I0886807

Geography Education and Explorations on Human Development and Culture

Cristina Honrubia-Montesinos
University of Castilla-La Mancha, Spain

Antonio Morcillo-Martínez
University of Castilla-La Mancha, Spain

IGI Global
Scientific Publishing
Publishing Tomorrow's Research Today

Vice President of Editorial	Melissa Wagner
Managing Editor of Acquisitions	Mikaela Felty
Managing Editor of Book Development	Jocelynn Hessler
Production Manager	Mike Brehm
Cover Design	Phillip Shickler

Published in the United States of America by
IGI Global Scientific Publishing
701 East Chocolate Avenue
Hershey, PA, 17033, USA
Tel: 717-533-8845
Fax: 717-533-8661
E-mail: cust@igi-global.com
Website: https://www.igi-global.com

Copyright © 2025 by IGI Global Scientific Publishing. All rights reserved. No part of this publication may be reproduced, stored or distributed in any form or by any means, electronic or mechanical, including photocopying, without written permission from the publisher.
Product or company names used in this set are for identification purposes only. Inclusion of the names of the products or companies does not indicate a claim of ownership by IGI Global Scientific Publishing of the trademark or registered trademark.

Library of Congress Cataloging-in-Publication Data

Names: Honrubia-Montesinos, Cristina, 1990- editor. | Morcillo-Martínez, Antonio, 1998- editor.
Title: Geography education and explorations on human development and culture / Edited by Cristina Honrubia-Montesinos, Antonio Morcillo-Martínez.
Description: Hershey, PA : IGI Global Scientific Publishing, [2025] |
Includes bibliographical references and index. | Summary: "The relevance that geography currently has from a social, cultural, political and economic point of view is unquestionable. This book aims to highlight the depth and extension of geography from different perspectives"--
Provided by publisher.
Identifiers: LCCN 2024054496 (print) | LCCN 2024054497 (ebook) | ISBN 9798369371855 (hardcover) | ISBN 9798369371862 (paperback) | ISBN 9798369371879 (ebook)
Subjects: LCSH: Geography--Study and teaching (Elementary) | Human geography--Study and teaching (Elementary) | Environmental education--Study and teaching (Elementary) | Sustainable development--Study and teaching (Elementary)
Classification: LCC G73 .G395 2025 (print) | LCC G73 (ebook) | DDC 372.89/1--dc23/eng/20241227
LC record available at https://lccn.loc.gov/2024054496
LC ebook record available at https://lccn.loc.gov/2024054497

British Cataloguing in Publication Data
A Cataloguing in Publication record for this book is available from the British Library.

All work contributed to this book is new, previously-unpublished material.
The views expressed in this book are those of the authors, but not necessarily of the publisher.
This book contains information sourced from authentic and highly regarded references, with reasonable efforts made to ensure the reliability of the data and information presented. The authors, editors, and publisher believe the information in this book to be accurate and true as of the date of publication. Every effort has been made to trace and credit the copyright holders of all materials included. However, the authors, editors, and publisher cannot assume responsibility for the validity of all materials or the consequences of their use. Should any copyright material be found unacknowledged, please inform the publisher so that corrections may be made in future reprints.

Table of Contents

Detailed Table of Contents

Section 1
Geography Education and Pedagogical Approaches for Socially Relevant Issues

The first section of the book offers educational approaches and practices focused on the current challenges and socially relevant problems that concern the field of geography, considering the contributions of specific didactics, the role of pedagogy and the importance of the teaching-learning process as an indispensable tool for achieving progress in societies. All these chapters are aimed at addressing these needs from an educational perspective, highlighting the promotion of geographical thinking and knowledge, the promotion of sustainability, the development of active citizenship, the implications of social and cultural transformations, the opportunities offered by the new intercultural scenarios, as well as the social and emotional literacy of these issues.

When geographic space is taught in Early Childhood Education, proper consideration should be given to the subjectivity of the concept itself, which is due to its different epistemological foundations. From the age of four, children start developing a sense of belonging and construction of identities linked to an understanding of geographic spaces and critical-thinking abilities (affective, conceptual, and interpretative). This study was developed as a pedagogical proposal with five-year-old students from a public school in the Valencian town of Benigànim (Spain) in the last quarter of the 2020-21 academic year. The teaching methods incorporated geographic literacy concepts and competences. The twelve activities included work with maps and with geography-related vocabulary, ranging from local geographic spaces to a more complex vision of global reality.

In this work, the Sendai Framework for Disaster Risk Reduction (2015-2030). It is the first major agreement of the post-2015 development agenda and offers Member States a series of concrete actions that can be taken to protect the benefits of development against the risk of disasters. Regarding the objectives pursued by this Framework, they are: the reduction of disaster risk and losses caused by disasters, both in lives, livelihoods and health as well as in people's economic, physical, social, cultural and environmental assets, companies, communities and countries. This is where the education factor would have its influential role. Although in the Spanish case there are still few works on experiences and didactic proposals regarding the teaching of natural risks from Geography, the Sendai Framework constitutes a first and fundamental step to integrate these phenomena in school, as stated in the LOMLOE (2020).

From the transition between the 18th and 19th centuries, the teaching of Geography emphasized the description of large global spaces, aligned with a rationalist perspective suitable for shaping expansionist elites. In the 19th century, this discipline fostered identification with the nation-state through the concept of territory, promoting a conformist form of citizenship. The following century, marked by two world wars and the Cold War, continued these educational traditions. The current century introduces the social appeal of participatory and multi-scalar territorial construction. The Nós Propomos!/ We Propose! Project embraces the challenge of fostering committed and proactive participation in territorial citizenship. Its methodological simplicity (problem identification by pupils, fieldwork, presentation, and sharing of proposals) ensures its applicability even in educational peripheries. The international expansion of the project demonstrates that it provides an alternative educational response, influencing educational practices beyond the boundaries of Nós Propomos! itself.

The inclusion of topics related to the postulates of sustainable development has gained special relevance in educational field. The current educational law, LOMLOE, includes principles and values related to the promotion of education for sustainable development pursuing the formation of citizens committed to the construction of a more sustainable world, being essential to involve students from an early age and develop educational actions aimed at addressing from the classroom the current problems and challenges of climate emergency. This chapter focuses on the treatment of environmental sustainability and aims to highlight the importance of addressing the sustainable development goals in the implementation of the educational curriculum in the 4th year of Primary Education, with the main objective of promoting awareness and sensitivity towards the marine world through the development and design of an innovative educational proposal focused on working the claims of sustainable development goal no. 14 (Life below water).

Primary education plays a fundamental role in shaping future citizens who are critical and engaged with their environment. In this context, the study of geographic space could become an essential tool for understanding the social and environmental realities of the world around us. When reviewing teaching practices, they are predominantly framed within a traditional approach that often focuses on memorization and abstract content. This makes it difficult for students to understand the relevance of geographic space in their daily lives and in the current society. To achieve these objectives, various didactic strategies are proposed throughout this chapter. These strategies are framed within a pedagogical approach that promotes knowledge construction through experience and critical reflection, enabling students to learn to understand social problems, investigating their causes and consequences, and participate in the search for solutions. This contributes to the construction of a more just, equitable, and sustainable society.

This chapter explores the importance of intercultural competence in today's education, emphasizing its role in promoting peaceful coexistence and social development in a multifaceted world. It analyzes pedagogical strategies that empower future educators to foster empathy, respect, and understanding of different cultures. Additionally, it identifies challenges in integrating interculturality into the educational curriculum, such as prejudice and lack of exposure to diversity. To address these obstacles, practical activities are proposed that integrate geography teaching, allowing students to contextualize their learning and build an inclusive environment. Ultimately, the purpose of the chapter is to help teachers become more aware, enabling them to act as agents of change in their communities

The natural environment, which has been damaged by human interventions, was initially able to tolerate human-induced negative impacts, but over time this destruction has reached the dimension of environmental disaster. Although steps were taken to address environmental problems from time to time, the measures taken were not sufficient and environmental problems continued to increase. Considering that the easiest, permanent and economical way to solve problems is through education, education should be given importance in the fight against environmental problems. The prerequisite for a livable environment is to raise an environmentally sensitive generation. The first target should therefore be the critically important early childhood period. In order to invest in our future, it is necessary to increase children's contact with nature, to ensure that they love and protect nature, and to develop their attitudes and skills to contribute to nature. So what are the duties of families and educators in this regard, what can we do for a more ecological, sustainable and livable environment?

Didactic itineraries serve as a valuable tool for teaching Social Sciences, particularly Geography. Therefore, a tour has been organized through the streets so that first-year students can learn about transportation systems and basic road safety rules. Thanks to the experience that provide this project, the benefits of didactic itineraries has been highlighted, as students acquire meaningful learning through them by directly engaging with their surroundings and classmates, which they find more engaging and motivating compared to traditional classroom lectures. Due to this, these itineraries are an ideal resource to incorporate into educational programs.

Section 2
Interdisciplinarity, Culture, and Geography

The social needs and requirements of the present requires multidisciplinary consideration that includes educational and pedagogical approaches involving other related disciplines that contribute to generate progress in these issues in a holistic way. For this reason, the second part of the book devotes efforts to present links with the contributions made by language and literature, the arts or the impact of technological advances.

Recovering Lucía Baquedano's work is essential for enriching juvenile literature and highlighting overlooked Spanish women authors. Cinco panes de cebada, Baquedano's novel, reflects rural Spanish life during the late Franco era and the country's modernization. The protagonist, Muriel, an urban teacher in a rural setting, undergoes a journey of adaptation and identity-building as she immerses herself in her new community. Over time, Muriel defies the stereotype of the traditional rural teacher by prioritizing her students' learning and well-being. The novel's confessional tone suggests an autobiographical quality and serves as a potential model for today's educators. The article ultimately argues for the renewed appreciation of works like this in educational contexts, as they address timeless themes such as teacher vocation and the transformative power of education for disadvantaged communities.

The aim of this paper is to revalue the Spanish writer Elena Quiroga, who received some recognition in her literary beginnings, winning prizes such as the Nadal, allowing her to insert herself into the narrative panorama of the 1950s. However, her figure was forgotten in Spain after her death. This writer stands out for her constant willingness to experiment in her narrative forms, which leads us to speak of a personal narrative universe. The great richness and variety of her work is the main reason why she deserves critical attention. For this reason, after studying the historical and literary panorama in which Elena Quiroga is situated, a corpus of three works has been selected, two short novels and a novel, to illustrate the evolution of her work by comparing them. These works are Trayecto uno, Plácida, la joven and Tristura.

Due to the increasing presence of foreign students in classrooms, Spanish educational institutions face the challenge of developing an intercultural educational model to provide quality education to all students, regardless of their geographic and social background. Specifically, in the subject of Spanish Language and Literature, incorporating texts from the Hispano-American literary tradition can facilitate the academic and social integration of many Latin American students enrolled in Spanish schools. This article reviews the didactic potential of the work of Horacio Quiroga, considered one of the great masters of the Hispano-American short story, for the simultaneous development of both literary and intercultural competencies. The article analyzes the constructive, formal, linguistic, and thematic characteristics of Quiroga's stories, which not only offer an artistically enriching reading experience but also provide an opportunity for our migrant and native students to engage in intercultural dialogue.

In the post-truth era, where fake news abounds, advancements in artificial intelligence (AI), particularly conversational AI, are producing tools capable of generating increasingly human-like text. This mixed-methods study explores future teachers' perceptions about the ability of conversational AI to write argumentative texts. To this aim, a group of 87 pre-service teachers compared two anonymised texts: an AI-generated one and a text written by a fellow student. Results indicate that most participants believed the AI-generated text would outperform the human-written one in an exam, though opinions were divided on which text conveyed more emotions. Participants generally associated better writing and information richness with AI, while they linked subjectivity and emotivity to the human author. However, not all correctly identified the author of each text, highlighting the need for further preparation of future teachers on current AI abilities and limitations.

Foreword

It has often been reiterated that geography serves as a bridge between the natural and social sciences, with geographic knowledge emerging from the interactions between humans, their activities, and the environment in specific places and distinct spaces. The centenary celebration of the International Geographical Union in 2022 underscored geography's indispensable role in analyzing the transformations impacting humanity's daily lives, as well as addressing the grand challenges societies and cultures will face in the coming decades.

In a finite world with limited surface and resources, population growth, economic expansion, urbanization, and the exponential rise in global exchanges—of goods, services, capital, information, and culture—have drastically transformed landscapes. These changes have not only resulted in artificialization but also in re-naturalization efforts. Climate change continues its relentless advance, risks and hazards have intensified, and environmental awareness has grown, along with a stronger emphasis on sustainability. However, these phenomena highlight that geography's core questions are inherently tied to human and social development, cultural diversity, global understanding, and the evolving geopolitical and geostrategic dynamics of our ever-changing world.

Recent developments in geographic thought and epistemology have reinforced the importance of (post-)cultural, (post-)humanist, and social geographies, partly as a reaction to the positivist resurgence brought about by advances in geospatial technologies and Geographic Information Systems. The United Nations' 17 Sustainable Development Goals (SDGs) all possess an undeniably spatial and multiscale dimension. Addressing these challenges requires more than an analytical approach; it necessitates fostering a culture of sustainability that integrates local spaces, national frameworks, and global processes.

In geographic education, this necessity is even more pronounced. Despite efforts to align geography with STEM subjects and the integration of geo-information into classrooms—introducing groundbreaking educational resources—geography remains predominantly categorized under the social sciences in primary and secondary education worldwide. Today's human development mirrors the characteristics of the modern world: changeable, uncertain, complex, interconnected, unequal, and diverse. Consequently, geographic education must embrace human, social, and cultural geography to remain relevant.

The *International Charter on Geographical Education* (2016) by the International Geographical Union and the *GEODEM Benchmark Statement for Teaching Europe in Geography* (2023) by EUROGEO emphasize the importance of geography in sparking intellectual curiosity and fostering spatial empathy. These principles are rooted in the understanding of Earth as a mosaic of cultures, societies, and communities—each possessing unique identities but also sharing common values of humanity.

The different chapters of this book reinforce these principles, but also the necessary claim that geographic education should be based on geographic intelligence (which goes beyond spatial intelligence, incorporating interpersonal or social intelligence) and specific geographic competencies (spatial thinking, geographic knowledge, and spatial citizenship) that transcends the contents and procedures based on geographic knowledge and information. Such an approach equips students to become responsible and informed citizens capable of navigating the complexities of modern societies.

Finally, as highlighted in this volume *Geography Education and Explorations on Human Development and Culture*, the challenges of geographic education extend beyond pedagogy and curriculum design. As UNESCO recently noted, the integration of artificial intelligence (AI) into education presents ethical considerations and is reshaping geographical teaching and learning practices. By promoting ethical awareness, AI applications in geography education can align with societal values like human development and spatial citizenship. Moreover, AI offers opportunities to refine curriculum design and lesson planning, ensuring they are attuned to increasingly diverse cultural contexts.

Rafael de Miguel González
University of Zaragoza, Spain, & European Association of Geographers, Belgium

Preface

Geography has long been recognized as a critical discipline for understanding the world around us, yet its educational role remains underappreciated in many contemporary contexts. This book, *Geography Education and Explorations on Human Development and Culture*, emerges as an essential response to this gap and pretends to highlight the relevance of Geography.

As editors, we aim to emphasize geography not merely as the study of physical landscapes but as a multifaceted area of inquiry that influences and is influenced by human development, culture, and education. From a global perspective, geography provides vital insights into social, political, economic, and environmental issues that shape our world. Yet, despite its significance, geography often does not receive the attention it deserves in educational curricula and research discussions. In fact, in recent decades, Geography has lost influence within the Social Sciences. However, the most recent perspectives published by important geographical associations, such as the American Association of Geographers, indicate that Geography is the only discipline that allows us to observe the world and make sense of it. This conception emphasizes that its subject matter—the world—is both motivating and exciting, and that Geography, of course, seeks to make sense of it (Lewis, 2020). Similarly, the Royal Geographical Society in the United Kingdom states that Geography is unique in bridging the social and natural sciences. On one hand, human Geography refers to understanding the dynamics of cultures, societies, and economies, while physical Geography focuses on understanding the dynamics of landscapes and the environment. Geography, therefore, in its broadest sense, is an education for life and for living. Learning through Geography helps us all to become more socially and environmentally aware, better informed, and more responsible citizens (Royal Geographical Society, 2020).

According to Carlo Manuel, president of the College of Geographers in 2021, Geography's primary interest lies in understanding the relationship between society and the natural world. The quest for answers to these relationships can be approached from multiple perspectives. This demonstrates that Geography straddles the natural sciences, social sciences, and the humanities. Furthermore, Geography analyzes territory, territorial sustainability, global change, and territorial information—key aspects in contemporary society. All of this underscores the importance of Geography, elevating it to the recognition it deserves as a science for and by society.

For students, the relevance of Geography lies in understanding social reality with spatiality as a conditioning factor. In other words, it involves examining the role of space in social practices and how these practices shape space. Geography can be recognized as a field of inquiry into the spatiality of individuals, groups, and broader social structures. It implies understanding the importance of these spatialities in comprehending the dynamics of social life, analysed from a multi-scalar dimension (Barbosa and Cavalcanti, 2011). As a scientific discipline, Geography, through its varied lines of research, contributes to developing spatial understanding that enables humans to cultivate their geographical thinking.

Additionally, Geography's importance lies in its connection to other disciplines by addressing socially relevant problems that affect both the environment and societies. In this regard, specific didactics play an essential role, particularly the didactics of Geography. Its objective is not only to teach about continents, mountains, or rivers but also to critically analyze the social, cultural, political, and environmental phenomena occurring in different territories. Thus, it is crucial that geography education goes beyond technical knowledge of maps and coordinates, focusing instead on current issues affecting both the environment and societies.

As a result, socially relevant problems such as climate change, global warming, depopulation, poverty, and social exclusion can and should be addressed through Geography, with an interdisciplinary approach that provides a holistic view of these issues. Geography can establish links with other areas and disciplines to offer a deep understanding of contemporary phenomena.

For instance, the link between Geography and interculturality is profound, as the territory and geographic space serve as the stage where cultures meet, interact, and sometimes intertwine or come into conflict. The study of human Geography allows us to understand how different societies have inhabited and transformed their environment, turning the natural landscape into a reflection of their customs, beliefs, and cultural practices. Through geographic analysis, we can understand the dynamics of mobility, migration, and displacement that characterize the contemporary world, where human and cultural flows cross borders, creating new contexts for interaction and coexistence (Meer and Modood, 2011). This crossing of geographic and cultural boundaries is a constant process that highlights the diversity of humanity and the need to build bridges of understanding between different ways of life.

Interculturality, in this context, is not limited to the coexistence of cultures but advocates for active and respectful interaction, where recognizing differences and fostering dialogue between peoples become fundamental principles for strengthening a plural society. Interculturality invites us to reflect on how different cultures, in their varied expressions, can contribute to mutual enrichment instead of being seen as isolated elements or sources of conflict. It is in the contact between cultures—in the exchange of ideas, beliefs, and customs—that new meanings and practices emerge, constructing a global collective identity that transcends the local or national.

Language, as an instrument of communication, plays a central role in this process. It is not merely a means of transmitting information but also a tool for creating and sustaining social relationships, expressing ideas and emotions, and articulating our worldview. Through language, cultures represent, preserve, and transform themselves. It is also through language that people can overcome geographic and cultural barriers, establishing connections that transcend physical borders. In this sense, literature stands out as one of the most powerful forms of language expression. Literary texts often reflect the historical, social, and cultural movements that shape societies and their territories. Each literary work is a manifestation of human experience, rooted in a specific geographic and cultural context yet possessing the power to transcend those boundaries and touch the hearts of other peoples, languages, and geographies.

Moreover, literature is a space where interculturality manifests tangibly. In literature, languages and cultures meet, mix, and transform, creating new forms of expression that reflect the richness of encounters and conflicts between peoples. From the oral narratives of early civilizations to contemporary works exploring globalization phenomena, literature has been and continues to be a mirror of human interaction—a place where identities are negotiated, questioned, and reinvented. Literary narratives allow us to delve into the lives of others, understand their struggles, dreams, and hopes, while simultaneously challenging our own worldviews.

The intertwining of these three fields of knowledge—Geography, interculturality, and language and literature—not only enriches our academic understanding but also invites us to adopt a broader and more compassionate perspective on the world. We live in an era of global interconnectedness where geographic and cultural barriers are fading, bringing us closer to one another like never before. This context demands new ways of understanding and valuing human diversity, recognizing the importance of cultural narratives, and fostering intercultural dialogue to live in harmony.

By exploring the geography of peoples and cultures, we study not only physical spaces but also the soul of the communities inhabiting those spaces. Language and literature become vehicles for expressing and transmitting that soul, and interculturality teaches us how these expressions can be shared, respected, and mutually enriched. Geography, in its capacity to study place and space, connects us to the very root of cultures; language and literature connect us to the stories those cultures have told and continue to tell; and interculturality invites us to live and think in ways that celebrate shared stories and the possibilities of a common future.

To fully grasp the global nature of Geography, it is essential to prioritize geography education and didactics from early stages to foster knowledge, skills, and attitudes. It is worth highlighting that geography education is currently thriving, with the implementation of innovative strategies and pedagogical models aimed not only at improving understanding of geographical knowledge but also at promoting active, participatory, and meaningful learning for students.

Technology has played a fundamental role as a resource that enhances motivation and the development of geographical knowledge. Specifically, geographic information technologies have revolutionized the teaching and learning process in Geography (Buzo Sánchez, 2016; De Miguel González et al., 2019). They provide opportunities for understanding both local and distant contexts, enabling work from the local to the global and vice versa, fostering a deeper comprehension of geographic phenomena in various spaces (Álvarez-Otero et al., 2022; Murga-Menoyo, 2020). The use of these technological resources not only enriches the teaching-learning process but also prepares students to face the challenges of an increasingly interconnected and globalized world.

Furthermore, current resources have fostered innovative geographic education, increasing students' interest (Lázaro-Torres and Puertas-Aguilar, 2022; Milson, 2011). These resources include geographic games, web-based GIS tools, and digital platforms (Crespo-Castellanos, 2019; De Miguel González, 2015; Souto González, 2013). The integration of research and education in Geography didactic innovation is opening new possibilities for teaching and learning the discipline. Advanced methodological strategies, pedagogical models based on TIG and GIS, and the creation of quality resources are shaping a new educational paradigm. This integrative approach not only improves Geography education but also prepares students to face future challenges with a solid and critical understanding of geographic space.

In geographic education, there is academic consensus on the importance of geographic thinking and world conceptualization. Geographic thinking is highly valued and considered powerful (Brooks et al., 2017) as it replaces the descriptive, fact-based image of Geography. As Jackson (2006) noted, demonstrating the power of geographic thinking helps dismantle outdated Geography content and reverse declining student interest and misconceptions about the discipline. In geographic education, geographic thinking refers to the cognitive process of thinking like a geographer, encompassing both concepts and procedures.

Starting with concepts, Cox et al. (2020) highlight that some aspects of geographic thinking involve interactions between humans and nature, scale, local-global relationships, geographic issues, and their links to personal choices. Similarly, Uhlenwinkel (2017) states that "people who think geographically use geographic concepts such as place, space, scale, interdependence, diversity, proximity, or distance to analyze phenomena in the world around them" (p. 45). According to Jackson (2006), thinking geographically involves a unique way of viewing the world, understanding complex problems, and thinking about interconnections at various scales (from global to local). This author emphasizes space and place, scale and connection, proximity and distance, and relational thinking as key concepts. Addressing these concepts provides excellent tools to communicate the true meaning of geographic thinking and how to implement it in schools. Similarly, the Geographical Association (2009) considers geographic thinking. as an essential educational outcome of learning geography: "Being able to apply knowledge and conceptual understanding about the changing world to new environments, that is, to think" (Geographical Association, 2009, p. 10). Lambert (2017) adds that while absorbing the vast amount of geographical data can be impressive, it is not, in itself, a sign of intellectual development. For this, the development of conceptual knowledge is needed, which involves a certain set of rules, procedures, and patterns

Continuing with the procedures, research and geographic inquiry are considered key means for acquiring geographic thinking (Bednarz et al., 2013; Favier & Van der Schee, 2014; Finlayson et al., 2017; Karkdijk et al., 2019; Howell & Maddox, 2022). According to the U.S. National Geography Standards (2012), the inquiry process within geographic thinking involves five main phases: formulating geographic questions, acquiring geographic information, organizing geographic information, analyzing geographic information, and responding to geographic questions.

Situated between these two positions is García de La Vega (2022), who approaches geographic thinking as a cognitive domain but defines it in relation to key concepts. This author emphasizes that teaching geography through these concepts helps organize and connect content with its broader context. This approach fosters meaningful learning, a deep understanding of geography as an intellectual core of knowledge, and the ability to think geographically (García de La Vega, 2022).

Similarly, though with greater conceptual emphasis, are the studies of Karkdijk et al. (2019) and Murphy and Hare (2016). On one hand, Karkdijk et al. (2019) highlight that geographic thinking is a reasoning process where students apply geographic knowledge to a regional context. On the other hand, Murphy and Hare (2016) develop the idea of cross-cutting themes. The connection between these studies lies in their shared focus on the necessary association between concepts and procedures for the acquisition of geographic thinking.

In summary, this preface highlights the importance of geography, interculturality, and language and literature as fundamental pillars for understanding the modern world from a holistic and plural perspective. Geography, by analyzing the interactions between society and the environment, as well as spatial phenomena at multiple scales, provides tools to understand and address global and local challenges. Interculturality invites us to value diversity and build bridges of dialogue and respect among cultures, while language and literature act as vehicles for expressing, preserving, and transforming cultural and human identities. Together, these disciplines foster critical, sensitive, and responsible thinking, capable of addressing contemporary issues through an interdisciplinary and innovative approach. Thus, it advocates for an integrative education that combines academic reflection, the use of advanced technologies and methodologies, and the construction of meaningful knowledge that promotes a more inclusive, supportive, and sustainable world.

The collection of chapters presented below brings together diverse perspectives from international scholars, aiming to reinvigorate the teaching and learning of geography, especially in early childhood and primary education settings. The contributions offered in this volume are rooted in a comprehensive understanding of geography as a tool for critical thinking, intercultural engagement, and sustainable development. Each chapter addresses key themes such as sustainability, cultural diversity, migration, environmental education, and the integration of modern technologies like Geographic Information Systems (GIS) in the geography classroom.

Our collective goal is to offer both theoretical and practical insights into how geography can be taught in a way that promotes not only knowledge but also social responsibility and active citizenship. By incorporating methodologies, resources, and innovative practices, we aim to equip educators with the tools to foster a deeper understanding of the world in their students and prepare them to engage with the complex challenges of the 21st century.

This book is structured to appeal to a broad audience, from early childhood and primary education teachers to researchers and geographers. The content spans diverse topics, ranging from the pedagogical approaches to human geography to the exploration of interculturality and the challenges of teaching geography in multicultural settings. Additionally, it highlights the role of education in fostering a deeper understanding of environmental issues and social justice.

We hope that the contributions in this volume will inspire educators to rethink their approach to geography teaching, encourage further research, and ultimately cultivate a generation of learners who are well-equipped to navigate the complexities of the world with an informed, compassionate, and globally conscious mindset.

Chapter overview

1. The Identities That Reveal Emotions: Geographical Thinking in Early Childhood Education

This chapter examines the development of geographic thinking in young children, emphasizing the importance of fostering a sense of belonging and identity through geographic spaces. It details a study involving five-year-old students, exploring how children's understanding of local and global spaces can enhance critical thinking and emotional development. The authors advocate for incorporating geographic literacy into early childhood education to nurture not only cognitive skills but also affective and interpretative abilities that will shape children's identities.

2. The Teaching of Physical Geography in Teacher Training: 20 Years of Educational Field Trips in Protected Natural Areas

This chapter reflects on the role of field trips in teaching physical geography to future primary education teachers. By focusing on the use of natural parks in Castilla-La Mancha as educational settings, the authors illustrate how immersive experiences can foster environmental awareness and a deeper understanding of the local geography. The chapter discusses the methodology behind these field trips, documenting over two decades of successful educational practices that integrate theory with practice in the teaching of geography.

3. Migrations and Interculturality: A Strategy for Early Childhood Education from Social Geography

With the rise of multicultural classrooms in current time, this chapter explores strategies for addressing the educational challenges posed by this demographic shift. It underscores the role of social geography in promoting intercultural understanding and democratic citizenship. The authors propose a cooperative learning model aimed at fostering social integration and solidarity among young students, preparing them to thrive in a multicultural society.

4. Sendai Framework for Disaster Risk Reduction (2015-2030): A Before and After for Teaching Natural Risk

In this chapter, the authors analyze the Sendai Framework for Disaster Risk Reduction and its implications for geography education. Focusing on Spain's implementation of this global initiative, the chapter explores how integrating natural risk education into school curricula can better equip students to understand and address disaster risks. The authors argue that the framework provides a critical opportunity for embedding disaster preparedness within educational systems.

5. Geographic Education and Citizen Participation: The Project We Propose!

This chapter discusses the evolution of geography education, emphasizing the shift towards participatory and multi-scalar territorial citizenship. The authors present the 'Nós Propomos!' project, which encourages students to engage in problem identification and fieldwork as part of a broader effort to foster active citizenship. This participatory approach is shown to be applicable even in educational peripheries, demonstrating its effectiveness in fostering critical thinking and territorial awareness among students.

6. Learn to Take Care of the Environment: An Innovative Proposal for Sustainable Development Goal 14 in Primary Education

Focusing on the importance of environmental sustainability, this chapter presents an innovative approach to teaching about marine conservation and the Sustainable Development Goals (SDGs) in primary education. By aligning the educational curriculum with SDG 14, the chapter proposes a set of practical activities designed to raise awareness and sensitivity towards the marine world, aiming to cultivate a generation of environmentally-conscious citizens.

7. The Study of Relevant Social Issues in Primary Geography Education

This chapter explores the relevance of geographic space in understanding social and environmental issues. The author critiques traditional teaching methods that emphasize memorization over experiential learning and proposes new didactic strategies aimed at promoting critical reflection and active participation. The chapter encourages students to investigate social problems and contribute to the search for solutions, fostering a more just and sustainable society.

8. Fostering Intercultural Skills Through Geography: A Proposal for Future Primary Education Teachers

This chapter highlights the importance of intercultural competence in geography education, particularly for future primary education teachers. The authors propose a series of pedagogical strategies to help teachers build empathy, respect, and understanding of diverse cultures. The chapter emphasizes geography's role in promoting an inclusive and respectful classroom environment, preparing future educators to be agents of change in multicultural societies.

9. Preparing Children for the Future with Environmental Education

In this chapter, the author emphasizes the critical role of early childhood education in addressing environmental challenges. The chapter explores how to cultivate environmentally-conscious attitudes in young children, encouraging them to protect nature and engage with sustainability practices. By focusing on the early years, the chapter advocates for a long-term investment in environmental education to ensure a more sustainable future.

10. Learning on the Move: Merging Geography with Road Safety Education

This chapter introduces the concept of didactic itineraries as a method for teaching geography and road safety. By incorporating field-based learning, the chapter highlights how these activities engage students and provide meaningful, contextual learning experiences. It emphasizes the importance of experiential learning in geography education and its ability to motivate students by connecting academic content with real-world applications.

11. The Revival of Lucía Baquedano's Work in Youth Literature

This chapter focuses on the literary contributions of Lucía Baquedano, a Spanish author whose work is often overlooked in educational contexts. Through an analysis of her novel '*Cinco panes de cebada*', the chapter explores the themes of teacher vocation, identity, and the transformative power of education. The authors argue for the inclusion of Baquedano's work in educational curricula to inspire future educators and promote a deeper understanding of rural Spanish life.

12. Elena Quiroga: A Writer to Rediscover

This chapter seeks to rediscover the works of Spanish author Elena Quiroga, whose contributions to literature have been largely forgotten. By analyzing a selection of her novels, the chapter sheds light on Quiroga's narrative innovations and their relevance to contemporary education. The authors advocate for a renewed appreciation of Quiroga's work, particularly for its exploration of themes such as identity, memory, and the human condition.

13. Literature that Dismantles Borders: Hispano-American Short Stories and Interculturality in the Classroom

This chapter discusses the potential of Hispanic-American literature, specifically the works of Horacio Quiroga, to promote intercultural understanding in the classroom. By analyzing Quiroga's short stories, the authors explore how literature can serve as a bridge for social and academic integration, particularly for Latin American students in Spain. The chapter argues for the inclusion of these texts to foster intercultural dialogue and literary appreciation.

14. Pre-service Teachers' Conceptions of AI-Generated and Peer-Written Texts

This chapter examines how future teachers perceive the writing capabilities of conversational AI in comparison to human authors. Through a study with pre-service teachers, the authors explore how AI-generated texts are viewed in terms of quality, emotion, and expressiveness. The chapter calls for greater awareness and understanding of AI's role in education, encouraging teachers to critically engage with technological advancements.

15. The Impact of Oral Transmission and Geolocation on El Señor Don Gato

This chapter explores the impact of oral transmission and geographical context on the evolution of the traditional Spanish romance 'El Señor Don Gato'. By analyzing different versions of the song from various regions, the authors demonstrate how geography, culture, and social interaction have shaped this enduring piece of folklore. The chapter highlights the adaptability of oral traditions and their role in preserving cultural heritage.

In bringing together the diverse contributions featured in this volume, we hope to have illuminated the multifaceted role that geography plays in shaping education and human development. From early childhood education to primary schooling and beyond, geography serves as a critical tool for understanding not just the world's physical spaces but also the intricate connections between people, cultures, and environments. Through the chapters presented here, we have witnessed the immense potential of geography to inspire critical thinking, foster intercultural empathy, and promote sustainable development.

The collective insights in this book offer a clear call to action for educators and researchers alike: to reconsider and revitalize geography's place in curricula, ensuring that it is approached not merely as a subject to be memorized, but as a dynamic discipline that connects students to global challenges and local realities. We believe that, through innovative pedagogies and the integration of new technologies like GIS, geography education has the power to cultivate informed, responsible citizens—ready to tackle the pressing issues of our time with a global and ethical mindset.

As editors, we are grateful for the diverse perspectives shared by scholars and practitioners from around the world, each of whom has demonstrated how geography can serve as a bridge between education and the world's complex challenges. The hope is that the lessons drawn from these chapters will inspire educators to take bold steps in reimagining geography education, making it an engaging and impactful subject that prepares future generations to navigate the complexities of the 21st century.

This book stands as both a resource and a catalyst for change, reminding us that geography education is not only about understanding the physical world but also about cultivating a deeper, more compassionate understanding of one another. Through this volume, we aim to spark ongoing conversations, inspire further research, and ultimately contribute to a more globally conscious, sustainable, and equitable future for all.

Section 1
Geography Education and Pedagogical Approaches for Socially Relevant Issues

The first section of the book offers educational approaches and practices focused on the current challenges and socially relevant problems that concern the field of geography, considering the contributions of specific didactics, the role of pedagogy and the importance of the teaching-learning process as an indispensable tool for achieving progress in societies. All these chapters are aimed at addressing these needs from an educational perspective, highlighting the promotion of geographical thinking and knowledge, the promotion of sustainability, the development of active citizenship, the implications of social and cultural transformations, the opportunities offered by the new intercultural scenarios, as well as the social and emotional literacy of these issues.

Chapter 1
The Identities That Reveal Emotions:
Geographical Thinking in Early Childhood Education

Diego García Monteagudo
https://orcid.org/0000-0003-0505-0608
University of Valencia, Spain

Clara Vañó Albert
https://orcid.org/0009-0009-7024-0847
Independent Researcher, Spain

ABSTRACT

When geographic space is taught in Early Childhood Education, proper consideration should be given to the subjectivity of the concept itself, which is due to its different epistemological foundations. From the age of four, children start developing a sense of belonging and construction of identities linked to an understanding of geographic spaces and critical-thinking abilities (affective, conceptual, and interpretative). This study was developed as a pedagogical proposal with five-year-old students from a public school in the Valencian town of Benigànim (Spain) in the last quarter of the 2020-21 academic year. The teaching methods incorporated geographic literacy concepts and competences. The twelve activities included work with maps and with geography-related vocabulary, ranging from local geographic spaces to a more complex vision of global reality.

DOI: 10.4018/979-8-3693-7185-5.ch001

Copyright © 2025, IGI Global Scientific Publishing. Copying or distributing in print or electronic forms without written permission of IGI Global is prohibited.

1. INTRODUCTION

This research demonstrates that teaching geographic space in Early Childhood Education (ECE) is based on a set of activities supported by emotions and the students' appreciation of their environment as a lived space. That appreciation of geographic space appears in each student's mind as something unique and can be didactically enriched to shape identities from the earliest years of school. The project was developed in the Valencian town of Benigànim, that established a link between emotions and geographic space (Vañó, 2021). On the one hand, the town offered both unique and common characteristics that aid the comparison with other geographic contexts. On the other hand, emotions can be diverse, and some are generated individually, while others are shared in a group. In that aspect, family is the main support for students in developing a broader sense of belonging that starts at home and extends to society and the local space (Ferrer & Riera, 2015). Geographical knowledge requires the connection of different elements, among which emotions, perceptions, and ideas stand out.

Teaching geographic space in the ECE can be based on two didactic models. From one point of view, understanding of geometric characteristics of reality is perceived via children's own bodies, which they measure through everyday objects, thus acquiring some knowledge of location, distance, and orientation. From another point of view, it is assumed that geographic space consists of mental images that are created in childhood through children´s interaction with reality. The spatial trilogy (the lived, the perceived, and the imagined space), derived from the juxtaposition of the geographical school of thought (geography of perception and behavior, humanistic geography, and postmodern geography), forms the epistemological basis for a didactic strategy that progressively shapes the learning process. Some previous studies show that from the age of four, boys and girls gradually develop a sense of belonging in which affective (geographical feeling), conceptual (basic concept of geography such as places and landscapes), and interpretive thinking about reality are linked together thus forming the basis for geographical thinking (Canet et al., 2019; García Monteagudo, 2022; Jaráiz et al., 2016).

Additionally, this case study is based on action research and demonstrates the results of a didactic intervention carried out in a public school in the Valencian town of Benigànim in the last quarter of the 2020-21 academic year. The didactic proposal consisted of twelve activities prepared for a group of ten five-year-old students, combining the research method, the discovery method, and the symbolic play. These strategies include structuring concepts, geographical thinking competences, and skills, which consolidate broader knowledge of geographic space as a socially contextualized space in which emotions construct the language expressing spatiality.

Three didactic objectives closely related to promoting geographic literacy among the ECE students were proposed. The first objective was aimed at bringing learners closer to their geographic space and the recognition of social and cultural elements of their immediate reality. To this end, a number of places in the town of Benigànim were carefully selected to be linked to other elements that would allow students to observe a more complex geographic space through its association with everyday moments, thus encouraging the perception of that town as a lived space.

The second objective was to develop a set of didactic activities that would introduce students to emotions in the context of their immediate environment. Specifically, twelve programmed sessions followed an ascending sequence that started from the students' preliminary comprehension and continued with a series of techniques that would broaden the learners´ knowledge about some locations based on emotional reflection. The third objective was to encourage the development of habits related to the society, the local culture, and the local identity. That meant demonstrating geographic spaces closely to students so that they would understand the concepts that intervene in the process of identity formation through craft activities, among other resources.

The activities promoted geographic literacy via work with maps, with verbal, iconic, and graphic language. Those facilitated understanding of local geographic spaces from geometric postulates to a social context. The initial perception became more complex through work with places in the neighborhood, dealing with names of streets associated with famous people and jobs, and the construction of mini models, which had to demonstrate urban and cultural elements of the lived space. The connection between society and natural environment was channeled through the study of ancient trades, in which geographical elements marked some patterns and trends that described landscapes through a holistic perspective. Toward the end of the teaching sequence, students had to develop a distinctive style and be able to understand the drawings of other boys and girls, thus reinforcing their skills and fostering coexistence in the classroom. Finally, the last activity showed that mental representation of geographic space favored the eventual introduction of knowledge via a multi-step perspective, allowing students to gradually acquire some features of geographical awareness and collectively reach geographical thinking abilities.

Apart from the introduction (1), this chapter contains four more sections (2-5). First (2), we outline the fundamental studies on geographic literacy and the benchmarks to develop geographical thinking in ECE. Second (3), we define the methodology, with a special focus on action research and the school context in which the didactic intervention took place. Subsequently (4), we demonstrate the results of activities that were used to promote geographic literacy among the ECE students. Finally (5), we compare the results with other studies, reflect on reaching the proposed objectives, and set lines of future research.

2. GEOGRAPHIC LITERACY IN CHILDREN'S EDUCATION: PRINCIPLES AND PEDAGOGICAL PROPOSALS

At the ECE stage, a series of factors may compromise children's literacy from a broad perspective, and those are not exclusively restricted to reading and writing development. Emotional and psychomotor skills, the acquisition of values, and the discovery of reality imply a more active relationship with the elements that create a context for those competences that are regulated by teaching decrees. In that context, equal opportunities and the development of full citizenship should be guaranteed, in line with the Convention on the Rights of the Child (Carmona, 2012; Martínez, 2020) and the Council of Europe (Kowalczyk, 2004; Requena, 2000).

The fulfillment of the above rights has been the focal point of geographies of childhood in the last four decades. In particular, cultural geography studies have revolved around the social experiences of boys and girls in spaces such as schoolyards, towns, or cities (Sáiz-Linares & Ceballos-López, 2022). Those areas can be used for recreational activities to encourage the generation called "the backseat generation" (Karsten, 2005) to get closer to the natural environment. According to Louv (2005), the increase in children's exposure to screens produces a nature deficit syndrome, that is, a behavioral disorder generated by people's detachment and disconnection from the natural world. Nevertheless, there are some educational resources and tools that bring the reality closer to the ECE students. They allow us to address multiculturalism, to get access to cultural heritage in other countries, to get closer to history through photographs, and to learn with interactive maps or 3D tools. In line with Alonso-Sainz (2022), those experiences would not have been appreciated in the classroom if they had not been technologically supported.

The organization of content or basic knowledge in the ECE curricula in multidisciplinary areas does not influence the disappearance of specific disciplines, but it does show an imbalance between languages and mathematics compared to social science. The interdisciplinary nature of geography justifies its inclusion in the ECE teaching and learning process, as it contributes to the development of identities through their integration with the context in which students live and gradually learn. For several decades, active pedagogy and pedagogical innovation movements have highlighted the potential of geographical education from the first years of school (González Iglesias, 2003; Souto, 2017).

In recent years, competence-based teaching has been promoted, and the subject of geography has enabled the development of a series of skills and objectives incorporated into the European Education and Training curricula (including geographic literacy). It is about learning location, distribution, and spatial relationships; teaching and learning spatial representation based on observation and mental maps;

developing cartographic skills and appreciation of territorial reality; and promoting education in values (Tonda, 2001).

Geographic literacy or geo-literacy (Turner & Leydon, 2012) is the ability to understand, process, and use spatial information, which encompasses two distinct but complementary skills. On the one hand, geographical (disciplinary) knowledge provides first-order concepts or meta-concepts, including geographic space, territory, place, region, landscape, and scale (Araya and Cavalcanti, 2018). On the other hand, geo-spatial identification combines geographical concepts and skills that involve spatial thinking and decision-making about geographic spaces (Herbst et al., 2022; Uz et al., 2021).

In the first place, (disciplinary) geographical knowledge requires a definition of geographic space, a process that addresses the polysemy of the term dating to different epistemological trends or schools. Traditionally, there are two basic and divergent models to understand geographic space. On the one hand, the concept of absolute or geometric space is understood as neutral, abstract, and objective and more typical of positivism. That concept was taken up by Piaget and Inhelder (1947) and Hannoun (1977), who explained that students learn topological, projective, and Euclidean categories of space. On the other hand, the relative, partial, concrete, and subjective space can be explained from the constructivist postulates (Sebastià and Tonda, 2015) and can be linked to the humanistic aspect of the Geography of Perception and Behavior (Marrón Gaite, 1999; Vara Muñoz, 2008).

Those two models are complemented by social representation (Souto, 2018), with different symbolic uses of geographic space (whether public or private, physical, virtual, or digital) in the modern world of extended media and artificial intelligence. That complex geographic space applies Soja's spatial trilogy from a socio-constructivist perspective which merges the lived spaces of humanistic geography and the perceived and conceptualized spaces of postmodern geography. Therefore, geographic space incorporates social representations and explains people's behavior and attitudes (Sammut et al., 2015). Those depend on a mental image in which personal experiences, other people's perceptions, and notions of territorial power are interwoven.

Geo-spatial identification begins based on those postulates, which combine disciplinary knowledge with theories of learning and pedagogy. As summarized in Table 1, geography has structuring concepts (Granados, 2019) linked to competences (Case & Wearing, 2016) that develop geographical thinking in the ECE (García Monteagudo, 2024).

Table 1. Concepts, competences, and purposes of geographic literacy

Structuring concepts (Granados, 2019)	Brief definition	Guiding questions	Geographical thinking skills (Case & Wearing, 2016)
Perception	Inner feeling, produced by real-world impressions, which creates a mental image.	What elements of the geographic space do students perceive?	Spatial factor, among others
Location	Find out where someone or something is. It informs us of its characteristics (location, establishment...).	Are they able to locate buildings, parks, and other elements on the map?	Spatial factor
Distribution	Learn the location of geographical elements on a map to get closer to reality.	Can we understand whether there are rules to explain the phenomenon in other geographical locations and its spatial effect?	Patterns and trends
Interaction	Analyze the relationship between the variables to explain their impact on reality.	Are the natural aspects linked to the social and cultural aspects of the place?	Interrelationship
Arrangement	Understand why a certain spatial reality exists and what contributions were made by the administration and the citizens.	Is it possible to explain a new layout of the place and to understand its social impact?	Evidence and interpretation Geographic analysis

Source: own elaboration based on indicated authors.

The selection of concepts and competences above explains that geographic literacy involves three progressive levels of cognitive complexity, the first two of which are accessible to the ECE students. The lower level deals with the tools to learn places and their location, a fundamental basis of geographical science. The middle level includes understanding the relationship between people and the environment and its causes and the ability to question, confirm, investigate, and analyze geographical information. The high level consists of a geographical approach that requires critical interpretation, an understanding of the relationship of geography with natural resources, and the complexity of developing citizenship (Memişoğlu, 2017).

To develop geographical thinking, students must have reached a mature level of geo-literacy, but in the case of ECE, this level should be adapted to the students' abilities. Geographical thinking is a way of understanding reality from a geographical perspective, i.e., thinking about space, moving in different spaces (multiscalarity), and being aware of this spatiality (knowing that personal and collective actions take place in a space that is changing and mobile). It requires a conceptual framework, geographical representation, and critical-creative thinking. Learners would understand that they are the main figures who make decisions within the space, with the space, and about the space (Ortega Rocha, 2020). The final stage of geographical thinking is geographical awareness because it means that the students' reflection process is transmitted into action, i.e., they are ready to make geographical choices. This includes values, attitudes, ethics, and memory to build an argumentative geographical discourse, which is considered as powerful geographical knowledge in the scientific literature (Bednarz & Lee, 2019; Flores et al., 2023; Maude & Caldis, 2019).

Consequently, teachers should promote strategies that would encompass geographical competences. For the ECE, this means that students would symbolize the world through various cartographic forms that would help them explore it, even if it is fictional (Musgrave, 2020). To achieve geographic literacy, three main elements must be combined (Welton, 1988, in Memişoğlu, 2017): first, knowledge of the physical environment; second, the ability to use maps; and third, knowledge of geographical concepts. In short, the ECE students become geo-literate when geographical knowledge helps them to move independently in the global context, which requires knowledge of geography-related language (Souto, 2018), which is transformed into skills (Dikmenli, 2013; Martin & Owens, 2005; Massey, 2005; Palmer & Birch, 2004).

The first geographic literacy studies in the ECE were developed in the late 1970s. Reading maps was one of the first skills to develop location and orientation at the age of three to five, showing that it is possible to organize trips from the family environment (home) to unfamiliar places such as parks, streets, and other public places. At the age of three, the floor plan can be read if the orientation coincides with that space; at the age of four, the floor plan can be viewed outside the space, but it should be placed in the same direction, while at the age of five, the floor plan can be manipulated inside or outside the space, regardless of its position (Bluestein & Acredolo, 1979). This progression demonstrates that different levels must be applied to learning symbols, which is why Piñeiro (1987) used to work out geographic space on a model representing a room. That study showed that children between the ages of three and six learn the position of objects using their own bodies and that their knowledge of reality is syncretic, considering a degree of abstraction adapted to their abilities.

As studies on the didactics of social science progressed, the subjective understanding of geographic spaces in childhood was achieved. The study by Piñeiro (1996) is an example of an analysis of town and city. It showed that between the ages of four and seven, both lived and perceived space contributed to a more complex representation. Didactic itineraries can complement that representation, as they allow students to interact directly with geographic space, acquiring civic and artistic skills, among others (Fernández Portela, 2016).

In recent decades, the value of kinaesthetic coding has been utilized to explain the orientation of boys and girls in space. The information that children receive through senses or movements aids the understanding of space, even without interpersonal communication (Rivero and Gil, 2011). In fact, during the first five months of life, babies build their first spatial images in the brain from the perception of light and shadow, lines, surfaces, volumes, and some colors. By the first year of life, sensory perception is combined with grasping and helps to construct mental images of objects, surfaces, and volumes. That is the moment when children orient themselves by crawling through some parts of their home, recognizing simple objects, and placing them in different positions, although they cannot yet verbalize their actions. Between the first and second year of life, children develop a sense of orientation through their movements around the house and nearby places based on the recognition of landmarks (benches, streetlights, trees, blackboards...) found in those spaces or on regular routes.

In this development of spatial thinking, children's personal experiences are gradually enriched, and they become familiar with geographic spaces, regardless of their physical proximity or distance (Comes, 2002). The proximity is mediated by travel, narratives, family stories, the Internet, and other recent means of communication. It is expressed as appropriation achieved through information coding, as cultural changes (social relations, travel…) expand the spatio-temporal context in children's minds. This means that various socialization factors explain that learning about geographic spaces occurs through relationships that become complex, independent of physical location and other qualities of spatial reasoning (Inarejos, 2018; Moreno and Gallego, 2013; Pena and Caamaño, 2019), but not of geographical reasoning. In this aspect, every boy or girl acquires geographical information from images and maps that simplify the lived or the perceived reality. Studies in cognitive psychology reveal that the emotional connection with the environment influences the representation of space and affects the activity carried out in that environment. That emotional connection would influence cognitive maps, with the common action patterns derived from the feelings experienced (Caballero, 2002).

During the ECE, boys and girls follow itineraries in which their geographical perception of distance is developed. The difference between the perceived distance and the real distance is quite confusing for three-year-old students, as they have difficulty discerning between near and far when the distance is from 1,500 to 2,000 meters. At four years old, the differentiation between near and far is more accomplished, but at five years old, there is some indecision (boys and girls point out as close something that reaches 2,000 meters and as far away something that is below 1,000 meters). The explanation for this indecision is that they travel more by car than on foot compared to four-year-olds. Therefore, distance is a quality that is understood by the time traveled and not so much by its spatial perception (Jaráiz et al. 2016); therefore, space and time are two inseparable variables in the acquisition of geographic literacy.

The complex geographic space described above as social representation is understood as a social construction that begins in the ECE. Students' experiences and imagination can act as explanatory elements for students' identities, emotions, and relationships projected between the local and the global spaces, achieving a certain notion of multiscalarity. Teachers promote those qualities through drawings, stories, and evaluative didactic projects that help children comprehend space and awaken the learners´ emotions about some subjective space (Canet et al., 2019). The ECE students acquire new concepts through a set of well-defined activities and connect their experiences with the local space. Moreover, teachers can promote a set of habits related to citizenship and social values by focusing on some basic rules of coexistence in different spaces. An example of this is the project on Syrian refugees in a school in Almería (Spain), in which contents on History (Spanish Civil War), Art History (the work El Guernica), and geographical competencies were covered. It combined historical (location of the conflict, causes, and consequences) and citizen education (working with NGOs and the search for solutions to the war conflict) applied to the early childhood stage. The five-year-old students' capacity for dialogue encouraged teachers to address aspects such as conflict resolution, solidarity, tolerance, respect, and geo-historical awareness of the multinational environment. With this, some behavioral problems during playground time were resolved, thus promoting social, geographical, and historical learning (López, 2017), which directly appealed to identities and the construction of a feeling of belonging to the global world.

Those qualities and skills develop three of the four dimensions of the feeling of belonging in students (Corona, 2020). First, there is a social-psychological dimension or the feeling of security that students obtain when they feel valued at school. It promotes the development of social and civic competence. Second, there is an affective dimension or the feeling of belonging to a group. In this case, class and school entail the students' first social relationships. Third, there is a physical dimension since identity construction unfolds a deeper connection between the students and the physical environment. It can be applied to a number of spaces, whether personal or common. Eventually, there is a link between emotional education, identities, and the development of a feeling of belonging, which influences the learning process and provides direct contact with some local places. According to García Monteagudo (2022), those are the key elements of geographical thinking that need to be introduced in ECE. He gives an example of understanding localities as complex geographical spaces in which other locations (parks, organizations, libraries, markets...) are marked with emotions and awaken a feeling of belonging. The latter consequently develops habits and social values in students. When the emotional dimension forms part of didactic activities, it fosters solid knowledge, which combines concepts (basic vocabulary), skills, and attitudes fundamental for the promotion of geographic literacy.

In short, stimulating students' imagination through drawings or other techniques provides teachers with information based on which they can ask questions to boost visualization of geographic spaces from an early age. Observing and analyzing students' work and asking open-ended questions is a clue to fostering creativity, which is more important at an early age than any mapped product. Among the first cartographies produced by children, we can find flora or fauna of the neighborhood garden, the schoolyard flora, the creation of a friendship tree, or even the cartographies of the future (Musgrave, 2020). Those contents are the basis for education in values that is oriented towards citizenship training. Moreover, some interdisciplinary proposals have turned their contents (for example, peace education) into curricular projects that have reached early childhood education classrooms (Palma, 2023).

3. METHODOLOGY

The implementation of this project in the ECE classrooms had a thorough preparation, and a pilot study was conducted by the authors of this research (Vañó, 2021). It consisted of checking whether the drawing instrument aimed at the representation of a locality was understood by the ECE students. To do this, 50 drawings were collected from boys and girls aged from 3 to 5 years in various public schools in the metropolitan area of Valencia and in the Vall d'Albaida region. After some adjustments, it was decided to use that instrument in the first phase of this study.

During four months, some studies on geographical thinking in early childhood education were reviewed, and some activities were designed to explore the didactic possibilities of the town of Benigànim, where the didactic project (that makes up this qualitative study) was subsequently developed. The intervention prepared by the author corresponded to the participatory action research (Pérez et al., 2010); it was carried out in a public school in Benigànim between February and May 2021. That public school was aware of the previous study by Vañó (2021) and demonstrated interest in connecting early childhood education students with their local space (Well & Ebans, 2003). Action research allowed in-service and pre-service teachers to take a leading role in the teaching and learning process and jointly outline the research problems around the didactics of geographic space in the ECE. The training model was based on a socio-critical paradigm that considers teachers as reflective thinkers and researchers in their performance (Pruzzo, 2014).

Ten students who participated in this educational project were six girls and four boys between the ages of four and five. The lessons were organized in small groups. Most learners were in the syllabic and alphabetic stage for reading and writing. The development of intuitive thinking and socio-constructivist learning were adjusted to the knowledge of reality to help children observe, investigate, and raise some hypotheses. In the first intervention phases, the perception of the environment, feelings, the clarification of ideas, and the elaboration of concepts were essential to lay foundations for some principles of geographic literacy.

Regarding methodology, elements of research and discovery methods were combined (Aranda, 2016) and complemented by symbolic representation or games (Ruiz de Velasco y Abad, 2011). With those methods and the drawing technique (Ortiz et al., 2012), educational objectives and contents were defined based on the curricular decree for the 2020-21 academic year (*Royal Decree 1630/2006, of December 29, which establishes the minimum teaching standards for the second cycle of Early Childhood Education*), according to which the intervention on geographic literacy was carried out (Table 2).

Table 2. Synthesis of the objectives proposed in the didactic project

Teaching objectives	Didactic contents
-Understand the place that students occupy in a local society. -Classify places by the emotions they generate. -Link activities and space: game areas, movement, and information. -Go from abstract to concrete through the study of space. -Build a social bond through the space a person is in - Use emotional language to complement verbal and gestural expression.	-Graphic representation: the map. -Senses: sight, touch, and listening. -Feelings: control, impulsivity, positive attitude, and respect. -Traditional jobs: the messenger and the night watchman. -Stories and characters: identification and representation. -Popular culture: instruments and activities. -Urban space: the street and the license plates. -Abstraction: geometric shapes (roundabouts) and silhouettes (lines)

Source: own elaboration

The didactic project was divided into twelve sessions (Table 3) in which students were expected to take an active role in their assignments. The aim was for students to acquire geographic literacy skills through research tasks and, therefore, develop geographical thinking by linking emotions, perceptions, and ideas on a local scale to their experiences.

Table 3. A set of activities that made up the didactic project on geographic literacy

Sessions	Activity title	Brief description
1	"Initial Drawings"	Presentation of the project, motivation, and initial ideas. Analyze the town (Benigànim) using a free-line drawing; compare the evolution of concepts with another representation at the end of all the sessions.
2	"A Different View"	Learning through forms. The outline of the population can be related to different didactic contents.
3	"Great Little Committee"	Fusion, perspectives, and spatial association. Ideas, characteristics, and definitions are associated with local spaces to define them.
4	"Unsuspected Signal"	Our names introduce us to others. How do we call our streets to recognize them? Knowledge of spaces by their proper names (toponymy).
5	"Hiding Places"	Dialogue, identification, and location of relevant areas. Knowledge and recognition of spaces locally.
6	"Circles with Stories"	The connection between historical narrative and spaces. Production based on history and learning through narrative.
7	"Water Resources"	Space and resources. Contact with resources and understanding their value and how they change over time. Resources mark our lives and history, defining the places we inhabit and why we do so.
8	"The Messenger"	Spatial thinking. Practicing concepts through symbolic play and three-dimensional maps. If we are able to find a place, can we do it for someone else? Creating a model in the classroom to make deliveries to different areas of Benigànim.
9	"My Space"	Own spaces and the relationship with the members of the group. The spaces are us and are part of our relationship with others. We understand the importance of space through relationships with others.
10	"The Robbers"	Individual graphic representation of the classroom. Knowledge of the classroom and classmates, dealing with a map. The game of hide-and-seek with objects: who has hidden it, and what are their preferences?
11	"The Night Watchman"	From territorial concepts to experiences. Real scale game. Jobs and real life. Recreation of streets and houses.
12	"Final Drawings"	Consolidation and participation in learning. Final comparison and analysis of the changes at the end of the teaching process. Evaluation through a free drawing with the same instruction and technique as in the initial activity.

Source: own elaboration

4. RESULTS OF THE DIDACTIC INTERVENTION: THE SET ON GEOGRAPHIC LITERACY

As indicated in the previous section, the set of activities started with brainstorming. It was necessary to discover the students' ideas to adapt us to their needs and to prepare the activities. This section summarizes the results by sessions and activities, contextualizing some of the objectives that aided in achieving the goals proposed by the project on geographic literacy.

At the beginning of the didactic project, the activity "First Drawings" was used to determine the students' mental image of the town of Benigànim. A worksheet was designed in which the following question was asked: How would you draw your town? More complex or abstract terms such as locality, geographic space, or habitat were omitted so that the students would understand the main objective. Analysis of all the drawings revealed a common pattern in eight of the ten examples: the reference to one's home, i.e. the closest place where a sense of belonging develops through the family.

Figure 1. Two initial representations of the town of Benigànim from the participants in this study.

With those first results showing that home was the most important lived space for students, the second activity began with a higher level of abstraction. The teacher presented a silhouette to the group, and some children associated it with animals, vehicles, and other elements. In a dialogue with the group, the teacher redirected the situation by asking learners to imagine that the silhouette was their room and asking them about the objects they could find there and their location. Consequently, learners discovered that the silhouette corresponded to the outline of the town (Figure 2). That comparative analysis helped students understand the locality concept through an abstract shape. The goal was to consolidate the concept of a map or a representation of space on paper.

Figure 2. Initial representation of a map of Benigànim.

In the third activity, "Small Big Committee," geographical location was taught through awareness of the presence of students in a specific area. The teacher prepared a series of objects for each learner based on their tastes, interests, and personality traits. Each object was meant to identify the meaning of some proverbs, which were recited to help the participants link that object with the location. Those objects led to some streets, and subsequently, a conversation was initiated between the students and the teacher to get to know the streets and other associated elements. With those connections, learners were able to detach themselves from their own homes and identify with other spaces, developing a certain empathy shared with their classmates who introduced them to other places. Those ideas were successfully shared because each student dealt with wrapped objects, which facilitated dialogue between the children and encouraged further interactions.

Figure 3. Image of the objects for the third activity.

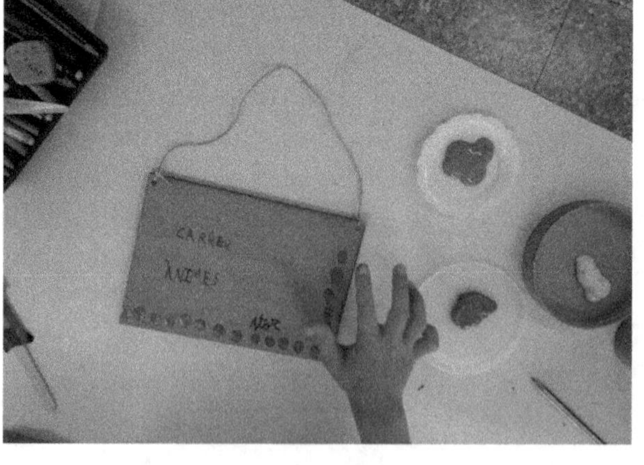

For the fourth session, "Unsuspected Signal," the teacher previously contacted the students' parents to collect information about the streets where they lived and their (possibly historical) relation with those families. Students learned to recognize their address in the context of the street map and learned about some aspects related to the nomenclature of that passageway (Figure 4). The brief presentation of ideas encompassed the children´s protagonism ("I" and "me") and extended to the immediate surroundings (other houses in their street) and the place itself through peer narratives and family stories. The teacher explained the meaning of the street names and provided children with other landmarks to expand the location in more detail.

Figure 4. Materials to work with street names

The next activity, "Hiding Places," required an interview with families about the places they visited in the neighborhood that they considered important for their children's education. Based on that data, the teacher created a collage of pictures of those town areas (Figure 5). A photo of each student was placed in the area he/she had not visited, thus arousing curiosity to find themselves. Students were able to recognize themselves without having been to that place, assuming that the collage situation was real. Eventually, the students recognized different places in the city without having made a trip; they correctly located those places and marked them using icons.

Figure 5. Collage with pictures of students and places in Benigànim

In the sixth session, "Circles with Stories," the teaching strategy took a turn compared to the previous activity. Until now, geographic space was represented by images that provided information about some spatial or temporal context. In this activity, however, models were created. Those contained information about the changes in selected places (Figure 6). That information included customs and cultural changes related to the local identities. Students felt identified with some of those activities that reinforced family ties (connection with their grandparents and parents). As we can see, meaning could be attributed to geometric shapes selected for the models, thus reinforcing emotional attachment and a sense of belonging. For example, a story can be linked to the circle representing a traffic roundabout. It helps to understand the geometric shape, its location, and its associated road safety.

Figure 6. Elements of models made by students

For the posterior activity, "Water Resources," students had already established an emotional attachment to the town. The session aimed at explaining the importance of wells as communal spaces, their role in the town development, and the use of those facilities. Therefore, the seventh activity focused on water supply and its spatio-temporal significance (Figure 7), such as exploitation, transportation, and access to that service in the past and nowadays.

Figure 7. Experiments with water to simulate water resources

The eighth session was related to the previous activity and it was dedicated to the profession of "The Messenger." That is a job where a person is responsible for taking deliveries from one place to another (nowadays similar to a "freelancer"). To do this, each student had to take on the role of the messenger; he or she had to find their way around a three-dimensional map of the town, which had already been practiced in the previous activities. The teacher used colored caps that the messenger had to carry from one place to another, while the other students described the places he or she passed by using the address and directions. The imaginary character (with which a certain empathy was developed) encouraged spatial thinking. On those routes, students had to name the elements they encountered (buildings, people, asphalt, trees, parks...) and understand the distance, urban morphology, and the period of time from one place to another. Pupils used simple sentences to point out that it took longer to get from home to another place because it was further away; they also indicated why some buildings were smaller than, for example, the church bell tower.

The ninth activity, "My space," was about relationships with others and alterity from a social and spatial perspective. To do that, the teacher used a bottle symbolizing the space each student inhabited, filled it with pictures representing the relationship with other students, and drew each child in the space of their choice. Memory and emotions associated with that space influenced the representation of events and places that the students described. They demonstrated extensive geographical knowledge and showed emotional closeness to things they expressed orally or through some drawings.

In the tenth session, "The Robbers," each student made a map of the classroom with its most important elements (door, furniture...), which he or she handed to the teacher. The teacher gave each child a sealed envelope containing a classroom map with an object they had to find. Each pupil had a different indication of where the object was located. The aim of the activity was to link the room to its graphic representation. It was demonstrated that some spatial corners were most frequently visited by the pupils. No questions were asked about the emotions those places triggered in the pupils; however, it was discovered which objects were hidden.

The penultimate activity recreated the job of a night watchman. It explored the idea of geographic space from imagination to experience. Until then, space had been approached through emotions, but social relationships and their spatial mediation had to be analyzed. For that purpose, houses were built out of cardboard to replicate the houses and streets of the town and to take on the role of a night watchman (Figure 8). The night watchman was a security guard who patrolled streets at night to ensure the safety of the neighborhood and its houses. The spatial categories ("near" and "far"), the knowledge about that profession, and some associated values were transmitted in class.

Figure 8. The houses and the model built by students

At the end of the project, the students were asked to create a new representation of the town of Benigànim (Figure 9). Their drawings showed more than one house, with significant changes in proportions and distances and new elements that were not present in the initial representations. Among those components, a total of sixteen people were displayed, most of whom were identified as family members.

Figure 9. Example of two representations at the end of the teaching project

The evaluation process was constant during the development of the didactic project. The teacher had a class diary, and the observation system was used to record progress and obstacles in the development of activities. Given the diversity of the tasks and their relationship with the construction of geographical thinking, specific evaluation criteria were designed, which corresponded to some techniques and instruments explained in Table 4.

Table 4. Proposed evaluation system to measure the learning process

Evaluation criteria	Techniques and instruments
Understand the place that each child occupies in the local society	Initial drawing and the silhouette of the town map (Activity 2)
Identify the places learned through the emotions produced	Photographs and cards (various activities)
Relate activities to places: locations, traffic, and information	Symbolic play (Activity 8)
Become aware of the transition from the abstract to the concrete (study of space)	Model of the town (Activity 8) Final drawing
Build a social bond through the place in which each child is located	Graphic representation of the classroom (Activity 10)
Add an emotional component to verbal and gestural language	Systematic observation. Teacher's diary. Final drawing

Source: own elaboration

5. DISCUSSION AND CONCLUSION

The main goal of the study (to teach geographic space at a local level) was achieved, but not without some difficulties related to the ECE content and the didactic method. Regarding content, its interdisciplinary base was necessary for students to interact and make links between the concepts proposed in different sessions of the didactic project. In terms of the method, students paid more attention to recreational tasks than to other activities linked to learning objectives or curricular skills. Although emotional learning was more a means than an end, it ensured that students developed a geographical awareness of the space where they lived based on multidisciplinary knowledge derived from geography.

Students formed a worldview of the complex geographic space that lies between the geography of perception, behavior, and social representations (Souto, 2018). In this spatial representation, social relations are a product of spatially contextualized information since understanding geographic space cannot be separated from socio-emotional content that provides students with a sense of belonging. The spaces evoke experiences that give meaning to the fundamental concepts interwoven; those could be revised through activities of the didactic project. The constant mention of local geographical knowledge awakened the emotional connection and gave students confidence when they participated in educational activities.

The result of the didactic project also involved other learning experiences associated with the ECE stage. The students acquired habits of personal hygiene, correct sitting posture, silence, and other similar habits, which, according to their families, were also trained outside school. The support of the families was an incentive and a challenge, especially because of the integration of cultural and linguistic diversity, which required geographical and pedagogical project management at the local level. In the didactic planning of the geographical content, students with auditory and visual deficit were taken into account; linguistic diversity beyond the local context (Valencian and Spanish) was also taken into consideration. To attend students with hearing impairment, we worked with digital whiteboards as a visual reinforcement. Lip-facial reading formed part of some activities that were developed within the classroom, fostering teamwork in small groups. Regarding visual impairment, the arrangement of the furniture was taken into account. Logical order with the predominance of sensory activities (especially touch) was applied to promote the teaching-learning process. It is relevant to mention that no integration problems were detected for those students. Regarding linguistic diversity, vocabulary and communication were partially worked in the Valencian language so that the pupils could identify some concepts (toponyms, street names...) in the local language.

Geographic literacy (with the comparative method at the forefront) required a progressive approach with guidelines to bring pupils closer to geographic space. On the one hand, geographic space was used as a primary concept (Araya and Cavalcanti, 2018). However, its polysemy was taken into account based on landscapes (lived space) and their perceptions (the perceived space); in other words, humanistic geography influences the sense of place. On the other hand, some geo-spatial identity was achieved by linking the structuring concepts to geographical competencies or skills. All of those formed the basis for students' socio-spatial thinking (Table 5).

Table 5.Complementary acquisition of structuring concepts and geographical competencies for each session and activity

Sessions	Activity title	Structuring concepts (Granados, 2019) and geographical competencies (Case & Wearing, 2019).
1	"Initial Drawings"	Location and spatial factor.
2	"A Different View"	Location, distribution, and spatial factor.
3	"Great Little Committee"	Location, perception, and geographic perspective.
4	"Unsuspected Signal"	Location; evidence and interpretation.
5	"Hiding Places"	Location, distribution, and spatial factor.
6	"Circles with Stories"	Location, distribution, patterns, and trends.
7	"Water Resources"	Distribution, interaction, and interrelationships.
8	"The Messenger"	Perception, organization, and geographic perspective.
9	"My Space"	Perception, organization, and geographic perspective.
10	"The Robbers"	Perception, organization, and geographic perspective.
11	"The Night Watchman"	Perception, organization, and geographic perspective.
12	"Final Drawings"	Perception, organization, and geographic perspective.

Source: own elaboration

The structuring concepts of location, distribution, and perception appeared in all the sessions, either separately or together. Those concepts were complemented by spatial-factor competences, patterns, trends, and geographic perspective, among others. This means that Piaget's psychoevolutionary stages were surpassed in the understanding of geographic space, and skills other than location and orientation were obtained, as found in the first studies on geographic literacy (Bluestein & Acredolo, 1979). The students started from an initial representation in which their home and immediate environment were considered as lived spaces (beyond the rooms or interior spaces that were not depicted). In this case, the progression in geographical awareness delved into the premises of Piñeiro's (1987) research. The geographic literacy achieved was less dependent on the children´s bodies, and a lower degree of syncretism was observed towards the end of the didactic project.

It can be stated that the first objective of the study, that was, bringing students closer to their geographic space based on the recognition of social and cultural elements of their immediate reality, was met. One of the relevant findings of this study was that we managed to conduct didactic work with fictitious spaces, especially in the fifth activity, in which the teacher linked a photograph of each student to a place in the town where it had never been taken. That decision was not a coincidence, but within the context of COVID-19, with its corresponding movement restrictions and several pandemic waves in the town during a short period of time (Vañó, 2021). This collage served as cartography to narrate geographic spaces, inspiring creativity, teaching socio-spatial relationships, and fostering geographical awareness, taking into consideration the students´ abilities. With that collage, an aspirational map was created (Musgrave, 2020), which enhanced the learning context of the fish morphology that appeared in the second activity, aiding to understand the boundaries in the area.

Model construction illustrated the students´ commitment to subjective spaces. It differs from Piñeiro´s (1987) idea to teach location and direction at the scale of some internal space, such as a household room or a school classroom. In this study, the models represented space outside school, i.e. geographic spaces to which students attributed codes that provided patterns and trends for territorial reality. Those patterns and trends served as a basis for developing interrelationships between people and the natural environment based on water resources in the seventh activity. In fact, reflection on the uses and functions of water in the study by Muñoz and García Monteagudo (2021) favored the development of historical empathy. It demonstrated that geographic space and historical time were two inseparable variables of teaching and learning at the ECE stage. That conclusion was also evidenced in the research by Jaráiz et al. (2016) on the movements of the ECE students.

Regarding geographic literacy, the didactic project was composed of progressive levels in the complexity of its activities. The results show that the basic-level skills were acquired, and the participants learned the names of the places and their location. Moreover, training in other structuring concepts, such as distribution, interactions, and organization, provided an optimal level of geographic literacy that went beyond basic spatial reasoning skills (Dikmenli, 2013; Martin & Owens, 2005; Massey, 2005; Palmer & Birch, 2004). Geographic competences formed the basis of structuring concepts, such as a spatial factor, patterns, trends, and a geographic perspective, which ensured reaching an intermediate degree of geographic literacy. That meant that the relationships between human beings and the environment were acquired (evidenced in the activities related to jobs and water resources). Other skills for processing geographical information were developed via photographs, models, and games. However, it cannot be claimed that a high level of geographic literacy was achieved, although a certain geographical awareness, typical of geographical analysis, was developed (Case & Wearing, 2016). That was confirmed via drawings in the final activity, where organization skills were displayed in line with the study by Canet et al. (2019). All that does not mean that geographical awareness related to a concerning issue of natural resources management was understood. It required a higher level of geographic literacy (Memişoğlu, 2017). Nevertheless, it does confirm that the students acquired a certain vision on sustainable citizenship (Benavent, 2022).

An intermediate level of geographic literacy means that students have developed a considerable base for geographical awareness. The local territorial reality was approached from a geographical perspective, particularly emphasizing on social relations and identity. The students reflected on space and were immersed in various spaces, thus acquiring an awareness of that locality. The majority of students understood that people belong to some geographic space, which they represented in a creative and critical way. It led them to perceive that they were important players in that local geographic space. That is how geographical awareness was established, which is, the final stage of geographical thinking according to Ortega Rocha (2020). As advocated in a number of curricular frameworks and the territorial innovation and citizenship project "We Propose" (Claudino & Mendes, 2021), students should finish the ECE stage with geographical knowledge that prepares them to make some territorial decisions in tune with cognitive maturity that leads to their personal autonomy. Therefore, geographical knowledge is based on values and attitudes that provide foundations for geographical narratives that empower students in the development of global citizenship (Bednarz & Lee, 2019; Flores et al. 2023; Maude & Caldis, 2019). Consequently, the second objective of this research was met, which was to develop a set of didactic activities that would motivate students to reveal and recognize emotions linked to their immediate space.

During the design and implementation of the project, teachers played a key role in setting basic foundations for geographic literacy. It has been manifested that the ECE reality can be represented via a number of cartographic sources, thus enriching the student's personal experiences in understanding space through participants´ emotional ties and in combination with physical proximity (Comes, 2002). In this research, three elements of geographic literacy outlined by Memişoğlu (2017) have been met. First, knowledge of the physical environment was achieved. The elements that build the reality of the local space were presented in a global and holistic way adapted for the ECE (Aranda, 2016). Second, map skills were accomplished. In addition to maps, we worked with photographs and models that made geographical knowledge more accessible to students. Third, concepts related to geography were trained. Along with the structuring concepts already mentioned above, other elements were included, such as work with maps, with water resources, with trades, with jobs, and with attitude-related concepts linked to education in values for the ECE stage. Having worked with those three elements of geo-literacy ensured competence-based learning for students (Massey, 2005; Palmer & Birch, 2005) who became familiar with the language of geography (Souto, 2018).

This study contributed to the (limited) research on geographic literacy in the ECE, which focused on the subjective understanding of geographic spaces. We did not distinguish between the morphological categories of town and city, as Piñeiro (1996) did, but the students understood that the town of Benigànim was the product of a complex representation in which the lived, the perceived, and the conceived spaces interacted. Due to COVID-19 restrictions, field trips could not be carried out, so we reflected on the teaching method and the didactic opportunities to teach geographic space. It allowed us to combine citizenship with other (artistic) skills, which occupy their place in the interdisciplinarity of geography (Fernández Portela, 2016).

The action research methodology allowed us to fill in some gaps in practice that were not foreseen in the previous study by Vañó (2021), which led us to this research, carried out in the second half of 2021 with a high participation rate (Pérez et al. 2010). Teamwork and continuous reflection stimulated by this methodology produced a curriculum adapted to the moment and the circumstances during different training sessions. The opinions and attitudes of the pupils, as well as the opportunities that presented themselves on site, determined the concrete measures for the content proposal during the project implementation (collection of information from various sources). Thus, when initial contact between students and the local environment was established (Well & Ebans, 2003), both pre-service and in-service teachers positioned themselves in a socio-critical paradigm, assuming a reflective and research-based function for their own practice (Pruzzo, 2014), as recommended by didactic innovation projects (González Iglesias, 2003; Jardón, 2019; Souto, 2017).

In consequence, this study attempted to provide a different perspective that encourages curiosity, discovery, and interest in geographical education from an early age. The link between family and school, the knowledge of society, and local spaces were the catalyst to expand socio-geographical knowledge since childhood by means of simplified language and emotions in order to reach an abstract territorial understanding. Reinforcement of geographical knowledge begins with a drawing and continues with the representation of shapes and silhouettes of animals, among other elements typical for the ECE. However, the teacher's goal is to constantly guide and realign that perception in order to establish a connection between previous and learned knowledge. Fantasy and imagination, as mentioned by Egan (1999), take on a pedagogical relevance that has been demonstrated with Benigànim students.

To conclude and to get back to the title of this study, emotions have been the most important means of communication in this project on geographic literacy. Language transmits information, but linguistic diversity in a context such as a small classroom demands a universal and inclusive language, which compensates for some sensory deficit to achieve an approach to geographic space. Emotions affect the development of identities that are essential to ensure personal autonomy, relationship with reality, and communication in all its forms. Therefore, the third objective has been met, which was to promote habits related to the appreciation of society and local culture, as well as to promote elements that contributed to the construction of identities. It should be kept in mind that the results of this case study were contextualized for a specific area and that the main goal was to construct the basis of geographical thinking performed in a small group of the ECE students.

Finally, the original focus of this study opens future prospects related to geography learning. Teachers and students should be trained in critical teaching of relevant social problems (from this perspective, skills-based learning is achievable from an early age). Furthermore, it is beneficial to introduce new methodologies such as, for example, service-learning, which favors the socialization of students and citizenship education. Overcoming the boundaries of early childhood education and seeking relationships with primary education could eventually aid in making citizens geographically literate. Those recommendations have a direct application in any school context whose teaching team seeks to promote geographical thinking.

FUNDING

This study is developed within the framework of the knowledge generation project I+D+i 'INCLUCOM-Curricular models and historical-geographical competences of teachers for the construction of inclusive identities (reference PID2021-122519OB-I00)', funded by the Spanish Ministry of Science and Innovation and by 'FEDER: A way of doing Europe'. It also belongs to the Emerging Educational Innovation Project (PIEE) of the University of Valencia: 'Teacher training and praxis in socio-environmental problems on an international scale: analysis and implementation of didactic proposals for sustainable citizenship (reference UV-SFPIE_PIEE-3328962)'.

6. REFERENCES

Agencia Estatal Boletín Oficial del Estado del Gobierno de España. (2024). *Real Decreto 1630/2006, de 29 de diciembre, por el que se establecen las enseñanzas mínimas del segundo ciclo de Educación infantil.* https://www.boe.es/eli/es/rd/2006/12/29/1630

Alonso-Sainz, E. (2022). Las TIC en la etapa de educación infantil: Una mirada crítica de su uso y reflexiones para las buenas prácticas como alternativa educativa. *Vivat Academia.Revista de Comunicación*, 155, 241–263. DOI: 10.15178/va.2022.155.e1371

Aranda, A. M. (2016). *Didáctica de las ciencias sociales en Educación Infantil.* Síntesis.

Araya, F., & Cavalcanti, L. (2018). Desarrollo del pensamiento geográfico: Un desafío para la formación docente en Geografía. *Revista de Geografía Norte Grande*, 70(70), 51–69. DOI: 10.4067/S0718-34022018000200051

Bednarz, R., & Lee, J. (2019). What improves spatial thinking? Evidence from the Spatial Thinking Abilities Test. *International Research in Geographical and Environmental Education*, 28(4), 262–280. DOI: 10.1080/10382046.2019.1626124

Benavent, M. (2022). Una mirada constructiva a la realidad: los ODS en educación infantil. Una propuesta de promoción de la conciencia socio-ambiental. En D. García Monteagudo, N. Palacios, L.A. Rodríguez y C. Fuster (coords.), *La enseñanza de problemas socio-ambientales para una praxis escolar crítica* (pp. 37-52). Nau Llibres.

Bluestein, N., & Acredolo, L. (1979). Developmental Changes in Map-Reading Skills. *Child Development*, 50(3), 691–697. DOI: 10.2307/1128934

Canet, S., Morales, A. J., & García Monteagudo, D. (2019). Pensar geográficamente en la educación infantil: De la imaginación a la construcción social del espacio concebido. *Didáctica Geográfica*, 19(19), 23–46. DOI: 10.21138/DG.415

Carmona, R. (2012). The obligations of States Parties under the Convention on the rights of the child: a focus on rights in childhood policies in Spain. *Educatio Siglo XXI, 30*(2), 69-88. https://revistas.um.es/educatio/article/view/153691

Case, R., & Wearing, J. (2016). *Teaching Geographical Thinking. Revised and expanded edition*. Royal Canadian Geographical Society.

Claudino Nunes, S., & Mendes, L. F. (2021). Project "We Propose!" Territorial citizenship and curricular innovation in portuguese geographical education. *Didáctica Geográfica*, 22(22), 47–71. DOI: 10.21138/DG.564

Comes, P. (2002). La interpretación de las imágenes espaciales. En P. Martínez e I. González (coords.), *La geografía y la historia, elementos del medio* (pp. 209-244). Ministerio de Educación, Cultura y Deporte del Gobierno de España.

Corona, A. (2020). El sentido de pertenencia, una estrategia de mejora en el proceso formativo en las artes. Estudio de caso en Danza en una universidad mexicana. *Páginas de Educación*, 13(2), 59–79. DOI: 10.22235/pe.v13i2.2172

Dikmenli, Y. (2013). Geographic literacy perception scale (GLPS) validity and reliability study. [MIJE]. *Mevlana International Journal of Education*, 4(1), 1–15. DOI: 10.13054/mije.13.43.4.1

Egan, K. (1999). *Fantasía e imaginación, su poder en la enseñanza primaria: una alternativa a la enseñanza y el aprendizaje en la educación infantil y primaria*. Morata.

Fernández Portela, J. (2016). El estudio del espacio geográfico cercano en educación infantil: itinerario didáctico para el conocimiento de la ciudad de Valladolid. In Alanís, L., Almuedo, J., de Oliveira, G., Iglesias, R., & Pedregal, B. (Eds.), *Nativos digitales y geografía en el siglo XXI: educación geográfica y sistemas de aprendizaje* (pp. 484–496). Grupo de Didáctica de la Geografía de la Asociación de Geógrafos Españoles y Universidad Pablo de Olavide.

Ferrer, M. J., & Riera, M. A. (2015). Relaciones y encuentros con las familias en Educación Infantil. *RELAdEI: revista latinoamericana de educación infantil, 4*(2), 27-42. https://revistas.usc.gal/index.php/reladei/article/view/4879

Flores, C., De la Cruz, G., Domínguez, E., & Pérez, B. D. (2023). Revisión teórica acerca del conocimiento geográfico potente. Un nuevo enfoque conceptual en la educación geográfica. *Didáctica Geográfica*, 24(24), 17–38. DOI: 10.21138/DG.666

García Monteagudo, D. (2022). De la representación del espacio a la construcción del pensamiento geográfico en Educación Infantil. En A. Paz, V. Figueora, E.J. Rodríguez y A.J. Montes (coords.), *Libro de Actas del 2º Congreso Caribeño de Investigación Educativa: Nuevos paradigmas y experiencias emergentes* (pp. 437-443). Instituto Superior de Formación Docente Salomé Ureña.

García Monteagudo, D. (2024). Formación inicial del profesorado de Educación Infantil en pensamiento geográfico: Paisaje e identidades territoriales. *ENSAYOS. Revista De La Facultad De Educación De Albacete*, 39(1), 57–73. DOI: 10.18239/ensayos.v39i1.3488

González Iglesias, E. (2003). El trabajo con mapas en Educación Infantil. Un recurso para la educación en valores. En M.J. Marrón Gaite, C. Moraleda y H. Rodríguez (coords.), *La geografía ante las nuevas demandas sociales* (pp. 89-101). Asociación de Geógrafos Españoles.

Granados, J. (2019). Definición y justificación de un marco conceptual para la didáctica de la geografía. In Hortas, M. J., Dias, A., & De Alba, N. (Eds.), *Enseñar y aprender didáctica de las ciencias sociales: la formación del profesorado desde una perspectiva sociocrítica* (pp. 41–49). Asociación Universitaria de Profesorado de Didáctica de las Ciencias Sociales., https://dialnet.unirioja.es/servlet/articulo?codigo=7609709

Hannoun, H. (1977). *El niño conquista el medio.* Kapelusz.

Herbst, E., Cruz, T., Bower, C. A., Hirsh-Pasek, K., & Michnick, R. (2022). Playing for the Future: Spatial Thinking Belongs in Preschools and Home Environments. In Lynn, A., & Thai, K. P. (Eds.), *Handbook of Research on Innovative Approaches to Early Childhood Development and School Readiness* (pp. 416–451). IGI Global., DOI: 10.4018/978-1-7998-8649-5.ch017

Inarejos, J. A. (2018). *Guía docente para la enseñanza de la geografía y la historia en educación infantil.* Servicio de Publicaciones de la Universidad de Extremadura. https://dehesa.unex.es:8443/handle/10662/7023

Jaráiz, F. J., Hernández, A. M., & Hernández, A. (2016). La percepción del espacio: estudio de caso. En L. Alanís, J. Almuedo, G. de Oliveira, R. Iglesias & B. Pedregal (coords.), *Nativos digitales y geografía en el siglo XXI: educación geográfica y sistemas de aprendizaje* (pp. 601-616). Grupo de didáctica de la Geografía de la Asociación de Geógrafos Españoles, Universidad Pablo de Olavide y Universidad de Alicante. DOI: 10.14198/2016-nativos-digitales-y-geografia

Jardón, P. (2019). El sentido de la enseñanza de la Geografía y la Historia. 30 años del proyecto GEA-CLÍO. *Ar@cne: revista electrónica de recursos en internet sobre geografía y ciencias sociales*, 235, 1-16. https://raco.cat/index.php/Aracne/article/view/361146

Karsten, L. (2005). It all used to be better? Different generations on continuity and change in urban children's daily use of space. *Children's Geographies*, 3(3), 275–290. DOI: 10.1080/14733280500352912

Kowalczyk, I. (2004). El Consejo de Europa: defensor de los derechos de las niñas y los niños. *Infancia en Europa: revista de una red de revistas europeas, 7*, 8-11.

López, M. J. (2017). Los problemas relevantes de la sociedad actual en un aula de 5 años: qué hacemos con las personas refugiadas sirias. *Investigación en la Escuela*, 92, 19-31. http://www.investigacionenlaescuela.es/articulos/R92/R92-2

Louv, R. (2005). *Last child in the woods: saving our children from nature-deficit disorder*. Algonquin Books of Chapel Hill.

Marrón Gaite, M. J. (1999). La geografía del comportamiento y de la percepción. Aportaciones a la investigación y a la enseñanza de la geografía. *Didáctica Geográfica*, 3, 85–108.

Martin, F., & Owens, P. (2005). Young children making sense of the world. In Scoffham, S. (Ed.), *Primary geography handbook* (pp. 63–73). Geographical Association.

Martínez, C. (2020). *Infancia, pandemia y derechos. Treinta años de la Convención de los Derechos del Niño en España*. Thomson Reuters Aranzadi.

Massey, D. (2005). *For space*. Sage Publications.

Maude, A., & Caldis, S. (2019). Teaching Higher-Order Thinking and Powerful Geographical Knowledge Through the Stage 5 Biomes and Food Security Unit. *Geographical Education, 32*, 30-39. https://eric.ed.gov/?id=EJ1238503

Memişoğlu, H. (2017). Opinions of Teachers and Preservice Teachers of Social Studies on Geo-Literacy. *Educational Research Review*, 12(19), 967–979. DOI: 10.5897/ERR2017.3315

Moreno, M. C., & Gallego, C. I. (2013). "Egipto en mi aula": didáctica de la geografía para educación infantil. En R. de Miguel, M.L. de Lázaro y M.J. Marrón (coords.), *Innovación en la enseñanza de la geografía ante los desafíos sociales y territoriales* (pp. 353-366). Diputación provincial de Zaragoza.

Muñoz, E. M., & García Monteagudo, D. (2021). El agua como propuesta didáctica piloto en el grado de maestro/a en Educación Infantil. En J. Granados y R.M. Medir (coords.), *Enseñar y aprender geografía para un mundo sostenible* (pp. 361-377). Octaedro.

Musgrave, R. A. (2020). Los niños no pueden explorar el mundo, pero sí cartografiarlo. *National Geographic*. https://www.nationalgeographic.es/family/2020/06/los-ninos-no-pueden-explorar-el-mundo-pero-si-cartografiarlo

Ortega Rocha, E. (2020). Concepciones del profesorado chileno sobre la conciencia geográfica. *REIDICS.Revista De Investigación En Didáctica De Las Ciencias Sociales*, 6, 126–143. DOI: 10.17398/2531-0968.06.126

Ortiz, A., Prats, M., & Baylina, M. (2012). Métodos visuales y geografías de la infancia: dibujando el entorno cotidiano. *Scripta Nova, Revista Electrónica de Geografía y Ciencias Sociales*, XVI (400). https://www.ub.edu/geocrit/sn/sn-400.htm

Palma, A. (2023). Didáctica de las Ciencias Sociales y Educación para la Paz. Valoraciones desde la práctica docente. *Revista UNES. Universidad. Escuela y Sociedad*, 14(14), 5–25. DOI: 10.30827/unes.i14.27121

Palmer, J., & Birch, J. (2004). *Geography in the early years*. Routledge.

Pena, M., & Caamaño, M. J. (2019). El libro informativo y juvenil: un recurso para la enseñanza y el aprendizaje de la geografía. En X.C. Macía, F.X. Armas y F.R. Lestegás (coords.), *La reconfiguración del medio rural en la sociedad de la información: nuevos desafíos en la educación geográfica* (pp. 581-590). Andavira.

Pérez, A. I., Soto, E., & Serván, M. J. (2010). Participatory action research and reconstruction of teachers' practical thinking: Lesson studies and central reflection. An experience in Spain. *Educational Action Research*, 18(1), 73–87. DOI: 10.1080/09650790903484533

Piaget, J., & Inhelder, B. (1947). *La représentation de l'espace chez l'enfant*. Presses Universitaires de France.

Piñeiro, M. R. (1987). La lectura del mapa en el niño de preescolar. *Revista interuniversitaria de formación del profesorado, 1*, 66-76. https://digitum.um.es/digitum/handle/10201/124224

Piñeiro, M.R. (1996). Representación del concepto de pueblo y ciudad en niños asturianos de 4 a 7 años. *Iber: didáctica de las ciencias sociales, geografía e historia, 9*, 27-36.

Pruzzo, V. (2014). La enseñanza entre la praxis y la poiesis. Aportes a la didáctica general. *Revista Pilquen. Sección Psicopedagogía, 11*(1), 1–13. https://revele.uncoma.edu.ar/index.php/psico/article/view/2103

Requena, M. (2000). La protección jurídica del menor en el ámbito del consejo de Europa: el Convenio Europeo sobre el ejercicio de los derechos del niño. En M. Calvo y N. Fernández (coords.), *Los derechos de la infancia y la adolescencia* (pp. 187-207). Editores Mira.

Rivero, M. ª.P., y Gil. A. (2011). Pensar y expresar el espacio en el aula de Infantil. En Mª P. Rivero (Coord.), *Didáctica de las Ciencias Sociales para Educación Infantil* (pp. 31-47). Editores Mira.

Ruiz de Velasco, A., & Abad, J. (2011). *El juego simbólico*. Graó.

Sáiz-Linares, A., & Ceballos-López, N. (2022). Una mirada a las geografías de la infancia desde el análisis de los patios escolares. *Educación Básica, Cultura, Currículo, 52*, e09886, 1-20. DOI: 10.1590/198053149886

Sammut, G., Andreouli, E., Gaskell, G., & Valsiner, J. (2015). Social representations: a revolutionary paradigm? In Sammut, G., Andreouli, E., Gaskell, G., & Valsiner, J. (Eds.), *Cambridge Handbook of Social Representations* (pp. 3–11). Cambridge University Press. DOI: 10.1017/CBO9781107323650.003

Sebastià, R., & Tonda, E. M. (2015). El concepto y representación del espacio geográfico en la enseñanza de la Geografía en los niveles educativos no universitarios. Análisis bibliométrico. In de la Riva, J., Ibarra, P., Montorio, R., & Rodrigues, M. (Eds.), *Análisis espacial y representación geográfica: innovación y aplicación* (pp. 1505–1514). Universidad de Zaragoza-Asociación Española de Geografía.

Souto, X. M. (2017). Propuestas para innovar e investigar sobre Conocimiento del medio en Educación Infantil y Primaria. *Biblio 3w: revista bibliográfica de geografía y ciencias sociales*, 22. https://raco.cat/index.php/Biblio3w/article/view/329017

Tonda, E. M. (2001). *La didáctica de las ciencias sociales en la formación del profesorado de Educación Infantil*. Servicio de Publicaciones de la Universidad de Alicante.

Turner, S., & Leydon, J. (2012). Improving Geographic Literacy among First-Year Undergraduate Students: Testing the Effectiveness of Online Quizzes. *The Journal of Geography*, 111(2), 54–66. DOI: 10.1080/00221341.2011.583263

Uz, C., Anteneh, M. R., & Thompon, M. (2021). What's So Special About Spatial?: A Review Study Joining Virtual Reality and Spatial Ability. In Russell, D. (Ed.), *Implementing Augmented Reality Into Immersive Virtual Learning Environments* (pp. 56–73). IGI GLOBAL., DOI: 10.4018/978-1-7998-4222-4.ch003

Vañó, C. (2021). *Las emociones del espacio vivido. La identidad desde el ámbito local.* Trabajo Final de Grado inédito. Grado en maestro/a en educación infantil. Universitat de València.

Vara Muñoz, J. L. (2008). Cinco décadas de Geografía de la percepción. *Eria*, 77, 371–384. DOI: 10.17811/er.0.2008.371-384

Wells, N., & Evans, G. (2003). Nearby nature. A buffer of life stress among rural children. *Environment and Behavior*, 35(3), 311–330. DOI: 10.1177/0013916503035003001

KEY TERMS AND DEFINITIONS

Alterity: the ability to expect and recognize differences in a common scenario (engagement) without oppression or empowerment. Those hinder the construction of flexible and multiple identities that serve as a key to citizenship.

Complex geographic space: concept linked to the mental representation that a person constructs about reality, a result of the juxtaposition of feelings, perceptions, and consciousness that make up that territory.

Early Childhood Education: in Spain, this is the school level with its own identity, aimed at a specific age group: girls and boys from birth to the age of six. Its purpose is to contribute to the holistic and harmonious development of students in all its dimensions: physical, emotional, sexual, affective, social, cognitive and artistic, as well as the education of civic values for living together. It is divided into two cycles, both of which are educational in nature: the first covers the age group up to three years, the second up to six years.

Geographic literacy: a set of skills to manage information and geographic language (verbal, graphic, iconic, and cartographic) in order to make decisions that affect citizenship. Those observation skills, analysis, interpretation of information, and geographical language can be learned within basic education system.

Geographical thinking: a way of understanding reality from a geographical perspective. It involves reflection on space and the development of an affective, conceptual, and critical view. It results in people´s awareness of their actions when they make decisions related to citizenship.

Identities: a dual construction process (individual and collective) that consists in the integral development of a feeling of multiple and multi-scalar belonging, which is required to achieve the notion of full and complex citizenship.

Map: a graphic representation of selected facts or aspects of the Earth's surface in two dimensions, drawn on a reduced scale and with a vertical perspective corresponding to that surface.

Mental map: a spontaneous depiction of images, some subjective space that a person records on a physical or electronic medium. That is a reflection, part of the mental processes that one acquires, captures, stores and identifies from spatial information which explains the reality to that person.

Sense of belonging: the notion of what a person experiences in relation to other groups, communities, and organizations. It allows people to be part of something and to feel accepted by others expressed through different forms of communication and via evaluation by others.

Chapter 2

The Teaching of Physical Geography in Teacher Training:

20 Years of Experiences in the Organization of Educational Field Trips in Protected Natural Areas

Óscar Jerez García
https://orcid.org/0000-0001-6020-5268
University of Castilla-La Mancha, Spain

Manuel Antonio Serrano de la Cruz Santos-Olmo
https://orcid.org/0000-0003-0122-4877
University of Castilla-La Mancha, Spain

ABSTRACT

Physical Geography has used educational itineraries in natural environments as a key tool to promote practical learning and environmental awareness. In Castilla-La Mancha, National and Natural Parks serve as ideal settings for training activities organized by the Department of Geography at the University of Castilla-La Mancha, primarily aimed at future Primary Education teachers. Over two decades, more than thirty field trips have been conducted with nearly 1.500 students, documenting and analyzing learning outcomes. The hypothesis is that students have limited knowledge of local natural areas, making it crucial to provide them with the means to know and appreciate these environments. A model has been developed that combines theoretical and methodological foundations to design school excursions, thereby promoting an understanding of landscapes and the ability to organize educational

DOI: 10.4018/979-8-3693-7185-5.ch002

Copyright © 2025, IGI Global Scientific Publishing. Copying or distributing in print or electronic forms without written permission of IGI Global is prohibited.

outings that integrate geographical knowledge with environmental awareness.

1. INTRODUCTION

Geography, and particularly Physical Geography, has employed educational itineraries in natural environments as an essential tool to promote spatial learning and environmental awareness among students at various educational stages, from Early Childhood Education to higher education, but especially in Primary and Secondary Education. These itineraries allow students to apply theoretical knowledge in a practical context, thereby promoting engagement and the appreciation of actions aimed at solving environmental and territorial problems. Castilla-La Mancha has a rich network of protected natural areas, highlighted by two National Parks and seven Natural Parks, which provide ideal settings for conducting educational activities in the field of Physical Geography.

The Department of Geography at the University of Castilla-La Mancha has organized various educational activities in these natural spaces over the past two decades, aimed especially at future Primary Education teachers, but also at students from other university degrees who may potentially become Secondary Education teachers. These experiences aim to document and analyze learning outcomes, based on course reports and organizing educational itineraries that integrate geographical knowledge from a didactic perspective.

It is postulated that students possess insufficient knowledge about the most relevant natural spaces in their surroundings, a deficiency evidenced in the initial assessments of the didactics of Geography courses in the initial teacher training programs. Consequently, it is imperative, first, to provide future educators with the necessary resources to acquire a deep understanding and a critical appreciation of these environments. Second, it is essential to develop and propose a pedagogical method that empowers trainee teachers to competently and effectively organize and carry out school field trips.

Considering both objectives, one conceptual and theoretical and the other methodological and practical, the main results obtained from more than thirty field trips organized during the period 2004-2024 are presented, in which nearly 1,500 students have participated. This extensive experience in organizing field trips has allowed for the development of a theoretical model: a didactic proposal for designing school excursions.

This design addresses the theoretical and methodological foundations necessary for organizing field trips and for identifying and utilizing places of didactic interest in the teaching of the natural environment and landscapes. The approach seeks to enrich the understanding of geographical landscapes and their dynamics through education, emphasizing the importance of selecting significant locations that facilitate contextualized learning and stimulate students' critical thinking.

Through this didactic model, a greater understanding of geographical processes and a deeper appreciation of the natural environment are promoted. Furthermore, it seeks for future educators to develop the ability to plan and execute educational excursions that not only convey geographical knowledge but also foster environmental awareness and commitment to the conservation of the environment. Therefore, the work of the Department of Geography at the University of Castilla-La Mancha, through the organization of educational itineraries in National Parks and Natural Parks, has proven to be an effective strategy for enhancing the training of future teachers.

1.1. Physical Geography, Landscape, and Education

In the teaching of Physical Geography, it is essential to address the importance of integrating scientific research within pedagogical practice, especially in the school context. It is important that teachers not only transmit knowledge but also apply the results of their research in the classroom, seeking to bridge the gap between research and educational practice, as Martínez (2000) points out. This disassociation limits educational effectiveness and overcoming it could significantly improve the didactic preparation of teachers.

The concept of "didactic transposition," proposed by Chevallard (2000), is central to this process. This transposition involves adapting and transforming scientific knowledge so that it can be taught within the school context, which not only entails a simple reduction of content but also a proper reworking for teaching purposes, as explained by Rodríguez (2000). In this regard, teacher-researchers must consider key questions about what to teach, why, and how to do it, linking science, pedagogical principles, and didactic methods with curricular objectives (Souto, 1999). This approach seeks to balance "scientific knowledge" with "knowledge to be taught" (González, 2002), promoting a dialectic between scientific research and its educational application, and avoiding both the absolutization of science and didactics devoid of content (Sánchez, 2005). Other approaches propose analyzing the origins, development, and challenges of problem-centered socio-environmental curricula as an alternative to disciplinary academicism, linked to the crisis of the liberal order and aspirations for democratic, inclusive, and general education, while reviewing its reactions, obstacles, and future projections (Romero, 2021). In other cases, the interdisciplinarity of environmental knowledge is addressed from various sciences and disciplines (Gavidia et al., 2015), or the heritage value of landscapes is examined (Gómez, 2013).

Physical Geography, as a scientific discipline, has multiple functions that encompass cultural, scientific, and practical aspects. Geography not only expands human spatial knowledge but also provides rational methods for studying this dimension and facilitates spatial planning. In addition to these general purposes, Physical Geography plays a crucial role in developing an ethic of conservation, promoting sensitivity and admiration for nature, which gives it significant social responsibility (Porrit, 1988, in Palacios, 1990).

In recent decades, the concept of natural landscape, a subject of study in Physical Geography, has gained importance, being regarded as both a resource and a heritage. Since the 1960s and 1970s, it has been recognized that the landscape is an essential environmental resource, comparable to other traditional resources such as water or vegetation (Cancer, 1999). This conception of the landscape is accepted by both scientists and society at large from two perspectives: as a resource that requires protection from degradation and as a resource that influences the development of territorial activities, such as recreational and tourism-related activities.

According to Cancer (1999), the landscape can be categorized as a natural, cultural, scientific, educational, or recreational resource, among others, and it often presents multiple qualities simultaneously. The increasing attention to applied landscape studies reflects this evolution. Many physical geographers have paid more attention to applied issues, which is evident in the inclusion of chapters dedicated to practical applications in various books and the rise of specific works on these topics (Unwin, 1995).

Within Physical Geography, several main areas of focus are identified: the chronological study of landscape and environment evolution; the analysis of natural processes; the effects of human activity; the concept of systems as an underlying paradigm; the temporal interpretation of the physical environment; and the application of geographical knowledge in environmental planning (Sala and Batalla, 1996). In this way, the landscape is increasingly approached as an economic, environmental, and educational resource, fitting within the realm of applied science. This approach involves studying the construction and evolution of landscapes, their current uses, their valuation, and their planning and management (Sanz, 2002). The landscape is a natural, historical, social, and cultural asset that, as part of our common heritage, requires both scientific and administrative attention. This trajectory is summarized in the activities of the Landscape Group of the Spanish Association of Geography (Alomar-Garau & Gómez-Zotano, 2022). Landscape studies, in addition to their scientific function, have an important applied role. These studies are essential for better grounding territorial planning tasks, contributing to the proper conservation of geographic spaces that present high-quality scientific, educational, and environmental values (García, 1995).

In short, the search for a connection that facilitates didactic transposition is proposed, allowing scientific knowledge of Physical Geography about the natural environment and the landscape to become effective and meaningful didactic knowledge for students.

1.2. Physical Geography in the School Context. The Natural Environment in the Spanish Curriculum

Physical Geography can be defined as the branch of this scientific discipline that deals with the physical or natural aspects of the Earth, encompassing the study of processes and patterns in the natural environment, such as relief, climate, waters, and natural landscapes (Aguilera et al., 2009). The right to know, study, and enjoy a healthy nature and environment is recognized by numerous national and international organizations. This right is acknowledged by the Convention on the Rights of the Child (UNICEF, 1989), Agenda 21, the Earth Charter, the Millennium Development Goals, the Sustainable Development Goals, and, in the case of Spain, by the Spanish Constitution itself in Article 45.

In the Spanish educational system (LOMLOE, 2020), there is no subject, matter, or area of knowledge called Geography within Basic Education, which includes the stages of Primary Education and Compulsory Secondary Education (between 6 and 16 years old). There is a subject, but it is optional, in the Bachillerato stage, specifically in the second year, called Geography (which is actually Physical and Human Geography of Spain). In Early Childhood Education (from 0 to 6 years), there is an area of knowledge and experience called Knowledge of the Environment, which includes a Geography adapted to this stage. In Primary Education, until the last educational reform of 2020 and its curricular development in 2022, the area of Social Sciences included a block of content on Physical Geography, another on Human Geography, and another on History. The significance of Physical Geography is noteworthy, as it held the same curricular weight as Human Geography and even History, but in any case, it was included in the area of Social Sciences, not in Natural Sciences. With the change in the curriculum, both areas disappear and are merged into a single subject called Knowledge of the Natural, Social, and Cultural Environment. This integrative conception of the sciences has led to a reduction in the curricular content of Physical Geography, which is now shared with content from other social and natural disciplines and must be specified, curriculum-wise, in each of the 17 regions or autonomous communities in Spain that have educational competencies.

For example, in Castilla-La Mancha, some specific curricular competencies in this area of knowledge, related to the natural environment and landscape, are (Decreto 81/2022):

- "Identify the characteristics of different elements or systems of the natural, social, and cultural environment, analyzing their organization and properties and establishing relationships between them to recognize the value of cultural and natural heritage, conserve it, enhance it, and undertake actions for its responsible use, especially that of our autonomous community".
- "Identify the causes and consequences of human intervention in the environment from social, economic, cultural, technological, and environmental perspectives, to improve the ability to face problems, seek solutions, and act individually and cooperatively in their resolution, and to implement sustainable lifestyles that respect, care for, and protect people and the planet".

These competencies are associated with basic curricular knowledge that develops content related to landscape and natural spaces. For the first cycle, within the block of Societies and Territory, ecosocial awareness, the basic curricular knowledge includes:

- "Knowledge of our environment. Natural landscapes and humanized landscapes, and their elements. Human action on the environment and its consequences".
- "Natural spaces of Castilla-La Mancha close to the reality of the students: characteristics and peculiarities".

In the second and third cycles, landscapes are related, especially, to climatic issues, also within this block of Societies and Territories, both in the section of Ecosocial Awareness and in the Challenges of the Current World:

- "The main ecosystems and their landscapes".
- "The natural environment. The geographical diversity of Spain and Europe".
- "Natural and cultural heritage as an asset and resource; its use, care, and conservation. Protected spaces of Castilla-La Mancha and UNESCO-designated assets".
- "Climate change. Introduction to the causes and consequences of climate change, and its impact on the landscapes of the Earth. Mitigation and adaptation measures".

1.3. Physical Geography for the Development of Environmental Education in Protected Natural Areas

Within the multiple disciplines that contribute to the implementation and fulfillment of the fundamental principles of Environmental Education (EE), Geography, in general, and Physical Geography, in particular, have played a prominent role, providing ongoing and transversal support to this purpose. Particularly relevant is its contribution to the teaching-learning process oriented toward social transformation, where activities in direct contact with nature, such as geographical didactic itineraries, play an essential role.

Protected Natural Areas (PNAs) emerge as key references in the planning of environmental training programs and strategies. These areas have facilitated the generation of environmental awareness in line with the principles of conservation and sustainable management of natural resources, although challenges persist regarding the achievement of sustainable development goals from both social and naturalist or conservationist perspectives (Cid, 1997; Merino et al., 1998; VV.AA., 1998; Ayuso, 2006; Corrales & Garrido, 2024). Despite these limitations, PNAs offer exceptional opportunities to develop activities related to EE that can engage a society that is often disconnected from local spatial and environmental realities. The most relevant activities are those planned in specific programs that utilize the infrastructures and facilities of the areas themselves, forming part of the administrative management of these sites (Henares, 2015). However, the lack of active programs or the absence of specific approaches for certain groups, such as university students, positions PNAs as ideal outdoor laboratories for the development of educational initiatives, such as didactic itineraries, whether proposed by the administration or adapted to specific educational interests.

It is evident that there is still much progress to be made in this area, facing significant challenges such as the need to reorient the perception of natural spaces, valuing them not only as reservoirs of biodiversity but also as sources of goods and ecosystem services. It is also essential to strengthen the role of EE as a tool for citizen participation and to encourage the use of these spaces as observatories and laboratories of natural processes (Cid, 1997; Henares, 2015), which is fundamental to promoting understanding and awareness of the importance of the relationships between humans and nature.

In Castilla-La Mancha, PNAs form an extensive network that encompasses a significant set of protected areas aimed at nature conservation. This network includes 114 Protected Natural Areas: 2 National Parks, 7 Natural Parks, 22 Natural Reserves, 5 River Reserves, 28 Natural Monuments, 48 Micro-reserves, and 2 Protected Landscapes (https://areasprotegidas.castillalamancha.es). These spaces harbor a significant portion of the region's natural values, both in terms of geodiversity and biodiversity.

The Network of Protected Natural Areas of Castilla-La Mancha was included in the Regional Plan for Environmental Education (JCCM, 2003), and in the recent Environmental Education Strategy for Castilla-La Mancha, Horizon 2030 (Government of Castilla-La Mancha, 2020), reference documents that establish a set of measures to unify criteria, define lines of action, and set achievable and evaluable objectives in the field of EE. In recent decades, various experiences have been implemented within these plans, particularly the chapters related to natural spaces, addressing in part the lines of action already outlined in the Conservation Plan for the Natural Environment of Castilla-La Mancha regarding Public Use and Environmental Education (JCCM, 1995).

1.4. Socio-Environmental education: The Teaching of Physical Geography in the Natural Environment through Didactic Itineraries

Socio-environmental education is an educational approach that promotes the understanding and analysis of the interactions between human societies and the environment, aiming to foster responsible attitudes, values, and actions towards environmental and social sustainability. It seeks to empower individuals to identify environmental issues, assess their social and cultural impact, and actively participate in finding solutions. In this process, the teaching of Physical Geography, the natural environment, or natural sciences (or science) from a spatial and territorial dimension is included in formal education and, therefore, in the Spanish curriculum, from the Early Childhood Education stage to the Baccalaureate stage. Some studies analyze the training levels of Early Childhood Education teachers in geographical thinking through landscape teaching and territorial identities, revealing that most of the trainee teachers reach declarative levels, which explains some of the difficulties in applying geographical thinking in early childhood classrooms (García-Monteagudo, 2024). Various studies highlight the importance of conceptualizing, understanding, and valuing the landscape from a socio-environmental perspective (Romero & Luis, 2008; Hernández & García, 2022; García, 2023). A classic method for studying landscape and the natural environment from Geography, but fully renewed conceptually and methodologically, consists of fieldwork outings, didactic itineraries, interpretive routes, school walks, or hiking, concepts with different connotations but whose common denominator is based on active teaching outside the classroom (Jerez, 2019).It has been emphasized and reinforced numerous times from the field of Geography that certain field practices in the form of didactic itineraries constitute one of the most suitable and productive methods or exercises for addressing geographic teaching (García and Martínez, 1988; Palacios, 1988; Sánchez, 1996; García, 2004 and 2012; Jerez et al., 2014; Sousa et al, 2016; Crespo et al., 2018; Liceras, 2018, among others). The analysis and interpretation of the territory in situ aid comprehension, facilitate awareness of the visited spaces, and generate direct awareness of their most significant issues. All this helps to consolidate these exercises as one of the main tools for developing different geographic skills in students. These practices also become particularly interesting in the field of Environmental

Education (EE), with which they share many points of interest, as they all converge in the geographical space where all relationships between human beings and the natural environment unfold.

The constructivist approaches to learning typically used in these initiatives are based on considering the student's prior knowledge as the foundation upon which they must build their own. This establishes a competency-based learning that involves the student acquiring geographic and environmental content, skills, and values. Current regulations (both national and regional) emphasize the need for students to be aware of their natural, cultural, and social heritage, recognizing its cultural, political, social, and educational significance. However, teaching experience warns us of the limited knowledge and esteem that young people generally have for it. On the other hand, many teachers are concerned that their teaching practice directly connects students to their heritage, so that they know, love, respect, and conserve it while serving as a stimulus for reflection on natural and cultural variability and respect for it. In light of this, we assert that EE should be based on a research methodology by students, where guided contact with nature is reinforced as one of the main strengths for acquiring knowledge and values through the acquisition of geographic competence and environmental awareness, concretized in:

- Completing theoretical training with practical study in direct contact with nature, relating to various theoretical contents studied in the Geography subjects present in the curricula of the mentioned degrees.
- Knowing and valuing the diversity, variety, and uniqueness of territorial systems, landscapes, and protected areas in the local environment.
- Raising awareness and understanding of the main natural values and significant issues present in Protected Natural Areas (PNAs) and their areas of territorial influence.

The geographic didactic itinerary thus emerges as an educational strategy that consciously brings the individual closer to reality, serving as a valuable teaching and learning opportunity for both the teacher and the student. In it, the process of observation, information collection, interpretation, conjecturing, explanations, and projections is enhanced, enabling them to read, think, and reconstruct their social environment. Moreover, fieldwork is simultaneously a social activity and an opportunity to learn about and recognize the characteristics of the environment. These activities can even serve as an excellent complement to the training capabilities, where participants can acquire the skills to demonstrate the basic techniques and procedures needed to be competent in designing, programming, and executing them. The interpretative itinerary thus becomes a reference resource for exploring, diagnosing, recognizing, and proposing possible solutions to various local environmental issues.

The scientific literature on didactic itineraries focused from a geographical perspective, and more specifically in the natural environment, is extensive in Spain (García Ruiz, 1997; García de la Vega, 2004; Ramiro, 2005; García Martín et al., 2018), particularly highlighting those planned and implemented in protected natural spaces (Santana et al., 2016; Serrano de la Cruz, García and Jerez, 2016) and specifically in National Parks (Jerez, 2009; Crespo, 2012, among others).

According to this, numerous educators have considered nature as a pedagogical resource, apart from the anthropocentric view of nature presented by formal education. Classical pedagogues such as Rousseau, Dewey, Fröbel, Montessori, Manjón, Giner de los Ríos, Cossío, etc., have highlighted the importance of education in nature (Negrín, 2005; Martínez, 2012). There is consensus demonstrating that, for several decades, there has been a distancing between schoolchildren, children, and also adolescents, from the natural environment. School-aged children are increasingly less engaged with nature, while outdoor play spaces are also diminishing in urban areas. In the last years of the 20th century and the early years of the 21st century, a decline has occurred in developed countries regarding the time children dedicate to outdoor activities such as hiking, walking, playing in the field, in gardens, in parks, etc. (Clements, 2004). Some studies have shown that unstructured free play outdoors has decreased by about 9 hours over a 25-year study period (Hofferth & Sandberg, 2001). Children spend less time outdoors than their parents did when they were children. A notable British study shows that children aged 8 in the UK, in the 1990s, were much better acquainted with Pokémon characters than with the wildlife species in their surroundings (Sebba, 1991), or a Spanish study showing that adolescents know the starting lineups of top football teams better than they know historical figures and geographical place names (Sánchez and Jerez, 2004).

To understand this contemporary phenomenon, according to Piulats (2017), three historical processes can be identified in which humanity has distanced itself from nature. The first was the Neolithic revolution, several millennia ago, when humanity began to domesticate plants and animals and distanced itself from the natural environment by living in sedentary settlements. The second was the industrial revolution, during which population explosion, increased consumerism, and resource exploitation improved human quality of life but also increased environmental problems. The third would be marked by the digital technological revolution, in which increased screen time and sedentary behavior have enabled real-time connections with any place on the planet, the ability to purchase anything without leaving home, studying at any educational institution in the world, and starting an online business from the comfort of home. However, all of this has led to a detachment from the environment, the real and tangible world, and especially from the natural environment.

It is undeniable that today's generations of children spend less time in contact with nature, and schools, within the formal and official educational context, do not particularly promote this contact, especially through field trips and school excursions. Unlike other surrounding countries, there is no legislation in Spain regarding schools that educate in nature. Nature schools are not officially recognized in Spain, although there are some ongoing homologations. They are recognized in countries such as Germany, Switzerland, Sweden, Norway, the United Kingdom, Denmark, the United States of America, Canada, Australia, New Zealand, Japan, and some Ibero-American countries (Hueso, 2021: 131). There are mixed educational models, such as those that dedicate some days to classroom learning and others to outdoor activities, or where students spend part of the week in the field, or where each day consists of hours in the classroom and hours outdoors. According to EDNA (National Association for Education in Nature), this type of educational process can be defined as learning based on regular, permanent, and direct contact with wild nature, utilizing the elements and spaces it offers from a respectful perspective (https://asociacionedna.wordpress.com/).

Groups that benefit from education in nature are minors, in general, especially preschool and primary school students. There is a logical reason for this, as during these stages there is greater sensitivity to all living things, termed biophilia. This concept was coined by Eric Fromm (Eckardt, 2006) but developed by Wilson (2021), who defined it as the innate affiliation of human beings with other living organisms. It is a sentiment common to all cultures and environments, explaining why we enjoy going to the countryside or living in a house with a garden or views of nature, among other things. American geographer Tuan (2007) coined a term with a similar analogy, which he called topophilia, referring to the affiliation to a physical or natural environment and the feeling of belonging to that place where we feel comfortable. However, we know that humans are born with this instinct, biophilia and topophilia, but it is also necessary to cultivate the means to maintain it: by being in contact with nature (Hueso, 2021: 95). Playing in nature provides physical strength and improves defenses, increases visual acuity and attention and observation skills, fosters creativity and autonomy. On an emotional level, it teaches us to regulate sensations and emotions, to cooperate rather than compete, to develop social skills, and to manage conflicts. Learning solidifies in a strong and lasting manner (Hueso, 2021: 104). Moreover, it is proven that people who spend time in nature and have a higher degree of connection to it exhibit pro-environmental behaviors in adulthood (Chawla, 2020). According to this author, in just one generation, children's lives have largely shifted indoors, leading to the loss of free exploration of the nearby natural world, even when research indicates that direct experiences of nature in childhood contribute to lifelong care for the environment. In response to this reality, many naturalist, conservationist, and educational organizations advocate for connecting children with nature, resulting in a growing interest in measuring young people's connection to nature, understanding how it relates to their well-being and behavior, and creating programs to enhance that connection (Chawla, 2020). Some studies have shown that individuals who grew up in families with more pro-environmental attitudes also exhibited more pro-environmental behavior as young adults (Evans et al., 2018). Prospective longitudinal analyses revealed a similar association between families' pro-environmental behaviors and their young adult children's pro-environmental behavior over a 12-year period, from ages 6 to 18. It was found that the family's educational level, but not political ideology, was also associated with more pro-environmental behavior as children matured. Additionally, the time spent outdoors in childhood was positively associated with greater environmentally responsible behavior in early adulthood (Evans et al., 2018). Caring for the natural environment is a logical consequence of the bond created in childhood. If connection fosters knowledge and affection, the result is a protection of what is loved, namely, nature (Hughes et al., 2019).

The current school and the educational system, in general, do not favor this necessary connection with nature. According to Kellert (2002), human-nature interactions can be grouped into three types: a) direct contact, when visiting a natural space, the countryside, a forest, a park, a garden, etc.; b) indirect contact, when a real interaction occurs, but with a part of nature artificially prepared for this purpose, such as in a zoo, farm school, school garden, etc.; c) vicarious or symbolic contact, which occurs when there is no real physical contact with nature, but through a medium, support, or intermediary, such as a book, images, photographs, videos, documentaries, movies, the Internet, etc. Following this typological classification, the magnitude, use, and time dedicated from the educational system would be exactly in reverse order: there is a greater vicarious or symbolic interaction with the natural environment (which we could also call digital or virtual), followed by indirect contact, and lastly, a more direct link. By analogy, taking this hypothesis (Wilson, 2021) as a reference and expanding the scope not only to the set of living beings but to the entire landscape, both the perceivable landscape or phenolandscape and the non-perceivable landscape or cryptolandscape, both to its biotic or living components and abiotic or inert ones, and both to its physical or natural elements and human or sociocultural ones, we can speak of the term "geophilia." We can define it as an innate tendency of human beings to seek connections with the landscape. Based on the taxonomy proposed by Kellert & Wilson (1993) based on the biophilia hypothesis, we can outline seven types of human values or interests that support this theory, according to which, despite the gradual distancing from our landscapes, especially the natural ones, we maintain that connection we can call "geophilia": a) practical interest: the landscape and its elements provide us with resources, food, shelter, medicines, raw materials, etc.; b) naturalistic interest: to know, explore, and engage in recreational activities in the natural environment; c) ecological and scientific interest: for the systematic, rigorous, and scientific study of the landscape through disciplines such as Geography, Biology, Ecology, etc.; d) aesthetic interest: to enjoy the beauty of landscapes, inherent to all cultures and epochs; e) symbolic interest: by assigning feelings, thoughts, metaphors, etc., attributed to nature, as seen in children's stories, fables, religions, myths, totems, etc.; f) moral interest: responsibility to defend, protect, and conserve landscapes; g) dominant interest: to control, manage, subordinate, and exploit nature. This geophilia is detected in the primary interest in certain types of landscapes.

Building on the work of Tuan (2007), the so-called Savannah Hypothesis (Balling & Falk, 1982) has been developed, suggesting an evolutionary trend in human environmental preferences, positing that humans show an inherent inclination toward savanna-type landscapes. According to studies conducted by these authors, people prefer landscapes characterized by open spaces, gentle reliefs, scattered trees, the presence of water, and uniform vegetation over other types of environments such as deserts, steppes, forests, jungles, or mountains. This preference is based on the theory that our ancestors, who lived in savannas, achieved significant survival advantages, which may have led to the development of a genetic predisposition toward these environments. As a result, the attraction to savanna-like environments has persisted over time and is currently manifested in the appeal of parks, resorts, and golf courses that resemble these landscapes. Empirical evidence supports this hypothesis, showing a clear trend in people's reactions to images of savannas compared to other natural environments. This suggests that, although humans have inhabited various types of environments, savannas have exerted a significant evolutionary influence that continues to affect our current landscape preferences. The application of this understanding has practical implications in the design of landscapes and recreational environments, where the incorporation of elements characteristic of this type of landscape could enhance the emotional and aesthetic connection with users (Balling & Falk, 1982). Moreover, it is observed that this preference is especially pronounced in young children, suggesting that designing play spaces and settings for children's stories with these characteristics could foster greater engagement and enjoyment, aligning with individuals' natural inclinations. In our environment within the interior of the Spanish plateau, the open Mediterranean forest, specifically the dehesa landscape, serves as an exemplary embodiment of this theory. Therefore, a significant portion of our field trips has been conducted in such landscapes, like those predominant in the Cabañeros and Tablas de Daimiel National Parks, or in the Natural Parks of the Lagunas de Ruidera and the Valle de Alcudia and Sierra Morena (Jerez et al., 2017).

In accordance with these postulates, we believe that from the initial training of teachers, this contact with the natural environment should be promoted, allowing for the development of feelings of biophilia, topophilia, and geophilia, from which knowledge, skills, values, and attitudes can be transmitted to future students of Basic Education. It is clearly evident that there is an increasing disconnection between younger students and the natural environment, to the extent that some authors consider this lack of relationship a physical and psychological affliction, which they term nature deficit disorder (Louv, 2018). Thus, for the past two decades, a series of complementary and extracurricular educational activities have been promoted from the Faculty of Education of Ciudad Real to immerse and reconnect university students with the natural environment, especially in Protected Natural Areas and, within them, particularly in National Parks and Natural Parks.

2. OBJECTIVES AND METHOD

The Department of Geography at the University of Castilla-La Mancha (UCLM) has been organizing a series of formal and informal educational activities in the Protected Natural Areas (PNA) of the province of Ciudad Real for several years, which has enabled the development of one of the most significant educational actions from Higher Education for the training of future teachers and geographers. The present contribution aims to collect and organize these experiences, outline the criteria for the development of the selected itineraries, present the method of implementing these activities, and analyze the learning outcomes of the various types of students to whom they have been directed and, ultimately, who have participated in them.

To carry out this work, different phases of study have been developed in which, on one hand, a documentation process has been undertaken through the compilation of these types of experiences, primarily collected from course reports organized within the Faculty of Education of Ciudad Real at UCLM by various professors from the aforementioned department. On the other hand, the itineraries and materials used in these activities, focused on the transmission of geographical knowledge for environmental educational purposes, have been organized in order to conduct a qualitative analysis of the frequency and destination of the developed didactic itineraries and the learning outcomes obtained, based on the fulfillment of the objectives and educational competencies proposed in these activities.

The courses have been developed according to a dual methodology: theoretical-expository and applied-participatory. The first part is dedicated to presenting, through a predominantly graphical geographical language, using images, photographs, maps, drawings, and brief texts, the main natural and socio-cultural characteristics of the previously selected space. The second part is dedicated to conducting a field trip, for which a didactic itinerary has been designed or selected, aiming to concretize the geographical elements explained in the classroom and analyze and discuss the most significant issues affecting the conservation and balance of the most characteristic natural components.

The methodological proposal is thus based on the preparation of a practical work centered on a natural space and its environmental issues, grounded in the principles of meaningful learning and active participation (Jerez et al., 2014). It consists of several phases related to the natural environment and its environmental issues: exploration (contact with nature), analysis (environmental diagnosis), and treatment (proposal of educational solutions). In this, the student first explores and analyzes the landscape; subsequently, establishes a diagnosis of the environment; and lastly, proposes a series of educational solutions, without delving into technical, political, or executive considerations.

As a complement to this part, the students must finally prepare a brief report or dossier as a fieldwork notebook demonstrating the assimilation and utilization of the main ideas developed during the course, practically applied to the different selected didactic itineraries, with the aim of promoting observation and reflection on the various themes introduced. The required notebooks are individual works articulated by a collection of photographs, properly organized and commented on, taken by each student during the explanations provided throughout the development of the itinerary on the field day. To this end, it is required that each of the selected images be representative of both the main natural components and the most notable interventions or impacts that shape the landscapes in the completed itineraries.

3. TRAINING ACTIVITIES FOR TEACHERS IN NATIONAL AND NATURAL PARKS OF CASTILLA-LA MANCHA (SPAIN)

The network of facilities and the social visibility of the Protected Natural Areas (PNAs) facilitate the development of a significant number of environmental education (EE) activities within them, though not as abundantly as one might expect. Additionally, the geographical and natural uniqueness of these protected enclaves makes them excellent external teaching laboratories. They allow for educational and training actions related to conservation criteria, linked both to the knowledge and commitment to their maintenance through the development of actions respectful of the environment, and to their consideration as essential spaces for professional learning.

The province of Ciudad Real, which covers an area of almost 20,000 km^2, has a considerable number of these PNAs under various types of protection, which can serve as fundamental resources in disseminating the main geographical values, as it encompasses outstanding examples of various natural components due to their uniqueness and state of conservation. Among them, this province has the privilege of being one of the few in the national territory to have National Parks within its administrative boundaries, specifically two: the National Park of Las Tablas de Daimiel and the National Park of Cabañeros, which were classified by the Ministry of Agriculture, Food and Environment, along with Monfragüe, as "Interior Rural Parks." Furthermore, there are also two significant Natural Parks: the Natural Park of Las Lagunas de Ruidera and the Natural Park of Valle de Alcudia and Sierra Madrona. In total, almost 11% of the provincial area falls under some form of protection; it comprises 40% of all PNAs in Castilla-La Mancha (CLM) and encompasses just over 36% of the protected area.

3.1. Initial Approach and Planning

The interest in offering complementary university education of a geographical nature, coupled with the concern to link these subjects to Environmental Education (EA), given their limited formal presence in the curricula, has enabled the development of a set of activities focused on organizing proprietary courses related to the wealth of Protected Natural Areas (PNAs) in the provincial context during the period from 2004 to 2024. These activities have been regarded as a priority for the voluntary supplementary training of students from the Faculty of Education and the Faculty of Humanities at the University of Castilla-La Mancha (UCLM) in the Ciudad Real Campus. Their development has pursued clear objectives centered on:

a) Establishing direct contact with nature.
b) Reinforcing theoretical content taught in the classroom.
c) Achieving real and committed engagement with the space.
d) Raising awareness of environmental and management issues.
e) Promoting awareness of good conservation practices.
f) Presenting the foundations of a professional ethics framework of an educational and geographical nature.

All these activities have been approached from a dual perspective: firstly, from the pedagogical intention to address theoretical conceptual content about the natural environment and its various elements that add value to these spaces; and secondly, from the didactic interest in tackling different attitudinal content related to the appreciation of the natural and cultural landscape combinations present in these enclaves. Both conceptions allow for the analysis, in-depth exploration, and reflection on the anthropogenic interventions that significantly condition the management and conservation of these spaces while exercising an appreciation of the meaning and utility of protecting natural spaces and their relevance or sense of uniqueness or exclusivity within the provincial or regional geographical framework.

These activities have been specifically directed at students of the former Diplomas and new Bachelor's degrees in Early Childhood Education, Primary Education from the Faculty of Education in Ciudad Real, and the Bachelor's degree in Geography from the Faculty of Humanities at UCLM. They have been conceived as a complement to the geographical content of these programs, while also reinforcing the limited presence of EA content within them (from 2010 onwards, the Geography program includes an elective course on EA in its curriculum).

The initiative to develop these proprietary courses also arises as a situational response to a dual opportunity: satisfying the interest of a portion of the student body in enhancing their academic offerings to complete their higher education and providing the faculty the opportunity to incorporate EA themes that are barely present in the official curricula for such important future professionals as teachers and geographers. Therefore, the initial design of these courses develops within the context of the former programs (diplomas and degrees), which had a lower degree of specialization (or mention, in the case of Education), maintaining a significant load of elective credits and freely configured options that students could freely cover with different complementary courses based on their interest in the offered topics. This created a curricular need on the part of the student that often outweighed the thematic interest in the course in question (in this case, EA), but allowed for the introduction of certain aspects with notable deficiencies in the design of the curricula.

3.2. Procedure for Organizing Excursions

One of the objectives of these activities is to train students, specifically future Primary Education teachers, to organize school excursions and didactic itineraries in the natural environment. To achieve this, they are guided with various theoretical and methodological studies, from which we have designed a model based on a series of methodological steps. In the realm of designing excursions and didactic itineraries in the natural environment, there has been limited methodological research. According to Professor Marrón (2000), the design of a didactic itinerary involves four phases: 1) Delimitation of the route; 2) Reasons for selecting the study area; 3) Establishment of objectives; and 4) Execution of activities at the stops.

On the other hand, Wass (1992) adds a classical methodological approach with six fundamental parts: 1) Teacher knowledge: content, maps, first aid, legislation, etc.; 2) Initial planning: time factors, group size, authorizations, etc.; 3) Subsequent planning: selection of the route, accommodation, preparatory work, etc.; 4) Direction of the excursion: organization of the group, travel, transport, etc.; 5) Activities during the excursion: study, observation, collection of samples, etc.; and 6) Follow-up: data processing, images, models, etc.

It is crucial for teachers to master these phases and other aspects such as legislation, materials for fieldwork, use of ICT, knowledge of the geographical environment, travel insurance, etc. According to Muñoz and Jerez (2018), every field trip must include: the theme of the trip, place of visit, objectives, distance and sequencing of the route, planned stops, and human and material resources.

Based on academic methodological proposals and personal teaching experience in organizing field trips and didactic excursions, a method for organizing excursions has been developed, which, with some variations, is the one we have followed during these activities. It is proposed that every field trip should be carefully planned following these five methodological steps:

1. Knowledge of the geographical environment of the place to be visited. Analysis of geographical sources, maps, information on the Internet, etc. a. Virtual study of the itinerary: using ICT, online cartography, and sources on the Internet. b. Preliminary reconnaissance trip with students to familiarize them with the terrain, interpretative stops, technical stops, distances of the motorized route and times, distances of the pedestrian route and times, accessibility, etc. c. Measure and delimit the distance of the motorized route and the walking distance. d. Measure and delimit the chronological time and sequencing. e. Plan the stops, their educational content, and duration. f. Consult the expected weather conditions for the day of the excursion and its time sequencing (high probability up to three days prior).
2. Select the theme of the field trip. Identify and relate the theme to the curriculum. Delimit work content in the field.
3. Establish the curricular and specific objectives of the excursion.
4. Prepare materials for fieldwork: maps, documents, texts, graphics, worksheets, field notebooks. a. Create a map of the itinerary and the necessary graphic information. b. Plan the human and material resources.
5. Other relevant information: a. Knowledge of basic legislation: easements, rights of way, public roads and paths, regulations in protected areas, etc. b. Application of environmental and social attitudes and values. c. Consideration of logistical aspects: travel and school insurance, first aid, parental permissions, inclusion of the activity in annual teaching plans, etc.

These guidelines ensure thorough and effective planning of field trips in various environments.

3.3. Organization and Selection of Itineraries

For the selection of itineraries, priority has been given to the areas represented by National and Natural Parks, which are especially relevant not only for their significant conservation importance but also for embodying characteristic geosystems linked to water (Tablas de Daimiel and Lagunas de Ruidera) and Mediterranean forests (Cabañeros and Valle de Alcudia and Sierra Madrona). These areas are particularly fragile and bear a significant common social responsibility for the proper management and conservation of their values.

For each of these areas, a didactic itinerary has been designed or selected, utilizing their public use zones, which have been implemented over various years. Sometimes one area has been prioritized over others, while in other instances, multiple itineraries have been developed in the same year (see Table 1). The criteria for developing or selecting itineraries were centered on identifying stops of didactic interest related to some of the main characteristics defining the visited protected natural area (ENP). To this end, sometimes existing official itineraries were used, while in other cases, itineraries were designed to cater to the specific interests of each course, always ensuring that a maximum number of stops included at least one feature of alteration or threat of disappearance of a significant natural characteristic. These principles allowed for the articulation of various explanations and interpretations of the different heritage components of the territory, supported by didactic materials previously discussed in class and during the development of the itinerary (maps, figures, photographs, etc.). This approach aimed to facilitate scientific information regarding theoretical content, encourage reflection and valuation of landscape components, and inspire potential frameworks for future didactic action for the students.

In total, 32 didactic excursions have been held between 2004 and 2024, with the participation of 1,450 students, who exhibited a remarkable level of interest and engagement. However, a clear trend shift occurred starting in 2010 due to the implementation of the new undergraduate degrees. The reduction in elective options (which are now covered by subjects to ensure the achievement of new specific "mentions" in certain degrees) and the practical disappearance of free choice have led to a decline (and in some cases, even the elimination) of such activities, which has resulted in decreased enrollment in the offered courses. Although these initiatives were revived in 2013, attendance has been considerably lower (see Table 1).

Table 1. Didactic Excursions and Itineraries Organized by the Department of Geography at the Faculty of Education of Ciudad Real (UCLM) from 2004 to 2024.

Año:	National Park Tablas de Daimiel	National Park Cabañeros	Other Protected Natural Areas *	Number of Participants
2004		⊗		50
2005	⊗	⊗		100
2006		⊗ ⊗ ⊗	⊗	200
2007		⊗	⊗	100
2008	⊗	⊗	⊗ ⊗	200
2009	⊗		⊗	100
2010				--
2011				--
2012				--
2013		⊗		30
2014	⊗		⊗	70
2015				
2016			⊗	50
2017	⊗	⊗		100
2018	⊗			50
2019	⊗ ⊗	⊗		150
2020	**Pandemic**			
2021				
2022	⊗	⊗	⊗	150
2023		⊗		50
2024		⊗ ⊗ ⊗		150
Total:	9	15	8	1.450

Source: Department of Geography (Faculty of Education of Ciudad Real, University of Castilla-La Mancha). Own elaboration.

* Other Protected Natural Areas: Natural Park of Lagunas de Ruidera; Natural Park of Valle de Alcudia and Sierra Madrona; Natural Monument of the Volcán de Piedrabuena..

3.4. Assessment and Learning Outcomes of the Participants

The learning outcomes from the various editions of the itineraries conducted in different courses, in relation to the percentage of the total enrolled students, are associated with the set of itineraries planned each year, which received certificates of achievement upon fulfilling the evaluation requirements after attending theoretical classes and completing practical fieldwork. These requirements included attendance on the field day and participation in over 80% of the planned activities for each course, as well as the proper completion of fieldwork notebooks. These notebooks were prepared within one week following the conclusion of the itinerary and were used to demonstrate the acquisition of theoretical knowledge related to recognizing the main components and natural values of the most representative landscapes of the visited areas, as well as procedural and attitudinal content expressed through the acknowledgment of environmental issues and their possible solutions in the visited protected natural spaces.

The learning outcomes of the participants, as evidenced by the required evaluation tests, show a high success rate in meeting academic objectives. The average percentage exceeds 96%, ranging from 100% achieved during several excursions (for example, in the years 2005, 2007, 2009, 2016, 2017, 2019, etc.) to a minimum percentage of 85% in 2006, although all excursions have fluctuated between 90% and 100%. However, the challenge of monitoring the knowledge acquired concerning behaviors or habits in the environment has not been addressed and must be considered in future training initiatives. Thus, the importance of establishing a dual evaluation is emphasized: one that focuses on short-term academic characteristics and another that must assess the true long-term success of these initiatives. The first concentrates on participatory actions developed throughout the itineraries, involving interaction with local agents, as well as on the preparation of field reports that include commented photographic materials based on the characteristics addressed during the excursions (main types of landforms, protected vegetation, ecosystem and geosystem valuation from a geomorphological and landscape perspective, etc.). The second, as yet unqualified, should focus on the continuity of these educational initiatives in future personal or professional actions of the participants.

3.5. Practical Results: The Field Notebook

For each field trip, students are required to submit a project demonstrating that they have acquired a range of competencies and skills. In summary, the process for developing these activities is organized into three phases: a) Presentation by the faculty, either in the classroom or through online resources, regarding the site to be visited and its natural and landscape elements; b) Conducting the field trip; c) Submission of a project. This project typically consists of designing a programmatic activity centered around an educational excursion, integrated into a teaching program, a teaching unit, or a part of it, incorporating educational materials created by the students. Among these materials, students are asked to design a field notebook based on what they learned during the excursion, organized according to structured guidelines.

Fieldwork is essential in the training process concerning activities developed during the excursion, as it allows for the collection of direct information about the study elements and their environment. In addition to validating, reinforcing, and specifying the theory acquired in the classroom, it facilitates the observation of natural and landscape realities and their relationship to conservation issues. It is fundamental to document data, even those that may seem irrelevant, in the field notebook, which is an indispensable tool in this training process. This notebook not only aids in data collection for hypotheses but also develops observation and description skills that are fundamental for any student.

There are various models for creating this type of agenda, diary, or report of the practices developed through fieldwork, which we generically refer to as a field notebook (Pinillas & Torralba-Burrial, 2021; Fuentes, 2015; Parra, Domínguez, & Caballero, 2008; Roa & Vargas, 2009). In our case, we distinguish two models based on whether they are created by preservice teachers in Early Childhood Education or in Primary Education. In both cases, they can be developed manually or using software programs. For Early Childhood Education teachers, a very simple field notebook is requested, one that can be used by students in this educational stage. Essentially, it consists of a sheet of paper, generally sized A4, folded in half to create four pages of A5 size. On the first page, students are asked to draw a landscape or element from the visited location that is significant and allows for the identification of representative elements of that landscape. The second page includes activities and exercises of a conceptual nature, procedural exercises on the third page, and attitudinal activities on the fourth page (Figures 1 and 2).

In the case of the field notebook created by Primary Education faculty, the philosophy for the cover is maintained; however, within the content of the notebook, students must draw a sketch of the route, excursion, or itinerary, indicating the stops made and, subsequently, activities and exercises related to those stops (Figures 3, 4, 5, and 6).

Figure 1. Example of the cover (right) and an activities page in a field notebook for Early Childhood Education.

Figure 2. Example of the two interior activity pages in a field notebook for Early Childhood Education.

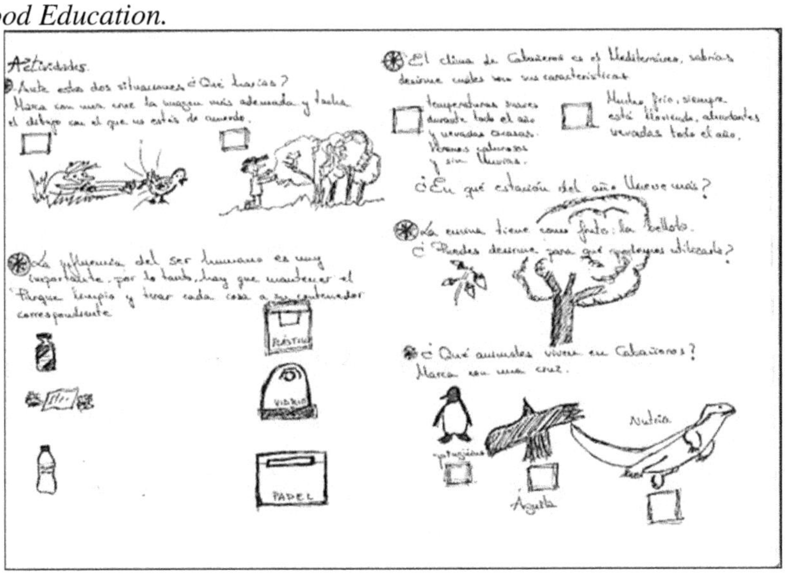

Figure 3. Example of the interior of a field notebook for Primary Education.

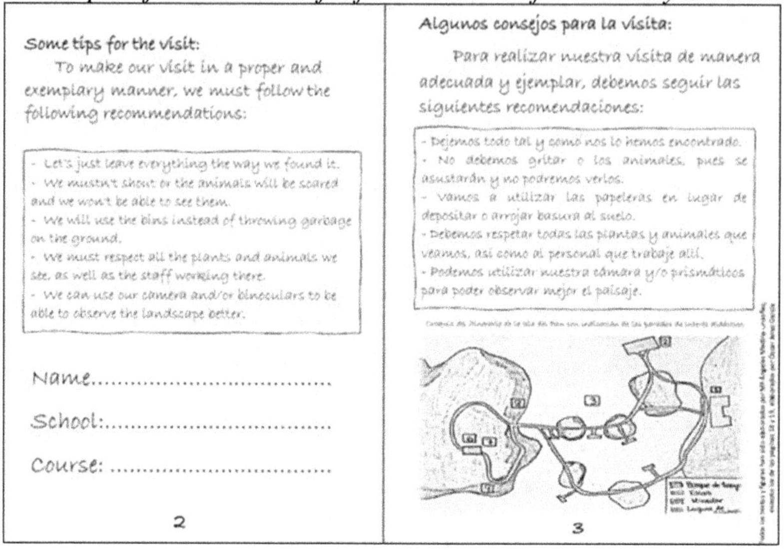

Some tips for the visit:

To make our visit in a proper and exemplary manner, we must follow the following recommendations:

- Let's just leave everything the way we found it.
- We mustn't shout or the animals will be scared and we won't be able to see them.
- We will use the bins instead of throwing garbage on the ground.
- We must respect all the plants and animals we see, as well as the staff working there.
- We can use our camera and/or binoculars to be able to observe the landscape better.

Name.................................

School:.................................

Course: ....

2

Algunos consejos para la visita:

Para realizar nuestra visita de manera adecuada y ejemplar, debemos seguir las siguientes recomendaciones:

- Dejemos todo tal y como nos lo hemos encontrado.
- No debemos gritar a los animales, pues se asustarán y no podremos verlos.
- Vamos a utilizar las papeleras en lugar de depositar o arrojar basura al suelo.
- Debemos respetar todas las plantas y animales que veamos, así como al personal que trabaje allí.
- Podemos utilizar nuestra cámara y/o prismáticos para poder observar mejor el paisaje.

3

Figure 4. Example of the interior of a field notebook for Primary Education.

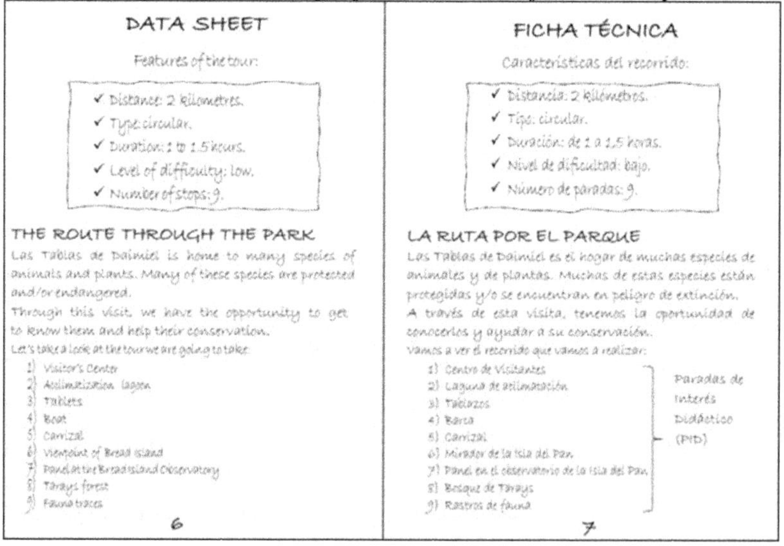

DATA SHEET

Features of the tour:

✓ Distance: 2 kilometres.
✓ Type: circular.
✓ Duration: 1 to 1.5 hours.
✓ Level of difficulty: low.
✓ Number of stops: 9.

THE ROUTE THROUGH THE PARK

Las Tablas de Daimiel is home to many species of animals and plants. Many of these species are protected and/or endangered.

Through this visit, we have the opportunity to get to know them and help their conservation.

Let's take a look at the tour we are going to take:

1) Visitor's Center
2) Acclimatization lagoon
3) Tablets
4) Boat
5) Carrizal
6) Viewpoint of Bread Island
7) Panel at the Bread Island Observatory
8) Tarays forest
9) Fauna traces

6

FICHA TÉCNICA

Características del recorrido:

✓ Distancia: 2 kilómetros.
✓ Tipo: circular.
✓ Duración: de 1 a 1,5 horas.
✓ Nivel de dificultad: bajo.
✓ Número de paradas: 9.

LA RUTA POR EL PARQUE

Las Tablas de Daimiel es el hogar de muchas especies de animales y de plantas. Muchas de estas especies están protegidas y/o se encuentran en peligro de extinción.

A través de esta visita, tenemos la oportunidad de conocerlos y ayudar a su conservación.

Vamos a ver el recorrido que vamos a realizar:

1) Centro de Visitantes
2) Laguna de aclimatación
3) Tabiazos
4) Barca
5) Carrizal
6) Mirador de la Isla del Pan
7) Panel en el observatorio de la Isla del Pan
8) Bosque de Tarays
9) Rastros de fauna

Paradas de Interés Didáctico (PID)

7

Figure 5. Example of the interior of a field notebook for Primary Education.

Figure 6. Example of the interior of a field notebook for Primary Education.

4. CONCLUSION

Fieldwork through a didactic itinerary serves as a formative tool for analyzing and diagnosing environmental problems, as well as for developing significant environmental values in the university education of future teachers and geographers. The planning and design of such activities in protected natural areas (PNAs) also respond to a method that sensitizes participants, particularly concerning the acquisition of formative competencies, as well as the development of skills and personal perceptions that are especially important for embracing the principles of environmental education (EE).

The results obtained have highlighted a relevant activity in the area of complementary training for future teachers and geographers at the University of Castilla-La Mancha (UCLM), centered on EE through the reinforcement of geographic competence. Over the past few decades, activities have been developed with the primary objective of providing interdisciplinary education from a geographical perspective through interpretive itineraries in contact with the natural environment of the main PNAs in the province of Ciudad Real. Additionally, these activities have promoted the development and dissemination of educational materials regarding the key natural elements of the selected areas (Serrano de la Cruz & Sánchez, 2007; Jerez, 2009a, 2009b, 2014; Jerez et al., 2014; García et al., 2022; Jerez et al., 2024, among others). These materials have been actively utilized in the field with students, establishing working dynamics focused on observation and the collection of written and photographic data to create personal products valuable for raising awareness about the main characteristics and issues of these protected areas and, at the same time, for evaluating these activities.

The results also indicate very high levels of participation and evaluation. From 2004 to the present (academic year 2023/2024), a total of 1,450 students have participated in this type of activity, achieving a high percentage of compliance with the proposed objectives. For EE, action and educational continuity are essential and should not rely solely on isolated educational content. It remains critical to understand, in a more applied and critical manner, the degree of subsequent influence that such activities may have had concerning the development of similar initiatives by individuals who acquired knowledge and skills in EE during these experiences. Such results could channel these principles more directly into broader citizen groups, whether in the educational sphere or in the more general social realm, where developing such proposals adapted to local realities is particularly important.

The advantages of these complementary training actions are clear: reinforcing environmental and geographical knowledge in the training of future education and territory professionals strengthens the foundations of sustainable and environmentally respectful principles supported by the main conservation figures as prominent symbols of our nature conservation efforts.

However, these activities appear to be insufficient. There is no comprehensive planning of a university or extracurricular nature that serves as a foundation for continuity and strengthening the transmission of this training and its associated values. Additionally, and this may be part of new research directives on these topics, there are no subsequent evaluation elements that allow for assessing the degree of transmission of these values in the personal and/or professional activities of future graduates. Finally, as has been insisted upon from critical positions in various forums on EE, there is a continuous focus on educational efforts possibly directed at the most educated and aware generations of recent decades, while it remains challenging to reach the most uninformed or least aware entities on these topics, whether in the context of areas used as laboratories for field training experiences or in the institutional or associative contexts of those responsible for the direct and daily management of these spaces.

The new curricula resulting from the European Higher Education Area have not favored the development of certain alternative training activities as much as previous ones. The organization of such activities to complement the training of future teachers and geographers has been sidelined as a consequence of the reorganization of subjects and credits in the curricula. By constraining the specialty and reducing the capacity for complementary free training, the topics developed in the self-training courses have been diminished, significantly reducing the transversal content related to EE that was previously present in these activities.

Thus, this type of activity raises, as a final reflective conclusion, the possibility of conducting future training and environmental awareness actions focused on the fundamental principles and foundations of EE. Among these, the interrelation of activities among different actors stands out, where the university and higher education can play a decisive role in promoting joint actions with local populations and institutions, while also addressing needs and opportunities to regional administrative bodies responsible for incentivizing the development of more global EE programs, which are often neglected in many Spanish regions.

5. REFERENCES

Aguilera, M. J., González, M. P., & Santos, J. M. (2009). *Geografía General I: Geografía Física*. UNED.

Alomar-Garau, G., & Gómez-Zotano, J. (2022). Del paisaje en la Geografía: Perspectivas y actividades del grupo de paisaje de la Asociación Española de Geografía. *Cuadernos de Geografia*, 108-109(108-9), 857–874. DOI: 10.7203/CGUV.108-9.24067

Ayuso, S. (2006). Educación ambiental en espacios naturales protegidos: Aprendiendo en la naturaleza. *Medio Ambiente Castilla-La Mancha*, 13, 22–27.

Bailey, P. (1981). *Didáctica de la geografía*. Editorial Cincel.

Balling, J. D., & Falk, J. H. (1982). Development of visual preference for natural environments. *Environment and Behavior*, 14(1), 5–28. DOI: 10.1177/0013916582141001

Cancer, L. A. (1999). *La degradación y la protección del paisaje*. Cátedra.

Chawla, L. (2020). Childhood nature connection and constructive hope: A review of research on connecting with nature and coping with environmental loss. *People and Nature*, 2(3), 619–642. DOI: 10.1002/pan3.10128

Chevallard, Y. (2000). *La transposición didáctica: del saber sabio al saber enseñado*. Aique.

Cid, O. (1997). Educación ambiental en los espacios naturales protegidos. En R. Mendoza Castellón (Coord.), *Actas de la VIII aula de ecología en educación ambiental: Almería, 14 de febrero de 1994* (pp. 51-60). Instituto de Estudios Almerienses.

Clements, R. (2004). An investigation of the state of outdoor play. *Contemporary Issues in Early Childhood*, 5(1), 68–80. DOI: 10.2304/ciec.2004.5.1.10

Consejería de Educación, Cultura y Deportes de la Junta de Comunidades de Castilla-La Mancha. (2022). Decreto 81/2022, de 12 de julio, por el que se establece la ordenación y el currículo de Educación Primaria en la comunidad autónoma de Castilla-La Mancha. *Diario Oficial de Castilla-La Mancha*, 135, 19860–19994.

Corrales, M., & Garrido, J. (2024). Enseñanza de la geografía empleando como recursos los parques naturales de Extremadura: experiencia práctica con docentes en formación. En J. Mora, J. Garrido, & R. A. Castanho (Dirs.), *Estrategias públicas en la ordenación del espacio geográfico* (pp. 551–564). Aranzadi.

Crespo, J. M. (2012). Un itinerario didáctico para la interpretación de los elementos físicos de los paisajes de la Sierra de Guadarrama. *Didáctica Geográfica*, 13, 15–34.

Crespo, J. M., Gómez, M. L., & Cruz, L. A. (2018). Una aproximación a los Parques Nacionales y sus paisajes a través de itinerarios didácticos. *Espacio, Tiempo y Forma. Serie VI, Geografía*, 11(11), 121–140. DOI: 10.5944/etfvi.11.2018.22359

Eckardt, M. H. (2006). Fromm y su concepto de "biofilia". En J. Silva García (Coord.), *El humanismo de Erich Fromm: Actualidad del autor de "El arte de amar" y "El miedo a la libertad"* (pp. 75-84). Universidad Autónoma de México.

Evans, G. W., Otto, S., & Kaiser, F. G. (2018). Childhood origins of young adult environmental behavior. *Psychological Science*, 29(5), 679–687. DOI: 10.1177/0956797617741894 PMID: 29447064

Fuentes, A. F. (2015). Cuaderno de campo. En P. M. Jorgensen, A. F. Fuentes, T. Miranda, & L. Cayola (Eds.), *Manual de Trabajo, Proyecto Inventario Botánico de la Región del Madidi* (Cap. 5).

García, A. (2004). El itinerario geográfico como recurso didáctico para la valoración del paisaje. *Didáctica Geográfica*, 6, 79–95.

García, A. (2012). El aprendizaje basado en problemas en el itinerario geográfico. *Educación y Futuro*, 27, 155–175.

García, A. (2023). Pensamento narrativo e construção de conceitos da paisagem. Oficina narrativa em Geografia. *Revista Signos Geográficos*, 5, 1–18. DOI: 10.5216/signos.v5.77428

García, A. L. (1997). El proceso de desarrollo de los Itinerarios Geográficos. *Didáctica Geográfica*, 2, 3–9.

García, A. L., & Martínez, J. M. (1988). Los itinerarios didácticos en la enseñanza de las Ciencias Sociales. *Revista del Departamento de Didáctica de las Ciencias Sociales de la Universidad de Granada, sep. 2º cuadernillo*, 107-113.

García, J. L. (1995). *El medio natural en los Montes de Ciudad Real y el Campo de Calatrava*. Diputación Provincial.

García, J. L., Serrano de la Cruz, M. A., & Jerez, Ó. (2022). *Guía del Parque Natural del Valle de Alcudia y Sierra Madrona (Ciudad Real)*. 3d3 Editores.

García, M.. (2018). Se hace geografía al andar: La salida de campo itinerante y senderista. *Didáctica Geográfica*, 19, 103–125.

García-Monteagudo, D. (2024). Formación inicial del profesorado de Educación Infantil en pensamiento geográfico: Paisaje e identidades territoriales. *ENSAYOS. Revista de la Facultad de Educación de Albacete*, 39(1), 57–73. DOI: 10.18239/ensayos.v39i1.3488

Gavidia, V., Solbes, J. A., & Souto, X. M. (2015). La supresión del área del conocimiento del medio en la educación primaria. *Aula*, 243-244, 31–36.

Gobierno de Castilla-La Mancha. (2020). *Estrategia de Educación Ambiental de Castilla-La Mancha. Horizonte 2030 (2020-2025)*. https://www.castillalamancha.es/sites/default/files/documentos/pdf/20200604/estrategia_ea-clm_h2030.pdf

Gómez Mendoza, J. (2013). Del patrimonio paisaje a los paisajes del patrimonio. *Documents d'Analisi Geografica*, 59(1), 5–20. DOI: 10.5565/rev/dag.48

González, I. (2002). La formación inicial de los profesores de ciencias sociales de Educación Secundaria: necesidades y proyectos. In *VV. AA. Nuevos horizontes en la formación del profesorado de Ciencias Sociales* (pp. 429–451). Asociación Universitaria de Profesores de Didáctica de las Ciencias Sociales.

Henares, I. (2015). La educación ambiental en los espacios naturales protegidos: retos. En *Blog Educación Ambiental 26 #EA26*. https://educacionambiental26.wordpress.com/2014/02/19/la-educacion-ambiental-en-los-espacios-naturales-protegidos-retos/

Hernández, A., & García, A. (2022). Del análisis del paisaje a la emoción del paisaje. Aportaciones a su didáctica. *REIDICS.Revista de Investigación en Didáctica de las Ciencias Sociales*, 10, 6–23. DOI: 10.17398/2531-0968.10.6

Hofferth, S. L., & Sandberg, J. F. (2001). How American children spend their time. *Journal of Marriage and Family*, 63(3), 295–308. DOI: 10.1111/j.1741-3737.2001.00295.x

Hueso, K. (2017). *Somos naturaleza. Un viaje a nuestra esencia*. Plataforma Editorial.

Hueso, K. (2021). *Educar en la naturaleza. Mejores personas para un planeta mejor*. Plataforma Editorial.

Hughes, J., Rogerson, M., Barton, J., & Bragg, R. (2019). Age and connection to nature: When is engagement critical? *Frontiers in Ecology and the Environment*, 17(5), 265–269. DOI: 10.1002/fee.2035

JCCM. (1995). *Plan de Conservación del Medio Natural de Castilla-La Mancha*. Consejería de Agricultura y Medio Ambiente, Junta de Comunidades de Castilla-La Mancha.

JCCM. (2003). *Plan Regional de Educación Ambiental de Castilla-La Mancha*. Junta de Comunidades de Castilla-La Mancha.

Jerez, Ó. (2009a). *El Parque Nacional de las Tablas de Daimiel. Itinerario didáctico por un área protegida "en peligro de extinción"*. Materiales didácticos del Dpto. de Geografía en la E. U. de Magisterio de Ciudad Real. Carranchín.

Jerez, Ó. (2009b). *El Parque Nacional de Cabañeros. Cuaderno de campo etno-botánico*. Materiales didácticos del Dpto. de Geografía en la E. U. de Magisterio de Ciudad Real. Carranchín.

Jerez, Ó. (Dir.). (2014). *Itinerarios didácticos por La Mancha Húmeda. Libro digital bilingüe*. Óptima diseño e impresión.

Jerez, Ó. (2019). La transferencia de conocimientos didáctico-geográficos en el medio rural. El senderismo didáctico como propuesta para un desarrollo rural sostenible. En X. C. Macía Arce, F. X. Armas Quintá, & F. Rodríguez Lestegás (Coords.), *La reconfiguración del medio rural en la sociedad de la información: nuevos desafíos en la educación geográfica* (pp. 1321–1333). Andavira Editora.

Jerez, Ó., García, J. L., & Serrano de la Cruz, M. A. (2024). *Guía de excursiones por los espacios naturales de Castilla-La Mancha: Los Montes de Toledo*. Universidad de Castilla-La Mancha.

Jerez, Ó., Rodríguez, M. Á., & Escobar, E. (2014). El itinerario didáctico como recurso educativo para el desarrollo de valores ambientales. In Sáez-Martínez, F. J., Guadamillas, F., & Martín, R. (Eds.), *Experiencias docentes en Energía y Medio Ambiente: University Vocational Training Network* (pp. 60–73). Servicio de Publicaciones de la Universidad de Castilla-La Mancha.

Jerez, Ó., Serrano de la Cruz, M. A., & García, J. L. (2017). *Los espacios naturales protegidos en la provincia de Ciudad Real*. Diputación Provincial de Ciudad Real.

Kellert, S. R. (2002). Experiencing nature: Affective, cognitive, and evaluative development in children. In Kahn, P. H.Jr, & Kellert, S. R. (Eds.), *Children and nature: Psychological, sociocultural, and evolutionary investigations* (pp. 117–151). MIT Press.

Kellert, S. R., & Wilson, E. O. (1993). *The biophilia hypothesis*. Island Press.

LOMLOE. (2020). *Ley Orgánica 3/2020, de 29 de diciembre, por la que se modifica la Ley Orgánica 2/2006, de 3 de mayo, de Educación*.

Louv, R. (2018). *Los últimos niños en el bosque. Salvemos a nuestros hijos del trastorno por déficit de naturaleza*. Capitán Swing.

Marrón, M. J. (2000). *Enseñar geografía. De la teoría a la práctica*. Síntesis.

Martínez, J. (2000). Educación ambiental y conservación de paisajes frágiles. En *Libro de actas de la Reunión Internacional de Expertos en Educación Ambiental*.

Martínez, J. M. (2012). De las escuelas al aire libre a las aulas de naturaleza. *Áreas. Revista Interna de Ciencias Sociales*, 20, 171–182.

Merino, C., Pérez, M., & Calaghan, P. (1998). Espacios naturales protegidos y educación ambiental. En P. Aznar (Coord.), *La educación ambiental en la sociedad global: actas* (pp. 292-298). Universitat de València, Servei de Publicacions.

Muñoz, E., & Jerez, Ó. (2018). El trabajo de campo como herramienta didáctica y su relación con el medio en el proyecto ¡Nosotros proponemos! En M. A. Rodríguez & S. Claudino (Coords.), *¡Nosotros proponemos! Ciudadanía, sostenibilidad e innovación geográfica ante los desafíos educativos de la sociedad* (pp. 41-65). Graó.

Negrín, O. (2005). *Teorías e instituciones contemporáneas de educación*. Editorial Universitaria Ramón Areces.

Palacios, D. (1988). La aplicación del método indagatorio a la enseñanza de la Geografía a través del trabajo de campo. En *Actas de las Primeras Jornadas de Didáctica de la Geografía* (pp. 124-130). A.G.E-Grupo de Didáctica.

Palacios, D. (1990). *Propuesta de un método para la didáctica de la geografía física a través del trabajo de campo. La indagación* (Tesis doctoral). Universidad de La Laguna.

Parra, M., Domínguez, G., & Caballero, P. J. (2008). El cuaderno de campo: Un recurso para dinamizar senderos desde la educación en valores. *Ágora para la Educación Física y el Deporte*, (7-8), 145–158.

Pinillas, S., & Torralba-Burrial, A. (2021). El cuaderno de campo como eje del aprendizaje de naturaleza cercana en Educación Infantil. *Revista Eureka sobre Enseñanza y Divulgación de las Ciencias, 18*(3), 3202.

Piulats, O. (2017). *La ciencia espiritual de la naturaleza: Seis meditaciones sobre ecología*. Ediciones Carena.

Ramiro, E. (2005). Ideas para preparar itinerarios de una forma amena y con buen humor. *Didáctica Geográfica*, 7, 463–472.

Roa, P. A., & Vargas, C. (2009). El cuaderno de campo como estrategia de enseñanza en el Departamento de Biología de la Universidad Pedagógica Nacional. *Bio-grafía: Escritos sobre la Biología y su Enseñanza, 2*(3), 1-41.

Rodríguez, F. (2000). *La actividad humana y el espacio geográfico*. Síntesis.

Romero, J., & Luis, A. (2008). El conocimiento socio-geográfico en la escuela: Las tensiones inherentes a la transmisión institucionalizada de cultura y los dilemas de la educación para la democracia en este mundo globalizado. *Scripta Nova, Revista Electrónica de Geografía y Ciencias Sociales, 12*(270). https://www.ub.edu/geocrit/sn/sn-270/sn-270-123.htm

Romero Morante, J. (2021). Apuntes para una sociogénesis del currículum centrado en problemas sociales. *REIDICS.Revista de Investigación en Didáctica de las Ciencias Sociales*, 9, 14–40. DOI: 10.17398/2531-0968.09.14

Sala, M., & Batalla, R. J. (1996). *Teoría y Métodos en Geografía Física*. Síntesis.

Sánchez, L. (2005). Enseñar Geografía: De la significatividad psicológica local a la significatividad lógica global. En E. Sande Lemos (Coord.), *Ensinar Geografia na sociedade do conhecimento* (pp. 429-451). APG-AGE.

Sánchez, L., & Jerez, Ó. (2004). El lugar en las ciencias sociales. In *Métodos para enseñar ciencias sociales: interacción, cooperación y participación* (pp. 69–80). Íber, Didáctica de las Ciencias Sociales, Geografía e Historia, Graó.

Santana, D., Morales, A. J., Colomer, J. C., Campo Pais, B., & Caurín Alonso, C. (2016). Parques naturales: La necesaria conceptualización transformadora en la educación primaria y secundaria. *Didáctica Geográfica, 16*, 73–94. https://didacticageografica.age-geografia.es/index.php/didacticageografica/article/view/296

Sanz, C. (2002). El paisaje como recurso. In *Estudios sobre el paisaje* (pp. 75–84). Fundación Duques de Soria. UAM.

Sebba, R. (1991). The landscapes of childhood: The reflection of childhood surroundings in adult memory and children's attitudes. *Environment and Behavior*, 23(4), 395–422. DOI: 10.1177/0013916591234001

Serrano de la Cruz, M. A., García, J. L., & Jerez, Ó. (2016). Propuesta preliminar para la identificación de lugares de interés didáctico en espacios naturales protegidos. *Didáctica Geográfica*, 17, 159–176.

Serrano de la Cruz, M. A., & Sánchez, L. (Coords.). (2007). *Educación geográfica a través de los paisajes de la provincia de Ciudad Real*. Ediciones de la Universidad de Castilla-La Mancha y Asociación de Geógrafos Españoles.

Sousa, S. A., García Monteagudo, D., & Souto, X. M. (2016). Educación geográfica y las salidas de campo como estrategia didáctica: un estudio comparativo desde el Geoforo Iberoamericano. *Biblio3W. Revista Bibliográfica de Geografía y Ciencias Sociales, 21*(1155). https://www.ub.edu/geocrit/b3w-1155.pdf

Souto, X. M. (1999). *Didáctica de la geografía. Problemas sociales y conocimiento del medio*. Serbal.

Tuan, Y. F. (2007). *Topofilia*. Editorial Melusina.

UNICEF. (1989). *Convención sobre los Derechos del Niño*. https://www.unicef.org/

Unwin, T. (1995). *El lugar de la Geografía*. Cátedra.

VV. AA. (1998). La educación ambiental en los espacios naturales protegidos. *Aula Verde: Revista de Educación Ambiental*, 16, 3–12.

Wass, S. (1992). *Salidas escolares y trabajo de campo en la educación primaria*. Ediciones Morata.

Wilson, E. O. (2021). *Biofilia: El amor a la naturaleza o aquello que nos hace humanos*. Errata Naturae Editores.

KEY TERMS AND DEFINITIONS

Biophilia: The innate inclination of humans toward nature and living beings. This connection, deeply rooted in our biology and psychology, reflects our evolution and the natural desire to be in contact with the natural environment, providing us with emotional and physical benefits.

Didactic Itinerary: A planned and structured route designed to facilitate learning in a specific context, such as a natural or cultural environment. It promotes the acquisition of knowledge, skills, and values through exploration and direct experience, integrating educational activities and didactic resources along the way.

Excursion: A journey, route, or temporary outing to a specific place for educational, recreational, or cultural purposes. Generally conducted in groups, it is organized to provide practical experiences and learning outside the usual environment, promoting exploration and direct knowledge of the visited location.

Field Notebook: A tool used to record observations, data, and notes during outdoor activities or in natural settings. It serves to document findings and make detailed notes about the terrain, relief, waters, vegetation, fauna, and other aspects of the natural environment, facilitating subsequent analysis and reflection.

Funding: This work has been developed within the framework of the call for funding applied research and innovation projects under the UCLM's Own Plan, specifically within the project 2022-GRIN-34383 titled "Evaluation of Landscape as a Tourist and Educational Resource in the Natural Parks of Castilla-La Mancha."

Geophilia: The love or attraction that individuals feel toward the earth and its landscapes. This concept describes the desire to connect with nature and natural environments, highlighting how landscapes and the earth can influence our emotional and psychological well-being. The Earth's surface and has been formed and shaped by natural processes, both abiotic and biotic, without significant intervention from human activities.

Natural Landscape: The morphology of a spatial or territorial system that is part of

Topophilia: The emotional affection or attachment that people develop toward certain places. This concept reflects how physical environments, such as landscapes or urban settings, can influence our experiences, feelings, and memories, creating a significant emotional connection to those locations.

Chapter 3
Migrations and Interculturality:
Design of a Strategy in the Degree of Early Childhood Education From Social Geography

Mercedes de los Ángeles Rodríguez-Rodríguez
https://orcid.org/0000-0003-1015-2027
University of Las Palmas de Gran Canaria, Spain

María Yazmina Lozano Mas
https://orcid.org/0000-0002-1599-1930
University of Las Palmas de Gran Canaria, Spain

Manuel Ramón González Herrera
https://orcid.org/0000-0002-2104-4702
Autonomous University of the City of Juárez, Mexico

ABSTRACT

In recent years, the Canary Islands have experienced a notable increase in the number of students from diverse international backgrounds entering school classrooms, fostering the emergence of a distinctly multicultural society. Addressing the educational needs arising from this demographic shift is crucial at the early childhood education level, where the foundation for mutual understanding in a multicultural context is established. In this regard, it is vital to underscore the role of Social Sciences education in cultivating democratic citizenship, with the overarching goal of equipping society to make collective, inclusive, and informed decisions that shape its actions. Achieving this requires educators proficient in these competencies, emphasizing the urgent need for strategies that support their professional develop-

DOI: 10.4018/979-8-3693-7185-5.ch003

Copyright © 2025, IGI Global Scientific Publishing. Copying or distributing in print or electronic forms without written permission of IGI Global is prohibited.

ment. This study aligns with that goal, advocating for cooperative learning, social integration, and solidarity.

INTRODUCTION

Today's globalized world is marked by the coexistence of individuals from diverse countries and cultures, the significant flow of social movements, continuous innovation, and the expansion of information and communication technologies. These new trends in mobility have "implications in political, social, economic, cultural, and consequently, educational life, underscoring the need to create new spaces for coexistence that foster understanding and social cohesion" (Rodríguez-García et al., 2019, p. 60).

School classrooms are the setting where these dynamics unfold. Thus, educators capable of guiding these groups of children, especially in early childhood education, toward healthy and enriching coexistence are essential. One of the key objectives of this formative stage in early childhood education is to contribute to the cultivation of civic values that promote student coexistence and the development of civic competence.

In this regard, the inclusion of Social Sciences in the formation of democratic citizenship is essential, as it fosters the development of a society capable of making decisions in a free and informed manner, collaboratively, with such decisions guiding collective actions (Benejam Arguimbau, & Pagès Blanch, 1997; Funes et al., 2019). With these two principles in mind, we emphasize the importance of addressing these issues from the earliest stages of education to prepare students for the processes of migration and the intercultural dynamics that emerge as a result. However, are future early childhood education teachers equipped to handle this responsibility? Do they possess the necessary tools to adequately develop these skills?

The phenomenon of migration has gained considerable importance in Spain, and particularly in the Canary Islands, over recent decades. Immigrants come from diverse regions, including Asia, Latin America, and Africa, arriving in the Canary Islands through various means and with life experiences. Thousands of children from these regions, fleeing poverty, conflict, and socioeconomic and political instability, enter Canary Islands' classrooms at the preschool level. The integration process is complex and challenging, not only for immigrant children, but also for their peers and, most importantly, for Early Childhood Education teachers, who require additional tools to effectively address migration and interculturality within their classrooms. This study focuses on these two concepts: migrations, defined as displacements involving a change of residence (Domínguez-Mujica, 2008), and interculturality, understood as a value that shapes personal behavior and promotes a lifestyle to be encouraged by educators (Hernández-Prados et al., 2021), who must first develop competence in intercultural literacy.

Therefore, this study presents an innovative educational proposal centered on the design of a strategy aimed at developing civic competence in future Early Childhood Education teachers through the course Didactics of Social Sciences in Early Childhood Education, offered in the first semester of the 4th year of the degree program at the University of Las Palmas de Gran Canaria. The objective is to enhance the preparation of future early childhood educators within university classrooms by addressing contemporary issues, thereby enriching the teaching and learning process.

The content is structured into several sections. First, a curricular framework is presented, followed by a discussion on the Sustainable Development Goals (SDGs) and the significance of initial teacher training. Next, the current socio-educational context in which the strategy is implemented is analyzed, followed by a comprehensive description of the strategy. Lastly, final considerations and definitions of key terms are provided from the authors' perspective, concluding with the bibliographical references that support the study.

1. CURRICULAR FRAMEWORK

Organic Law 3/2020, of December 29, which amends Organic Law 2/2006, of May 3, on Education (hereinafter LOMLOE), is a comprehensive legal reform comprising a single article with ninety-nine sections that either rewrite or modify the previous Organic Law on Education (hereinafter LOE). It includes eleven additional provisions, five transitional provisions, one repeal provision, and six final provisions (Montero Caro, 2021).

The LOMLOE incorporates the Sustainable Development Goals (hereafter SDGs), particularly Goal 4, into its Preamble. According to Montero Caro (2021), the integration of the five fundamental approaches upon which the law is based reflects a strong commitment to shaping education through the lens of sustainability. These approaches are as follows:

1. Children's rights serve as one of the guiding principles of the system.
2. Promotion of gender equality through coeducation.
3. Mainstreaming to ensure that all students have opportunities for success.
4. Focus on sustainable development as outlined in the 2030 Agenda.
5. Emphasis on digital transformation (LOMLOE, 2020).

The promotion of democratic values and a culture of sustainable development are emphasized both in the Preamble and throughout the articles of the law (Montero Caro, 2021).

In the context of Early Childhood Education, the LOMLOE redefines its purpose, focusing on contributing to the development of cognitive, emotional, psychomotor, and volitional capacities, integrating various areas of learning through a holistic approach (Pallares Jiménez & Lleida Añón, 2023). The goal is to cultivate an autonomous, imaginative, and critical childhood, with a strong sense of personal identity (Resa Ocio et al., 2023).

Royal Decree 95/2022, of February 1, outlines the organization and minimum educational requirements for Early Childhood Education and incorporates education for responsible and sustainable consumption as one of its pedagogical principles (art. 6.4).

Among the various key competencies, this study will focus on civic competence, which seeks to lay the foundations for practicing democratic citizenship. It emphasizes models that promote learning attitudes grounded in values such as respect, equity, equality, inclusion, and coexistence, alongside the peaceful and dialogued resolution of conflicts. This competence establishes the conditions for fostering behaviors that are respectful of oneself, others, and the environment, while also preventing discriminatory actions of any kind.

2. SDGS IN THE FRAMEWORK OF EARLY CHILDHOOD EDUCATION

At a global level, societies, institutions, and organizations acknowledge the necessity of advancing toward sustainable development, a process that entails profound transformations, including shifts in values and principles (García Esteban & Murga Menoyo, 2015). UNESCO advocates for values associated with human dignity, fundamental freedoms, human rights, equity, and environmental stewardship, while endorsing the Earth Charter as a declaration of fundamental ethical principles that serve as the foundation for constructing a just, sustainable, and peaceful global society (García Esteban & Murga Menoyo, 2015).

In this context, in December 2002, the Decade of Education for Sustainable Development (2005-2014) was declared. The goal of this Decade was to refocus education and training toward sustainability, acknowledging the transformative power of education at all levels, from basic education to university instruction.

In 2015, the United Nations (UN) adopted the 2030 Agenda for Sustainable Development, which encompasses 17 objectives and 169 goals forming a global action plan focused on three key axes: economic, social, and environmental. This agenda aims to foster the development of a global citizenship committed to sustainable development. Education is positioned as one of the fundamental pillars of societal transformation within the 2030 Agenda (Guardeño Juan & Monsalve Lorente, 2024).

Education for sustainability must commence at the early childhood education stage, as it is during this period that the foundational skills are cultivated, enabling students to acquire more advanced competencies in the future (Alacid de Pascual, 2021).

At the national level, the Ministry of Science and Innovation incorporated the achievement of the SDG objectives into its priorities for the 2021-2027 period. Similarly, the Conference of Rectors of Spanish Universities (hereafter CRUE) has committed its resources, expertise, and influence to support the 2030 Agenda for Sustainable Global Development (Pallarés Jiménez & Lleida Añón, 2023).

Students at this stage can and should be introduced to the learning process of the various SDGs, contributing from the outset of their schooling to the development of critical, active, and participatory citizenship.

In general, education for sustainability is approached through a gender perspective (SDG 5) or equality (SDG 10) at both the primary and early childhood education stages (Romero Claudio et al., 2023). However, the proposed experience seeks to advance education in values that foster interculturality and respect for diverse individuals and societies (SDG 10 and 16). This aligns with an approach that advocates for expanding freedoms, enabling individuals to take advantage of opportunities they deem most valuable, thereby contributing to overall well-being (Murga Menoyo, 2020).

Table 1. SDGs to work on from this experience

Goal	Issue
4	Quality Education
5	Gender Equality
10	Reduced Inequalities
16	Peace, Justice and Strong Institutions

Source: United Nations. Sustainable Development Goals (n.d.)

The citizenship that will engage with the tasks related to the SDGs is of critical importance, as it compels us to reconsider educational approaches in order to train students with a planetary consciousness. This consciousness is rooted in the ethics of responsibility and care, structured around five fundamental pillars:

a. Universal human rights.
b. Democracy and participation.

c. Equity.
d. Protection of minorities.
e. Peaceful resolution of conflicts.

This framework is grounded in a supportive and inclusive vision of the world (Murga Menoyo, 2020).

Furthermore, SDG 4 emphasizes the significance of striving for an Education for Sustainable Development that ensures students gain both theoretical and practical knowledge. This knowledge enables them to promote sustainable development through elements such as human rights, fostering a culture of peace and non-violence, appreciating cultural diversity, and recognizing the role of culture in contributing to sustainable development (Murga Menoyo, 2020).

This type of education encompasses three key areas of learning: cognitive, socio-emotional, and behavioral. In addition, two more areas are included: behavioral skills and non-cognitive skills, which cover social abilities such as empathy, conflict resolution, communication, and the capacity to work in networks and engage with individuals from diverse backgrounds, cultures, and perspectives (Murga Menoyo, 2020). However, these approaches show a significant deficit, particularly in the training of future Primary Education teachers, and especially in Early Childhood Education, highlighting the need to promote strategies that address and overcome these barriers.

3. WORKING WITH SOCIALLY RELEVANT PROBLEMS IN THE SECOND CYCLE OF EARLY CHILDHOOD EDUCATION

The teaching of Social Sciences aims to cultivate students' social thinking, enabling them to understand and actively participate in society while confronting contemporary issues, all the while considering both past and future contexts. In this regard, it is essential to integrate the realities in which students exist as part of society, ensuring that the issues relevant to that society are reflected in the classroom (López et al., 2022). Various authors assert that school instruction in Social Sciences should be based on Relevant Social Problems (López, 2017), which have been referred to by different names, each with slight variations in definition—social controversy topics, controversial issues, pressing social conflicts, socially relevant problems, or socially alive issues, among others (Díaz & Felices, 2017). However, they all stem from real societal problems and conflicts that generate ongoing debates without consensus, distinguishing them from traditional, agreed-upon school subjects (González, 2011). In fact, some Didactics of Social Sciences scholars argue that teacher training focused on critical understanding of reality and social intervention would be ineffective without the inclusion of social problems. Introducing such problems allows for the development of students' critical and social thinking, fosters their ability to make autonomous, constructive, and responsible decisions, and enhances their dialogue and consensus-building skills (Oller, 1999; in Díaz & Felices, 2017).

The teaching of Social Sciences based on social problems, socially relevant issues, or controversial topics aligns directly with the approach discussed here, as it fosters the learning of social justice and encourages active, engaged, and responsible participation through the development of responses or alternatives (Ortega & Olmos, 2018). Teaching based on real-world social issues in Social Sciences enables students to work with concepts, methodologies, and interpretive frameworks that help them navigate the complexities of reality. In order to understand social phenomena, such as migration and the cultural processes that unfold in host societies, a multidisciplinary approach is essential; we must draw on geography, history, sociology, and economics. Only by doing so can we comprehensively analyze the social phenomena associated with migration (Ocampo & Valencia, 2019). Addressing socially relevant topics allows students at this educational stage to integrate multiple aspects of learning, thereby moving toward the globalized practices needed in Early Childhood Education. Nevertheless, the Socially Relevant Problems approach remains closely associated with the field of Social Sciences (Ocampo & Valencia, 2019).

The introduction of controversial topics such as the environmental crisis, war, xenophobia, or poverty into the classrooms of the second cycle of Early Childhood Education has been explored by various authors, yielding positive outcomes for students, teachers, and the broader community (López, 2019; López et al., 2022). In this regard, the proposed topic aligns with the premise of being controversial: immigration and the tensions it generates when students from diverse backgrounds encounter differing cultural patterns within the classroom.

4. INITIAL TEACHER TRAINING

Initial teacher training in Spain, at both the primary and early childhood education levels (although not compulsory), has been shaped in accordance with the country's social, economic, and political realities across various historical periods (Luzón & Montes, 2018).

In this context, the training of future teachers at the Early Childhood Education stage across the various disciplines comprising the curriculum is essential for competently approaching teaching. In the case of Social Sciences, this training is crucial for understanding their role as agents of change and the perspective from which the teaching-learning process is approached. Initial university training is a critical period, particularly when preparing to teach in a world that is constantly evolving.

Pedagogical work with future teachers is essential for them to acquire the epistemological foundations that support the development of their critical thinking, making them receptive to the social and cultural diversity of contemporary society (Farias, 2009 in Pallares Jiménez & Lleida Añón, 2023).

From a critical perspective, the curriculum serves as a document that teachers rely on, adapted to the social context (Hernández, 2021). In this sense, we agree that university education equips future teachers with the necessary skills to design proposals that enable students to transform knowledge into an educational tool (Blanco Lozano et al., 2008).

Following the proposal put forth by various authors within the framework of the previous legislation, and while awaiting a new White Paper, it is crucial for students to recognize and work with key concepts related to the social environment. This experience emphasizes concepts such as identity/otherness and diversity/inequality (Benejam Arguimbau, 2002 in Blanco Lozano et al., 2008).

In the university setting, it is important not only to teach content but also to demonstrate how to apply it to the socio-educational context we currently face. To achieve this, students must be able to identify the Social Sciences content within the curriculum, taking into account the globalizing nature of the work in the Early Childhood Education stage. By doing so, diversity can be treated as a thematic core, addressing the need to incorporate other curricular elements. This enables students in training to acquire the necessary skills to plan and organize lessons around the environment as a central theme, helping children initially understand their immediate social surroundings, before integrating this knowledge into broader social systems (Blanco Lozano et al., 2008).

Among the various aspects that are covered in Social Sciences, both generally and within Early Childhood Education, we will focus on two key areas:

- Firstly, children's acquisition of social knowledge, which involves understanding both themselves and other individuals or groups, as well as the relationships that can be formed between them. This is a key aspect of their socialization process and is a fundamental objective of the Early Childhood Education stage (Blanco Lozano et al., 2008).
- Secondly, when addressing the social dimension, the importance of education in values cannot be overlooked. Values education is crucial not only for individual development but also for life in society. According to Blanco Lozano et al., modern society increasingly demands the formation of an integral, critical, participatory, tolerant, and supportive citizenry capable of meeting the challenges posed by globalization. From this perspective, the goal is to cultivate a society that promotes social justice, equity, inclusion, and interculturality. To achieve this, universities must identify and nurture

the factors that promote social and civic competence (Blanco Lozano et al., 2008).

As teachers, our commitment is to foster a society that advances democracy, equity, gender equality, tolerance, and respect for others. Schools are the ideal setting for facilitating these types of learning.

Despite this, recent research indicates that social content is not given the emphasis it deserves, and the same applies to Social Sciences within university curricula. This results in a gap in the preparation of newly graduated teachers. Studies point to several contributing factors, such as the globalizing approach to education at this stage, the prioritization of areas like literacy and numeracy, and deeply rooted conceptions about children's cognitive development. Additionally, the early childhood stage is often viewed more in terms of caregiving than as a period for civic education (Morales Hernández & López Martínez, 2022). Integrating education for sustainable development into teacher training is essential to meet current educational demands. While CRUE has already made plans to address this, research continues to highlight gaps in the training received by university students. However, project- and workshop-based learning has demonstrated a positive overall impact (Gómez Gómez & García Lázaro, 2023). In this context, it is crucial to consider both the discipline under discussion and the need to introduce active methodologies and competency-based learning approaches.

5. SOCIO-EDUCATIONAL CONTEXT IN CANARY ISLANDS CLASSROOMS REGARDING MIGRATORY PROCESSES AND INTERCULTURALITY

5.1. Current Socio-Educational Context

Migrations are a phenomenon as old as humanity and continue to be present in today's societies. Migrations remain an unresolved issue in the global order, with migrant movements now more frequent than at any point in history. These migratory processes bring significant economic, social, cultural, and security changes, which are often heavily politicized, particularly with the involvement of the media (Muñoz Jumilla, 2002). The media tends to simplify global migration trends, their evolution, historical and contemporary contexts, and their contributions to both sending and receiving societies, often proliferating myths.

Thus, human movements, beyond the social issues they raise, have the power to bring different cultures closer together. However, the question of the other, the foreigner, or otherness, challenges our ability to coexist. In recent decades, migration has become increasingly significant in Spain, particularly in the Canary Islands. In 2024, the number of residents in Spain born abroad reached its highest point, with 8.8 million people, according to data from the Continuous Population Statistics (INE, 2024). This figure represents 18.1% of the population, with uneven distribution across the autonomous communities. The highest percentages are found in the Balearic Islands (27%), Catalonia and Madrid (24%), and the Region of Valencia and the Canary Islands (both, 22%) (INE, 2024). This immigrant population, arriving from diverse regions such as Asia, Latin America, and Africa, enters Spanish territories through various means and with different life experiences. Regarding those arriving from Latin America (Domínguez et al., 2020), the following question was raised:

Following Spain's integration into the European Union, Latin American immigration has been interpreted within the framework of push-pull factors that arise from the differing economic levels of countries around the world, based on the theory of sociodemographic complementarity, as well as the favorable Spanish legal and institutional framework for the Latin American community (Prieto Rosas & López Gay, 2015). In this context, three elements have been highlighted to explain the rise of Latin American immigration: a) the increased intensity and diversification of destinations for Latin American international migration; b) changes in the demographic structure of both Spain and Latin American countries; and c) the educational expansion in Latin America. Additionally, these factors are complemented by the level of integration that this community has achieved in Spain (Izquierdo & Martínez, 2014, p. 39).

Moreover, Spain ranks among the top countries receiving immigrants by sea from Africa, alongside Italy and Greece. Its geographical proximity and lower travel costs, compared to countries like France, the United Kingdom, or the United States, make Spain an attractive destination for those seeking to emigrate illegally (Mbaye, 2014). Additionally, non-Spanish Europeans, including Italians, Germans, Russians, and Ukrainians, have a strong presence in Early Childhood Education classrooms.

This migration process has a direct impact on the educational system and immigrant students, who, like all children, must be educated under Article 27 of the Spanish Constitution (1978), which guarantees the right to education. However, the question arises: Are Spanish classrooms prepared to accommodate these students? Are teachers equipped to lead the education of children from different cultural backgrounds, sharing the same space?

These challenges must be handled with care and efficiency to avoid, as suggested: "The dichotomous vision between the group of "natives" and the group of "foreigners" arises because this discourse, originating from the political class, spreads, multiplies, and amplifies through the media, thereby colonizing the common thought of our societies" (Van Dijk, 2010, p. 84).

Although many Spaniards maintain that "others are racists" (Calvo Buezas, 1989, p. 53), there is clear evidence of "an increase in xenophobia and subtle, low-intensity racism. This type of racism is expressed in indirect ways, often characterized by cultural and symbolic elements and a hidden but persistent resistance to diversity" (Díez Gutiérrez, 2020, p. 84).

Therefore, the school must be equipped to address the diversity of its school community, particularly when the needs arise from social and cultural backgrounds and individual capacities. In this context, the diversification, personalization, and universal design of curricula and teaching processes are established as key strategies for managing the diversity present in classrooms (Casanova Rodríguez & Rodríguez, 2009; UNESCO, 2017; Luis Rico, 2020).

However, the reception processes differ significantly depending on the educational stage. In compulsory education, different personal and administrative resources are mobilized to ensure that students meet stage objectives appropriately.

Nevertheless, as Luis Rico (2020, p. 72) asks: "What happens in the Early Childhood Education stage? Due to its non-compulsory nature and the developmental level of students, this stage generates ideas that directly shape how Early Childhood Education students are received".

The number of foreign-born children living in Spain is notably significant within the school system. Over the last decade, this figure has increased by 45%, surpassing one million for the first time during the 2023-2024 academic year. Specifically, according to statistics from the Ministry of Education, Vocational Training, and Sport, 1,066,875 students were enrolled in Early Childhood Education, with 16.6% in Spain and 12.5% in the Canary Islands. Socio-political and demographic indicators suggest that in the coming years, thousands of people from groups, countries, and cultures outside Europe will settle in Spain, seeking a better quality of life and a more secure future for themselves and their families. It is essential to recognize that they are undoubtedly part of our community. Immigrant students present a variety of needs (Fernández, 2004), making it crucial to train teachers who will be responsible for their education in Spanish classrooms and particularly in the Canary Islands.

5.2. Current Intercultural Context

Spanish society, and specifically Canarian society, has undergone significant structural changes over the past thirty years, which have had profound effects on cultural management, a responsibility shared between the State and the Self-Governing Regions. As Briones Gómez (2007, n.p.) observes, "We have transitioned from a culturally and religiously almost homogeneous and monochromatic society to a plural or multicultural one. Moreover, this transformation is evident in all aspects of the Spanish socio-cultural system". This shift has had a strong impact on the teaching-learning process, where children from various socio-cultural backgrounds now share educational spaces.

As a result, educational centers have become contexts where highly dynamic social interactions take place, requiring that institutions and educators implement strategies to effectively address cultural diversity. In light of this scenario, it is essential to reflect on the educational implications and take proactive steps to manage diversity, fostering values rooted in equality and the acceptance of multiculturalism in future generations. Rodríguez et al. emphasize the importance of this approach:

The justification for intercultural education today, and the consequent development of intercultural competence, is strongly supported by the historical gaps that have been imposed upon us, as the many acts committed over time have resulted in deep sources of exclusion and social marginalization (2019, p. 63).

Interculturalism goes beyond a mere multicultural perspective, as it "proposes a proactive model of relationships between cultures, where cultural interaction is viewed as an educational process in itself" (Fernández, 2004, p. 42). Expanding on this idea, other authors argue that "interculturality is founded on interdependence, reciprocity, and symmetry between cultures, which require dialogue and communication" (Ávila Dávalos, 2022, p. 12), with a focus on the acceptance and recognition of cultural differences (Alavés Ruíz, 2014). This framework calls for the establishment of cultural relations and the integration of diverse cultures. Thus, intercultural education must be embedded within the overall educational framework, specifically the school project, which must take into account the particular context of each educational center. Likewise, the autonomy afforded to schools in the design and implementation of programs allows for the creation of initiatives tailored to the needs of both the center and its students, ensuring intentional and effective attention to diversity in intercultural education (Fernández, 2004; Tebár Cuesta, 2018; Mansilla Sepúlveda et al., 2022).

6. DESCRIPTION OF THE PROPOSED STRATEGY

The early childhood stage (0-6 years) is crucial for children's development. It is essential to provide motivating experiences that promote meaningful and innovative learning. Through this kind of educational approach, children can engage with important content and values related to culture, equality, tolerance, and solidarity in a playful, natural, and engaging manner (Rodríguez-García et al., 2019).

Taking this into account and recognizing the critical role teachers play in the formative experiences of young children, this educational strategy has been designed to convey knowledge to students in a more practical and effective manner, particularly on contemporary and controversial issues such as migration and interculturality.

6.1. Methodological Proposal

This strategy will be implemented as an educational experience within the course Didactics of Social Sciences in the Early Childhood Education Degree at the University of Las Palmas de Gran Canaria (ULPGC). This is the only course on Social Sciences taught in the degree, and it is a compulsory course offered during the first semester of the fourth year, with a workload of 4.5 credits (45 contact hours and 66.5 non-contact hours).

In addition to the general and transversal competencies, the course includes the following specific competencies:

- ED2: Understand the scientific, mathematical, and technological foundations of the curriculum for this stage, along with the theories of acquisition and development of relevant learning.
- ED3: Foster interest in and respect for the natural, social, and cultural environment through suitable educational projects.
- ED4: Acquire knowledge about the evolution of thought, customs, beliefs, and social and political movements throughout history.

As for these competencies, this experience allows us to address all three specific competencies. First, students must base their proposals on the curriculum of the Self-Governing Region of the Canary Islands (ED2). Secondly, the experience encourages students to reflect on their surrounding social and cultural environment and learn how to develop suitable educational projects rooted in that context (ED3). Lastly, through this proposal, students will gain an understanding of how society, thought, and related elements have evolved over time within this particular subject (ED4).

According to the *Verifica* teaching project, the objectives to be met are:

1. To prepare future Early Childhood Education graduates from a professional perspective in relevant subject areas.
2. To introduce basic ideas that should be included in the Early Childhood Education curriculum related to Social Sciences.
3. To provide tools, methods, and techniques for the effective implementation of teaching practices.

As with the competencies, we can affirm that this experience will contribute to achieving the course's objectives, as it will be approached from both an innovative and professional perspective for teachers in initial training in the field of Social Sciences education. It will address a highly relevant topic, immigration, which is particularly significant in Spain and the Canary Islands, offering novel methodologies to engage with this issue.

Finally, we must focus on the content outlined in the teaching project. This project includes five topics, and we believe that the most appropriate for this experience is Topic 4: First contact with the social world, the development of social knowledge, and the understanding of society and its elements. Elements of representation of the social world. The construction of economic, political, artistic, and philosophical notions. Procedures and resources for their teaching.

Upon completing the course, students are expected to achieve the following Learning Results (RA):

- RA1: Understand and apply the scientific and methodological foundations necessary to address the curriculum for this stage, designing experiences related to natural and social sciences using a scientific methodology.
- RA2: Place relevant scientific and social events in their historical context and assess their impact on social progress.
- RA3: Design teaching proposals that are integrated into the school curriculum and connect classroom activities with various perspectives of the natural, social, and cultural environment, with a focus on respect for the environment and sustainable development.

The proposed experience will contribute to achieving the learning outcomes, as it requires students to approach the curriculum from a Social Sciences perspective (RA1). They will need to contextualize the phenomena to be integrated into the curriculum within the sociocultural environment in which they occur (RA2), and design proposals that fit into the curriculum with a focus on contributing to sustainable development. The objectives of this experience include:

- To bring Early Childhood Education students closer to practical and real-world experiences with second-cycle Early Childhood Education students (4-5 years old).
- To propose an innovative strategy that incorporates quality education initiatives focusing on values such as equality and gender.
- To identify cultural diversity as a key element of 21st-century society in the Canary Islands.

6.1.1. Methodology

The proposed work methodology is project-based learning (PBL), an alternative methodology to articulate the content addressed in the designed educational strategy. This methodology is selected because research indicates that PBL tends to increase motivation and improve learning outcomes. It achieves this by allowing students to apply the knowledge they have gained to specific products or processes, putting their conceptual frameworks into practice to solve tangible problems (Medina-Nicolalde, & Tapia-Calvopina, 2017; Rodríguez-Rodríguez & Parreño Castellano, 2023). This approach enhances the relevance of the subject matter for students. Additionally, PBL is expected to contribute to greater thematic coherence by organizing the course content around a project. Furthermore, the collaborative nature of PBL encourages students to work in teams, fostering a questioning of individualism in the classroom while promoting respect for diversity.

6.2. Development of Strategies

As mentioned in previous sections, our proposal is included in Topic 4, titled "First contacts with the social world: the development of social knowledge, the understanding of society and its elements. Elements of representation of the social world. The construction of economic, political, artistic, and philosophical notions." According to the teaching plan, this unit comprises ten hours—four hours dedicated to theoretical content and six hours focused on practical work, which includes the presentation of the final projects by the student groups.

6.2.1. First Part

Before starting the PBL, a two-hour classroom activity is proposed, focusing on family geographies, based on Bale's (2012) proposal. The task will involve students mapping their place of birth, as well as the birthplaces and residences of their parents and grandparents, including their current places of residence. The activity will be prepared in advance through family investigation, and students will then work on it in class. The objective of this activity is to analyze geographical mobility within families, examining why populations move from one place to another, and to raise awareness of the factors driving migration. Initially, students will complete the task individually, using information gathered at home, and later they will share their findings in class. This sharing phase is intended to foster reflection on their personal migration stories and those happening in the broader context of their environment. Bale suggests two actions for this activity:

- Worksheet: Students will record the various places of residence in a table, listing them from most recent to oldest.
- Geography Tree: Instead of a traditional family tree that focuses on names, this geography tree will focus on the places of birth of family members.

This activity aims to identify the history of mobility or lack thereof within each student's family and to draw a parallel between "immigrant Spain" and "new emigration." Through this process, students will gain an awareness of mobility patterns, whether from their own personal experience or that of their classmates. They will come to understand that migration is often more prevalent than initially perceived. For instance, approximately 4 out of 10 Canarian students in the Early Childhood Education Degree at ULPGC have family migration histories, with key destinations being the Americas, particularly Cuba, Venezuela, and Argentina (Domínguez et al., 2002; Monzón, 2022).

In the second part of the work, students are guided to undertake a collaborative project, which they will develop in teams. The project's objectives and expected competencies are introduced and discussed in class, along with the various stages involved in the project. This includes a collaborative approach, the level of commitment required, and the evaluation method, including the use of rubrics. A key point is the importance of the materials prepared for the final presentation. These materials will not only be used for the project presentation but will also play a role in the Practicum III course, which takes place in Early Childhood Education classrooms during the second semester of the 4th year. The materials will contribute to the final evaluation in that phase, as there will be coordination among the teachers evaluating the different subjects. The presentation phase is critical because, without a solid understanding and commitment from the students at the outset, it will be difficult to achieve satisfactory results.

The students will form work groups, with a maximum of five students per group, creating six teams in the classroom. The central theme of the project—interculturality and migration in the Canary Islands—will be defined with the teacher's guidance. Each team will explore this topic from the perspective of a specific territory assigned to them. The overarching project has been titled "Equal and Different," although each team will approach it from the territory it is to defend. These two adjectives, according to Hegel, represent opposing concepts, and from their conflict, a third arises: the synthesis. This synthesis, as Gadamer (2020) notes, holds more truth than either of the original opposing terms.

At this preparatory stage, the teacher selected a Peters map, which was divided based on the number of groups (6) and the regions that send immigrants to the Canary Islands: North and Central America (isthmic and insular), South America, Europe (excluding Spain), Asia, North Africa, and sub-Saharan Africa. Each group was assigned a section of the map, which would serve as the focus of their analysis. The groups were tasked with analyzing their assigned region from various perspectives, including its geographical location, sociodemographic and economic characteristics, cultural features, historical context, and migratory connections with the Canary Islands. These aspects would be incorporated into the final report that each group would submit as part of their project work.

The project was organized into several phases following this initial stage, which focused on information, organization, and motivation (Table 2). The first phase aimed to help the groups explore different perspectives on the topic they were assigned, based on bibliographic and electronic resources. This allowed them to more accurately define the project's objectives and, in some cases, identify key hypotheses for analysis. In this phase, guiding questions were formulated, and the necessary tools to carry out the tasks were identified.

Table 2. Phases of the project-based learning model

<table>
<tr><th>Nº</th><th>Description of the phases of PBL</th><th>Observations</th></tr>
<tr><td>1</td><td>After assigning the territory to each team, they will work on the topic of migration and interculturality using the PBL methodology and will address different aspects:
• Conducting a bibliographic search and developing the theoretical framework of the topic, with each team highlighting the sociodemographic, historical, and economic aspects of its assigned territory.
• Each team will create a database of Early Childhood Education students born outside the Canary Islands and enrolled in the last three academic years. Each team will focus on their assigned territory. For example, teams working on South America will select children born in countries from that region.
• Justification of the interest in the topic to be investigated.
• Development of the guiding questions to be posed to the interviewees.</td><td rowspan="4">The course teachers monitored and facilitated the development of the various phases throughout the face-to-face practical activity sessions. Also, as the project progressed, the workgroups were required to prepare evaluative materials.
In this way, no group was left behind, and continuous work was guaranteed.</td></tr>
<tr><td>2</td><td>• Prepare and design a semi-structured interview guide in the classroom, with each team contributing ideas to create a unified model.
• Selection of experts to be interviewed. These experts will include staff from NGOs that work with immigrants in the Canary Islands and those managing reception centers for unaccompanied migrant minors. Proposed organizations include the International Red Cross, the Coliseum Association, and Quorum Social 77, among others. Additionally, one team must interview representatives from the Island Council or Town Councils involved in addressing migrant issues in the Canary Islands. Experts will also include a primary school director and an early childhood education teacher who works in classrooms with migrants from various nationalities. Each team is required to conduct at least three interviews, with one interview being the teacher.</td></tr>
<tr><td>3</td><td>• Prepare and implement a discussion group for each project (one per work team). Example: “Equal and Different: Latin American Students in Early Childhood Education Classrooms in Gran Canaria.”</td></tr>
<tr><td>4</td><td>• The teacher will observe each discussion group using an observation guide. Additionally, the discussion group will be recorded and later analyzed to gather information.</td></tr>
<tr><td>5</td><td>• Finally, each team will then give an oral presentation of the results obtained, supported by a written presentation highlighting the key findings and proposing personalized strategies tailored to the needs identified by each group.
• During the final presentation, a roundtable will be organized, with each team's moderator and representative participating.
• The moderator will offer a general conclusion, summarizing the most relevant results and the most effective strategies developed by the teams.</td><td>This phase consisted of two stages: the first took place in the classroom during the practical sessions and tutorials with the various groups, and the second occurred when the results were presented in a plenary session. During this session, the subject teachers served as a panel, and each student participated in assessing the work of their classmates.</td></tr>
</table>

Source: Own elaboration

The activities will be conducted both in person, utilizing a computer room or classroom, and remotely, including fieldwork. The use of the virtual classroom and ICT tools within the course will be highly beneficial for ensuring coordination and adherence to the planned schedule. The teaching staff will serve as facilitators throughout all phases, providing resources and assisting in problem-solving. The final phase of the project will involve the preparation and presentation of the main results to the class. A conference will be organized to showcase these presentations, coinciding with the Educational Innovation Week held annually at ULPGC, which typically takes place in the second half of November. The conference will be open for registration, with special invitations extended to Early Childhood Education teachers and organizations working with migrant children to support their integration into Canarian society.

This final phase also includes tutoring for the face-to-face activities and culminates in the continuous assessment of this part of the course. Students will present how they developed their projects, outlining the steps they followed, the challenges they encountered, how they resolved them, and the key results achieved.

6.3. Expected Results

The Project-Based Learning strategy offers an alternative and a key opportunity to implement experiences that strengthen civic education within educational communities, particularly through the lens of Critical Didactics of Geography, History, and Social Sciences (Abarca, 2017). This approach was chosen in this research with the goal of fostering civic education and developing 21st-century thinking skills in Early Childhood Education degree students, who are future teachers. The strategy aims to help students connect real-world experiences and outcomes with the school community and the local social environment. A fundamental principle of this approach is the presentation of results to public audiences beyond the classroom and the teacher, thus broadening the impact of the students' work (Abarca, 2017).

The application of the PBL methodology is expected to contribute to the development of various general and specific competencies within the degree program and the course. Among the general competencies, the didactic approach directly stimulates the ability to apply the scientific method, as well as skills in analysis and synthesis, problem-solving, decision-making, teamwork, information management, and the application of theoretical knowledge in practice (Parreño & Rodríguez, 2021). Additionally, it fosters autonomous learning and promotes awareness of human rights and the reduction of all forms of inequality, aligning with the Sustainable Development Goals (SDGs). As for specific competencies, studies have demonstrated the effectiveness of this didactic approach (Abarca Salina, 2017; Parreño-Castellano & Rodríguez-Rodríguez, 2021) in developing knowledge in areas such as human geography, particularly social geography. It enhances the ability to use geographic information to address social issues, integrate temporal and spatial dimensions in the analysis of socio-territorial processes, and relate and synthesize cross-sectional territorial information (Parreño, 2021).

The active nature of PBL facilitates the achievement of additional learning outcomes that may not be explicitly expressed through competencies or objectives, such as the development of self-esteem, self-recognition, (partial) social immersion, and leadership skills. While these competencies are not formally included in the course curriculum, they will become evident through the work students engage in. It is common for these types of skills to emerge in PBL environments (Vergara Ramírez, 2016; Rodríguez & Parreño, 2023).

In addition, it is important to note that the development of transversal competencies is closely linked to this applied methodology. It enables students to communicate with diverse audiences, effectively utilize media and communication channels—particularly those related to new information and communication technologies—and apply their professional skills and knowledge to address the interests, needs, and concerns of their specific communities or regions.

7. FINAL CONSIDERATIONS

Through the design and future implementation of the project "Equals and Different," students of the Degree in Early Childhood Education will have the opportunity to engage in a practical teaching experience, one that they initiate themselves, while working with second-cycle Early Childhood Education students on content related to intercultural education and migration.

The training of future teachers presents a challenge that must be addressed. Although many professionals have expressed disappointment with the current study plans, particularly due to the omission of Social Sciences content, this should not be an obstacle. We must continue working with the necessary responsibility and determination to produce highly qualified Early Childhood Education professionals. Social Geography phenomena deserve a prominent place in teacher training, and therefore, introducing current social issues into the subject content is a priority task.

There is a stark contradiction between the significant value attributed to geographical knowledge and Social Sciences in children's education and the limited number of credits allocated to these subjects in teacher training programs. This is particularly concerning when considering that the Early Childhood Curriculum includes an area dedicated to "Discovering and exploring the environment," which could serve as an integrating core for the other two areas (growth in harmony and communication, and representation of reality).

It is now essential to consider how to integrate and sustain the learning of young children within the cultures of educational centers. These are the environments where our future Early Childhood and Primary Education teachers will complete their practicum and, eventually, perform professionally. This commitment to blending academic and practical knowledge reflects the ongoing effort to dismantle the traditional hierarchy between university learning and social issues, aiming to form "a more equal and dialectical relationship between academic and practical knowledge" (Zeichner, 2010, p. 132).

8. REFERENCES

Abarca Salinas, M. E. (2017). El aprendizaje basado en proyectos, una estrategia didáctica desde historia y geografía para intencional la participación ciudadana en las comunidades educativas. *Revista Andamio, 4*(2), 53-71, https://revista-andamio.cl/index.php/revista/article/view/83

Alacid de Pascual, S. (2021). La sostenibilidad dentro del aula de Educación Infantil. *Quest Journals. Journal of Research in Humanities and Social Science*, Volume 9-Issue 10, 74-83. https://www.questjournals.org/jrhss/papers/vol9-issue10/Ser-5/K09107483.pdf

Alavés Ruíz, A. (2014). *Interculturalidad: conceptos, alcances y derecho*. Centro de Producción editorial del GCPRD. https://rm.coe.int/1680301bc3

Ávila Dávalos, H. (2022). Multiculturalidad e interculturalidad: El papel de la educación superior para generación de competencias interculturales para el contexto organizacional. *Educación y Humanismo*, 24(43), 13–34. DOI: 10.17081/eduhum.24.43.4838

Bale, J. (2012). *Didáctica de la Geografía en la Educación Primaria*. Morata.

Benejam Arguimbau, P. (2002). La oportunidad de identificar conceptos clave que guíen la propuesta curricular de ciencias sociales, AAVV. *Las ciencias sociales: concepciones y procedimientos*, Graó, 11-19.

Benejam Arguimbau, P.; & Pagès Blanch, J. (coord.) (1997). *Enseñar y aprender Ciencias Sociales, Geografía e Historia en la Educación Secundaria*. ICE/HORSORI.

Blanco Lozano, P., De la Calle Carracedo, M., Fernández Rubio, C., Molero Otero, B., Ortega Sánchez, D., & Sánchez Agustí, M. (2008). La formación en didáctica de las Ciencias Sociales del maestro de Educación Infantil. Una propuesta adaptada al Espacio Europeo de Educación Superior. En R.M. Ávila Ruíz, M. Alcázar Cruz Rodriguez y M.C. Díez Bedmar (Coords.), *Didáctica de las Ciencias Sociales, currículo escolar y formación del profesorado. La didáctica de las Ciencias Sociales en los nuevos planes de estudio* (pp.113-134). Universidad de Jaén.

Briones Gómez, R. (2007). La integración del 'otro 'en la emergente España multicultural. *Gazeta de Antropología, 23*. https://docs.repositoriobiocultural.org/2021/08/03/la-integracion-del-otro-en-la-emergente-espana-multicultural/

Calvo Buezas, T. (1989). *Los racistas son los otros: gitanos, minorías y derechos humanos en los textos escolares*. Editorial Popular.

Casanova Rodríguez, M. A., & Rodríguez, H. J. (2009). *La inclusión educativa, un horizonte de posibilidades*. Editorial La Muralla.

Díaz Moreno, N., & Felices de la Fuente, M. M. (2017). Problemas sociales relevantes en el aula de primaria: la "cartografía de la controversia" como método, *REIDICS. Revista de Investigación en Didáctica de las Ciencias Sociales, nº 1*, 24-38, https://dehesa.unex.es/handle/10662/6610

Díez Gutiérrez, E. J. (2020). Migraciones y refugiados: el papel de los centros educativos ante la situación actual. En M. Fernández-Hawrylaw, D. Heras Sevilla y J.A. Gómez Monedero. (Coords.), *Ampliando horizontes en Educación inclusiva* (pp. 81-88). Universidad de Burgos.

Domínguez Mujica, J. (2008). Un nuevo diálogo migratorio: Canarias, Latinoamérica y el Caribe en la Era de la Globalización. *Anuario de Estudios Atlanticos*, 1(54), 469–509. https://dialnet.unirioja.es/ejemplar/216156

Domínguez Mujica, J., Díaz Hernández, R., & Parreño Castellano, J. M. (2002). Canarias, puerta de Europa. ¿Preferido destino de iberoamericano? *Coloquios de historia canario americana*, 57-77, https://accedacris.ulpgc.es/bitstream/10553/12033/4/Canarias_%20puerta_de_Europa.pdf

Domínguez-Mujica, J., López de Lera, D., Ortega-Rivera, E., & Pérez-Caramés, A. (2020). El sistema migratorio de Latinoamérica-España: ¿Ha sido la crisis económica un paréntesis? *Cuadernos Geográficos*, 59(3), 37–57. DOI: 10.30827/cuadgeo.v59i3.9223

Farias, F. (2009). La epistemología de las Ciencias Sociales en la formación por competencias del pregrado. *Cinta de Moebio*, 34(34), 58–66. DOI: 10.4067/S0717-554X2009000100004

Fernández Batanero, J. M. (2004). La presencia de alumnos inmigrantes en las aulas: un reto educativo. *Educación y educadores*, (7), 33-44, https://www.redalyc.org/pdf/834/83400705.pdf

Funes, A. G., Cerdá, M. C., & Jara, M. A. (2019). Las finalidades de la enseñanza de las ciencias sociales y la historia para la educación ciudadana del siglo XXI. Experiencias en Argentina, Balbé, M., & González Monfort, N., Santisteban, A. (Eds). *Quin professorat, quina ciutadania, quin futur?* 227-231, https://ddd.uab.cat/record/220367

Gadamer, H. G. (2000). *La dialéctica de Hegel*. Ediciones Cátedra.

García Esteban, F.E.; & Murga Menoyo, M.A. (2015). El profesorado de Educación Infantil ante el Desarrollo Sostenible. Necesidades formativas, *Enseñanza &Teaching*, 33-1, 121-142. DOI: DOI: 10.14201/et2014331121142

García Monteagudo, D. (2024). Formación inicial del profesorado de Educación Infantil en pensamiento geográfico: paisaje e identidades territoriales: Initial training of Early Childhood Education teachers in geographical thinking: landscape and territorial identities. *ENSAYOS.Revista De La Facultad De Educación De Albacete*, 39(1), 57–73. DOI: 10.18239/ensayos.v39i1.3488

Gómez Gómez, M., & García Lázaro, D. (2023). Concienciación y conocimientos sobre los Objetivos de Desarrollo Sostenible en la formación del profesorado, *Profesorado. Revista de currículum y formación del profesorado, vol. 27, Nº 3*, 243-364. DOI: 10.30827/profesorado.v27i3.27948

González Monfort, N. (2011). Reflexions al voltant de les questions socialment vives i l'ensenyament de les ciències socials. Conclusiones de les VII Jornades Internacionals de Recerca en Didàctica de les Ciéncies Socials, Pagès Blanch, J., Santisteban, A. (Coords): *Les questions socialment vives i l'ensenyament de les ciències socials* (pp. 153-161). https://ddd.uab.cat/record/197111

Guardeño Juan, M., & Monsalve Lorente, L. (2024). Educación para el desarrollo sostenible en el currículum de España e Irlanda. *Universidad y Sociedad*, 16(1), 30–44. http://scielo.sld.cu/scielo.php?script=sci_abstract&pid=S22183620202 4000100030&lng=es&nrm=iso

Hernández Cervantes, L. (2021). La formación del profesorado como técnico y reproductor social. La práctica docente de Ana Lorenia Hernández Cervantes. En L.M. Aguayo Rendón, L.M. López Velarde, J.C. (Coords). *Ser profesor y doctorantes en tiempos de intertidumbre. Aportes al desarrollo profesional desde y sobre la pandemia* (pp. 175-202). Taberna Libaria editores.

Hernández Prados, M. A., Carbonell Bernal, N., & Polanco Mora, B. (2021). Diseño de una experiencia para trabajar la interculturalidad en Educación Infantil. *Revista Electrónica en Educación y Pedagogía*, 5(9), 165–178. DOI: 10.15658/rev.electron.educ.pedagog21.11050912

Izquierdo Escribano, A., & Martínez-Buján, R. (2014). From Favourites to 'With No Return': Permanence and Mobility of Latin American Immigration in Spain during the 21st Century. *Canadian Ethnic Studies*, 46(3), 103–130. DOI: 10.1353/ces.2014.0038

López Martínez, M. J. (2017). Los problemas relevantes de la sociedad actual en el aula de 5 años: qué hacemos con las personas refugiadas sirias. *Investigación en la escuela nº 92*, 19-31, https://dialnet.unirioja.es/servlet/articulo?codigo=6275111

López Martínez, M. J. (2019). La enseñanza de las ciencias sociales a partir de problemas sociales relevantes en un aula de infantil. Estado de la cuestión y resultados de una investigación. *El futuro del Pasado: Revista electrónica de Historia*, nº 10, 57-59, https://dialnet.unirioja.es/servlet/articulo?codigo=7105038

López Torres, E., Carril Merino, M. T., & Alonso Neila, E. (2022). ApS con fuentes orales para hacer frente a la España vaciada en Educación Infantil. *Didáctica de las Ciencias Experimentales y Sociales*, (43), 53–70. DOI: 10.7203/dces.43.25270

Luis Rico, M. I. (2020). La acogida del alumnado de familia inmigrada en la etapa de educación infantil. In Ampliando horizontes en educación inclusiva. [Recurso electrónico]: XVII Congreso Internacional y XXXVII Jornadas de Universidades y Educación Inclusiva. 21, 22 y 23 de septiembre de 2020 (aplazado en abril por Covid-19) (pp. 71-75). Servicio de Publicaciones e Imagen Institucional.

Luzón-Trujillo, A., & Montes Moreno, S. (2018). Perspectiva histórica de la formación inicial del profesorado de Educación Infantil y Primaria en España. Una tarea inacabada. *Historia Caribe*, 13(33), 121–152. DOI: 10.15648/hc.33.2018.6

Mansilla Sepúlveda, J. M., Villalba, K., Mora García, J. P., & Turra Díaz, O. (2022). Tensión epistémica entre multiculturalismo anglosajón e interculturalidad latinoamericana: Sus proyecciones en educación. *Revista de Filosofia (Venezuela)*, 39(100), 132–143. DOI: 10.5281/zenodo.5979768

Mbaye, L. M. (2014). "Barcelona or die": Understanding illegal migration from Senegal. *IZA Journal of Migration*, 3(1), 1–19. DOI: 10.1186/s40176-014-0021-8

Medina-Nicolalde, M. A., & Tapia-Calvopina, M. P. (2017). El aprendizaje basado en proyectos una oportunidad para trabajar interdisciplinariamente. OLIMPIA. Revista de la Facultad de Cultura Física de la Universidad de Granma, 14(46), 236-246. https://dialnet.unirioja.es/servlet/articulo?codigo=6220162

Montero Caro, M. D. (2021). Educación, gobierno abierto y progreso: Los Objetivos de Desarrollo Sostenible (ODS) en el ámbito educativo. Una visión crítica de la LOMLOE. *Revista de Educación y Derecho*, (23), 1–26. DOI: 10.1344/REYD2021.23.34443

Monzón Perdomo, M. E. (2019). «Mujeres solas» luces y sombras de la emigración canaria a América (siglos XVIII-XIX). *Anuario de Estudios Atlánticos*, (65), 1-24. Cabildo Insular de Gran Canaria.

Morales Hernández, A. J., & López Martínez, M. J. (2022). Didáctica de las Ciencias Sociales en Educación Infantil: Expectativas curriculares y formación docente. *Revista Educatio Siglo XXI*, 40(3), 61–82. DOI: 10.6018/educatio.487691

Muñoz Jumilla, A. R. (2002). Efectos de la globalización en las migraciones internacionales. *Papeles de Población*, 8(33), 9–45. https://www.scielo.org.mx/pdf/pp/v8n33/v8n33a2.pdf

Murga Menoyo, M.A. (2020). El camino hacia los ODS: Conformar una ciudadanía planetaria mediante la educación. *Comillas Jorunal of International Relations*, núm 19: Los desafíos de la Agenda 2030. Educación y ciudadanía global, 01-11, DOI: 10.14422/cir.i19.y2020.001

Ocampo Ospina, L. F., & Valencia Carvajal, S. (2019). Los problemas sociales relevantes: enfoque interdisciplinar para la enseñanza integrada de las ciencias sociales, *REIDICS. Revista de Investigación en Didáctica de las Ciencias Sociales* n ° 4, 60-75, https://dehesa.unex.es/bitstream/10662/9328/1/2531-0968_2019_4_60.pdf

Oller, M. (1999). Trabajar problemas sociales en el aula, una alternativa a la transversalidad. En T, García Santa María. (Coord.), *Un currículum de ciencias sociales para el siglo XXI: qué contenidos y para qué*, (pp. 123-131)., Díada Editora S.L.

Ortega Sánchez, D., & Olmos Vila, R. (2018). Los problemas sociales relevantes o las cuestiones socialmente vivas en la enseñanza de las ciencias sociales. En M.A. Jara y A. Santisteban. (Coords). *Contribuciones de Joan Pagés al desarrollo de las ciencias sociales, la historia y la geografía en Iberoamérica* (pp. 203-214). Universidad Nacional del Comahue. https://ddd.uab.cat/pub/llibres/2018/196869/jara-santisteban-2018-bariloche.pdf

Pallares Jiménez, M. A., & Lleida Añón, A. (2023). Diseño de actividades en el rincón de ODS en el aula de Educación Infantil. En M.E. Cambil Hernández, A.R. Fernández Paradas, N. De Alba Fernández. (Coords). *La didáctica de las Ciencias Sociales ante el reto de los ODS* (pp. 675-682). Narcea S.A. Ediciones.

Parreño-Castellano, J. M., & Rodríguez Rodríguez, M. A. (2021). *Aprendizaje basado en proyectos colaborativos. Una experiencia con estudiantes del Grado en Geografía y Ordenación del Territorio* (No. COMPON-2021-CINAIC-0060). Universidad de Zaragoza, Servicio de Publicaciones. DOI: 10.26754/CINAIC.2021.0060

Prieto Rosas, V., & López-Gay, A. (2015). Push and Pull Factors of Latin American Migration. In Domingo, A., Sabater, A., & Verdugo, A. (Eds.), *Demographic Analysis of Latin American Immigrants in Spain: From Boom to Boost* (pp. 1–28). Springer., DOI: 10.1007/978-3-319-12361-5_1

Resa Ocio, A., Lucas Palacios, L. & Diez Ros, R (2023). Hacia unas competencias coeducativas en Educación Infantil para la formación docente desde la Didáctica de las Ciencias Sociales. *Clío. History and History teaching 49*, 209-233, DOI: 10.26754/ojs_clio/clio.2023499540

Rodríguez García, A. M., Hinojo-Lucena, F. J., & Ágreda-Montoro, M. (2019). Diseño e implementación de una experiencia para trabajar la interculturalidad en Educación Infantil a través de realidad aumentada y códigos QR. *Educar, 55*(1), 59-77, https://raco.cat/index.php/Educar/article/view/348881

Rodríguez Rodríguez, M. Á., & Parreño Castellano, J. M. (2023). Aprendizaje activo en el aula universitaria actual: una experiencia de aprender haciendo. *Didáctica geográfica.* nº 24-2023, 39-61. 39. DOI: 10.21138/DG.663

Romero Claudio, C., Heredia Ponce, H., & Romero Oliva, M. F. (2023). Formación inicial de las docentes y los docentes y biografías de mujeres en libros ilustrados de no ficción, *Cadernos CEDES*, Vol. 44. *Cadernos CEDES*, 122(122), 106–120. DOI: 10.1590/cc271187

Tebár Cuesta, F. (2018). Autonomía de los centros educativos (intervención de la Inspección de Educación). *Avances en supervisión educativa*, (29). DOI: 10.23824/ase.v0i29.617

Van Dijk, T. (2010). Racismo y la Prensa: Análisis Crítico del Discurso. Ponencia en el Auditórium del Edificio Centro de Innovación Docente de la Facultad de Filosofía y Humanidades de la Universidad Austral de Chile el 3 de enero de 2010.

Vergara Ramírez, J.J. (2016). Aprendo porque quiero. El Aprendizaje Basado en Proyectos (ABP), paso a paso. SM.

Zeichner, K. (2010). Nuevas epistemologías en formación del profesorado, Repensando las conexiones entre las asignaturas del campus y las experiencias de prácticas en la formación del profesorado en la universidad, *Revista Interuniversitaria de Formación del Profesorado*, Vol. 24, núm. 2, 123-149, https://www.redalyc.org/articulo.oa?id=27419198007

KEY TERMS AND DEFINITIONS

Equity: A principle that ensures fair treatment for all individuals, acknowledging and respecting differences, diversity, and specific needs. Unlike equality, which gives everyone the same resources, equity provides individuals with what they need to achieve the same outcomes, considering their unique circumstances.

Interculturality: A process of interaction, coexistence, and dialogue between different cultures, rooted in mutual respect and recognition of cultural diversity. Unlike multiculturalism, which simply acknowledges the presence of multiple cultures in the same space, interculturality encourages active, equitable relationships where cultures exchange ideas, knowledge, and values, leading to mutual enrichment.

Migrations: The movement of people from one location to another, either within the same country (internal migration) or across borders (international migration), to settle temporarily or permanently. Migration has been a continuous phenomenon throughout history, driven by economic, social, political, and environmental factors. People arriving in a new location are called immigrants, while those leaving are emigrants.

Second cycle of early childhood education: This educational stage covers ages 3 to 6, focusing on promoting the comprehensive development of children in cognitive, emotional, social, and physical aspects. It is considered one of the most critical phases in a person's life, fostering the development of language, personal autonomy, creativity, and foundational cognitive skills such as problem-solving and artistic expression. Children engage in more structured activities, such as learning numbers, letters, and social skills, through playful and participatory methods.

Social geography: A branch of geography that examines how human societies organize, structure, and experience space. It explores interactions between people and their environments, focusing on social, cultural, economic, and political factors that shape how human groups occupy and modify geographic spaces. Social geography aims to understand the relationship between social dynamics and territory, and how these affect the spatial distribution of populations, resources, and cultural phenomena.

Social justice: This term refers to the fair distribution of resources, opportunities, and rights across society, ensuring that all individuals, regardless of background, social class, gender, ethnicity, religion, sexual orientation, or abilities, can live dignified and fulfilled lives. It is rooted in equality, equity, respect for diversity, and solidarity, and it aims to eliminate structural inequalities that perpetuate poverty, exclusion, and discrimination.

Spatial justice: This concerns the equitable distribution of resources, opportunities, and services across geographic spaces, ensuring that all communities, regardless of location, have access to what they need to thrive. This concept focuses on how decisions about space usage and territorial planning affect different social groups, and how these decisions can either perpetuate or reduce inequalities.

Chapter 4

Sendai Framework for Disaster Risk Reduction (2015–2030):

A Before and After for Teaching Natural Risks

Álvaro Francisco Morote
https://orcid.org/0000-0003-2438-4961
University of Valencia, Spain

ABSTRACT

In this work, the Sendai Framework for Disaster Risk Reduction (2015-2030). It is the first major agreement of the post-2015 development agenda and offers Member States a series of concrete actions that can be taken to protect the benefits of development against the risk of disasters. Regarding the objectives pursued by this Framework, they are: the reduction of disaster risk and losses caused by disasters, both in lives, livelihoods and health as well as in people's economic, physical, social, cultural and environmental assets, companies, communities and countries. This is where the education factor would have its influential role. Although in the Spanish case there are still few works on experiences and didactic proposals regarding the teaching of natural risks from Geography, the Sendai Framework constitutes a first and fundamental step to integrate these phenomena in school, as stated in the LOMLOE (2020).

DOI: 10.4018/979-8-3693-7185-5.ch004

Copyright © 2025, IGI Global Scientific Publishing. Copying or distributing in print or electronic forms without written permission of IGI Global is prohibited.

1. INTRODUCTION

Climate change is one of the most significant socio-environmental challenges facing current society (Jacobs et al., 2022; Martí et al., 2021; Muñoz et al., 2020). The 21st century is the century of environmental sustainability and the fight against climate change (Marques et al., 2022; Romero & Olcina, 2021). It is also worth noting that the current process of this phenomenon on a global scale is supported by scientific data (Intergovernmental Panel on Climate Change, [IPCC], 2022).

The relationship between the increasing presence of greenhouse gases in Earth's troposphere and the progressive rise in air temperature is undeniable (Choulga et al., 2021). This is supported by the fact that these anthropogenic gases are altering the planet's energy balance, which is a key component of the Earth's climate system (Krauss, 2021). This situation, far from stabilizing or improving, continues to worsen, as fossil fuel consumption remains on the rise, despite international protocols (Kyoto, Paris) and, consequently, the proportion of CO_2 in Earth's atmosphere continues to increase (Butler & Montzka, 2020).

The latest international summits held have failed to reach an agreement to limit CO_2 emissions on a global scale. Therefore, the goal of achieving a global temperature increase of only 1.5°C or 2°C at most by the end of this century is increasingly compromised. Only the agreement to financially support the Climate Compensation Fund established in the Paris Agreement (2015) by developed countries has been reached, although without detailing timelines and procedures for obtaining the planned $100 billion annually.

There are a series of atmospheric processes related to the main climatic elements (temperatures, precipitation, winds) that are already manifesting the effects of global warming in the Mediterranean region (Cramer et al., 2018; Muñoz et al., 2020; Olcina & Biener, 2019; Olcina & Moltó, 2019). Regarding precipitation: 1) a general decrease in quantity, although with regional nuances; 2) changes in seasonality with less abundant rainfall in spring and more abundant in autumn; and 3) an intensification of hourly rainfall, turning these precipitations into rains that cause damage wherever they occur, as has been observed in recent years (De Luis et al., 2010; Miró-Pérez & Estrela, 2019; Olcina, 2017; Serrano, 2017).

Likewise, regarding the frequency of development of extreme weather events, it is worth highlighting torrential rains, windstorms and coastal storms, heatwaves, hailstorms, and shorter but more intense droughts (Muñoz et al., 2020). This last aspect has a more direct impact on people's lives and has an immediate economic cost. Additionally, it represents the main problem for the coming decades if the frequency of these events increases as indicated by climate models (State Meteorological Agency [AEMET], 2021).

Global warming has been exacerbated by human actions as indicated in the Sixth IPCC Report (2022). This report warns of the impact that this global phenomenon can have if mitigation and adaptation actions are not carried out in the short to medium term. As explained by Arnell et al. (2019) and Pausas and Millán (2019), this phenomenon will have various effects depending on the territories and receiving societies (Pérez-Morales et al., 2022). It is a global risk that requires concerted action from society as a whole (González-Gaudiano et al., 2020; Olcina & Morote, 2023a), and has managed to capture the interest of society, and perhaps, increasingly, humans should be aware of the need for action, for example, in the educational field (Gutiérrez- Pérez et al., 2020; Morote & Hernández, 2024).

In this sense, in recent years, different contributions have been made in which it is evident that one of the measures to mitigate the effects of climate change may involve raising awareness among humans (Cruz & Páramo, 2020; Eilam, 2022; Méndez et al., 2020; Morote & Olcina, 2023a). This interest was already recognized in the Kyoto Protocol (1997) (Article 6), and reaffirmed in Article 12 of the Paris Agreement (2015) where the importance of education to change long-term habits and promote a better understanding and training to understand this phenomenon and its effects is highlighted (UNESCO & UNFCCC, 2016).

Similarly, the IPCC, in its Fifth Report (2014), also noted this, as did the United Nations (UN, 2015a) in Agenda 2030 (Sustainable Development Goals [SDGs]) where it is explained that education is one of the most effective tools for reducing the effects of climate change (Goal n.13, "Climate Action").

The interest in studying climate change and its associated effects, as well as its treatment from an educational perspective, is the objective of this work, which focuses on the analysis of actions within the educational context of the Sendai Framework for Disaster Risk Reduction (2015-2030) (UN, 2015b).

The aim of this paper is to present the Sendai Framework for Disaster Risk Reduction (2015-2030) (UN, 2015b) and the actions proposed in this document for society's adaptation to natural hazards. All of this within a context of climate change. This is the first major agreement of the post-2015 development agenda and offers member states a series of specific actions that can be taken to protect development gains against disaster risk. The Sendai Framework is linked with other agreements of the 2030 Agenda, such as the aforementioned Paris Agreement (2015), the Addis Ababa Action Agenda on Financing for Development, the New Urban Agenda, and the SDGs.

2. THEORETICAL FRAMEWORK. A BRIEF REVIEW

2.1. Education and Climate Change in the International Context

Firstly, it is important to take in to account the document of the United Nations Framework Convention on Climate Change (1992). The issues related to the education of this phenomenon are addressed in Article 6: "Education, Training, and Public Awareness". This article establishes that, in order to achieve its objectives, the signatory parties will promote and facilitate: 1) the development and implementation of education and public awareness programs about climate change and its effects; 2) public access to information on climate change and its effects; 3) public participation in the study of climate change and its effects, as well as in the development of appropriate responses; 4) the training of scientific, technical, and managerial personnel; and 5) international cooperation in these fields.

Likewise, it is important to highlight that the implementation of the actions outlined in this Article 6 has been specified over the past decade in different programs: the Delhi Work Programme (2003-2007), the amended Delhi Work Programme (2008-2012), and the Doha Work Programme (2013-2020), which is currently in effect. The Doha Work Programme (2013-2020) recognizes that a key goal of education is to promote changes in lifestyles, attitudes, and behaviors to foster sustainable development. It is also reaffirmed that public participation and access to information are crucial for developing effective policies, as well as for actively engaging all stakeholders in their implementation.

The most notable educational guidance can be found in Table 2. The program, specifically in the field of education, aims to disseminate and publicize climate change, such as: 1) creating active networks of organizations and individuals to develop activities in thematic areas related to Article 6 of the Framework Convention; 2) informing the public about the causes of climate change, sources of greenhouse gas emissions, and actions that can be taken at all levels to address climate change; and 3) disseminating relevant information that is generated in relation to global warming. Among other suggestions, the preparation and dissemination of popular versions of key documents, such as the assessment reports of the Intergovernmental Panel on Climate Change, are included.

Table 1. Main guidelines of the Doha Work Programme (2013-2020)

-Designate a national focal point for Article 6, to which specific tasks will be assigned. -Prepare needs assessments concerning the topics related to Article 6, utilizing, among other methods, social research techniques. -Develop national strategies for addressing the issues included in Article 6 of the Convention. -Create communication strategies on climate change aimed at achieving behavior change. -Strengthen educational institutions that implement educational actions on climate change.

Source: UN (2024).

A third report and the analyzed documentation have been the reports published by the IPCC since 1990. The IPCC prepares comprehensive assessment reports on knowledge about climate change (its causes, potential impacts, and response options). Additionally, it produces special reports, which are assessments on specific topics, and methodological reports, which provide practical guidelines for the preparation of greenhouse gas inventories.

Since 1990, six reports on the state of the climate have been published, which show an evolution in the treatment of education as a measure to combat climate change (Table 3). So far, no special report has been dedicated to the relationship between climate change and education, although it is true that, as these studies have been developed, education has gained a larger presence in the overall text, especially starting from the Fourth Report (IPCC, 2007), and reaching its greatest significance in the Sixth Report (IPCC, 2021, 2022), which emphasizes the importance of educational actions as a tool for achieving resilient, gender-equal societies through the design of public education programs aimed at explaining environmental changes and their effects, as well as early warning systems for atmospheric hazards that states must implement (IPCC, 2021, 2022). The volumes dedicated to analyzing the impact and adaptation measures (Volume 2), as well as to mitigation (Volume 3), include references to the educational sector in actions for reducing the effects of climate change on an international scale.

In the specific context of the European Union, the main source of information consulted has been the European Commission, the legislative and executive body that is responsible for directing and managing it. Among its functions are daily management, proposing legislation, and upholding the Union's treaties. Comprising 28 countries, it decides who handles various policies (trade, environmental issues, competition, economy, security, etc.). Regarding climate change, a section on this phenomenon is dedicated on its website ("Energy, Climate Change and Environment"), where the heading titled "Climate Action and the Green Pact" stands out, outlining various measures for action.

In this context, those documents in which the education factor should be present have been consulted, such as the European Framework for achieving climate neutrality (2021), the European Union's Climate Change Adaptation Strategy (2021), and the Strategic Plan 2000-2024. Concerning the first, its aim is to ensure compliance with the measures established in the Paris Agreement (2015). In this regulation, the only reference linked to education is point 34, and with regard to the articles, it is Article 5 (see Table 3).

Table 2. Education and climate change in the European Commission

European Framework for Achieving Climate Neutrality (2021) - Point 34: "When adopting the relevant measures at the national and Union level to achieve the goal of climate neutrality, Member States, the European Parliament, the Council, and the Commission must take into account, among other issues [...] the need to make the transition fair and socially equitable through appropriate education and training programs [...]". - Article 5, "Adaptation to Climate Change", point 3: "The relevant institutions of the Union and the Member States will also ensure that the adaptation policies of the Union and the Member States are coherent, mutually reinforcing, provide co-benefits to sectoral policies, and are aimed at a systematic and more appropriate integration of climate change adaptation across all policy areas, including, where appropriate, relevant socioeconomic and environmental policies, as well as the Union's external action. They will pay special attention to the most vulnerable populations and economic sectors and will identify shortcomings in this regard through consultations with civil society".
European Union Climate Change Adaptation Strategy (2021) Point 2, "Building a Climate-Resilient Union", paragraph 2.2.2. - "Foster local, individual, and equitable resilience". - "In order to engage and empower European citizens to take direct adaptation actions, the Commission will use the European Climate Pact initiative and the Climate Education Coalition to inform, inspire, and connect". - "Increasing support is needed for education, training, and labor reskilling initiatives that lead to green jobs". - "It will support the reskilling and upskilling of workers for equitable and just resilience through education and training via the ESF+, the Erasmus+ program, and the European Solidarity Corps".
Strategic Plan (2000-2024) -Objective Number 4, "Communication and Dialogue": - The population places great importance on actions to combat climate change and has a fundamental role to play in the transition to a climate-neutral economy. - It is also explained that this plan will bring together regions, local communities, civil society, industry, and schools to design and commit to promises and ensure that everyone is part of this behavior change.

Source: European Comission (2024). Own elaboration.

Regarding the European Union's Climate Change Adaptation Strategy (2021), education and training issues in this area are addressed in point 2 ("Building a resilient Union to climate change"). This plan is interesting because it states that local action is the cornerstone of adaptation, and that EU support should help increase local resilience. This will build on existing initiatives and instruments, such as the smart cities market, the Digital Europe program, the Horizon Europe Framework Program, and the smart cities challenge. It is also important to note that this same point emphasizes training for climate change for workers directly affected by this phenomenon (job losses in sectors such as agriculture, fishing, and tourism) (see Table 3). A third document analyzed has been the Strategic Plan (2000-2024). In relation to global warming, objective number 4 ("Communication and Dialogue," has been reviewed, focusing on the importance of combating climate change through raising awareness and educating citizens") (Table 3).

The second organization consulted at the European level has been the European Environment Agency (EEA), which has a section on its website dedicated to climate change and education. The objective of this agency is to promote sustainable development and contribute to achieving significant and measurable improvements in the European territory by providing reliable, relevant, specific, and timely information to policymakers and society at large. According to the latest report from 2021, "EEA and Eionet Strategy for 2021-2030", the importance of current socio-environmental challenges such as climate change and COVID-19 is highlighted, but primarily, the measures focus on reducing greenhouse gas emissions and achieving the Sustainable Development Goals (SDGs) (Agenda 2030). Regarding education, this term is included in strategic objective number 5 ("Providing resources for our common ambitions") (Table 4).

Table 3. Strategic objective number 5 ("Provision of resources for our common ambitions")

- "Making the transitions to sustainability a reality will require all political areas and government levels to work together to drive changes and actions throughout society. To achieve this, coherent contributions must be made from various political spheres, ranging from research and innovation, the economy, industry, competition, and trade to employment, education, and well-being". - The five strategic objectives will be applied in the following areas: 1) biodiversity and ecosystems; 2) climate change mitigation and adaptation; 3) human health and the environment; 4) circular economy and resource use; and 5) trends, perspectives, and responses in sustainability matters. However, there is no further information on concrete measures to address climate change from the educational sphere.

Source: EEA (2024).

2.2. Sustainable Development Goals (SDGs)

In the field of education and relationship with climate change, undoubtedly, the approval of the SDGs has had a great impact (Granados & Medir, 2022). The 2030 Agenda, under the slogan "Transforming our world: the 2030 Agenda for Sustainable Development", was signed on September 25th (2015). It is, to date, the largest international commitment reached for the protection of the planet and its inhabitants and constitutes a roadmap with 17 goals, 169 targets, and 232 monitoring indicators. The general objectives pursued by the SDGs are: ending poverty in all its forms;

reducing inequality; creating an inclusive future; creating a resilient future, and achieving sustainable and equitable economic growth.

Regarding climate change issues, specific reference should be made to goal n. 13 ("Climate Action"), which refers to taking urgent measures to combat climate change and its effects. It is worth noting that there are several targets related to education and public awareness, such as goal n. 1 ("resilience and capacity to adapt to climate-related risks and natural disasters in all countries"), and goal n. 3 ("improve education, awareness, and human and institutional capacity regarding mitigation of climate change, adaptation to it, reduction of its effects, and early warning") (see Table 5).

Table 4. Goals of SDG 13 ("Climate Action")

- Goal 1. Strengthen resilience and adaptive capacity to climate-related risks and natural disasters in all countries. - Goal 2. Incorporate measures related to climate change in national policies, strategies, and plans. - Goal 3. Improve education, awareness, and human and institutional capacity regarding climate change mitigation, adaptation, reducing its effects, and early warning. - Goal 4. Fulfill the commitment of developed countries that are parties to the United Nations Framework Convention on Climate Change to achieve jointly mobilizing $100 billion annually by 2020 from all sources, in order to address the needs of developing countries for the adoption of specific mitigation measures and transparency in their application, and to fully operationalize the Green Climate Fund, capitalizing it as soon as possible. - Goal 5. Promote mechanisms to increase capacity for effective planning and management in relation to climate change in the least developed countries and small island developing states, with a particular focus on women, youth, and local and marginalized communities.

Source: UN (2015a). Own elaboration.

2.3. Recent Publications about Climate Change. A Geographic Approach

Authors such as González-Gaudiano et al. (2020), Méndez et al. (2020), and Rudd (2021) have been explaining and demonstrating the importance of Teaching-Learning (T-L) on climate change for years, not only in the younger population, but also in the general population. This is something that has been worked on both in Spain and internationally (Canaza et al., 2021; Eilam, 2022; Opuni-Frimpong et al., 2022; Puttick et al., 2022; Rudd, 2021; Sezen-Barrie & Marbach-Ad, 2021; Zhong et al., 2021).

The teaching of climate change is a topic that has been covered in an international context: Central America and North America (McWhirter & Shealy, 2018; Li et al., 2021; Sezen-Barrie & Marbach-Ad, 2021); South America (Da Rocha et al., 2020); Europe (Jeong et al., 2021; Kurup et al., 2021; León et al., 2021); Africa (Anyanwu

& Le Grange, 2017; García-Vinuesa et al., 2020); Asia and Oceania (Haq & Ahmed, 2020; Li & Liu, 2021; Liao et al., 2021; Pfautsch & Gray, 2017; Yu et al., 2020).

Within geography, the specialized discipline of climatology at non-university education levels has had an extensive scientific production for decades in Spain from the Didactics of Geography (García de la Vega, 2016; Martínez-Romera, 2013; Marrón, 2011; Tonda & Sebastiá, 2003; Valbuena & Valverde, 2006). However, with regards to publications on works about teaching climate change from the Spanish Geography field are relatively recent (during the last decade), as it is considered a "new" content (Morote, 2021). This author has analyzed publications on research and teaching of climatology carried out in Spain (1980-2020) and shows that in four decades studies on this topic amount to 89, with those related to teaching climate change making up 46.1% (mostly in the last decade). Additionally, three themes have been identified: 1) works on experiences and didactic proposals; 2) analysis of social representations; and 3) analysis of climate change in school textbooks.

According to Morote and Olcina (2023a), in the educational field, climate change faces several challenges for its proper teaching in the classroom (see Table 6). Additionally, it is worth noting that the main information received about climate change comes from digital media and social networks (Morote et al., 2021). Kažys (2018) highlights this issue (falseness and manipulation of information received from media), while Allen and McAleer (2018) add concern about the risks posed by so-called fake news. Another added problem is the little importance given by teachers to education as a measure to mitigate the risks associated with climate change, as Morote and Hernández (2020) have found. Even this information is being reproduced without scientific rigor in school textbooks (Olcina, 2017). For example, in the case of Geography textbooks (Secondary Education and baccalaureate), this last author has detected significant deficiencies: lack of rigorous information, excessive catastrophism, and an almost total absence of real data from academic work.

Table 5. Challenges in teaching about climate change

1. The difficulty of teaching content that even the scientific community finds complex due to the amalgamation of factors involved in them (Eilam, 2022; Ferrari et al., 2019; Olcina, 2020). 2. The need to teach this topic with scientific rigor (Caride & Meira, 2019; Escoz et al., 2020; Ferrari et al., 2019; Jeong et al., 2021; Kurup et al., 2021; Masters, 2020; Nelles and Serrer, 2020; Rudd, 2021). 3. The limited training of teachers on this topic (Morote & Olcina, 2021). 4. Stereotypes and inaccurate information spread in the media that are influencing both students and teachers in their thoughts and perceptions about this phenomenon (León et al., 2021; Morote et al., 2021; Rudd, 2021).

Source: Morote and Olcina (2023a). Own elaboration.

2.4. Recent Publications about Natural Hazards. A Geographic Approach

At the international level, there are few studies that have analyzed the state of the art concerning the teaching of natural hazards. Attention could be drawn to the work of Saregar et al. (2022), which aims to analyze the international literature on natural disasters in schools. These authors have identified a total of 216 works, highlighting that since 2015 (as this study has confirmed), the number of these contributions has increased significantly. Among the total studies, those published in the U.S. (20.8%) and Japan (14.3%) particularly stand out.

In Cuba, Muñoz and Otaño (2014) propose educational software utilizing multimedia resources on natural disasters, which assist the population with protective measures to confront them. This includes documents, images, and videos that can be highly useful for teachers when working with students. Additionally, in Andorra, Brugulat et al. (2021) present an educational project conducted by the Institute of Andorran Studies (IEA) for the schools of the Principality of Andorra, aimed at raising awareness of existing natural hazards and encouraging reflection and sensitivity towards these phenomena. This project has been established as a basic training course in civil protection, which is taught from the 5th year of Primary Education to the 4th year of Secondary Education. These authors highlight that between the academic years 2018-2019 and 2020-2021, 2,450 students and more than 40 teachers have been trained in knowledge about natural hazards. Meanwhile, Li et al. (2022), in China, present the ArcGis Online resource for analyzing and interpreting natural hazards, like what Morote et al. (2022) have done in Spain regarding floods in the Júcar and Segura River basins.

In the international landscape, the works related to natural hazards or disasters in general particularly stand out. However, some recent studies on floods include the one conducted by Nguyen and Pham (2018). These authors have verified the impact of training on this phenomenon across Ethiopia, India, Peru, and Vietnam on students' awareness. Other studies have focused on the theoretical and methodological foundations for education on natural hazards, which, in the Spanish context, correspond in this research to works linked with "conceptual analysis". For example, in Cuba, the study by Jiménez et al. (2017) should be highlighted, in which the authors review the foundations that support education for perceived disaster risks in the context of school education processes.

Moreover, there are authors who have analyzed the social representation and training of both teachers and students. For instance, the work of Suazo and Torres (2021) in Honduras examines the perceptions, knowledge, and teaching of climate change and disaster risk among 457 teachers. These authors found that teachers have an overall perception of 2.06 (out of 5), with a tendency to overestimate variables (immediacy, catastrophism, voluntariness, role of the press, involvement, and past history). Additionally, 60% report not having received training, while 59% indicate that they did not teach topics on this phenomenon to their students. These data could be compared to those obtained by Morote and Souto (2020) in Spain, where they calculated that only 12.1% of teachers had received training on floods.

In the U.S., in Colorado, Boyd et al. (2022) found, based on a survey of 279 teachers, that 70% indicate they teach content related to natural hazards at school. Regarding the reasons offered by teachers who do not teach this content, the lack of training and specificity of these contents in the curriculum should be noted. For those who do teach, they do so use readings, field trips, and class discussions.

Another study, by Gavari and Pastrana (2018), provides an analysis of education for natural risk reduction in Japan. The authors describe the Japanese model based on the legal framework adopted in 1961, which regulates the participation of all administrative levels in the matter, with well-defined functions within an integrated system. This model is complemented by the introduction of two key strategies in the school curriculum: the incorporation of disaster risk management content into core subjects and a specific study schedule designed by teachers. In this sense, as the authors point out, teachers play a fundamental role in risk reduction, not only due to the training they receive but also because of their connection with the community.

Also, finally, it is worth mentioning works published in recent years in Spain that are related to specific natural risks. For example, the most studied natural risk from the perspective of Geography education is flooding. This is a recent line of work approached from various perspectives: 1) analysis of the social representations of students and teachers (Morote & Gómez-Trigueros, 2023; Sánchez-Almodóvar et al., 2023); 2) analysis of school textbooks (Morote et al., 2022); and 3) didactic proposals (Cuello & García, 2019; Morote & Olcina, 2023b; Olcina et al., 2022; Zaragoza, 2022).

3. SENDAI FRAMEWORK FOR DISASTER RISK REDUCTION (2015-2030)

3.1. Background

The Sendai Framework for Disaster Risk Reduction 2015-2030 was adopted at the Third UN World Conference in Sendai, Japan, on March 18th (2015). It is the outcome of stakeholder consultations initiated in March 2012 and inter-governmental negotiations from July 2014 to March 2015, supported by the United Nations Office for Disaster Risk Reduction at the request of the UN General Assembly (UN, 2015b).

The Sendai Framework is the successor instrument to the Hyogo Framework for Action (HFA) 2005-2015: Building the Resilience of Nations and Communities to Disasters. The HFA was conceived to give further impetus to the global work under the International Framework for Action for the International Decade for Natural Disaster Reduction of 1989, and the Yokohama Strategy for a Safer World: Guidelines for Natural Disaster Prevention, Preparedness and Mitigation and its Plan of Action, adopted in 1994 and the International Strategy for Disaster Reduction of 1999.

The objectives pursued by this Framework are: the reduction of disaster risk and losses caused by disasters, both in terms of lives, livelihoods, and health, as well as economic, physical, social, cultural, and environmental assets of individuals, businesses, communities, and countries. In addition, it recognizes that the primary responsibility for reducing disaster risk lies with the State, but it is a responsibility that must be shared with other actors, such as local governments, the private sector, and other stakeholders. This is where the education factor would play its influential role (Table 7).

Table 6. Opportunity for countries

- To adopt a concise, focused, forward-looking and action-oriented post 2015 framework for disaster risk reduction. - To complete the assessment and review of the implementation of the Hyogo Framework for Action 2005–2015: Building the Resilience of Nations and Communities to Disasters. - To consider the experience gained through the regional and national strategies/ institutions and plans for disaster risk reduction and their recommendations, as well as relevant regional agreements for the implementation of the Hyogo Framework for Action. - To identify modalities of cooperation based on commitments to implement a post 2015 framework for disaster risk reduction. - To determine modalities for the periodic review of the implementation of a post 2015 framework for disaster risk reduction.

Source: UN (2015b). Own elaboration.

The Sendai Framework is built on elements which ensure continuity with the work done by States and other stakeholders under the HFA and introduces several innovations as called for during the consultations and negotiations. Many commentators have identified the most significant shifts as a strong emphasis on disaster risk management as opposed to disaster management, the definition of seven global targets, the reduction of disaster risk as an expected outcome, a goal focused on preventing new risk, reducing existing risk and strengthening resilience, as well as a set of guiding principles, including primary responsibility of states to prevent and reduce disaster risk, all-of-society and all-of-State institutions engagement. In addition, the scope of disaster risk reduction has been broadened significantly to focus on both natural and man-made hazards and related environmental, technological and biological hazards and risks. Health resilience is strongly promoted throughout.

Also this Framework articulates the following: the need for improved understanding of disaster risk in all its dimensions of exposure, vulnerability and hazard characteristics; the strengthening of disaster risk governance, including national platforms; accountability for disaster risk management; preparedness to "Build Back Better"; recognition of stakeholders and their roles; mobilization of risk-sensitive investment to avoid the creation of new risk; resilience of health infrastructure, cultural heritage and work-places; strengthening of international cooperation and global partnership, and risk-informed donor policies and programs, including financial support and loans from international financial institutions. There is also clear recognition of the Global Platform for Disaster Risk Reduction and the regional platforms for disaster risk reduction as mechanisms for coherence across agendas, monitoring and periodic reviews in support of UN Governance bodies.

In relation with the Priorities for action (block IV) the article n.20 explain that "Taking into account the experience gained through the implementation of the Hyogo Framework for Action, and in pursuance of the expected outcome and goal, there is a need for focused action within and across sectors by States at local, national, regional and global levels in the following four priority areas: Priority 1: Understanding disaster risk; Priority 2: Strengthening disaster risk governance to manage disaster risk; Priority 3: Investing in disaster risk reduction for resilience; and Priority 4: Enhancing disaster preparedness for effective response and to "Build Back Better" in recovery, rehabilitation and reconstruction (see Table 8). Also, the Sendai Framework has different guiding principles (see Table 9).

Table 7. Priorities for action

Priority 1: Understanding disaster risk -Disaster risk management needs to be based on an understanding of disaster risk in all its dimensions of vulnerability, capacity, exposure of persons and assets, hazard characteristics and the environment.
Priority 2: Strengthening disaster risk governance to manage disaster risk -Disaster risk governance at the national, regional and global levels is vital to the management of disaster risk reduction in all sectors and ensuring the coherence of national and local frameworks of laws, regulations and public policies that, by defining roles and responsibilities, guide, encourage and incentivize the public and private sectors to take action and address disaster risk.
Priority 3: Investing in disaster risk reduction for resilience -Public and private investment in disaster risk prevention and reduction through structural and non-structural measures are essential to enhance the economic, social, health and cultural resilience of persons, communities, countries and their assets, as well as the environment. These can be drivers of innovation, growth and job creation. Such measures are costeffective and instrumental to save lives, prevent and reduce losses and ensure effective recovery and rehabilitation.
Priority 4: Enhancing disaster preparedness for effective response and to "Build Back Better" in recovery, rehabilitation and reconstruction -Experience indicates that disaster preparedness needs to be strengthened for more effective response and ensure capacities are in place for effective recovery. Disasters have also demonstrated that the recovery, rehabilitation and reconstruction phase, which needs to be prepared ahead of the disaster, is an opportunity to «Build Back Better» through integrating disaster risk reduction measures. Women and persons with disabilities should publicly lead and promote gender-equitable and universally accessible approaches during the response and reconstruction phases.

Source: UN (2015b). Own elaboration.

Table 8. Guiding principles

- Primary responsibility of States to prevent and reduce disaster risk, including through cooperation. - Shared responsibility between central Government and national authorities, sectors and stakeholders as appropriate to national circumstances. - Protection of persons and their assets while promoting and protecting all human rights including the right to development. - Engagement from all of society. - Full engagement of all State institutions of an executive and legislative nature at national and local levels. - Empowerment of local authorities and communities through resources, incentives and decision-making responsibilities as appropriate. - Decision-making to be inclusive and riskinformed while using a multi-hazard approach. - Coherence of disaster risk reduction and sustainable development policies, plans, practices and mechanisms, across different sectors. - Accounting of local and specific characteristics of disaster risks when determining measures to reduce risk. - Addressing underlying risk factors cost-effectively through investment versus relying primarly on postdisaster response and recovery. - "Build Back Better" for preventing the creation of, and reducing existing, disaster risk. - The quality of global partnership and international cooperation to be effective, meaningful and strong. - Support from developed countries and partners to developing countries to be tailored according to needs and priorities as identified by them.

Source: UN (2015b). Own elaboration.

3.2. Education in Risk

Regarding aspects related to the educational context and the teaching of natural hazards, these aspects are reflected in different sections and articles. For example, in article n. 14 it is indicated that there is a need to address existing challenges and prepare for future ones by focusing on monitoring, assessing and understanding disaster risk and sharing such information and on how it is created; strengthening disaster risk governance and coordination across relevant institutions and sectors and the full and meaningful participation of relevant stakeholders at appropriate levels; investing in the economic, social, health, cultural and educational resilience of persons, communities and countries and the environment, as well as through technology and research; and enhancing multi-hazard early warning systems, preparedness, response, recovery, rehabilitation and reconstruction. To complement national action and capacity, there is a need to enhance international cooperation between developed and developing countries and between States and international organizations.

Regarding to n. 17, this explains that, to attain the expected outcome, the following goal must be pursued: prevent new and reduce existing disaster risk through the implementation of integrated and inclusive economic, structural, legal, social, health, cultural, educational, environmental, technological, political and institutional measures that prevent and reduce hazard exposure and vulnerability to disaster, increase preparedness for response and recovery, and thus strengthen resilience.

Furthermore, n. 18 (d) is linked to the "educational facilities": "Substantially reduce disaster damage to critical infrastructure and disruption of basic services, among them health and educational facilities, including through developing their resilience by 2030". Meanwhile the article n. 19 (k) it is indicated that in the post-disaster recovery, rehabilitation and reconstruction phase, it is critical to prevent the creation of and to reduce disaster risk by "Building Back Better" and increasing public education and awareness of disaster risk.

Another notably interesting article for risk reduction is n. 24, which includes several sections (see Table 10), while the article n. 25 (f) is dedicated to environmental campaigns: "to develop effective global and regional campaigns as instruments for public awareness and education, building on the existing ones (for example, the "One million safe schools and hospitals" initiative; the "Making Cities Resilient: My city is getting ready" campaign; the United Nations Sasakawa Award for Disaster Risk Reduction; and the annual United Nations International Day for Disaster Reduction), to promote a culture of disaster prevention, resilience and responsible citizenship, generate understanding of disaster risk, support mutual learning and share experiences; and encourage public and private stakeholders to actively engage in such initiatives and to develop new ones at the local, national, regional and global levels".

Table 9. Article n. 24

(d) To systematically evaluate, record, share and publicly account for disaster losses and understand the economic, social, health, education, environmental and cultural heritage impacts, as appropriate, in the context of event-specific hazard-exposure and vulnerability information. (g) To build the knowledge of government officials at all levels, civil society, communities and volunteers, as well as the private sector, through sharing experiences, lessons learned, good practices and training and education on disaster risk reduction, including the use of existing training and education mechanisms and peer learning. (k) To promote investments in innovation and technology development in long-term, multihazard and solution-driven research in disaster risk management to address gaps, obstacles, interdependencies and social, economic, educational and environmental challenges and disaster risks. (l) To promote the incorporation of disaster risk knowledge, including disaster prevention, mitigation, preparedness, response, recovery and rehabilitation, in formal and non-formal education, as well as in civic education at all levels, as well as in professional education and training. (m) To promote national strategies to strengthen public education and awareness in disaster risk reduction, including disaster risk information and knowledge, through campaigns, social media and community mobilization, taking into account specific audiences and their needs.

Source: UN (2015b). Own elaboration.

Continuing with the analysis of the Sendai Framework, other articles worth noting are number 30 (j) "To strengthen the design and implementation of inclusive policies and social safety-netmechanisms, including through community involvement, integrated with livelihood enhancement programmes, and access to basic health-care services, including maternal, newborn and child health, sexual and reproductive health, food security and nutrition, housing and education, towards the eradication of poverty, to find durable solutions in the post-disaster phase and to empower and assist people disproportionately affected by disasters"; and the n. 33 (c): "To promote the resilience of new and existing critical infrastructure, including water, transportation and telecommunications infrastructure, educational facilities, hospitals and other health facilities, to ensure that they remain safe, effective and operational during and after disasters in order to provide live-saving and essential services".

Likewise, it is worth noting the "Role of stakeholders". Thus, for example, these actions can be found in article n.36 (a) and (d). For instance, in section (a) it is indicated that: "Civil society, volunteers, organized voluntary work organizations and community-based organizations to participate, in collaboration with public institutions, to, inter alia, provide specific knowledge and pragmatic guidance in the context of the development and implementation of normative frameworks, standards and plans for disaster risk reduction; engage in the implementation of local, national, regional and global plans and strategies; contribute to and support public awareness, a culture of prevention and education on disaster risk; and advocate for resilient communities and an inclusive and all-of-society disaster risk management that strengthen synergies across groups, as appropriate. On this point, it should be noted that: (ii) "Children and youth are agents of change and should be given the space and modalities to contribute to disaster risk reduction, in accordance with legislation, national practice and educational curricula".

And regarding the section (d): Media to take an active and inclusive role at the local, national, regional and global levels in contributing to the raising of public awareness and understanding and disseminate accurate and non-sensitive disaster risk, hazard and disaster information, including on small-scale disasters, in a simple, transparent, easy-to-understand and accessible manner, in close cooperation with national authorities; adopt specific disaster risk reduction communications policies; support, as appropriate, early warning systems and life-saving protective measures; and stimulate a culture of prevention and strong community involvement in sustained public education campaigns and public consultations at all levels of society, in accordance with national practices.

4. DISCUSSION AND CONCLUSIONS

The interest in studying climate change and its associated effects, as well as its treatment from an educational perspective, is the objective of this work, which focuses on the analysis of actions within the educational context of the Sendai Framework for Disaster Risk Reduction (2015-2030). Education is a fundamental tool in the fight against climate change and associated natural hazards. It is an element of social awareness generation through accurate messages on this issue, which must always be supported by scientific research. It is a very effective tool for implementing individual and collective actions for mitigation and adaptation to this phenomenon. However, as a means of disseminating science and culture, it requires careful and prolonged programming over time (years of educational training) that only yields results in the medium and long term. Therefore, despite being recognized by organizations and administrations at all levels of work (international to local), it does not find a development of concrete actions in line with its social importance.

The previous studies about education and natural hazards focus on how students and teachers understand, interpret, and address the risk of flooding, as well as on the evaluation of educational materials, such as school textbooks, and innovative pedagogical proposals to improve teaching about this type of natural disaster. By linking this approach to the Sendai Framework for Disaster Risk Reduction (2015-2030), we can identify several relevant connections that reflect the integration of disaster risk management in the educational field.

1. Focus on awareness and education about risk. The Sendai Framework emphasizes the importance of education and awareness about risk as a fundamental pillar for disaster reduction. In this regard, the studies mentioned in the text, which

address the social representations of students and teachers regarding flooding, are directly aligned with Priority 3 of the Sendai Framework, which aims to "increase public education and awareness on disaster risk reduction". Through research on how students and teachers perceive flooding, we can identify what knowledge and attitudes prevail in school communities and how they can be modified to improve resilience to these phenomena. The studies also explore how these educational approaches influence the preparation of new generations for natural risks, which is an essential component in building a culture of safety and prevention.

2. Evaluation of educational materials and school curricula. The analysis of school textbooks (Morote et al., 2022) is particularly relevant within the context of the Sendai Framework, which highlights the need to improve educational policies to integrate disaster risk into curricula. In Priority 2 of the Framework, there is an emphasis on "strengthening governance and the integration of risk into public policies", which includes, among other things, the review and improvement of available educational resources. The textbooks analyzed in the studies by Morote et al. (2022) can reveal the extent to which flooding content is present in secondary education programs, how complete these approaches are, and whether they adequately address issues related to prevention, mitigation, and adaptation to risk. Proper integration of these topics in textbooks, along with effective pedagogical strategies, can play a key role in disaster risk reduction by training students who are better equipped to manage flooding threats in their communities.
3. Innovative didactic proposals. The didactic proposals (such as those by Cuello & García, 2019; Morote & Olcina, 2023) are another critical aspect that directly connects to the Sendai Framework, as this document advocates for the promotion of participatory and practical approaches to risk education. The Sendai Framework encourages the adoption of educational approaches that are accessible, participatory, and that empower students and communities to act effectively in the face of risks. The didactic proposals explored in the mentioned studies, such as interactive activities or awareness projects based on local flooding experiences, can contribute to the creation of skills and capacities that improve community preparedness for disasters. By providing students with the tools to understand, assess, and make decisions about flood risk, these initiatives contribute to the broader goals of the Sendai Framework, particularly in reducing human, economic, and infrastructure losses.
4. Social representation and decision-making. The analysis of social representations regarding flooding risk is also closely related to Priority 4 of the Sendai Framework, which promotes the reduction of vulnerabilities through an inclusive and participatory approach. Social representations play a crucial role in how

people perceive risks and, therefore, in how they make decisions to protect themselves. If students and educators view flooding only as inevitable natural phenomena, they may not adopt effective preventive measures. However, if they are made aware of the human and structural factors contributing to floods, as well as ways to mitigate them, they are more likely to make informed decisions at both individual and collective levels. Analyzing how students perceive and react to risks helps shape educational interventions that foster a culture of prevention and a stronger community response to natural disasters.

In relation with the LOMLOE (2020), it represents a significant reform of the education system in Spain, aiming to guarantee a more inclusive, equitable, and quality education. This new law includes aspects that can directly contribute to increasing students' resilience to natural hazards. Below are some ways in which the LOMLOE can contribute to this goal:

1. Integration of education in values and transversal competencies. One of the key pillars of the LOMLOE is education in values and transversal competencies, such as social responsibility, sustainability, cooperation, critical thinking, and solidarity. These values are essential for resilience to natural disasters, as they foster a culture of prevention, collective action, and solidarity in emergency situations. The law promotes the inclusion of these values across all educational stages, which can raise students' awareness of natural hazards and prepare them to act in an informed and responsible way in the face of potential disasters.
2. Focus on environmental education and sustainability. The LOMLOE reinforces environmental education as a fundamental component of students' education, emphasizing the importance of sustainability and environmental protection. This environmental education can include content related to natural hazards, such as flooding, wildfires, earthquakes, among others. By integrating knowledge about natural risks into the curriculum, students develop greater awareness of natural phenomena and their impacts, enabling them to adopt more resilient behaviors and take preventive actions in their homes and communities.
3. Development of risk management and decision-making skills. The LOMLOE promotes the development of competencies through problem-based and project-based learning. In this context, students can be involved in emergency simulation and planning activities, addressing natural hazard scenarios such as floods or wildfires. By working practically on evaluating and making decisions on how to act in crisis situations, students not only acquire knowledge but also develop key skills to manage and face risk situations effectively. This also fosters the ability to quickly adapt to unexpected situations and reduce the impact of disasters.

4. Strengthening global citizenship education. The LOMLOE emphasizes the importance of global citizenship education, which includes aspects related to international cooperation, disaster management, and solidarity in the face of global challenges, including natural risks. By promoting a global perspective on disasters, students can understand that risks are not only local but have global implications, and that international cooperation is key in managing emergencies. Additionally, global citizenship education enables students to reflect on the interdependence of human communities and ecosystems, as well as the need to act collectively in the face of global challenges related to natural risks.
5. Inclusion of risk management in interdisciplinary projects. The LOMLOE encourages interdisciplinary education, encouraging teachers to integrate different areas of knowledge (natural sciences, geography, technology, etc.) to address complex problems in a holistic way. This offers an excellent opportunity for students to gain an integrated understanding of how different factors (environmental, social, economic, etc.) are interconnected in natural hazards and how effective risk management requires the collaboration of various disciplines.

Education is an essential commitment of any society. It should aim to educate citizens from teaching basic content on the subjects that make up the necessary training to thrive in the world. Current climate change is an essential issue of present and future on our planet, which will guide policies and practices of institutions and administrations, if its unwanted effects are to be effectively limited. Teaching about climate change, in the classroom (at all educational levels) and among society in general, requires more attention from organizations and governments on this topic, to make it an effective tool for mitigation and adaptation.

Regarding natural risks, initiatives for training and awareness raising are being carried out in Spain through the teaching of Geography, as well as the publication and development of a very interesting line of work. As for the Sendai Framework, it is a significantly important document that establishes a series of recommendations, many of them related to the teaching of natural risks. Undoubtedly, it is a document that has set a milestone in the field of risk education and must be taken into account in the future.

The teaching of natural hazards is essential for developing a society that is more aware and prepared to face the challenges posed by our natural environment. As communities confront the increasing frequency and intensity of events such as earthquakes, hurricanes, and floods, it is crucial for citizens, from an early age, to understand the mechanisms behind these phenomena and the associated risks. This education not only allows for the identification of potential threats but also fosters a culture of prevention that can save lives and reduce material losses.

Moreover, the teaching of natural hazards contributes to building stronger community resilience. By understanding the vulnerability of their environments, students learn to value proper urban planning, resilient infrastructure, and the importance of being prepared for disasters. This knowledge enables them not only to protect themselves and their families but also to actively participate in creating safer and more sustainable communities. Education on natural hazards, therefore, becomes a vital pillar for adaptation and mitigation in a world increasingly affected by climate change.

Finally, educating about natural hazards promotes a shared responsibility for environmental protection and collective well-being. By raising awareness among younger generations about how their actions can influence the severity of natural disasters, they are instilled with a proactive attitude towards sustainability and community cooperation. In this way, the teaching of natural hazards not only prepares people to face emergency situations but also contributes to the construction of a more balanced and environmentally conscious future.

REFERENCES

Agencia Estatal de Meterología (AEMET). (2021). *Proyecciones Climáticas para el siglo XXI en España*.https://www.aemet.es/es/serviciosclimaticos/cambio_climat

Allen, D. E., & McAller, M. (2018). Fake news and indifference to scientific fact: President Trump's confused tweets on global warming, climate change and weather. *Scientometrics*, 117(1), 625–629. DOI: 10.1007/s11192-018-2847-y

Anyanwu, R., & Le Grange, L. (2017). The influence of teacher variables on climate change science literacy of Geography teachers in the Western Cape, South Africa. *International Research in Geographical and Environmental Education*, 26(3), 193–206. DOI: 10.1080/10382046.2017.1330039

Arnell, N. W., Lowe, J. A., Challinor, A. J., & Osborn, T. J. (2019). Global and regional impacts of climate change at different levels of global temperature increase. *Climatic Change*, 155(3), 377–391. DOI: 10.1007/s10584-019-02464-z

Boyd, K. J., Gold, A. U., & Littrell, M. K. (2022). Teaching practices around natural hazards and community resilience in Colorado. *Journal of Geoscience Education*, 70(4), 460–473. DOI: 10.1080/10899995.2021.1964319

Brugulat, M., Margalef, A., & Echeverria, A. (2021). Educación en riesgos naturales y autoprotección en Andorra. *Geotemas, 18* (ejemplar dedicado a: *X Congreso Geológico de España*), 631.

Butler, J. H., & Montzka, S. A. (2020). *The NOAA annual greenhouse gas index. Global Monitoring Laboratory*. https://www.esrl.noaa.gov/gmd/aggi/aggi.html

Canaza, F., Escobar-Mamani, F., & Huanca-Arohuanca, J. (2021). Reconocer a la bestia: Percepción de peligro climático en estudiantes de educación secundaria. *Revista de Ciencias Sociales*, XXVII(2), 417–434.

Caride, J. A., & Meira, P. Á. (2019). Educación, ética y cambio climático. *Innovación Educativa (México, D.F.)*, 29(29), 61–76. DOI: 10.15304/ie.29.6336

Choulga, M., Janssens-Maenhout, G., Super, I., Solazzo, E., Agusti-Panareda, A., Balsamo, G., Bousserez, N., Crippa, M., Denier van der Gon, H., Engelen, R., Guizzardi, D., Kuenen, J., McNorton, J., Oreggioni, G., & Visschedijk, A. J. H. (2021). Global anthropogenic CO2 emissions and uncertainties as a prior for Earth system modelling and data assimilation. *Earth System Science Data*, 13(11), 5311–5335. DOI: 10.5194/essd-13-5311-2021

Cramer, W., Guiot, J., Fader, M., Garrabou, J., Gattuso, J. P., Iglesias, A., Lange, M. A., Lionello, P., Llasat, M. C., Paz, S., Peñuelas, J., Snoussi, M., Toreti, A., Tsimplis, M. N., & Xoplaki, E. (2018). Climate change and interconnected risks to sustainable development in the Mediterranean. *Nature Climate Change*, 8(11), 972–980. DOI: 10.1038/s41558-018-0299-2

Cruz, N., & Páramo, P. (2020). Educación para la mitigación y adaptación al cambio climático en América Latina. *Educación y Educadores*, 23(3), 469–489. DOI: 10.5294/edu.2020.23.3.6

Cuello, A., & García, F. F. (2019). ¿Ayudan los libros de texto a comprender la red fluvial de la ciudad? *Revista de Humanidades*, 37, 209–234.

Da Rocha, Tibola, V., Brandli, L.L., Mazutti, J., Dal Moro, L., Gasperina, L. D. & Locatelli, R.M. (2020). Teacher's Approach on Climate Change Education a Case Study. *World Sustainability Series*, 617-642. DOI: 10.1007/978-3-030-30306-8_37

De Luis, M., Brunetti, M., González-Hidalgo, J. C., Longares, L. A., & Martin-Vide, J. (2010). Changes in seasonal precipitation in the Iberian Peninsula during 1946–2005. *Global and Planetary Change*, 74(1), 27–33. DOI: 10.1016/j.gloplacha.2010.06.006

Eilam, E. (2022). Climate change education: The problem with walking away from disciplines. *Studies in Science Education*, 58(2), 231–264. DOI: 10.1080/03057267.2021.2011589

Escoz, A., Gutiérrez-Pérez, J., & Meira, P. Á. (2020). Water and climate change, two key objectives in the agenda 2030: Assessment of climate literacy levels and social representations in academics from three climate contexts. *Water (Basel)*, 12(1), 1–33. DOI: 10.3390/w12010092

European Comission. (2024). https://ec.europa.eu/info/index_es

European Environment Agency (EEA). (2024). *Adaptación al cambio climático.* https://www.eea.europa.eu/es/themes/climate-change-adaptation

Ferrari, E., Ballegeer, A. M., Fuertes, M. Á., Herrero, P., Delgado, L., Corrochano, D., Andrés-Sánchez, S., Bisquert, K. M., García-Vinuesa, A., Meira, P. Á., Martínez, F., & Ruiz, C. (2019). Improvement on Social Representation of Climate Change through a Knowledge-Based MOOC in Spanish. *Sustainability (Basel)*, 11(22), 1–21. DOI: 10.3390/su11226317

García de la Vega, A. (2016). Análisis curricular, rigor científico en los textos literarios sobre las observaciones del tiempo y clima: Propuesta didáctica con fuentes primarias. In Sebastiá, R., & Tonda, E. M. (Eds.), *La investigación e innovación en la enseñanza de la Geografía* (pp. 211–236). Servicio de Publicaciones de la Universidad de Alicante. DOI: 10.14198/GeoAlicante2015.15

García-Vinuesa, A., Rui, S. A., Azeiterio, U. M., Meira, P. Á., & Pereira, M. (2020). Mozambican students' knowledge and perceptions about climate change: An exploratory study in Pemba City. *International Research in Geographical and Environmental Education*, 0(0), 1–17. DOI: 10.1080/10382046.2020.1863671

Gavari, E., & Pastrana, J. (2018). Evolución del caso japonés como referente internacional en la educación para la reducción del riesgo de desastres. *Revista Española de Educación Comparada*, 32(32), 52–67. DOI: 10.5944/reec.32.2018.22319

González-Gaudiano, E., Meira, P. Á., & Gutiérrez-Pérez, J. (2020). ¿Cómo educar sobre la complejidad de la crisis climática? *Revista Mexicana de Investigación Educativa*, XXV(87), 843–872.

Granados, J., & Medir, R. M. (2022). *Enseñar y aprender geografía para un mundo sostenible*. Octaedro.

Gutierrez-Pérez, J., Meira, P. Á., & Gonzalez-Gaudiano, E. (2020). Presentación. *Revista Mexicana de Investigación Educativa*, 87, 820–842.

Haq, S. M. A., & Ahmed, K. J. (2020). Perceptions about climate change among university students in Bangladesh. *Natural Hazards*, 103(3), 3683–3713. Advance online publication. DOI: 10.1007/s11069-020-04151-0

Intergovernmental Panel on Climate Change. (2014). *Climate Change 2013 and Climate Change 2014* (3 vols.). https://www.ipcc.ch

Intergovernmental Panel on Climate Change. (2022). *Climate Change 2022: Impacts, Adaptation and Vulnerability*.https://www.ipcc.ch/report/ar6/wg2/

Jacobs, K., Holway, J., Hanak, E., Quay, R., Sternlieb, F., & Udall, B. (2022). Impacts and Opportunities at the Climate-Land Use-Energy-Water Interface: An Urgent Call for Dialogue. *Journal of the American Water Resources Association*, 58(3), 411–419. DOI: 10.1111/1752-1688.13014

Jeong, J. S., González-Gómez, D., Conde-Núñez, M. C., Sánchez-Cepeda, J. M., & Yllana-Prieto, F. (2021). Improving climate change awareness of preservice teachers (Psts) through a university science learning environment. *Education Sciences*, 11(2), 1–17. DOI: 10.3390/educsci11020078

Jiménez, O., Villalón, G., & Évora, O. E. (2017). La educación para la percepción de riesgos de desastres como prioridad del trabajo educativo en la escuela cubana. *Educare (San José)*, 21(3), 1. Advance online publication. DOI: 10.15359/ree.21-3.20

Kazys, J. (2018). Climate change information on internet by different Baltic Sea Region languages: Risks of disinformation & misinterpretation. *Journal of Security and Sustainability Issues*, 7(4), 685–695. DOI: 10.9770/jssi.2018.7.4(6)

Krauss, L. W. (2021). *El cambio climático. La ciencia ante el calentamiento global.* Ed. Pasado y Presente.

Kurup, P. M., Levinson, R., & Li, X. (2021). Informed-Decision Regarding Global Warming and Climate Change Among High School Students in the United Kingdom. *Canadian Journal of Science, Mathematics and Technology Education = Revue Canadienne de l'Enseignement des Sciences, des Mathématiques et de la Technologie*, 21(1), 166–185. DOI: 10.1007/s42330-020-00123-5

León, B., Boykoff, M. T., & Jordán, C. R. (2021). Climate change perception among Spanish undergraduates: A reception study on the combination of the local, global, gain and loss frames. *Communicatio Socialis*, 34(1), 57–75. DOI: 10.15581/003.34.1.57-75

LI, Y.Y. & Liu, S. C. (. (2021). Examining Taiwanese students' views on climate change and the teaching of climate change in the context of higher education. *Research in Science & Technological Education*, 1–14. DOI: 10.1080/02635143.2020.1830268

Li, C. J., Monroe, M. C., Oxarart, A., & Ritchie, T. (2021). Building teachers' self-efficacy in teaching about climate change through educative curriculum and professional development. *Applied Environmental Education and Communication*, 20(1), 34–48. DOI: 10.1080/1533015X.2019.1617806

Li, J. Q., Xia, H. M., & Zhao, X. Y. (2022). Web GIS for sustainable education: Towards natural desastres education for high school students. *Sustainability (Basel)*, 14(5), 2694. Advance online publication. DOI: 10.3390/su14052694

Liao, W., Yang, L., Zhong, S., Hess, J. J., Wang, Q., Bao, J., & Huang, C. (2019). Preparing the next generation of health professionals to tackle climate change: Are China's medical students ready? *Environmental Research*, 168, 270–277. DOI: 10.1016/j.envres.2018.10.006 PMID: 30342323

Marques, E., Nossa, P., & de Guimarães, R. B. (2022). Saúde e território em contexto de pós-pandemia: Secção temática. *Finisterra, 57*(121) (AOP). DOI: 10.18055/Finis29127

Marrón, M. J. (2011). Enseñar y aprender Geografía a partir de distintos lenguajes: Una experiencia de trabajo por proyectos para estudiar el clima de España a partir de los refranes. *Annales de Géographie*, 31(1), 107–123.

Martí, J., Amor, J. A., Giménez-García, R., Ruiz-Álvarez, V., & Biener, S. (2021). Episódios de chuvas torrenciais de 11 a 15 de setembro de 2019 no sudeste da Península Ibérica: Análise meteorológica e consequências das transformações nos usos do solo. *Finisterra*, 56(117), 151–174. DOI: 10.18055/Finis18993

Maters, M. (2020). *123 curiosidades que todo el mundo debería conocer sobre el clima*. Geoplaneta.

McWhirther, N., & Shealy, T. (2018). Case-based flipped classroom approach to teach sustainable infrastructure and decision-making. *International Journal of Construction Education and Research*, 1–21. DOI: 10.1080/15578771.2018.1487892

Méndez, M., Fernández, A., Cruz, A., & Bueno, P. (2020). De la representación social del cambio climático a la acción: El caso de estudiantes universitarios. *Revista Mexicana de Investigación Educativa*, 87, 1043–1068.

Miró-Pérez, J., & Estrela, M. (2016). Reconstrucción de la señal térmica local en la Comunidad Valenciana entre 1948 y 2011 a partir de un downscaling estadístico mediante una red neuronal artificial: Detección de patrones locales de cambio. *Boletín de la Asociación de Geógrafos Españoles*, 70(70), 113–147. DOI: 10.21138/bage.2165

Morote, Á. F. (2021). Investigación y enseñanza en Didáctica de la Climatología: Estado de la cuestión (1980-2020) (España). *Espacio, Tiempo y Forma. Serie VI, Geografía*, 14(14), 89–120. DOI: 10.5944/etfvi.14.2021.31332

Morote, Á. F., Campo, B., & Colomer, J. C. (2021). Percepción del cambio climático en alumnado de 4º del grado en Educación Primaria (Universidad de Valencia, España) a partir de la información de los medios de comunicación. *Revista Electrónica Interuniversitaria de Formación del Profesorado*, 24(1), 131–144. DOI: 10.6018/reifop.393631

Morote, Á. F., & Gómez-Trigueros, I. M. (2023). La brecha digital de género y enseñanza de los riesgos naturales en la formación del profesorado de Ciencias Sociales. *Research in Education and Learning Innovation Archives*, 30(30), 60–75. DOI: 10.7203/realia.30.24712

Morote, Á. F., & Hernández, M. (2020). Social Representations of Flooding of Future Teachers of Primary Education (Social Sciences): A Geographical Approach in the Spanish Mediterranean Region. *Sustainability (Basel)*, 12(15), 1–14. DOI: 10.3390/su12156065

Morote, Á. F., & Hernández, M. (2024). Knowledge and perception of Spanish school children of climate change. *Children's Geographies*, 22(3), 465–479. DOI: 10.1080/14733285.2024.2303581

Morote, Á. F., & Olcina, J. (2021). La importancia de la enseñanza del cambio climático: Propuestas didácticas para la geografía escolar. *Estudios Geográficos*, 82(291), e078. DOI: 10.3989/estgeogr.202189.089

Morote, Á. F., & Olcina, J. (2023a). Cambio climático y educación. Una revisión de la documentación oficial. *Documents d'Analisi Geografica*, 69(1), 107–134. DOI: 10.5565/rev/dag.749

Morote, A. F., & Olcina, J. (2023b). Educando en el riesgo. Formación y propuestas didácticas para incrementar la resiliencia socio-territorial de la población escolar. *Anales de Geografía de la Universidad Complutense*, 43(2), 413–434. DOI: 10.5209/agud.90582

Morote, Á. F., Olcina, J., & Hernández, M. (2022). How is flood risk explained in the subject of Geography in Spanish schools? An approach based on Social Science textbooks (Primary Education). *International Research in Geographical and Environmental Education*, 32(2), 124–139. DOI: 10.1080/10382046.2022.2133955

Morote, Á. F., & Souto, X. M. (2020). Educar para convivir con el riesgo de inundación. *Estudios Geográficos*, 81(288), 1–14. DOI: 10.3989/estgeogr.202051.031

Muñoz, C., Schultz, D., & Vaughan, G. (2020). *A Midlatitude Climatology and Interannual Variability of 200- and 500-hPa Cut-Off Lows*. American Meteorological Society.

Muñoz, J. A., & Otaño, Y. (2014). Software educativo: "a prepararnos para enfrentar los desastres naturales". *Ejemplar dedicado a: Memorias del XXVII Fórum Científico Nacional de Estudiantes Universitarios de Ciencias Pedagógicas, 14*, (39), 1-13.

Nelles, D., & Serrer, C. (2020). *El pequeño manual del cambio climático*. Grijalbo.

Nguyen, C. V., & Pham, N. M. (2018). The impact of natural disasters on children's education: Comparative evidence from Ethiopia, India, Peru, and Vietnam. *Review of Development Economics*, 22(4), 1561–1589. DOI: 10.1111/rode.12406

Olcina, J. (2017). Incremento de episodios de inundación por lluvias de intensidad horaria en el sector central del litoral mediterráneo español: Análisis de tendencias en Alicante. *Rev. Semata*, 29, 143–163.

Olcina, J. (2020). Clima, cambio climático y riesgos climáticos en el litoral mediterráneo español: Oportunidades para la geografía. *Documents d'Analisi Geografica*, 66(1), 159–182. DOI: 10.5565/rev/dag.629

Olcina, J., & Biener, S. (2019). Está cambiando el clima valenciano? Realidades e incertidumbres. [Is the Valencian climate changing? Realities and uncertainties] In Olcina, J., & Moltó, E. A. (Eds.), *Climas y Tiempos en el País Valencià* (pp. 162–170). Publicaciones de la Universidad de Alicante.

Olcina, J., & Moltó, E. A. (Eds.). (2019). *Climas y Tiempos del País Valenciano*. Publicaciones de la Universidad de Alicante.

Olcina, J., & Morote, A. F. (2023). Cambio climático y Objetivos de Desarrollo Sostenible (ODS). Adaptación y enseñanza como retos principales en el ámbito mediterráneo. *Finisterra. Revista Portuguesa de Geografía, 58* (122), 39-61. DOI: 10.18055/Finis28898

Olcina, J., Morote, Á. F., & Hernández, M. (2022). Teaching Floods in the Context of Climate Change with the Use of Official Cartographic Viewers (Spain). *Water (Basel)*, 14(21), 1–20. DOI: 10.3390/w14213376

Opuni-Frimpong, N. Y., Essel, H. B., Opuni-Frimpong, E., & Obeng, E. A. (2022). Sustainable Development Goal for Education: Teachers' Perspectives on Climate Change Education in Senior High Schools (SHS). *Sustainability (Basel)*, 14(13), 8086. DOI: 10.3390/su14138086

Pausas, J. G., & Millán, M. (2019). Greening and browning in a climate change hotspot: The Mediterranean Basin. *Bioscience*, 67(2), 143–151. DOI: 10.1093/biosci/biy157

Pérez-Morales, A., Gil-Guirado, S., & Olcina, J. (2022). La geografía de los riesgos en España (1992-2022). Cambios y oportunidades en una temática de trabajo consolidada y en alza. In *La Geografía española actual. Estado de la cuestión* (pp. 183–202). Comité Español de la UGI.

Pfautsch, S., & Gray, T. (2017). Low factual understanding and high anxiety about climate warming impedes university students to become sustainability stewards: An Australian case study. *International Journal of Sustainability in Higher Education*, 18(7), 1157–1175. DOI: 10.1108/IJSHE-09-2016-0179

Puttick, S., Chandrachud, P., Chopra, R., Robson, J., Singh, S., & Talks, I. (2022). Climate change education: Following the information. *Children. Education and Geography: Rethinking Intersections*, 30, 168–181. DOI: 10.4324/9781003248538-15

Romero, J., & Olcina, J. (Eds.). (2021). Cambio climático en el Mediterráneo. Procesos, riesgos y políticas. Ed. Tirant lo Blanc.

Rudd, J. A. (2021). From Climate Change Ignorant to Climate Change Educator. *Chemistry (Weinheim an der Bergstrasse, Germany)*, 27(20), 6107–6111. DOI: 10.1002/chem.202004414 PMID: 33704844

Sánchez-Almodóvar, E., Gómez-Trigueros, I. M., & Olcina, J. (2023). Climate Change and Extreme Weather Events in the Education of the Citizens of the Twenty-First Century: The Perception of Secondary Education Students. *Social Sciences (Basel, Switzerland)*, 12(1), 27. DOI: 10.3390/socsci12010027

Saregar, A., Sunyono, H. E. Y., Hasan, H., Putra, F. G., & Diani, R. (2022). Natural disaster Education in School: A bibliometric analysis with a detailed future insight overview. *International Journal of Educational Methodology*, 8(4), 743–757. DOI: 10.12973/ijem.8.4.743

Serrano, R. (2017). *Reconstrucción climática instrumental de la precipitación diaria en España: ensayo metodológico y aplicaciones*. Tesis doctoral. Universidad de Zaragoza. http://geografia.unizar.es/noticias/tesisrobertoserrano

Sezen-Barrie, A., & Marbach-Ad, G. (2021). Cultural-Historical Analysis of Feedback from Experts to Novice Science Teachers on Climate Change Lessons. *International Journal of Science Education*, 43(4), 497–528. DOI: 10.1080/09500693.2020.1870759

Suazo, L. E., & Torres, A. (2022). Perceptions, Knowledge, and Teaching of Climate Change and Natural Disaster Risk at Honduran Universities. *Journal of Higher Education Theory and Practice*, 22(3), 118–130. DOI: 10.33423/jhetp.v22i3.5090

Tonda, E. M., & Sebastiá, R. (2003). Las dificultades en el aprendizaje de los conceptos de tiempo atmosférico y clima: La elaboración e interpretación de climogramas. *Revista de Educación de la Universidad de Granada*, 16, 47–69.

UNESCO & UNFCCC. (2016). *Action for CLIMATE EMPOWERMENT. Guidelines for accelerating solutions through education, training and public awareness* (Issue October). UNESCO. https://unfccc.int/files/cooperation_and_support/education_and_outreach/application/pdf/action_for_climate_empowerment_guidelines.pdf

United Nations. (2024). *Programa de trabajo de Doha (2013-2020).*https://www.bing.com/search?q=organización+de+las+naciones+unidas&form=ANNTH1&refig=4e33c3212b664ce6ad8bfec203d421b4&sp=1&qs=AS&pq=organización+de+las+&sk=PRES1&sc=8-20&cvid=4e33c3212b664ce6ad8bfec203d421b4

United Nations (UN). (2015a). *Objetivos de Desarrollo Sostenible.* https://www.un.org/sustainabledevelopment/es/objetivos-de-desarrollo-sostenible/

United Nations (UN). (2015b). *Oficina de Naciones Unidas para la Reducción del Riesgo de Desastres. 2015. Marco de Sendai (2015-2030).*https://www.undrr.org/es/implementando-el-marco-de-sendai/que-es-el-marco-de-sendai-para-la-reduccion-del-riesgo-de

Valbuena, M., & Valverde, J. Á. (2006). La climatología local: Procedimiento para su enseñanza y aprendizaje. *Didáctica Geográfica*, 8, 93–108.

Zaragoza, A. (2022). *La importancia del factor educación como medida preventiva ante el riesgo de inundación. Análisis y propuestas para la Educación Secundaria en Orihuela.* Trabajo de Fin de Máster. Máster en Planificación de Riesgos Naturales, Universidad de Alicante.

KEY TERMS AND DEFINITIONS

Climate Change: A factor that influences the frequency and intensity of many natural hazards. The teaching of climate change focuses on how global climate changes can increase the risk of extreme natural events and how this should be integrated into planning and preparedness.

Disaster Preparedness: This concept encompasses the measures and strategies implemented before a natural disaster occurs to minimize its impact. In education, it includes drills, emergency plans, and awareness of disaster warning signs.

Mitigation: Refers to actions taken to reduce or eliminate the negative impacts of natural hazards. The teaching of mitigation includes urban planning, the construction of resilient infrastructure, and the promotion of sustainable practices to minimize damage.

Natural Hazards: Refers to natural events such as earthquakes, tsunamis, hurricanes, floods, and landslides that have the potential to cause harm to people, property, and the environment. In education, it addresses how these events occur, their frequency, and their impacts.

Resilience: The ability of a community, system, or individual to quickly recover after a natural event. The teaching of resilience involves educating about the importance of adaptation, safe rebuilding, and community support after a disaster.

Risk Maps: Visual tools that represent the probability and severity of natural hazards in a specific region. In education, they are used to teach students how to interpret these maps and understand the spatial distribution of hazards.

Vulnerability: This term describes the degree of susceptibility of a community or system to natural hazards. In an educational context, students are taught to identify factors that increase vulnerability, such as geographical location, population density, and disaster preparedness.

Chapter 5 Geographic Education and Citizen Participation: The Project We Propose!

Sérgio Claudino
https://orcid.org/0000-0001-6987-4812
University of Lisbon, Portugal

ABSTRACT

From the transition between the 18th and 19th centuries, the teaching of Geography emphasized the description of large global spaces, aligned with a rationalist perspective suitable for shaping expansionist elites. In the 19th century, this discipline fostered identification with the nation-state through the concept of territory, promoting a conformist form of citizenship. The following century, marked by two world wars and the Cold War, continued these educational traditions. The current century introduces the social appeal of participatory and multi-scalar territorial construction. The Nós Propomos!/ We Propose! Project embraces the challenge of fostering committed and proactive participation in territorial citizenship. Its methodological simplicity (problem identification by pupils, fieldwork, presentation, and sharing of proposals) ensures its applicability even in educational peripheries. The international expansion of the project demonstrates that it provides an alternative educational response, influencing educational practices beyond the boundaries of Nós Propomos! itself.

DOI: 10.4018/979-8-3693-7185-5.ch005

Copyright © 2025, IGI Global Scientific Publishing. Copying or distributing in print or electronic forms without written permission of IGI Global is prohibited.

1. ON CITIZENSHIP AND THE NÓS PROPOMOS!/ WE PROPOSE! PROJECT

The educational systems that emerged in the 19th century are linked to the construction of liberal citizenship. The disciplines of Social Sciences have always had a particular responsibility regarding the values they convey, with Geography playing a key role in promoting the construction of territorial identity among the youth.

In 2011/12, at the Institute of Geography and Territorial Planning of the University of Lisbon (IGOT-ULisboa), Portugal, the project Nós Propomos!/We Propose! Citizenship and Innovation in Geographical Education was created with a clear civic purpose: to challenge young people to identify problems in their community and to develop and share solutions for them. We Propose! expanded within Portugal and, starting in 2014, became an international initiative. Today, it is the leading international project for experimental Geography education.

This text reflects on the evolution of the citizenship challenges posed to the educational system and to the Geography discipline, influenced not only by different scientific perspectives but also by the dominant political-social projects from the 18th and 19th centuries to the present day. It discusses the need to value the local scale, in line with the changing scientific paradigms in Geography and the public demand for citizen participation and governance in territorial development. The We Propose! Project was created in response to a curricular requirement in the early 2000s Portuguese Geography programs and aims to address these demands. The text also discusses the concept of territorial citizenship, which was developed within the project and has become its main purpose, as the debate around geographical education increasingly focuses on building a democratic and participatory society.

In describing We Propose! Project, the principles and methodology are outlined, and the main phases of the project in Portugal are identified, which also inspire its operation in other countries. Some examples of projects implemented in the countries where the project is most established are given. The text then addresses the trend toward diversification of activities, influenced by the different educational levels, countries, and school contexts in which the project is developed, but also because the project organizers work in teams and take on the broader challenge of fostering debate and training around citizenship in schools. Initial teacher training is also becoming more closely associated with the project, especially in Brazil. Throughout the text, examples are provided to illustrate the discussion and characterization. Finally, the text reflects on the significance of We Propose! Project in renewing both the school system and geographical education itself, as the concerns with territorial citizenship, which characterize the project, are increasingly permeating the school environment, as seen in Portugal. Thus, the aim is not only to discuss the civic contribution of the school and the Geography discipline, including from a temporal perspective, but also to establish a framework for the principles and functioning of We Propose! Project - for those who wish to implement the project at their universities, other higher education institutions, schools, and with their pupils.

2. THE DEVALUATION OF THE LOCAL SCALE

Eighteenth-century rationalists, such as the encyclopedists Diderot and D'Alembert, focused their attention on the world as a whole. Fascinated by the planet, they revealed a world of systems and interconnections. At the turn of the 18th to the 19th century, Alexander von Humboldt traveled through the Americas, and his monumental work *Cosmos* reflected this universalist perspective.

In the 18th century, the first geography textbooks—often translated and adapted from French works for other countries—focused on nations considered most important, particularly European ones (Dufresnoy, 1757). This was the geography of the early 19th century, shaped by the *late Enlightenment* (Claudino, 2015). The Enlightenment paradigm aligned with the aspirations of hegemonic classes: there was a clear interest in teaching the children of elites about the contours of the world, its principal spaces, and their resources from an expansionist political perspective. Discussing geography education in the early 19th century, Horácio Capel et al. (1983) described it as a *science for the bourgeoisie*.

The 19th century was marked by Latin American independence movements, the triumph of liberal revolutions, and the unification of major European countries, all of which contributed to the construction of nation-states. The generalist tradition of the late 18th and early 19th centuries gave way to a nationalist cycle. Geography played a vital role in constructing new national identities: school chorographies traversed national territories through descriptions of mountains, rivers, cities, and innovative railways. Some authors emphasized the religious, judicial, and administrative organization of emerging liberal regimes. However, 19th-century citizenship discourse was not about civic participation but rather about acceptance and conformity to the new political order (Claudino, 2015).

The coexistence of universalist and nationalist discourses extended into the 20th century. The valuation of local realities had already been proposed in the 17th century by Comenius (1985) and later by pedagogical reform movements such as the New School (Vargas Silva, 2019). Nevertheless, these ideas never gained widespread or effective adoption in increasingly massified education systems. The two world wars and the Cold War's subsequent division into major political-military blocs sustained the worldview of nations. Geography curricula reflected and legitimized North-South contrasts, often explained by unfavorable environmental conditions and divergent legacies of economic and social development. In Europe, the expansion of the European Union sought to foster young people's identification with a broader homeland (Claudino, 1990) rather than encouraging participation in distant political mechanisms dominated by supranational entities. Thus, while the turn of the 18th/19th centuries brought a focus on understanding the world, and the 19th century prioritized nationalist discourse, the 20th century failed to adequately integrate the local scale and community experiences into school curricula.

The late 20th and early 21st centuries have been marked by calls for public participation and territorial governance. For example, the 1992 Rio de Janeiro Earth Summit on Environment and Development and subsequent summits emphasized that environmental challenges—especially those related to climate—demand the participation of all inhabitants of the planet, not just political and technical elites. Public approval of territorial planning instruments now requires participation and consensus-building among stakeholders. Urban spaces increasingly seek to mobilize inhabitants to adapt cities to the needs of groups and individuals. Local development depends on active citizen participation (Cristóvão & Baptista, 2023). Territorial governance requires educating the population to use the territory responsibly and effectively (Gonçalves & Claudino, 2021).

In 1990, the Educating Cities Network was established, aiming to make cities inclusive and sustainable spaces for educating their citizens (Osborne, Kearns & Yang, 2013). This initiative deepened with efforts to build learning societies (Ra, Jagannathan & Maclean, 2021). Recognizing the urgency of sustainable development, combating climate change, and fostering more just and inclusive societies, the United Nations introduced the Millennium Development Goals (2000) and later the Sustainable Development Goals (2015).

Despite calls for public participation in democratic governance and territorial decision-making, these principles have had only a limited impact on school curricula, including geography. Throughout its bicentennial tradition, geography has often lacked a social, political, and democratic commitment (Gaudelli & Heilman, 2009). In schools rooted in hierarchical, lecture-based traditions—contradictory to democratic practices—territorial issues are often presented as content lists in programs and textbooks rather than as challenges for pupils to identify and discuss. Political participation—whether voting, holding office at different scales, or engaging in direct democratic mechanisms such as participatory budgets—is conspicuously undervalued in school curricula, including geography. Yet this participation is central to democratic decision-making regarding territory.

Educating for democratic participation and governance in territorial decision-making presents a challenge for the educational system and geography as a discipline.

3. EDUCATING IN TERRITORIAL CITIZENSHIP, WITHIN A MORE HUMANISTIC AND CRITICAL GEOGRAPHY

The *We Propose!* Project arises in this context, positioning schools as privileged spaces for socialization (Tedesco, 2008) and civic education, as increasingly recognized by educational authorities, including in Portugal (Direção-Geral de Educação, 2017). Schools have differentiated facilities and equipment, bring together a group of qualified individuals—teachers—and educate the youngest members of the community. It is unreasonable for them to distance themselves from the problems of their surroundings. The school's contribution to the community, through discussion and the development of proposals to solve its problems, is also a way of giving back to the community for the investment it makes in the school. *We Propose!* seeks to stimulate the participation of schools—and Geography in particular—in improving the community.

Educating for citizenship means engaging with community problems (Claudino, 2019), being attentive to how its members perceive these problems, and participating in decisions that must be made. This project develops through the interaction between schools and the community. Within the framework of *We Propose!*, the concept of territorial citizenship emerges. The territory is a space for exercising power but also for fostering communities' identification with the land they inhabit (territorial identity). Ultimately, the territory is the portion of Earth's surface socially appropriated and shaped by human societies (Salgueiro, 2023). By linking the territory with the identification and spatial construction by a community, the concept of territorial citizenship was adopted to define the goals of *We Propose!*. Territorial citizenship represents the personal (and institutional) commitment to informed and proactive participation in resolving spatially-based social and environmental issues (Claudino, 2019), starting at the local and regional scales but necessarily adopting a multi-scalar approach. Territorial citizenship embodies a civic Geography that emphasizes people's and communities' participation in shaping their community's territory on various scales. Recognizing its undervaluation, asserting territorial citizenship helps recenter the debate in Geography education on activism and spatial advocacy (Mendes, 2024).

In practice, the *We Propose!* Project also signifies a paradigm shift in Geography and geographic education. It rejects a neo-positivist and culturalist approach focused on describing territories and the communities inhabiting them, which often accepts the inevitability of subpar living conditions for many groups due to physical constraints and traditional social organization. Instead, it aligns with Humanistic Geography, which values lived and experienced spaces (Cavalcanti, 2019), addressing the interests, concerns, and valuations of individuals (Bailly, Beguin & Scariati, 2016). This perspective also incorporates elements of Perceptual Geography, focusing on the representations we construct. The *We Propose!* Project is equally sensitive to Radical Geography's critique of social inequalities (Johnston, Gregory & Smith, 2009) and Critical Geography's denunciation of inequalities linked to marginalizing globalization processes (Santos, 2000). Poverty, inequality, and environmental issues must be central to geographical science (Souto & Claudino, 2019). In *We Propose!*, a grassroots Geography emerges, shaped by the perspectives and concerns of local residents (in schools, pupils and teachers) and a civic Geography deliberately focused on informed intervention in inhabited territories.

4. IMPLEMENTING THE CASE STUDY THROUGH A METHODOLOGICALLY ACCESSIBLE PROJECT

The *We Propose!* Project, as mentioned, challenges participants to identify local social and environmental issues that are meaningful to them, rooted in their daily lives (Neto, Silva, Alencar & Araújo, 2024), conduct fieldwork on these issues, and present solution proposals to share with the community.

The immediate motivation for We Propose! was the implementation of the Case Study requirement in the Portuguese 11th-grade Geography curriculum for pupils aged 16–17. Introduced in 2001 and implemented from 2003/4, this curriculum made the Case Study mandatory, assigning it significant prominence. While it did not necessarily need to focus on regional or local scales, it was strongly oriented toward them, with one of its objectives being "to critically analyze problems affecting the region where [the student] lives, reflecting on possible solutions to identified problems" (Alves, Brazão & Martins, 2001, p. 57). Despite its educational potential—encouraging pupils to address their community's issues and propose solutions—day-to-day interactions with schools revealed that the Case Study was rarely implemented. National exams in the subject overlooked it, and the routines of school practices (Souto & García-Monteagudo, 2023) appeared to stifle educational innovation, even when such innovations were formally included in the curriculum.

With the aim of fostering an "education for participatory democracy" (Bazolli, 2017, p. 20), the project focuses on developing attitudes of civic participation in addressing local problems. It directly associates the concept of citizenship in education with action (Moreno-Fernández, 2013), rejecting a more ambiguous and culturalist discourse limited to understanding problems at different scales.

Acknowledging the role of IGOT-ULisboa and its Center for Geographic Studies in research and the initial training of Geography teachers, the We Propose! Citizenship and Innovation in Geographic Education Project was launched. The intention was to create a network of schools that, with the support of the University of Lisbon, would promote the implementation of the 11th-grade Case Study.

According to its regulations (IGOT-ULisboa, 2024), the *We Propose!* Project has the following objectives:

i) to promote territorial citizenship within the school community;
ii) to bring schools and local governments closer to their communities through youth participation in solving local problems;
iii) to highlight the Case Study as an experimental project aimed at improving social and environmental conditions;
iv) to promote innovative methodological approaches in Geography education or other educational fields;
v) to stimulate the development of research, information processing, and presentation competences among young people, particularly in addressing local issues;
vi) to engage pupils and teachers in digital literacy for practical studies;
vii) to contribute to the sustainable development of localities and municipalities where the project is implemented;
viii) to foster networks of cooperation among local actors, such as universities, schools, municipalities, local associations, and companies;
ix) to contribute to the creation of a national and international network of pupils, teachers, schools, and universities that, through We Propose!, drive educational innovation and improvements in their communities.

From a methodological perspective, the project adopts a socioconstructivist learning approach that values pupils' interests (Díaz-Barriga, 2014; García-Monteagudo, Mendes & Lastória, 2024). The topics of study are defined by the pupils themselves, respecting their lived experiences in urban spaces (Gamalho, 2023), undoubtedly in dialogue with teachers, whose motivation to participate in the project is also considered—not as a sufficient condition but as a necessary one for engaging pupils (Veiga, García & García, 2019). Additionally, there is a deliberate emphasis on methodological simplicity to ensure feasibility in all schools, particularly those in more remote areas. The core phases of We Propose! include identifying and selecting the most significant problems, conducting fieldwork, and finally, developing and presenting the pupils' proposals. At the same time, the project acknowledges the diversity of experiences in each school and region while maintaining a focus on discussing and addressing community problems.

The project adopts a reformist perspective on education. As demonstrated by the limited implementation of the Case Study (Claudino & Coscurão, 2022), educational innovation does not arise from top-down reforms imposed by educational authorities, which rarely lead to "profound, lasting transformations" (Canário, 2005, p. 93). Instead, We Propose! initially focused on a single grade level, aims to influence a gradual shift in school practices over time.

5. PRINCIPLES OF WE PROPOSE! PROJECT

Based on the context and objectives outlined earlier, the We Propose! adheres to the following principles:

1. Territorial Citizenship

The project promotes a culture of civic engagement in territorial issues. While the concrete proposals pupils present are central to We Propose!, the primary output is fostering a critical perspective on the local environment. This includes identifying spatial problems and developing attitudes toward community participation aimed at addressing these issues.

2. Human Rights

Alongside respect for fundamental rights, all individuals are entitled to dignified living conditions, regardless of where they live or their social status. This principle is explicitly stated in the *International Charter on Geographical Education* of 1992 (Commission on Geographical Education/International Geographical Union, 2016) and reaffirmed in the revised Charter of 2016 (Commission on Geographical Education/International Geographical Union, 2016).

Some school projects We Propose! may not have an immediate connection to defending living conditions for populations, such as the development of local tourism itineraries or virtual museums, the harnessing of energy generated in gyms, or the beautification of already well-maintained public spaces. However, most student projects are directly related to human rights, the right to the city, and better living conditions for all, often emphasizing social inclusion. Examples include improved sanitation systems, more efficient waste collection and treatment, promotion of renewable energy sources, repurposing of food waste, rationalized water usage, better transportation infrastructure, enhanced public transportation, accessibility for people with disabilities, new and improved public facilities, such as schools and healthcare centres and support for elderly and isolated individuals.

Education, and Geography in particular, must be at the forefront of the fight for human rights (both environmental and social rights). The We Propose! Project makes a significant contribution to this effort.

3. Trust in Pupils

We Propose! centers on pupils taking responsibility for identifying territorial problems, collecting information, analyzing it, and developing intervention proposals. Teachers play a crucial role in facilitating and organizing the project, but the initiative places trust in pupils, rejecting infantilization (Loizaga, 2023). It seeks to validate pupils' insights and thinking processes (Sbardelotto & Francischett, 2021, p. 24), believing in their capacity to reinvent their city and community.

4. **Methodological Simplicity**

The project is intentionally designed with a simple methodology, ensuring that it can be implemented in all schools and engage the broadest range of pupils. This methodological feasibility is undoubtedly one of the reasons for the project's widespread adoption.

5. **Limited Financial Resource Requirements**

Consistent with the previous point, We Propose! operates with minimal financial resources. Pupils can take photos, conduct surveys, or perform other fieldwork activities within their local area and prepare their presentations without needing significant additional funding. This low-cost approach extends to schools as well.

6. **School and Social Inclusion**

The methodological simplicity and cost-efficiency also aim to ensure participation from schools and student populations with fewer resources, including those with learning difficulties. The project is particularly vital in schools located at the periphery of the educational system. Similarly, working on community problems is especially relevant in less privileged neighborhoods.

7. **Constructivism**

The project emphasizes pupils defining the themes or problems they wish to address, ensuring these topics are meaningful to them. This approach is reflected in the diversity of issues pupils explore, often closely linked to local and regional characteristics. For younger pupils, however, teachers often select the class project's theme.

8. **Competences Development**

The project helps pupils develop a wide range of competences, including research, information analysis, proposal development, and sharing results. It also encourages young people to (re)discover their local community, fostering a critical perspective. The thematic competitions associated with the project promote various competences, depending on the competition's scope. Pupils who participate in successive years show significant improvement in identifying problems and completing research phases, reflecting the growth in competencies gained through the project.

9. Interdisciplinarity

The Project originated in Geography and has primarily been implemented by those responsible for teaching Geography and with Geography students. In a way, this identification between We Propose! and Geography has been the Project's *strength*, granting additional social visibility to the field of Geography.

While it is true that Geography has a greater capacity to "understand and develop, within schools, the content related to the city" (Portela, 2024, p. 71) and serves as a facilitator for the dialogue of knowledge (Pulgarín Silva, Ramírez Acosta & Vélez Varga, 2019), the Project has a clear interdisciplinary vocation by addressing community issues. This interdisciplinary focus is now more evident in its implementation in Spain.

10. Dialogue and Horizontal Relationships

Universities or higher education institutions typically coordinate the project. It develops through close dialogue between universities and non-higher-education schools.

11. Affection

Like many other initiatives, We Propose! creates a network of emotional connections. The relationships built among participants, particularly between teachers across educational levels, are vital to constructing these networks.

Figure 1. Main partnerships of the We Propose! Project in Portugal

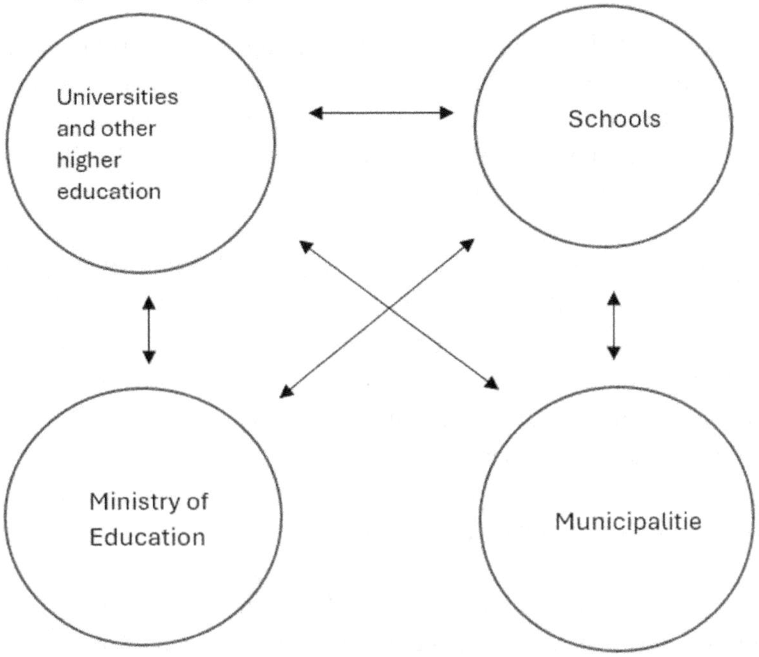

Figure 2. Pupils present their proposals to municipal officials and the community

12. **Partnerships**

The We Propose values partnerships between higher education institutions and schools, but a decisive third pillar is local government, particularly municipalities (Figure 1). The Ministry of Education and other educational authorities, depending on the country, also support the project initiative. Occasionally, partnerships are formed with companies, local associations, and other entities, which can play a significant role.

Municipalities can provide pupils with information about local issues, urban planning guidelines, and other data relevant to the projects. Local political engagement also raises awareness among municipalities about pupils' proposals. At the end of the academic year, sessions are often held where pupils present their proposals to local public officials and the community (Figure 2). The We Propose focuses on improving the local territory rather than merely criticizing existing problems. This constructive approach may explain the almost complete absence of conflicts between municipalities and participating pupils, who not only identify issues but also propose solutions.

The relationship with municipalities is highly varied and shaped by the attitudes of their political leaders. A key challenge is that municipal authorities often find it hard to believe that children or adolescents can develop relevant local action proposals. However, throughout the project, many have learned to trust the ideas of these young participants. In this sense, We Propose! has played a role in *educating* several municipalities.

Figure 3. Municipal plaque acknowledging the We Propose! Project for the installation of recycling stations at a school, in partnership with various entities

In some cases, public officials adopt proposals from We Propose! pupils but avoid publicly acknowledging their inspiration, which can be disheartening for both pupils and teachers. On the other hand, there are municipalities where local politicians and urban planning professionals committed to public participation encourage schools to join the project and provide full support.

Figure 4. Municipality honors the project for its educational contribution

Some municipalities publicly credit specific initiatives to proposals developed within the framework of the project (Figure 3). Others have recognized the project for its sustained contribution to fostering civic education among young people (Figure 4). Additionally, in several countries, the project has received numerous awards, granted either to the project itself or to its participants, particularly teachers.

In a paradoxical conclusion, one might say that We Propose! must both earn the support of local political authorities and, at times, allow itself to be embraced by them.

13. **Network**

As previously mentioned, and in alignment with the Second International Charter on Geographical Education emphasizing the importance of teacher networks for didactic innovation (Commission on Geographical Education/International Geographical Education, 2016), We Propose! is grounded in the belief that changing school practices requires a networked approach. This involves contributions from various actors who complement and inspire one another. Nationally, the project connects universities, schools, and municipalities. Internationally, it fosters dialogue between universities and other higher education institutions. In its Ibero-American expansion, the We Propose Network has benefited from the support of the Geoforo - Ibero-American Forum on Geography, Education, and Society, which emphasizes the importance of people's connections and feelings toward the places they inhabit and their daily social (and environmental) challenges (Rodríguez Moreno & Souto González, 2019). In some ways, the Geoforo has adopted We Propose! as its flagship school project. The We Propose! exists only because it is developed as a network.

14. **Flexibility**

The project accommodates the specific contexts of each school and higher education institution in their modes of operation, which becomes increasingly important as the network expands internationally. The project respects this diversity while ensuring that local problems are identified, researched, and addressed through intervention proposals.

15. **Diversification**

Directly related to the principle of flexibility, there is a growing diversification of activities under the We Propose! framework. These involve institutions, teachers and pupils. Increasingly, various activities aligned with the project's objectives are organized, contributing to its implementation.

16. **Evaluation**

Evaluations by teachers and pupils during the project, as well as a final evaluation, are significant milestones. These reviews help refine the project's objectives and its overall operation.

17. **Auto-Emulation**

Throughout the project, especially in its final phase, pupils feel their work is recognized. This acknowledgment largely stems from presenting and publicly sharing their proposals.

18. **Communication**

Communication of We Propose! activities gives the project visibility and strength. Additionally, it engages civil society, drawing attention to the pupils' proposals (Figure 6).

Figure 5. Students in initial teacher training for Geography (wearing black t-shirts) assist secondary school pupils in a workshop on research methodology

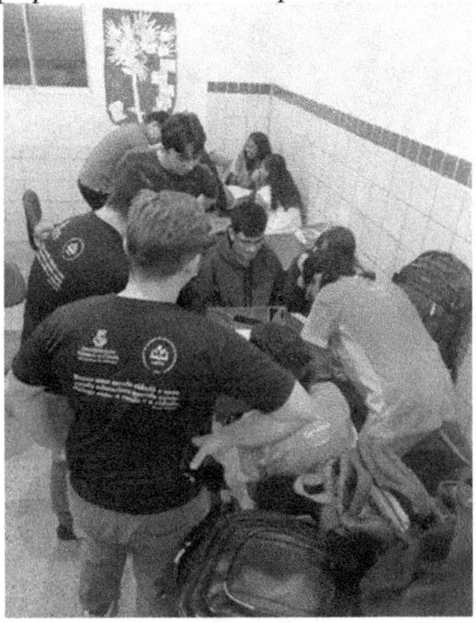

Figure 6. Pupils present the projects they developed on a radio station

Figure 7. Some of the printed books about the We Propose! Project

19. **Initial Teacher Training**

As previously mentioned, the We Propose! has primarily been organized through universities. In Brazil, the project is increasingly integrated into initial teacher training programs. In some cases, participation in project is included in the university curriculum, with teacher trainees accompanying schools' work on the project or even taking on its coordination themselves. The integrated mobilization of knowledge from the field of specialization (Geography, in general) and didactics, when working with young people on community issues, is, moreover, crucial for overcoming the divide between more theoretical education and more practical training (Gomes, 2023).

20. **Research**

The We Propose! has led to the publication of numerous scientific papers, some of which are books related to the work developed at a university (Figure 7). Additionally, there are multiple master's theses and doctoral dissertations focused wholly or partially on We Propose! Research on how pupils learn about their communities, share these learnings, and how this influences territorial development is an essential area of study.

6. THE MAIN PHASES OF THE PROJECT

The main phases of the We Propose! Project, as developed in Portugal and partially inspiring its implementation in other regions, are as follows:

a) Meeting with teachers

At the start of the academic year, a meeting is held to integrate teachers and schools participating in the We Propose! for the first time. During this meeting, the following are addressed the discussion of the project's objectives, a brief review of its performance in the previous year and an examination of the national regulations and the schedule of activities for the new academic year. As the project has expanded to schools nationwide, and many schools participate repeatedly, this initial discussion has become less focused on individual cases and is increasingly conducted via videoconference to accommodate all participants.

Figure 8. A Student Reads Aloud the Protocol Established Between IGOT-ULisboa and the School

b) Signing of cooperation protocols

At the core of the Project is the establishment of partnerships with various stakeholders, as previously mentioned. Collaboration protocols are signed between IGOT-ULisboa and schools, and, when possible, municipalities. These agreements include a general text on the cooperation between the involved entities, but they hold symbolic significance in the commitment to cooperation among the various institutions (Figure 8). At the international level, similar protocols have been signed between IGOT-ULisboa and higher education institutions in other countries.

c) Mobilization of schools, teachers, and pupils

The announcement of the new edition of the project is made at the end of the previous academic year and/or at the beginning of the new school year, through institutional websites, social media, and, naturally, through closer contact with already participating schools. In its first year of operation, the We Propose! project was limited to the 11th grade, with a program focused on the Case Study. However, it has gradually expanded to other education cycles. Since 2023/24, the following modalities have been in place:

Great citizens – for young people in the final cycle of Portuguese secondary education (ages 15 to 18)
Young citizens – for young people in the final years of basic education (ages 10 to 15)
Small great citizens – for children aged 6 to 10, in the early years of schooling

The number of younger pupils participating has increased over the years. The project has also included elderly people, from universities for the elderly, who discussed issues relevant to them. Its expansion to other countries demonstrates the universality and cross-cutting nature of the project.

d) Registration of schools, teachers, and pupils

The registration of schools and teachers is done through the project's website or the IGOT website. This process includes an optional request to indicate a municipal representative who will accompany the project, aiming to establish a relationship with local authorities from the outset. Once classes are underway, student registration is also requested, with two main objectives: to hold pupils accountable and engage them with We Propose!, and to establish direct contact with them throughout the project via email.

e) Selection of local issues by the pupils

Figure 9. Identification of local issues highlighted by pupils, prior to the selection of project themes

Figure 10. During the project theme definition phase, pupils explore the locality to identify relevant problems

The problems selected by the pupils mainly focus on local issues, while necessarily incorporating other scales. Their selection is usually conducted in the classroom through dialogue among the pupils and between pupils and teachers. Some teachers implement a survey to raise awareness of local issues, while others opt for brainstorming sessions (Figure 9). Additionally, some teachers organize field trips with pupils in the vicinity of the school to observe and raise awareness of local problems (Figure 10).

This phase of We Propose! is often time-consuming: many pupils are not accustomed to being asked to identify local issues. Frequently, it is parents and teachers—adults—who point these issues out to them. To some extent, much of the educational value of the project lies in this phase, as it prompts pupils to critically examine a community where they have almost naturalized or legitimized its problems. The role of teachers in provoking pupils to define topics is greater in the classroom when pupils have less autonomy. Throughout the We Propose!, pupils' perspectives on their community evolve through their own work and that of their peers, becoming more observant and critical of the socio-spatial challenges in their community.

Generally, pupils from more rural areas prefer to focus on the town where the school is located, which has a higher urban hierarchy than the localities where many of them live. This is partly due to the greater ease of group work, as they all converge daily in the same locality. However, this also occurs because they often feel constrained when identifying issues in the rural areas where they live. This runs counter to the spirit of We Propose!, which aims to bring visibility to the challenges of more peripheral areas. For many young people, however, urban spaces represent the idealized environment in which they aspire to live, and these are the areas they prefer to focus on.

f) Formation of work groups and definition of project themes

The formation of work groups and the definition of project themes often occur simultaneously. In the classroom, a general survey of problems is conducted, followed by the selection of a central theme for each group. Groups are either pre-formed or formed based on the pupils' shared interests.

Projects always carry a sense of idealism (the belief that it is possible to improve the community) combined with realism, ensuring that no implausible proposals are made. Regardless of the approach, projects should align with the pupils' interests and foster their intrinsic motivation. In cases where a particularly pressing issue dominates the local context (such as the use of a central urban plot that is currently or soon to be vacant), it is common for multiple groups to want to work on the same topic. Experience shows that even when addressing the same issue, different groups often propose diverse solutions, demonstrating their creative freedom.

Not all pupils are initially enthusiastic about the challenge of identifying community problems and creating solutions. For many, this process falls outside their personal and academic comfort zones. Some pupils require encouragement or pressure to participate, just as they do with other tasks. However, the overall experience of We

Propose! has been extremely positive, with high levels of student engagement and participation. Even initially reluctant pupils often become deeply involved, and it is common for teachers, public officials, and other stakeholders to be surprised and impressed by the pupils' commitment and performance.

The greatest resistance to implementing We Propose! typically comes from teachers. In schools with rigid schedules and routines, some view the project as an extracurricular activity rather than an integral part of the curriculum, even though it promotes a wide range of competencies outlined in educational programs. The innovative group work approach disrupts the daily rhythm of the school, presenting challenges for teachers as well. This reveals a gap between school discourses on citizenship and actual educational practices—a gap that We Propose! and similar projects seek to bridge.

The project often spreads to other schools when teachers who have participated in We Propose! move to new institutions, showcasing its acceptance and the positive experiences of the majority of educators who implement it.

g) Work session at the school with the coordination team

Figure 11. Work session at a school: dialoguing with pupils about the projects

Each academic year, one or more members of the coordination team visit each participating school (Figure 11). These work sessions are held with the classes, and during them: i) the objectives of the project are reviewed, ii) the phases of the project are explained, and iii) the national and international network of schools involved in the project is presented. Often, these sessions include an initial review of the projects that pupils intend to develop. Sometimes, pupils make a first presentation of the work they are developing, which is highly motivating for the entire class.

These visits to the schools also serve as an opportunity to engage in dialogue with all project participants, including school management teams and, occasionally, municipal political representatives. This personal contact is highly significant. With the expansion of the project, it has sometimes been challenging to hold these sessions at every school, but the team has managed to ensure that the majority of schools are reached.

h) Work Sessions with Municipal Technicians

Whenever possible, municipal technicians visit the schools and meet with pupils and teachers. They share the main concerns related to municipal planning instruments, such as the *Plano Diretor Municipal* (Municipal Master Plan), or, at times, pupils visit municipal government offices to discuss the municipality's issues (Figure 12). The goal of connecting schools and pupils, in particular, with local government authorities is, in fact, one of the core objectives of the We Propose!

Municipalities are generally cooperative, as is the case in Portugal. However, the relationship between municipalities and schools varies from one municipality to another and, even, from one country to another.

Figure 12. A class visited municipal facilities to engage in dialogue with a technician about their project issues

i) Conducting fieldwork

Figure 13. Pupils visit a wastewater treatment plant

Figure 14. A student interviews a local resident

Fieldwork is central to the project. Pupils take to the streets, visiting the spaces or facilities they want to address in their projects (Figure 13), and engage with local populations through small surveys (Figure 14) or interviews with local stakeholders. In We Propose!, fieldwork primarily involves listening to the community about the problems they experience.

As expected, there is an increasing use of social media networks for conducting surveys about the pupils' projects. This digital shift helps broaden the reach and inclusivity of data collection.

The very act of young people approaching the local population to gather their opinions on a community issue often brings that issue into the public sphere and encourages broader discussion. It is fair to say that the process associated with fieldwork in identifying local issues is, in itself, highly relevant, regardless of the final results.

j) Development of Proposals by Pupils

The proposals are presented using multimedia applications, and sometimes on posters. In much of the educational literature, it is argued that pupils follow/should follow the phases of the scientific research method in their projects (problem definition, hypothesis formulation, hypothesis testing, and conclusions). However, this has not been observed in the We Propose! Project over the years.

When pupils identify a problem to address (e.g., the revitalization and reactivation of a neglected and abandoned city square), they already have preliminary ideas about the measures that should be taken. They do not select a topic/problem without having some expectation of the actions needed to resolve it. As their fieldwork progresses, they may adapt or modify their initial proposals. It might be more appropriate to describe this as a school-based problem-solving method.

As mentioned earlier, it is common for younger pupils to work on tasks aligned with a broader proposal defined by their teacher. The work produced by these *young big citizens* often demonstrates remarkable creativity (Figure 15).

Figure 15. Proposal for a multi-function playground by a child

k) Intermediate Feedback

At an advanced stage of project development, the student groups complete an online questionnaire on the We Propose! project website. In this questionnaire, they identify the tasks they have completed, the ones still pending, and the challenges they are facing. This serves as an important self-assessment exercise for the groups.

At IGOT, a team of university pupils provides individualized feedback to each group of school pupils. They typically emphasize the positive aspects of the work completed so far and offer suggestions for the next steps, often focusing on the fieldwork still to be carried out. This phase is crucial as it provides pupils with their first external feedback on their project from individuals outside their school—from the *university*.

This intermediate feedback is not implemented by other higher education institutions, which oversee fewer schools and pupils, but it remains a valuable practice.

Additionally, this activity has been an important way to involve IGOT university students in the We Propose!, fostering collaboration and mutual learning between different educational levels.

l) Participation in Photography, Video, Text, Drawing, Advertisement Spot, and Poster Competitions

Figure 16. "Straight Street, an Abandoned Space" – highlighting an urban issue

In the early years of the We Propose!, photographs, texts, drawings, videos, and other by-products of the pupils' projects began to emerge, showcasing significant creativity and relevance. Thus, in the 2014/15 academic year, a photography competition for pupils was introduced. Due to its success, additional competitions were launched in subsequent years, including video, text, drawing, posters, advertisement spots, and songs (the latter particularly embraced by younger children). The themes of the pupils' works must align with the topics of their respective projects (Figure 16). These competitions engage pupils with a wide range of competences and bring renewed vibrancy and dynamism to the We Propose!

m) Sharing Pupils' Proposals

The pupils develop projects directly related to their community, and it is highly relevant for them to share their proposals. This is primarily done in seminars held at the end of the academic year, either for pupils from the same school or by bringing together colleagues from various schools. These seminars can take place at the school, the university, or another venue, such as a municipal building, in front of local politicians and other authorities. The pupils present their projects to their peers, possibly from other schools, as well as to public or private authorities, and also to the community and their families. The broader the audience, the better. These are very significant moments for the pupils, where they feel natural nervousness due to the public exposure, but which also motivate them to do their best work and make them feel gratified for the work they have done. These seminars are often covered in the local press.

At IGOT, a National Seminar is held at the end of the academic year. In the morning, all groups present their proposals (the work of all groups is valued!) – Figure 17. In the afternoon, there is a plenary session in a festive atmosphere (Figure 18), where some interventions are made by university, municipal, and school representatives. Afterward, small prizes are awarded to the pupils (Figure 19). The dissemination of the proposals often extends through schools and local authorities.

The question of whether or not to award distinctions for the best projects has been widely discussed, as there is a strong emphasis on valuing the work done by all the groups. In Portugal, several solutions have been tried, but it was finally decided to grant distinctions, aligning with the sensibilities of students and teachers. The voting for the best projects is done by a jury composed of teachers from each participating school, in the *Grand Citizens* category. In the project competitions for *Small Great Citizens* and *Young Citizens*, the voting is conducted by a jury from IGOT. The voting for awards in thematic competitions is also done by juries composed of IGOT students.

Figure 17. Pupils presenting their proposals at the National Seminar We Propose!

Figure 18. Plenary session of the National Seminar (opening artistic performance)

Figure 19. Pupils Displaying the Diploma Awarded to Their Project

n) Project Evaluation

At the end of the academic year, a survey is conducted to evaluate the project, involving both pupils and teachers. The pupils often express criticism regarding the collaboration of local government and other entities—they are working on solving local problems and feel frustrated by not receiving the support they believe they are entitled to. They often mention the difficulty of developing the project alongside other school activities. They consistently acknowledge that the project has helped them better understand their city/community and become more aware of its issues. As a Portuguese student mentioned in 2024, We Propose! allowed him to discover certain problems in the city that he wouldn't have sought to learn about if it weren't for this project. On the other hand, and perhaps understandably, they often fail to fully appreciate the competences they develop through their participation in Project. As for the teachers, there is a general consensus in applauding precisely the development of competences by the pupils.

In Portugal, the project follows an annual cycle. In other countries, such as Spain and sometimes Brazil, the We Propose! project is developed over one academic semester, which has the advantage of allowing the dissemination of results throughout the following semester.

7. SOME EXAMPLES OF IMPLEMENTED PROJECTS

The proposals from the We Propose! project have an impact that is relatively difficult to measure. The fact that they are often shared publicly helps build awareness about the existing problems, which is in itself a positive outcome. In the short term, the problems identified by the pupils and the proposals they suggest are not valued or resolved—leading to frustration among the pupils, who expect a more or less immediate resolution of the issues. In one year, a class received an award for the overall quality of their projects, which was undoubtedly also due to the methodological support provided by the teacher. After the award was announced, recognizing the merit of their proposals, the same teacher said she was often approached by pupils asking if their proposals, which were praised at a national level, would be implemented. However, in reality, that municipality traditionally did not value the pupils' proposals.

In any case, the public dissemination of local problems and related proposals creates pressure on decision-makers, leading to the resolution of many of these issues in the medium term—though this is usually monitored by the teachers, rather than the pupils, who are typically former pupils by then. On the other hand, it is sometimes difficult to categorically state that a particular initiative stemmed from a proposal made by the pupils, due to the time gap between the public announcement of the proposal and its implementation, which often involves adaptations made by the municipalities to the original proposal. It is also worth noting that schools do not always inform the public about the initiatives being carried out by local authorities. For instance, a Portuguese school held an exhibition showcasing the pupils' proposals along with newspaper articles reporting on their implementation, without the national coordination of We Propose! being informed beforehand about the realization of any of those proposals.

Although it is difficult to establish a clear pattern, it can be said that the greater the involvement of municipalities in launching the project, the higher the likelihood that they will implement the pupils' proposals. On the other hand, it is acknowledged that We Propose! needs to place more emphasis on presenting its proposals to local participatory budgets (which has happened, but not systematically when such budgets exist locally).

Figure 20. Class of little great citizens inaugurates a pedestrian crossing near the school, which they requested from the authorities

Figure 21. Inauguration of an eco-water fountain at the school, proposed by a group of pupils

Figure 22. Road with a sidewalk and lighting due to a proposal from We Propose! pupils

Figure 23. Vertical garden on the wall of the townhall buiding

Proposals from schools at the more elementary levels generally focus on initiatives shared between the teacher and the entire class, as previously mentioned. Those related to school equipment or accessibility to the school have been the most frequently implemented (Figure 20). With older pupils, the proposals are formulated by groups of pupils and cover a wide range of topics, such as the installation of an eco-water fountain in a school (Figure 21) or lighting and sidewalks for school access (Figure 22), the construction of a vertical garden (Figure 23), the creation of a sports field, the design of murals in public spaces, or the operation of a distribution vehicle for delicate goods, such as medicines, to isolated rural populations.

8. BUILDING PARTNERSHIPS AND REGIONAL NETWORKS

Figure 24. Countries with the We Propose! Project in 2024

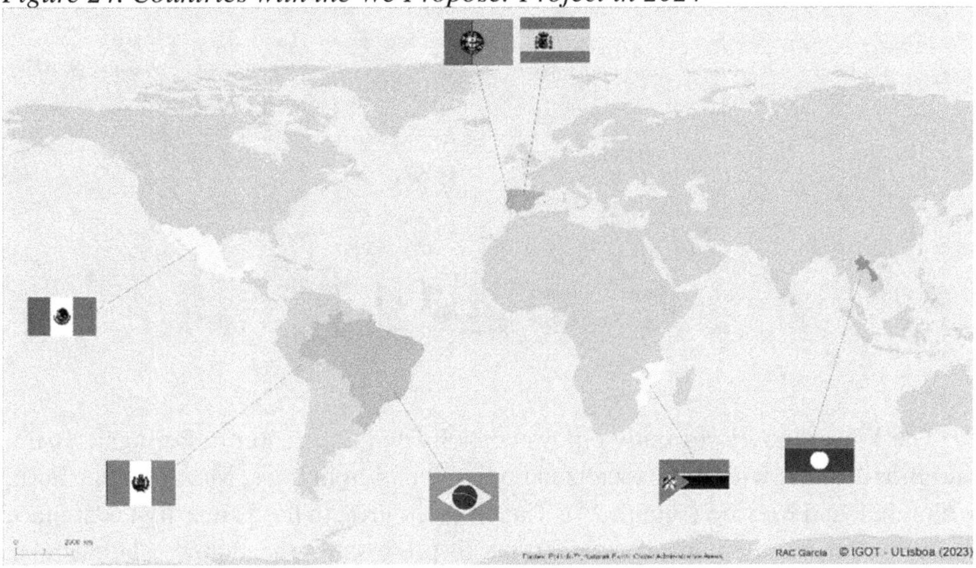

Figure 25. The We Propose! International Congresses are opportunities for sharing experiences, consolidating the network, and reinforcing the project's institutional presence

Figure 26. In Brazil, regional meetings of the project are taking place (2024)

The We Propose! currently enjoys consolidated expansion in Portugal, Spain, and Brazil, along with more localized implementation in Laos, Mozambique, Peru, Colombia, and Mexico (Figure 24). The project's growth has primarily taken place in Ibero-American countries, largely due to pre-existing academic relationships, the role of the Geoforo – Ibero-American Network of Geography, Education, and Society, as well as cultural and linguistic affinities.

The limited adoption outside the Ibero-American sphere may be attributed to this region's particular socio-environmental challenges, which make the project initiative especially relevant. In 2024, the project launched the *Manifesto We Propose! São Paulo. New Impetus*, at Brazil's National Meeting of Teaching Practices. This manifesto aims to expand the project to unrepresented regions within Brazil and to additional Ibero-American countries.

Since 2018, three major congresses have been held: the first in Lisbon; the second in Rio de Janeiro (2022, online); and the third in Ciudad Real, Spain, in 2024. The first two were categorized as Ibero-American, while the most recent was deemed international due to the diversity of participating countries (Figure 25). These congresses provide crucial opportunities for sharing experiences among educators and researchers from various nations and serve as important forums for bolstering the project's visibility and credibility among academic, educational, and political authorities.

The regional strengthening of the We Propose! network is particularly evident in Brazil, where regional meetings are being organized (Figure 26). These gatherings bring together different higher education institutions and schools that participate in the project, fostering collaboration and knowledge exchange.

9.THE RECENT EVOLUTION OF THE WE PROPOSE!: DIVERSIFICATION

In Portugal, the project has seen remarkable expansion. It began in 2011/12 with 400 pupils, and by 2018/19, it had grown to include more than 2,000 pupils and about 100 teachers. In recent years, the number of participants has stabilized at just under 3,000 pupils, with the growth primarily seen in elementary school levels.

The We Propose appears to be influencing Geography education beyond its own scope. Since 2018, when the project already had significant visibility and public recognition, national exams for secondary education have consistently included questions asking pupils to propose solutions for specific problems—something that can likely be linked to the influence of the project. The presence of such questions in exams undoubtedly influences school practices, encouraging discussions around solutions to territorial issues. Additionally, there has been a growing trend in activities where pupils are challenged to find solutions for geographic problems—a development that we have referred to as the *new geographic culture* (Claudino, Mendes & Coscurão, 2023), and which could perhaps be more accurately described as a new school culture of territorial citizenship. Another sign of the influence the We Propose is having on geographic education is the adoption of its concept within academic discussions, as well as by Portuguese socio-professional organizations in Geography (such as the Association of Geography Teachers and the Portuguese Association of Geographers, 2019). The 2023 Iberian Congress on Geography Didactics (organized by the Association of Geography Teachers and the University of Coimbra) was centered around the theme *Geography Didactics for Territorial Citizenship*...

In Spain, there is generally a strong partnership between universities, schools, and municipalities, which have supported the project since its inception and recognize its educational value. The project is mainly focused on engaging pupils in the appropriation and intervention in urban spaces. It involves pupils up to the age of 15, partly due to the limited presence of Geography as a subject in the *Bachillerato* (16-17 years old). As a result, the project's evaluation heavily emphasizes understanding how the pupils' knowledge of the city has evolved through their participation. Continuous teacher training has also been a key concern, supported both by the university and the municipalities (Figure 27).

Figure 27. Workshop for teachers on collaborative design of learning spaces, organized by a municipality

Figure 28. Lecture on citizenship, youth and territory organized as part of the project

Figure 29. Pupils identify various problems in their city and propose solutions for them, without fieldwork

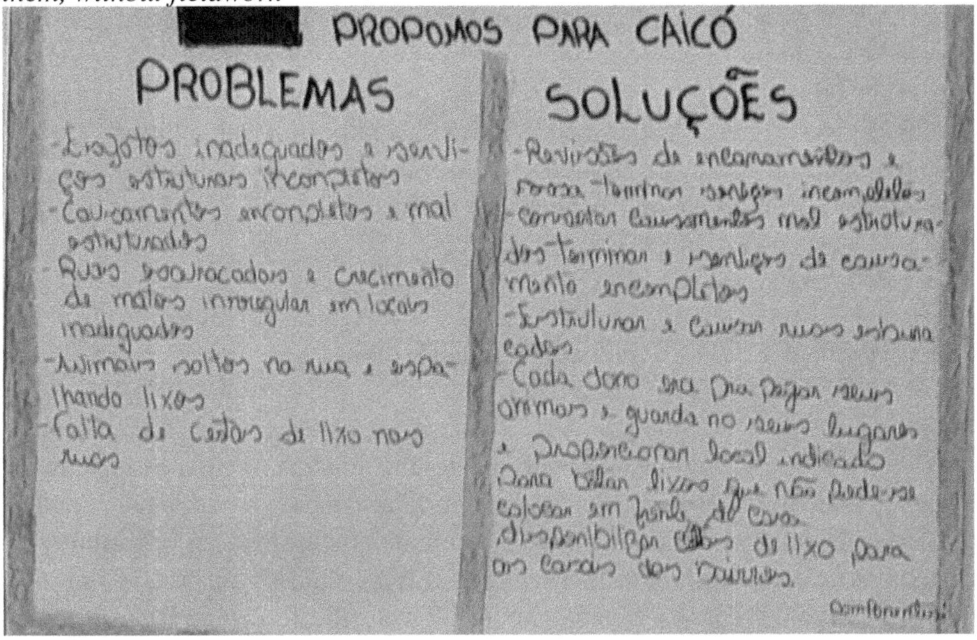

In Brazil, the project's organizers promote lectures that don't specifically address the project but help frame and highlight its value, as mentioned earlier (Figure 28). At times, during the implementation phase, study visits are conducted to an area where pupils observe and discuss a specific problem. In other cases, pupils do not conduct fieldwork: instead, in groups, they identify the main problems in their city and propose solutions (Figure 29). In both cases, even though the methodology of the We Propose isn't fully followed, these activities provide an important first critical look and reflection on community issues. Often, projects focus on the schools themselves, aiming to improve their conditions. As the project deepens, workshops are being developed to raise student awareness about urban issues, before fieldwork takes place, including educational games (Figure 3). The problems selected by pupils are often similar across schools, focusing on basic needs such as sanitation, mobility, and transportation.

As mentioned earlier, in Brazil, the involvement of schools is often linked to the direct support from university students who drive the We Propose! Many of them receive scholarships for assisting teachers in schools, while others attend Geography teacher training courses, where participation in the project is also incorporated into their initial training.

In Portugal, the project's expansion to other educational levels also involves adaptations. The Little Great Citizens classes typically adopt a theme suggested by the teacher (e.g., building a playground near the school, creating a tourist guide for the city, etc.). Pupils then work on the theme by creating texts, drawings, songs, dances, or other activities. Few experiences involve pupils exploring a theme in groups. However, this is a common practice in Spain. In Brazil, children tend to conduct a general survey of problems they identify in their city and then suggest solutions for them. As previously noted, the project aims to align with the sensitivities of teachers and schools.

10. THE SIMPLICITY AND STRENGTH OF A TRANSFORMATIVE PROJECT

The We Propose! was created with the goal of implementing a case study on local socio-environmental problems, attempting to counteract the undervaluation of the local scale, as mentioned earlier. The case study offers an opportunity to develop critical citizenship, fully aware that it is an almost utopian goal, but one that is absolutely necessary (Souto, García-Monteagudo & Campo País, 2024): in Geography, it presents the necessary utopia of building territorial citizenship.

The project aimed to create a proposal that could be shared by the largest number of schools and pupils. The focus was on the methodological feasibility of the project, which involves identifying local problems, conducting fieldwork on one of them, constructing solutions, and sharing them. Ultimately, the We Propose! develops a simple yet powerful idea: universities, schools, and education in Geography, in particular, have an effective commitment to the community in which they are embedded, a commitment they must take on in discussions about local issues and in creating proposals to solve them. The creation of concrete proposals by pupils for their communities represents an important educational contribution, but the project's value lies primarily in developing attitudes of civic participation—territorial citizenship.

The We Propose! has a key purpose that is repeated every year: universities and schools must mobilize their pupils to serve the community. Its unique feature, its signature, lies in not being just another well-intentioned but unimplemented educational proposal; it takes pupils outside the walls of schools, turning this goal into reality, and continues to encourage more universities, schools, and pupils to do the same.

The project has welcomed and encouraged diverse school experiences that share the goal of fostering a critical and interventive perspective among young people on their communities. Thousands of pupils and hundreds of teachers have participated in this project (over 20,000 pupils in Portugal alone), which continues to influence school practices, especially in Geography.

Due to its duration and international diffusion, the We Propose! constitutes a movement for the renewal of geographic education, giving local scale the centrality that Comenius, in the 17th century, and many others have advocated, often in speeches disconnected from real schools.

It is clear that educational innovation occurs through networks, not isolated teachers or institutions; that schools, and Geography specifically, have a genuine commitment to education for citizenship, which should not be an *extraordinary* addition to the school curriculum; that we must reach out to diverse educational systems and schools; and that we must embrace the diversity of paths towards building effective territorial citizenship. This represents both a legacy of the project and a challenge as the second quarter of the 21st century approaches. Education and Geography acquire new relevance and renew themselves in the challenge of fostering critical and active citizenship, as the We Propose! project demonstrates.

REFERENCES

Alves, M. L., Brazão, M., & Martins, O. S. (2001). *Programa de Geografia A*. Ministério da Educação.

Associação de Professores de Geografia & Associação Portuguesa de Geógrafos. (2019). *Educação Geográfica e Ordenamento do Território*. https://www.aprofgeo.org/wp/?p=1361

Bailly, A., Beguin, H., & Scariati, R. (2016). *Introduction à la géographie humaine* (9th ed.). Armand Colin. DOI: 10.3917/arco.baill.2016.01

Bazolli, J. A. (2017). Nós Propomos! e a busca inovação no campo da extensão universitária. In J. A. Bazolli, S. Claudino, M. V. C. Silva, S. F. R. Viana, & W. C. Silva (Org.), *A Extensão Universitária como indutora à cidadania: a experiência do "Nós Propomos"* (pp. 13-27). EDUFT.

Canário, R. (2005). *O que é a Escola?* Porto Editora.

Capel, H.. (1983). *Ciencia para la burguesía: renovación pedagógica y enseñanza de la geografía en la revolución liberal española, 1814-1857*. Universitat de Barcelona, Edicions i Publicacions.

Cavalcanti, L. S. (2019). *Pensar pela Geografia. Ensino e relevância social*. C&A Alfa Comunicação.

Claudino, S. (1990). Norman Graves: La Grande Patrie est la Communauté Européenne - un entretien avec Norman Graves. *The European Geographer*, 2, 51–54.

Claudino, S. (2015). A Educação Geográfica em Portugal e os Desafios Educativos. *Giramundo. Revista de Geografia do Colégio Pedro II*, 2(3), 7–19. DOI: 10.33025/grqcp2.v2i3.204

Claudino, S. (2019). Project We Propose! Building Territorial Citizenship from School. In J. A. Pineda-Alfonso, N. de Alba-Fernandez & E. Navarro-Medina, *Handbook of Research on Education for Participative Citizenship and Global Prosperity* (pp. 350-382). IGI Global. DOI: 10.4018/978-1-5225-7110-0

Comênio, J. A. (1985). *Didáctica Magna - Tratado da Arte Universal de Ensinar Tudo a Todos* (3rd ed.). Fundação Calouste Gulbenkian.

Comission on Geographical Education/International Geographical Education. (1992). *International Charter on Geographical Education*. Internacional Geographical Education. https://www.igu-cge.org/wp-content/uploads/2018/02/1.-English.pdf

Comission on Geographical Education/International Geographical Education. (2016). *International Charter on Geographical Education* (pp. 1–20). Internacional Geographical Education. https://www.igu-cge.org/wp-content/uploads/2019/03/IGU_2016_eng_ver25Feb2019.pdf

Coscurão, R., & Claudino, S. (2022). O desenvolvimento de competências de pesquisa na disciplina de Geografia do ensino secundário: uma experiência didática. *Finisterra – Revista Portuguesa de Geografia*, 57(119), 193-211 DOI: 10.18055/Finis24884

Cristóvão, A., & Baptista, A. (2023). Desenvolvimento em contexto rural. In Monteiro, A. A., & Fragoso, A. (Eds.), *Desenvolvimento local em Portugal* (pp. 37–54). Edições Afrontamento.

Díaz-Barriga, A. (2014). Construcción de programas de estudio en la perspectiva del enfoque de desarrollo de competências. *Perfiles Educativos*, XXXVI(143), 142–162. DOI: 10.22201/iisue.24486167e.2014.143.44027

Direção-Geral de Educação. (2017). *Estratégia Nacional de Educação para a Cidadania*. República Portuguesa.

Dufresnoy, L. (1757). *Descripçao da Terra, ou Methodo Breve da Geographia, Dividido em Liçoes por perguntas e respostas*. Officina de Joseph da Costa Coimbra.

Gamalho, N. P. (2023). Juventudes e as periferias. In V. H. N. Oliveira (org.), Geografias das Juventudes (pp. 39-59). GEPJUVE.

García-Monteagudo, D., Mendes, L. F. G., & Lastória, A. C. (2024). Percepciones docentes sobre ciencias sociales: El caso de Nosotros Proponemos. *Revista Colombiana de Educación*, 90(90), 9–34. DOI: 10.17227/rce.num90-14391

Gaudelli, W., & Heilman, E. (2009)... *Reconceptualizing Geography as Democratic Global Citizenship Education Teachers College Record*, 111(11), 2647–2677.

Gomes, M. F. V. B. (2023). Contribuições do Projeto Nós Propomos! Guarapuava à formação inicial dos professores de Geografia. In M. F. V. B. Gomes, C. L. Silva, A. Roik & E. Y. Yamamoto (org.), *Formação de professores no Projeto Nós Propomos! Guarapuava. Contribuições pedagógicas para a Agenda da ONU 2030*. C&A Alfa Comunicação.

Gonçalves, C., & Claudino, S. (2021). Governance, Civic and Geographic Education and Local Development. In S. Saúde, M. A. Raposo, N. Pereira, A. I. Rodrigues, *Teaching and Learning Practices That Promote Sustainable Development and Active Citizenship* (pp. 236-253). IGI Global.

IGOT-ULisboa. (2024). Regulamento do Projeto Nós Propomos! cidadania e Inovação na Educação Geográfica https://www.igot.ulisboa.pt/projeto-nos-propomos

Johnston, R. J., Gregory, D., & Smith, D. M. (2009). *The Dictionary of human geography* (3rd ed.). Blackwell Reference.

Loizaga, V. B. (2023). Ambitos sociolinguísticos en la ESO. El professorado ante el reto de la LOMLOE. *Íber.Didáctica de las Ciencias Sociales. Geografía e Historia*, 113, 29–33.

Mendes, L. (2024). Pensamento espacial crítica, activismo e cidadania territorial: introdução a uma trialéctica. In M. J. M. Gaité (Ed.) *Geografía, Educación y Innovacionnes y Didácticas* (pp. 146-153). Asociación Española de Geografía (AGE). Grupo de Didáctica.

Moreno Fernández, O. (2013). *Educación ambiental y educación para la ciudadanía desde una perspectiva planetaria. Estudio de experiencias educativas en Andalucía* [Environmental education and education for citizenship from a planetary perspective. Study of educational experiences in Andalucía]. Universidad Pablo de Olavide.

Neto, M. S., Silva, F. D. P. R., Alencar, J. J., & Araújo, R. L. (2024). Transformações espaciais e o Projeto Nós Propomos! em Campo Maior/Piauí/Brasil. In M. A. Rodríguez-Domenech (Ed.), *El pensamiento geográfico y su enseñanza en el siglo XXI: Tendencias y perspectivas a través del proyecto de participación ciudadana ¡Nosotros Proponemos!* (pp. 49-59). Dykinson Ebook.

Osborne, M., Kearns, P., & Yang, J. (2013). Learning cities: Developing inclusive, prosperous and sustainable urban communities. *International Review of Education*, 59(4), 409–423. DOI: 10.1007/s11159-013-9384-y

Portela, M. O. B. (2024). A cidade e o urbano no ensino de Geografia: possibilidades temáticas para o trabalho em sala de aula. In G. A. Teles, J. F. Sobrinho & S. Claudino, *Educação goegráfica e formação de professores no Brasil e Portugal: trajetórias e atualidades* (pp. 71-88). Sertão Cult.

Pulgarín Silva, R., Ramírez Acosta, J. I., & Vélez Venegas, C. (2019). La espacialidad y los derechos básicos de aprendizage de Ciencias Sociales en Colombia. In L. A. Rodríguez Pizzinato, N. Palacios Mena & X. M. Souto González, *La construcción global de una enseñanza de los problemas sociales desde el Geoforo Iberoamericano* (pp. 21-34). Nau Llibres.

Ra, S., Jagannathan, S., & Maclean, R. R. (Eds.) (2021) - Powering a Learning Society During an Age of Disruption. Springer. DOI: 10.1007/978-981-16-0983-1

Rodríguez Moreno, E. A., & Souto González, X. M. (2019). Síntesis del Geoforo en los ãnos 2008 a 2018. In L. A. Rodríguez Pizzinato, N. Palacios Mena & X. M. Souto González, *La construcción global de una enseñanza de los problemas sociales desde el Geoforo Iberoamericano* (pp. 9-20). Nau Llibres.

Salgueiro, T. B. (2023). Pensar o território através do território. In T. B. Salgueiro, *Temas e reflexões. Breve Antologia* (pp. 57-67). Associação Portuguesa de Geógrafos.

Santos, M. (2000). Por uma outra globalização: Do pensamento único à consciência universal. *Record.*

Sbardelotto, V. S., & Francischett, M. N. (2011). "Nós Propomos!" O conhecimento geográfico e a formação pelo ensino. In M. N. Francischett, A. C. Biz & G. Toffolo (orgs), "Nós Propomos!" Ressignficar o ensino na educação geográfica (pp. 23-37). Edunioeste.

Souto, X. M., & Claudino, S. (2019). Construímos uma Educação Geográfica para a Cidadania Participativa. O caso do Projeto Nós Propomos! *Revista Signos Geográficos – Boletim NEPEG de Ensino de Geografia*, 1, 16 p.

Souto, X. M., & García-Monteagudo, D. (2023). Formación docente y cultura profesional. In I. Bellatti, C. Fuentes Moreno, P. Miralles Martínez & L. Sánchez (coord.), *Las ciencias sociales y su didáctica pensamiento histórico y educación democrática: Homenaje a Joaquín Prats Cuevas.* Horizontes Universidad, Octaedro.

Souto, X. M., García-Monteagudo, D., & Campo País, B. (2024). Formando ciudadanos críticos en el Geoforo Iberoamericano. *Geo UERJ*, 46, 26 p. DOI: 10.12957/geouerj.2024.86466

Tedesco, J. C. (2008). *O novo pacto educativo. Educação, competitividade e cidadania na sociedade moderna* (3rd ed.). Fundação Manuel Leão.

Vargas Silva, F. A. (2019). *Tendencias, corrientes y modelos pedagógicos: relaciones, semejanzas y diferencias (una mirada desde occidente).* Sello Editorial Universidad del Tolima.

Veiga, F. H., García, O. F., & García, F. (2019). Envolvimento dos alunos na escola: conceptualização, avaliação e promoção. In *F. H. Veiga, Psicologia da Educação. Temas de aprofundamento científico para a Educação XXI* (pp. 37–58). Climepsi Editores.

KEY TERMS AND DEFINITIONS

Basic Geography: Geography constructed from the perspective of the territory and the concerns of its inhabitants—at school, from the perspective of pupils and teachers.

Citizen Geography: Geography that focuses on the participation of people and communities in the construction of their communal territory, at various scales.

Competence: A combination of interrelated knowledge, abilities, and attitudes that are required to solve problems.

New School Culture of Territorial Citizenship: The widespread adoption of activities that challenge pupils to find solutions to the problems of the spaces they inhabit.

School Method of Problem Solving: The definition and resolution of a research problem by pupils, with the expectation of finding a solution when beginning their research, while adapting their proposals based on the results from fieldwork.

Territorial Citizenship: A personal and institutional commitment to informed and proactive participation in the resolution of social and environmental community problems with a spatial basis, primarily at local and regional scales, but also at other scales.

Chapter 6
"Learn to Take Care of the Environment": An Innovative Proposal for Working on Sustainable Development Goal 14 "Life Below Water" in Primary Education

Nuria Cuartero Ruiz
https://orcid.org/0009-0001-9719-9612
Independent researcher, Spain

Antonio Morcillo-Martínez
https://orcid.org/0000-0002-8901-5406
University of Castilla-La Mancha, Spain

Isabel María Gómez-Barreto
https://orcid.org/0000-0002-0964-0923
University of Castilla-La Mancha, Spain

ABSTRACT

The inclusion of topics related to the postulates of sustainable development has gained special relevance in educational field. The current educational law, LOMLOE, includes principles and values related to the promotion of education for sustainable development pursuing the formation of citizens committed to the construction of a more sustainable world, being essential to involve students from an early age and develop educational actions aimed at addressing from the classroom the current problems and challenges of climate emergency. This chapter focuses on the treatment of environmental sustainability and aims to highlight the importance of addressing

DOI: 10.4018/979-8-3693-7185-5.ch006

Copyright © 2025, IGI Global Scientific Publishing. Copying or distributing in print or electronic forms without written permission of IGI Global is prohibited.

the sustainable development goals in the implementation of the educational curriculum in the 4th year of Primary Education, with the main objective of promoting awareness and sensitivity towards the marine world through the development and design of an innovative educational proposal focused on working the claims of sustainable development goal no. 14 (Life below water).

1. INTRODUCTION

This chapter aims to address the inclusion and treatment of pro-environmental issues based on the approaches of the Sustainable Development Goals (hereinafter, SDGs) in the educational curriculum. For this reason, the main objective of this work is to promote awareness and sensitization towards the marine world through the design of an innovative didactic proposal that is cross-sectional to the Natural, Social and Cultural Knowledge area of the curriculum in the 4th grade of Primary Education. In order to develop this didactic proposal, SDG 14 has been chosen, aimed at the conservation and sustainable use of oceans, seas, and marine ecosystem resources. The choice of this topic is motivated by the notable interest that the integration of these issues in the classroom awakens in students, as well as the current need to influence and act on the problems of this essential part of the planet (O'Flaherty and Liddy, 2018).

The convenience of working on aspects directly linked to sustainability and, in this particular case, the SDGs from an early age is a consequence of the primordial role played by the field of education in the promotion, acquisition and development of responsible values, attitudes and skills in accordance with the requirements of today's society (Cernev and Fenner, 2020). In this sense, it is essential to raise awareness and sensitize future generations about the conservation of natural resources, including the seas and oceans (SDG 14). Primary education becomes an ideal stage that allows the progressive acquisition and internalization of knowledge, behaviors and sustainable habits that will accompany students throughout their lives, contributing to the creation of future citizens with values of respect, responsibility and sustainable use of the environment, establishing themselves as future agents of change and having a great capacity for adaptation and potential for learning.

On the other hand, according to educational legislation, Organic Law 3/2020, of December 29, amending Organic Law 2/2006, of May 3, on Education (LOMLOE) and Decree 81/2022, of July 12, establishing the organization and curriculum of Primary Education in the autonomous community of Castilla-La Mancha (Spain) includes numerous allusions and approaches aligned with sustainability and sustainable development, establishing the need to promote and develop didactic proposals to work on them in the classroom.

In turn, a set of specific objectives that have served as a guide to achieve the main objective are set out, being these: 1) To investigate the importance of education for sustainable development and environmental sustainability in the Primary Education stage; and 2) To design an innovative didactic intervention aimed at addressing SDG 14, 'Life Below Water,' with the objective of promoting environmental sustainability by fostering progressive awareness, consciousness, and responsibility among students. This will be achieved through a set of strategies, resources, and activities planned for development in the area of Natural, Social, and Cultural Environmental Knowledge.

The structure of this work follows a sequential development process, divided into several parts. First, a theoretical framework will be established to address issues related to sustainability and sustainable development, along with their current implications. This will be followed by an exploration of the SDGs and the 2030 Agenda, with a special emphasis on SDG 14, 'Life Below Water.' Additionally, the current relevance of addressing the SDGs from early childhood will be discussed, focusing on the role of education for sustainable development, particularly in Primary Education classrooms. This first part, referring to the theoretical framework, concludes with a review of current legislation on sustainable development. Next, we present the methodology of the didactic intervention and the in-depth development of a didactic innovation proposal aimed at students in the 4th year of Primary Education, which is composed of 16 sessions in which a set of strategies, resources and activities focused on working on the goals established for SDG 14 have been planned.

2. THEORETICAL FRAMEWORK

2.1. Sustainable Development. Conceptualization and Current Implications

Currently, the concepts of sustainability and sustainable development have gained great relevance in relation to the treatment and awareness of environmental issues and are a top priority for action (Sheehy and Farneti, 2021). Sustainability has become a critical issue in today's world as the impact of human activities on the environment has become increasingly evident (Eyring et al., 2021). Climate change, dwindling natural resources and the increasing volume of waste are just a few examples of the challenges we face. The consequences of not addressing these issues are far-reaching and could have devastating results for both people and the planet (Tsalis et al, 2020).

In this regard, numerous supranational organisms highlight the importance and implications of sustainability in today's world (United Nations [UN], 2023). On the one hand, climate change is having an increasingly significant impact on our planet.

The average global temperature has increased by about 1.1 degrees Celsius since pre-industrial times, leading to melting glaciers, rising sea levels and more frequent natural disasters. Natural resources are finite and we are consuming them faster than they can be replenished (Chams and García-Blandón, 2019). Climate change and environmental degradation can damage our health, and sustainable practices can help protect the health and well-being of future generations.

Although the use of this term has experienced a notable increase in recent years, the concept of sustainability is actually not new. Some authors (López, 2015) have pointed out that the emergence of interest in protecting the planet dates back to the early 1970s, and with particular intensity in the 1980s, when different works warned of the need to establish limits to the western development model (Mebratu, 1998). At that time, a reflective process began on a global scale, promoted by the United Nations and the European Union, to address the issue of sustainability and sustainable development on our planet, which became one of the main challenges to be faced by the society of the 21st century. The debate was motivated by a progressive awareness of the limitations and incompatibilities between economic growth and environmental protection and the possibilities of maintaining the current level of growth of countries with more developed economies. As such, the first appearance of the concept of sustainable development occurred in the 1983 Brundtland Commission, whose report "Our Common Future" published by the World Commission on Environment and Development, defined the concept of sustainable development as "development that meets the needs of the present without compromising the ability of future generations to meet their own needs" (WCED, 1987, p. 8). The report warned about the negative impacts of economic development on the environment with the aim of finding possible solutions to the problems caused by industrialization and population growth.

The concepts of sustainability and sustainable development are often referred to and used interchangeably, although it is worth mentioning that both concepts, although similar at first glance, are different and should not be confused as their implications and purposes are slightly different (Olawumi and Chan, 2020; Ruggerio, 2021). On the one hand, considering the contributions of these authors, sustainability is about finding methods to ensure that the decisions and actions we take now do not have a detrimental influence in the future, implying the ability to maintain or preserve the conditions necessary for continued existence or success over time. Given the current rate of resource depletion and the growing concern about how human activity affects the environment, this idea is crucial. The concept of sustainability is composed of three fundamental pillars: environmental, economic and social, arguing that in order to achieve true sustainability it is necessary to balance the factors of economic, social and environmental sustainability in equal harmony (Murga-Menoyo and Novo, 2017).

In the environmental context, sustainability involves interacting responsibly with the planet to maintain sustainable development, aiming to preserve natural resources and biodiversity, maintain ecological systems in balance, reduce pollution and greenhouse gas emissions, among others. Economic sustainability involves ensuring that communities can maintain their independence, access the resources they need and secure sources of livelihood, while social sustainability focuses on creating communities that are fair, diverse, connected and offer a high quality of life. It also ensures that future generations benefit from at least the same quality of life that we enjoy, which means protecting universal human needs, well-being and rights (Boeve-de Pauw et al., 2015).

Sustainable development, on the other hand, is a broader concept that encompasses sustainability. Achieving sustainability is the primary goal of sustainable development, whose main objective is to achieve an optimal quality of life for all people, respecting the environment without abusing the limited resources that the planet offers us (Sachs et al., 2022). In other words, while sustainability is an ultimate goal, sustainable development is a process to achieve that goal. So much so that it has become the conceptual basis for theoretical approaches such as the green economy and the circular economy (Suárez-Eiroa et al., 2019). It has even become part of the common sense of a large proportion of the world's population and of political slogans of environmental advocacy (Alvarado-Herrera et al., 2017).

2.2. The Role of Education in Sustainable Development

As Barnosky and Hadly (2016) point out, global environmental conditions continue to deteriorate at an alarming rate due to a complex web of socio-ecological challenges that include, among others, climate change, air and water pollution, ocean acidification, land degradation and biodiversity loss. Numerous authors, such as Alacid (2021), argue that the treatment and incorporation of issues related to sustainability should begin in childhood, as this is the time when the development of crucial life competencies and skills begins. The Primary Education stage is a fundamental period to forge the formation and generation of active, responsible and committed citizens, becoming a privileged space to foster a more sustainable culture pursuing the promotion of positive environmental attitudes and values in students towards the care and preservation of nature and the environment that surrounds them (Castro and Renés, 2018).

The current implications and purposes of Sustainable Development represent the idea that it is necessary to pursue a change and transition from the current society to a more respectful and environmentally conscious one, an aspect that involves and concerns different areas, including education. From this premise, different approaches and strategies have emerged aimed at integrating the treatment of environmental issues in the educational curriculum in an intentional and systematic way. In this sense, the Education for Sustainable Development (hereinafter, ESD) proposed by the United Nations (Pauw et al., 2015), is an example of the paradigm shift initiated from the educational field to address current and future problems, as well as to promote an increase in environmental awareness focused on reversing these situations at an early age.

As Pauw et al. (2015) argue, Development Education involves an educational process that increases understanding and awareness of a rapidly and unevenly changing world, while ESD aims for education to instill a responsibility to create and enjoy a sustainable future (UNESCO, 2020). ESD starts as an idea, but in a short time it becomes a global movement (Hopkins, 2012) of importance for the educational field, putting the focus on the challenges to which the planet is subjected and which need to be urgently addressed. The accelerated rhythms of human beings have considerably modified terrestrial ecosystems to the point that our very survival in the future seems to be at risk due to changes that are increasingly costly to reverse. Implicit in this materialization of reality is the idea that in order to halt this environmental drift before it reaches catastrophic levels, environmental, social and economic problems must be addressed in a holistic manner. In this sense, ESD implies:

The inclusion of key issues related to sustainable development in teaching and learning; for example, climate change, disaster risk reduction, biodiversity, poverty reduction and sustainable consumption. It also requires participatory teaching and learning methods that motivate and empower learners to change their behavior and act in favor of sustainable development. Consequently, Education for Sustainable Development fosters competencies such as critical thinking, imagining future scenarios and collaborative decision making. (UNESCO, 2012, p. 2).

On the other hand, environmental education also involves a pedagogical approach, aligned with the claims of sustainable development, which seeks to foster awareness and knowledge about environmental issues, promoting responsible and sustainable behavior aimed at building an environmentally literate citizenry (Ardoin and Bowers, 2020). As such, environmental education fosters commitment to improving environmental sustainability and the effective use of environmental resources, pursuing the development and promotion of pro-environmental skills, values and attitudes to understand current and future challenges (Cernev and Fenner, 2020). Despite the above, authors such as Vilches et al. (2010), argue that environmental education aims to go beyond knowledge and understanding of the environment and the challenges it poses, encompassing a set of claims among which stand out: awareness and sensitization to the environment and the promotion of respect for it; the adoption of proactive attitudes and participation that lead to changes and improvements in the practices carried out; the promotion of the development and monitoring of sustainable habits; the development of problem-solving skills that necessarily involve reflection and critical thinking with subsequent decision making aimed at overcoming or reversing the problems. In this sense, both pedagogical approaches highlight the importance of the educational sphere in generating a basis for environmental sensitivity, interest and behaviors that allow understanding the connection and interdependence between human beings and the environment that surrounds them and the promotion of sustainable practices (Mastrángelo et al., 2019).

To highlight the importance and need to work on ESD in classrooms, previous research has addressed the implementation of educational programs aimed at addressing environmental sustainability issues and examined the effects they have had considering different aspects and areas of sustainable development. The studies of Estrada et al. (2020) and Aibar (2023) analyzed the changes in the development of environmental awareness in a sample of elementary school students following the implementation of an environmental care program and an environmental conservation program. The results indicated that in both cases the students significantly improved their levels of environmental awareness and sensitization, demonstrating the effectiveness of these programs for the improvement and promotion of environmental awareness. In the study developed by Velásquez (2017), the results also revealed that the application of a proposal aimed at improving environmental responsibility in students in the first year of Primary Education was effective and contributed to improving environmental commitment; students belonging to the experimental group significantly improved their levels of environmental responsibility compared to those in the group that did not receive any intervention.

Considering the impact and results obtained from the implementation of programs focused on the treatment of environmental and sustainability issues, the importance of continuing to advance in proposals aimed at achieving improvements in students in the primary education stage is evident, since, as evidenced by Ramos and Torralba (2020), there is a shortage of proposals in the educational field.

2.3. The Sustainable Development Goals and Agenda 2030

In 2000, the United Nations approved the Millennium Declaration, which a year later led to the proposal and adoption of the Millennium Development Goals, also known as MDGs, consisting of eight objectives to be developed between 2000-2015. The establishment of these goals was intended to address a wide variety of problems and challenges that generated concern worldwide, such as overcoming poverty, hunger, education, health or advancing in the change towards a more sustainable development, among others. By 2015, the date on which compliance was established, some progress had been made in reducing extreme poverty, and advances in medicine and education, although in general terms the eight goals had not been achieved as expected. This scenario led to the establishment of new goals and a new agenda for development, thus giving rise to the 2030 Agenda and the SDGs, whose proposals represent a new tool to increase global commitment and entrepreneurship to face the future (Hernando and Dikam, 2016). In this sense, the end of the MDG cycle represented a benchmark for the construction of the current 2030 Agenda (Gil, 2018). This global agenda represents a turning point, being the most ambitious one approved to date by the international community as a whole, and is made up of 17 goals and 169 targets (Sanahuja and Tezanos, 2017).

The SDGs are a set of 17 global goals that were agreed in 2015 to respond to the challenges of the 21st century, in accordance with the objectives set by the European Union and UNESCO for the decade 2020-2030. Being a set of global goals to address challenges such as poverty, inequality, eradicating wars, achieving quality education, caring for the planet and the environment, as well as other aspects of sustainable development (United Nations [UN], 2024). These goals integrate three fundamental dimensions of sustainable development, comprising the economic, social and environmental spheres.

The 2030 Agenda is described as "an Agenda of the people, by the people and for the people, and this will ensure its success" (United Nations, 2015, p. 12), establishing a roadmap and action by responding to five thematic points, known as the "five Ps": planet, prosperity, peace, people and partnership (Chams and García-Blandón, 2019).

The formation of the SDGs was an opportunity for the 193 countries that make up the United Nations and their societies to embark on a new path, protecting and improving the planet and human life. The Goals that make up the agenda, state that poverty eradication must go hand in hand with strategies that foster economic growth and address a range of social needs such as education, health, social protection and employment prospects, while combating climate change and protecting the environment (Gámez, 2019).

The SDGs are so many and so different from each other that, due to the need to address them today, a brief description of the purpose of each is set out in Table 1 below.

Table 1. Sustainable Development Goals and their purpose

Sustainable Development Goal	**Purpose**
No poverty	Ending poverty in all its forms worldwide.
Zero hunger	To end hunger, achieve food security and improved nutrition and promote sustainable agriculture.
Good health and well-being	To ensure healthy living and promote wellness for all at all ages.

continued on following page

Table 1. Continued

Sustainable Development Goal	Purpose
Quality education	Ensure inclusive, equitable and quality education and promote lifelong learning opportunities for all.
Gender equality	Achieving gender equality and empowering all women and girls.
Clean water and sanitation	Ensure the availability of water and its sustainable management and sanitation for all.
Affordable and clean energy	Ensure access to affordable, secure, sustainable and modern energy for all.
Decent work and economic growth	Promote sustained, inclusive and sustainable economic growth, full and productive employment and decent work for all.
Industry, innovation and infrastructure	Building resilient infrastructure, promoting inclusive and sustainable industrialization, and fostering innovation.
Reduced inequalities	Reducing inequality within and between countries.
Sustainable cities and communities	Making cities and human settlements inclusive, safe, resilient and sustainable.
Responsible consumption and production	Ensuring sustainable consumption and production patterns.
Climate action	Adopt urgent measures to combat climate change and its effects.
Life below water	Conserve and sustainably use the oceans, seas and marine resources for sustainable development.
Life on land	Sustainably management of forests, combat desertification, halt and reverse land degradation and halt biodiversity loss.
Peace, justice and strong institutions	Promoting just, peaceful and inclusive societies.
Partnerships for the goals	Revitalize the Global Partnership for Sustainable Development.

Source: United Nations [UN] (2024).

In line with UNESCO (2020), special mention should be made of SDG 4 "quality education", which is considered a key driver for the development of all the other goals. In view of this circumstance, the educational community as a whole acquires a set of responsibilities, among which are the transmission of values and the preparation of students to face such worrying phenomena as climate change, as well as to raise awareness of the environmental difficulties caused by humans to generate a change in people and awaken in them the desire to protect, conserve and maintain nature by proposing solutions to such problems (Canto, 2022). In this sense, environmental education and education for sustainable development approaches aim to involve students and teachers in the teaching-learning process to develop awareness about environmental conservation (Pulido-Capurro and Olivera-Carhuaz, 2018).

2.3.1. Sustainable Development Goal 14 “Life Below Water”

Given the breadth of purposes and goals involved in addressing the issues encompassed by sustainable development, this chapter is designed to have an impact on environmental sustainability, specifically focusing on the claims and implications of SDG 14 “Life below water”. Seas and oceans contain 97% of the total available water and occupy 71% of the planet's surface, so meeting the aforementioned goal is essential. Thus, using the simile given by Cicin-Sain (2015), “just as a person cannot live without a healthy heart and lungs, the Earth cannot survive without healthy oceans and seas” (p. 32).

The choice of this SDG responds to the need to raise awareness about the proper use, preservation and care of water (oceans, seas, rivers and lakes), marine ecosystems, and the species that inhabit them. Therefore, the school context must involve students in this responsibility, instilling from an early age actions and attitudes conducive to the care and protection of the environment in which we live, in order to reduce the impact of pollution and climate change, among others, through quality education. As pointed out by Sánchez et al. (2021), the intentional treatment of climate action and life below water (SDG 13 and SDG 14) will help to form critical citizens willing to change the current reality. In line with the above, it is essential to highlight the leading role of education in raising awareness of the marine environment, as well as in generating consensus and active involvement in the rational and sustainable use of marine resources (Negrín and Marrero, 2021).

As previously introduced, the incorporation of this SDG aims to protect, conserve and sustainably use the oceans, seas and marine resources. It should be emphasized that the oceans are considered a fundamental part of our planet and of vital importance for the survival of life on it, since among their main functions are climate regulation and mitigation of the effects derived as a result of climate change, the contribution of oxygen to the atmosphere, the supply of a relevant source of natural resources and food, constituting the largest ecosystem in the world made up of almost one million known species (United Nations [UN], 2015).

On the other hand, human intervention and activities derived from it exert a great environmental influence and impact that cause negative effects on the oceans, marine ecosystems and their biodiversity. Marine pollution is reaching very high levels, the most harmful of which is plastic. The oceans are also acidifying, endangering the survival of marine species and disrupting food chains. The existing problem is due to the fact that the oceans absorb heat from climate change, causing the temperature of the water to rise, in addition to the large amount of waste found in them. In order to protect marine biodiversity, we must reduce plastic consumption, consume only the food we need and regulate fishing (United Nations [UN], 2024).

Taking into consideration the above, a set of specific goals have been established to achieve this objective considering the various ocean problems, such as preventing and reducing pollution, protecting ecosystems, minimizing acidification, regulating fishing, conserving coastal areas, combating illegal and unregulated fishing, increasing economic benefits from the sustainable use of the sea and research to improve the health of the oceans and facilitate access for small fishermen, and implementing conventions on the law of the sea (United Nations [UN], 2024).

2.4. Current Educative Legislation and Sustainable Development

Taking into account the current need to raise awareness and work on the claims of Sustainable Development, the Organic Law 3/2020, of December 29, which amends the Organic Law 2/2006, of May 3, on Education (hereinafter, LOMLOE) establishes the need to focus on these issues. Among the multiple purposes of the LOMLOE is to promote the inclusion of education for sustainable development and global citizenship as part of the educational curriculum, with the firm purpose of contributing to the formation of citizens committed to achieving a fairer and more sustainable world.

The following is a summary of this law highlighting the most relevant information to work in relation to sustainability considering the environmental aspect. First, the preamble of said law recognizes the importance of addressing sustainable development and its cross-cutting nature in accordance with the provisions of the 2030 Agenda when it refers that Education for sustainable development includes the knowledge, skills, values and attitudes necessary to make decisions and actively participate in the solution of problems shared by all people in the world. It also states that education for sustainable development and ecological transition is essential to respond to the climate crisis, so that students understand the consequences of actions on the planet and develop empathy for their natural and social environment (LOMLOE, 2020).

The LOMLOE promotes principles and values related to education for sustainable development and global citizenship. Some of the key elements in relation to education and sustainability in the LOMLOE are the following:

– Article 1, focused on the educational principles that guide the law, is education for ecological transition with social justice criteria to contribute to economic, social and environmentally sustainable development (LOMLOE, 2020).
– Article 2, in relation to the purposes of the educational system, it is established to achieve values that promote respect for living beings, animal rights and the environment, especially the value of forest areas and sustainable development (LOMLOE, 2020).

– Title IV, about school centers, recognizes that the education system cannot be oblivious to the challenges posed by global climate change. Schools must become a place of stewardship and care for our environment in the different educational stages (LOMLOE, 2020).
– Articles 19 (pedagogical principles) and 121 (educational project): cross-sectionality is raised at the time of developing education for responsible consumption and sustainable development, and to include the cross-sectional treatment of sustainability in the educational projects of the center (LOMLOE, 2020).

Likewise, the inclusion of environmental topics has been reviewed considering the provisions of Decree 81/2022, of July 12, which establishes the organization and curriculum of Primary Education in the autonomous community of Castilla-La Mancha. Based on what is established in said curricular concretion at the autonomous level, we find multiple references related to sustainable development. In this sense, it should be highlighted that in article 6 (pedagogical principles), sustainable development is promoted as a cross-sectional element that appears in the curriculum permeating all educational work from any area; Article 7, which establishes the general objectives for the Primary Education stage, establishes an objective directly linked to the treatment of environmental issues and sustainable development:

ñ) To know the limits of the planet in which they live and the means available to them to ensure that resources prevail in time and space for as long as possible, abandoning the linear economy model followed so far and acquiring habits of conduct and knowledge of a circular economy. (Decree 81, 2022, p. 5)

Regarding the achievement of the key competences, which are the set of performances considered essential so that students can progress, with guarantees of success, in their training itinerary and face the main global and local challenges, we can find evident links with the following:

– Mathematical competence and competence in science, technology and engineering which implies understanding the world with scientific methods to transform the environment in a decisive, responsible and sustainable way.
– Citizenship competency as it helps students to carry out responsible citizenship and to actively commit to sustainable development and to the achievement of global citizenship.

The above has shown that the LOMLOE introduces into the Spanish education system the spirit of the 2030 Agenda, and with it the importance and unavoidable need for education (SDG 4) to play a central role in promoting sustainable development. Although it is true that the legislation does not incorporate the SDGs explicitly in its structure, it promotes environmental education and awareness for sustainability that is aligned with the values and principles of sustainable development established by the United Nations.

Since the present work is focused on the specific treatment of environmental sustainability and, specifically, centered on the approaches of SDG 14 "Life below water", we will now proceed to make a table extracting the relevant information from the curriculum in the area of Natural, Social and Cultural Environment Knowledge for the second cycle of Primary Education, which will allow us to detect those aspects of sustainable development that can be worked on in the 4th grade. Thus, Appendix 1 shows the interrelation of the different curricular elements: specific competences, evaluation criteria and basic knowledge, as well as their link with the operative descriptors of the competences.

However, it should be noted that, despite focusing on the specific area of Natural, Social and Cultural Environment Knowledge for the realization of this work because it presents a more direct link, given the cross-cutting nature of the subject matter in question and the globalizing approach that characterizes the teaching, SDG 14 could be worked from any area of knowledge of the curriculum and even involving interdisciplinarily several of them.

After highlighting the information extracted from both Organic Law 3/2020, of December 29, which modifies Organic Law 2/2006, of May 3, on Education, and Decree 81/2022, of July 12, which establishes the organization and curriculum of Primary Education in the autonomous community of Castilla-La Mancha, the importance of working on sustainability issues in the classroom becomes evident, as a consequence of the great climate emergency that we are currently experiencing and the need to act in order to reverse this situation.

3. METHODOLOGY

The modality followed in the development of this chapter is of a professional nature, consisting of the elaboration of a didactic innovation proposal with the approach and design of a complete and contextualized intervention. Some authors such as Pascual and Navío (2018) point out that educational innovation implies introducing a change to achieve improvements in educational processes, highlighting that in the classrooms it is necessary to work actively to generate a pedagogy of change that favors student learning. Rieckman (2018) establishes that, for ESD to be effective, the use of active methodologies is needed. The present work includes an in-depth practical design, planned to be implemented in a classroom belonging to the 4th year of Primary Education in the area of Natural, Social and Cultural Environment Knowledge. The aforementioned proposal focuses on the treatment of contents and topics related to environmental sustainability and specifically focused on the marine environment and ecosystems, based on the purposes set forth in Sustainable Development Goal 14 “life below water”, which seeks the conservation and use of oceans, seas and marine resources in a sustainable manner.

In order to determine the most appropriate way of pedagogical action in the classroom, it is important to know the peculiarities and psycho-pedagogical characteristics of the students to whom it is addressed. Placing the students in the stage of development in which they find themselves allows focusing and adapting the teaching action based on their needs, interests and learning styles. Therefore, the elaboration of the didactic proposal has been designed considering the psycho-evolutionary characteristics of the students of the fourth grade of Primary Education, which are determined by evolutionary features that affect the cognitive, socio-affective and psychomotor development of the children.

– Cognitive development: characterized by the evolution of thinking and the development of logical reasoning as a consequence of the progressive acquisition of skills such as conservation and classification, overcoming egocentrism and animism. Likewise, they carry out processes of analysis and abstraction, managing to solve problems of greater complexity. In turn, there is also an evolution of language linked to their greater capacity for imagination and creativity (Martín and Navarro, 2015).

– Affective and social development: characterized by the important role acquired by peer relationships and social norms, resulting in the construction of the basis of their personality and ethical system. The development of autonomy poses a greater protagonism of the students in their learning process, which can be exploited through the approach of active methodologies. In this sense, they show a greater interest in working and collaborating in groups, an ideal moment to implement cooperative learning dynamics (Pujolas, 2004). Finally, the social sphere is of special importance, a crucial moment in which to work on values education and student behavior (Puig and Martín, 2015).

– Motor development: characterized by the progressive evolution of movements, being more capable of performing in more complex and precise situations. There are improvements in body control, developing gross and fine psychomotor skills (Rigal, 2006), as well as the expansion of the spatial plane of students.

4. PROPOSAL DESIGN

The following section includes the development of the didactic innovation proposal designed under the title “Learning to take care of the environment”.

4.1. Justification

Once the importance of developing the SDGs in Primary Education classrooms has been presented, the need for an innovative proposal to improve attitudes and behaviors in terms of environmental sustainability considering aquatic ecosystems and the problems derived from them becomes extremely important. To this end, the following set of strategies, resources and activities focused on working on SDG 14 "life below water" is proposed.

Throughout the proposal, a set of activities are included that seek to generate in students an interest in caring for the environment, respect for marine biodiversity, recycling awareness, knowledge of reuse methods, among other issues. The set of proposed activities that make up the battery is divided into three clearly defined phases:

– First phase (Awareness): includes activities that aim to make students aware of an objective reality and to make the relevance of the issue latent. It implies that students become aware and acquire clear and reflective knowledge of reality.
– Second phase (Sensitization): includes activities that seek to generate an emotional impact. They consist of ensuring that students reach the ability to value something or respond emotionally to it moved by feelings.
– Third phase (Action-taking): includes activities aimed to foster a change of behavior in students and to get them actively and participatively involved in making decisions to put them into practice.

4.2. Didactic Objectives of the Proposal

This proposal aims to achieve the following didactic objectives in the students, organized according to the three previously established phases. Regarding the first level (awareness-raising), the objectives are: (1) to learn about Sustainable Development Goal 14, its main functions and purposes; (2) to analyze and reflect on the importance of marine sustainability, fostering critical thinking and environmental awareness; (3) to understand the importance and role of marine ecosystems and the need to preserve the oceans for human life; (4) to learn about the diversity of aquatic ecosystems, through the identification of various marine species and habitats; and, finally, (5) to raise awareness of the problems derived from marine pollution, the main sources of pollution and its effects on life below water.

On the other hand, the objectives of the second level (raising awareness) are: (6) to develop attitudes of respect and care towards the marine environment; (7) to reflect on the main dangers suffered by life below water (loss of biodiversity, degradation of natural habitats, endangered animals...); and (8) to explore interactions with the ocean, the influence of negative human activities and actions in affecting marine ecosystems and promote critical and reflective thinking about possible solutions.

Finally, the third level (action taking) objectives are: (9) to promote environmental responsibility through proactive participation for the conservation of marine ecosystems; (10) to promote sustainable habits and practices with the seas, oceans and the resources derived from them by making responsible decisions; (11) to acquire habits and customs that contribute to responsible management of the waste generated.

4.3. Description of the Proposal

The proposal is planned to be carried out through the use of active methodologies in which students can participate during the teaching-learning process and become the main protagonist and responsible for it. Specifically, cooperative learning is proposed as the main methodology, thus provoking the union of the group-class and the participation of all members as a whole, promoting learning from each other and enhancing learning outcomes (Juárez-Pulido et al., 2019). The didactic innovation proposal has been planned to be developed in the area of Natural, Social and Cultural Environment Knowledge, based on the treatment of basic knowledges and specific competences directly involved with environmental sustainability, selected after the educational curriculum review process (see Appendix 1).

4.4. Timing

The timeline for this proposal is designed to span approximately two months. It is structured into a total of 16 sessions, each lasting 45 minutes. Since this is a cross-curricular topic, it will be addressed in two weekly sessions throughout the school year.

4.5. Proposal Development

The didactic proposal is detailed from Table 2 to Table 14 of this section. The tables include the set and variety of strategies, resources and materials, student groupings and the breakdown of activities designed and proposed to promote comprehensive and meaningful learning in students. Each of the sessions, and the activities that comprise them, are linked to the achievement of the specific didactic objectives previously defined, which in turn are associated with a set of achievement indicators designed to evaluate the progress and advances of the students in each of the aspects that are addressed. Likewise, the specific topics worked on in each session are linked to the curricular elements previously reviewed and selected (available in Appendix 1).

Table 2. Development of the didactic proposal: session 1.

SESSION 1. WHERE ARE THE FISH?			
Curricular linkage	Basic knowledges: Block C. Societies and territories. 4. Ecosocial awareness. - Climate change. Introduction to the causes and consequences of climate change. - The transformation and degradation of natural ecosystems due to human action.		
	Specific competencies: 6	**Evaluation criteria:** 6.1	
Didactic objectives: 1, 2, 3 and 5	**Timing:** 45 minutes.	**Spaces:** classroom	
	Groupings: As a whole the entire class, except where specified.		
	Materials: story "Where are the fish?", index cards.		
Description of activities			
Activity 1. What do they know? (10')	As an initial test, each student will take a sheet of paper and individually answer the questions: How are the seas? Does a dirty sea make marine survival difficult? What is the sea for? What do you know about sustainable development goal 14?		
Activity 2. Story (15')	The story *Where are the fish?* is read. This story is about a fisherman who goes to the sea and catches everything except fish, giving visibility to the amount of garbage in the sea. With this story you can work on pollution, which is what SDG 14 intends. It also serves to reflect on environmental awareness and gives rise to talking about the importance of the ocean for human beings.		
Activity 3. Reflection (20')	Taking advantage of the number of things that the story allows to work on, reflections are made by answering questions such as, Is SDG 14 necessary? Why? How do you think all that garbage reaches the ocean? Does it really happen or is it just fiction? What benefits do you think the ocean brings to humans? Are we trained to take care of the seas? How can they be taken care of? Is it possible to meet the goal until 2030? Among others that arise as a result of the students' interventions.		
Evaluation			
Indicators	**Achieved**	**Not achieved**	**In process (observations)**
To recognize the purposes of SDG 14.			
To analyze the exposed environmental situation.			
To understand the role of the oceans for human life.			
To know marine pollution.			

Source: own elaboration.

Table 3. Development of the didactic proposal: session 2.

SESSION 2. MARINE SPECIES			
Curricular linkage	Basic knowledges: Block A. Scientific culture. 2. Life on our planet. - Characteristics of animals that allow their classification. - Ecosystems as a place where factors intervene. - Relationship between human beings and ecosystems.		
	Specific competences: 5	**Evaluation criteria:** 5.1, 5.2	
Didactic objectives: 3, 4, 7 and 8	**Timing:** 45 minutes.	**Sapces:** classroom.	
	Groupings: Groups of 4 except where specified.		
	Materials: digital images. **Resources:** computers.		
Description of activities			
Activity 1. Brainstorming (5´)	The students are asked what marine animals they know and a list is made on the blackboard. Together, a relatively long list should come out, with approximately one species per student.		
Activity 2. Search (30')	They are placed in groups of 4 people and each group chooses 4 species from those written on the board. If there are not enough, some can be repeated. They must search for information on the computer about the species they have chosen, areas where they live, danger of fishing or not, danger of extinction or not, a brief description of each one and their food chain, all of this must be written in a technical sheet.		
Activity 3 (10´) Creating a website	A website will be created featuring the different animals that have been investigated with the information that the students themselves have gathered so that it can be consulted quickly and everything is organized and classified.		
Evaluation			
Indicators	**Achieved**	**Not achieved**	**In process (observations)**
To know the role of marine ecosystems.			
To recognize the diversity of marine species.			
To reflect on the dangers of marine diversity.			
To understand human damage to marine diversity.			

Source: own elaboration.

Table 4. Development of the didactic proposal: session 3.

SESSION 3. IMAGES			
Curricular linkage	Basic knowledges: Block C. Companies and territories. 4. Ecosocial awareness. - The transformation and degradation of natural ecosystems due to human action. Conservation and protection of nature.		
	Specific competences: 6	**Evaluation criteria:** 6.1	
Didactic objectives: 2, 5, 7 and 8	**Timing:** 45 minutes.	**Spaces:** classroom.	
	Groupings: Groups of 4 except where specified.		
	Materials: digital images (anexo 4). **Resources:** projector.		
Description of activities			
Activity 1. Viewing (25')	Different images that show marine pollution scenarios will be selected by the teacher. Each group must analyze them, extracting a small synthesis to understand the current problem of garbage in the ocean.		
Activity 2. Podcast (15')	In groups they must make a short podcast lasting approximately a minute and a half, in which they must talk in general terms about the current problems found in the marine ecosystem. Once they have recorded it, a folder will be generated on Spotify where they will all be published. together so that everyone has access to them.		
Activity 3. Listening (5´)	All podcasts will be played in class, so everyone will be heard and a little feedback can be given to each group about what is good and what can be improved.		
Evaluation			
Indicators	**Achieved**	**Not achieved**	**In process (observations)**
To analyze the problems of the marine ecosystem.			
To know the main sources of marine pollution.			
To reflect on the dangers of the marine biosphere.			
To explore negative human actions for the marine ecosystem.			

Source: own elaboration.

Table 5. Development of the didactic proposal: session 4.

<table>
<tr><th colspan="4">SESSION 4. NEWSCAST</th></tr>
<tr><td rowspan="2">Curricular linkage</td><td colspan="3">Basic knowledges: Block C. Companies and territories.
4. Ecosocial awareness.
- The transformation and degradation of natural ecosystems due to human action. Conservation and protection of nature.</td></tr>
<tr><td>Specific competences: 6</td><td colspan="2">Evaluation criteria: 6.1</td></tr>
<tr><td rowspan="3">Didactic objectives: 1, 2 and 3</td><td>Timing: 45 minutes.</td><td colspan="2">Spaces: classroom.</td></tr>
<tr><td colspan="3">Groupings: Groups of 4 except where specified.</td></tr>
<tr><td colspan="3">Materials: cardboard, scissors. Resources: projector, video (link).</td></tr>
<tr><td colspan="4">Description of activities</td></tr>
<tr><td>Activity 1. Viewing (5')</td><td colspan="3">It starts with watching the video. This video is a recapitulation of what students must have learned from the previous sessions.</td></tr>
<tr><td>Activity 2. The argument (30')</td><td colspan="3">After watching the video, they must invent a story in which they say that they have gone to the beach and the reality that they have encountered and they must propose a solution to remedy it.</td></tr>
<tr><td>Activity 3. We break the news (10')</td><td colspan="3">Each group chooses a member of the group to, with the cardboard previously cut into a frame, simulate television and the news that all the group participants have written.</td></tr>
<tr><td colspan="4">Evaluation</td></tr>
<tr><td>Indicators</td><td>Achieved</td><td>Not achieved</td><td>In process (observations)</td></tr>
<tr><td>To recognize the purposes of SDG 14.</td><td></td><td></td><td></td></tr>
<tr><td>To analyze the problems of the marine ecosystem.</td><td></td><td></td><td></td></tr>
<tr><td>To understand the importance of the oceans for humans.</td><td></td><td></td><td></td></tr>
</table>

Source: own elaboration.

Table 6. Development of the didactic proposal: session 5.

<table>
<tr><th colspan="4">SESSION 5. VISIT FROM AN EXPERT</th></tr>
<tr><td rowspan="2">Curricular linkage</td><td colspan="3">Basic knowledges: Block A. Scientific culture.
2. Life on our planet.
– The functions and services of ecosystems.
– Relationship between human beings and ecosystems.</td></tr>
<tr><td>Specific competences: 5</td><td colspan="2">Evaluation criteria: 5.1</td></tr>
<tr><td rowspan="3">Didactic objectives: 3, 4, 7 and 8</td><td>Timing: 45 minutes.</td><td colspan="2">Spaces: classroom.</td></tr>
<tr><td colspan="3">Groupins: the entire class as a whole.</td></tr>
<tr><td colspan="3">Materials: depending on the manager. Human resource: guest.</td></tr>
<tr><td colspan="4">Description of activity</td></tr>
<tr><td>Activity 1. Expert (45')</td><td colspan="3">An expert in the field is invited to work, in this case, he is an expert in marine life, he knows a large part of the species that inhabit the oceans and seas, he knows their relationships between them and of course the danger they face today. in day. You have the entire session to work however you want. He usually asks what they know during the first few minutes, then goes on to share his knowledge in an attractive way and finally plays a game with the students.</td></tr>
<tr><td colspan="4">Evaluation</td></tr>
<tr><td>Indicators</td><td>Achieved</td><td>Not achieved</td><td>In process (observations)</td></tr>
<tr><td>To understand the role of marine ecosystems.</td><td></td><td></td><td></td></tr>
<tr><td>To identify different marine species.</td><td></td><td></td><td></td></tr>
<tr><td>To reflect on the dangers suffered by the marine ecosystem.</td><td></td><td></td><td></td></tr>
<tr><td>To explore negative human actions in the marine ecosystem.</td><td></td><td></td><td></td></tr>
</table>

Source: own elaboration.

Table 7. Development of the didactic proposal: sessions 6 and 7.

SESSIONS 6 AND 7: STORY COMPETITION			
Curricular linkage	Basic knowledges: Block A. Scientific culture. 2. Life on our planet. - The kingdoms of nature from a general perspective. - Ecosystems as a place where factors intervene. - Relationship between human beings and ecosystems. Block C. Companies and territories. 4. Ecosocial awareness. - Climate change. Introduction to the causes and consequences of climate change. - Ecosocial responsibility		
	Specific competences: 5 and 6	**Evaluation criteria:** 5.3 and 6.1	
Didactic objectives: 6, 7 and 8	**Timing:** 2 sessionss of 45 minutes each.	**Spaces:** classroom.	
	Groupings: Individually except where specified.		
	Materials: story "Where are the fish?". **Resources:** computer.		
Description of activities			
Activity 1. Session 6. Story (45')	A reminder is given to the entire class, with a primary focus on the story Where Are the Fish?, as it sets the direction they should follow. Additionally, the objectives of SDG 14 are reviewed. Each student individually begins to write their own story, inspired by the story mentioned above, being able to use everything that has been worked on during the interventions. We discuss how they carry out the story, to get a general idea of whether they should be allowed more time for creation or, on the contrary, they should finish it at home.		
Activity 1. Session 7. Illustrations (40')	With the cutout.pro application, the illustrations for the story are created. Each student will have a computer and they must make illustrations to give more life and meaning to the story, finally everything will be united in the same document.		
Activity 2. Session 7. Printing (5')	All documents are sent to be printed and bound, and placed in the classroom library so that all classmates can read them.		
Evaluation			
Indicators	**Achieved**	**Not achieved**	**In process (observations)**
To develop attitudes of respect for the marine environment.			
To reflect on the dangers that marine life suffers.			
To know the interactions with the marine ecosystem environment.			

Source: own elaboration.

Table 8. Development of the didactic proposal: sessions 8 and 9.

SESSIONS 8 AND 9. WE RECYCLE			
Curricular linkage	Basic knowledges: C. Societies and territories. 4. Ecosocial awareness. - Ecosocial responsibility. - Conservation and protection of nature.		
	Specific competences: 6	**Evaluation criteria:** 6.1	
Didactic objectives: 6, 9 and 11	**Timings:** 2 sessions of 45 minutes each.	**Spaces:** classroom, except where specified.	
	Groupings: It is specified in each activity.		
	Materials: cardboard boxes, cardboard, colors, collected waste, etc.		
Description of activies			
Activity 1. Session 7. How do we classify? (10')	As a whole, we talk about the different waste in the class, in this way we know if the students are familiar with the different types of containers that exist.		
Activity 2. Session 7. We create containers (35')	In small groups of 4 people, the different containers begin to be created. Each group will be in charge of a different container, it is free, each group can make the container as they see fit, respecting the colors that correspond to them.		
Activity 1. Session 8. What do we have? (fifteen)	With gloves and together with the whole class, the waste that was obtained on the beach in complementary activity 2 will be distributed, each group of those who formed the containers will take a bag.		
Activity 2. Session 8. We classify. (25')	The time comes to recycle, each group with its bag will classify the different waste, the rest of the class observes to also learn where everything goes.		
Activity 3. Session 8. Distribution (5')	Once the classification is finished, the containers are taken and taken to the center hall, so that all students can use them.		
Evaluation			
Indicators	**Achieved**	**Not achieved**	**In process (observations)**
To develop caring attitudes towards the marine environment.			
To participate in cleaning the marine ecosystem.			
To acquire responsible waste management habits.			

Source: own elaboration.

Table 9. Development of the didactic proposal: session 10.

SESSION 10. THE 3 Rs			
Curricular linkage	Basic knowledges: C. Societies and territories. 4. Ecosocial awareness. - Ecosocial responsibility. - Conservation and protection of nature. - Sustainable lifestyles. Responsible consumption and production.		
	Specific competences: 6	**Evaluation criteria:** 6.1	
Didactic objectives: 6 and 11	**Timing:** 45 minutes.	**Spaces:** classroom.	
	Groupings: Grouping of the entire class, except where specified.		
	Resources: projector, computer, photography camera.		
Description of activities			
Activity 1. Do you reuse? (15´)	In this activity the 3 "Rs" (reuse, reduce and recycle) are explained. Recycling has been worked on before, so this activity focuses on the importance of reducing and reusing.		
Activity 2. Game (15´)	In small groups they must order where they think each waste goes. Finally, together with the entire class, the game will be solved, in this way the errors it has are known and if the concept is clear.		
Activity 3. Photo-essay (15´)	To end the session, a photo-essay will be created. First, images of all the poorly recycled, reused and reduced garbage will be posted, and at the end, images will be posted showing how to do it correctly. The images can either be made by themselves or found online. AI, Google, etc.		
Evaluation			
Indicators	**Achieved**	**Not achieved**	**In process (observations)**
To develop attitudes of care for the environment.			
To acquire habits that contribute to correct waste management.			
To acquire habits that contribute to responsible waste management.			

Source: own elaboration.

Table 10. Development of the didactic proposal: session 11.

<table>
<tr><th colspan="4">SESSION 11. WE REUSE</th></tr>
<tr><td rowspan="2">Curricular linkage</td><td colspan="3">Basic knowledges: C. Societies and territories.
4. Ecosocial awareness.
- Ecosocial responsibility
- Conservation and protection of nature.
- Sustainable lifestyles. Responsible consumption and production.</td></tr>
<tr><td>Specific competences: 5 and 6.</td><td colspan="2">Evaluation criteria: 5.3 and 6.1</td></tr>
<tr><td rowspan="3">Didactic objectives: 6 and 11</td><td>Timing: 45 minutes.</td><td colspan="2">Spaces: classroom.</td></tr>
<tr><td colspan="3">Groupings: individual.</td></tr>
<tr><td colspan="3">Materials: socks, buttons, pieces of foam, cloth rags, wool, glue, etc.</td></tr>
<tr><td colspan="4">Description of activities</td></tr>
<tr><td>Activity 1. Story (5')</td><td colspan="3">The story that was created in previous sessions is reread individually, this is to remember the characters that made it up and to be able to make the puppets.</td></tr>
<tr><td>Activity 2. Puppets (40')</td><td colspan="3">With an old sock that is no longer used, either because it is torn, or because the rubber no longer fits or it has become too small, we are going to make our puppets for a theater. The puppets will be the characters from the story previously prepared and the way to create it is free, the objective of the activity is for them to learn to reuse and give things a second life.</td></tr>
<tr><td colspan="4">Evaluation</td></tr>
<tr><td>Indicators</td><td>Achieved</td><td>Not Achieved</td><td>In process (observations)</td></tr>
<tr><td>To develop attitudes of respect for the environment.</td><td></td><td></td><td></td></tr>
<tr><td>To acquire habits that help responsible waste management.</td><td></td><td></td><td></td></tr>
<tr><td>To acquire habits that collaborate with responsible waste management.</td><td></td><td></td><td></td></tr>
</table>

Source: own elaboration

Table 11. Development of the didactic proposal: session 12.

<table>
<tr><th colspan="4">SESSION 12. STORIES</th></tr>
<tr><td rowspan="2">Curricular linkage</td><td colspan="3">Basic knowledges:
Block A. Scientific culture. 2. Life on our planet.
- The kingdoms of nature from a general perspective.
- Ecosystems as a place where factors intervene.
- Relationship between human beings and ecosystems.
Block C. Companies and territories. 4. Ecosocial awareness.
- Climate change. Introduction to the causes and consequences of climate change.
- Ecosocial responsibility.</td></tr>
<tr><td>Specific competences: 5 and 6</td><td colspan="2">Evaluation criteria: 5.2 and 6.1</td></tr>
<tr><td rowspan="3">Didactic objectives: 3, 5, 7 and 8</td><td>Timing: 45 minutes.</td><td colspan="2">Spaces: classroom</td></tr>
<tr><td colspan="3">Groupings: Small groups, except where specified.</td></tr>
<tr><td colspan="3">Materials: puppet theater, puppets, story.</td></tr>
<tr><td colspan="4">Description of activities</td></tr>
<tr><td>Activity 1. Placement (5')</td><td colspan="3">They are placed in small groups, but individually each student prepares his puppets and his story. Blank sheets of paper are distributed so that in groups they can evaluate their partner about their story and their puppets. The entire group must agree. This procedure is so that they are attentive and do not get distracted.</td></tr>
<tr><td>Activity 2. Stories (35')</td><td colspan="3">We proceed to listen to all the classmates' stories, everyone will use the same puppet theater, but using their puppets, each story will last approximately one minute. They must evaluate once the partner finishes the story while preparing the next one, quickly and dynamically.</td></tr>
<tr><td>Activity 3. Evaluation (5')</td><td colspan="3">During the last minutes the notes will be distributed, not only those of the classmates will be taken into account, but it is a way to involve them in the process and give them their importance.</td></tr>
<tr><td colspan="4">Evaluation</td></tr>
<tr><td>Indicators</td><td>Achieved</td><td>Not achieved</td><td>In process (observations)</td></tr>
<tr><td>To understand the need to conserve the oceans.</td><td></td><td></td><td></td></tr>
<tr><td>To know the problems of marine pollution.</td><td></td><td></td><td></td></tr>
<tr><td>To reflect on the dangers that marine life suffers.</td><td></td><td></td><td></td></tr>
<tr><td>To promote critical thinking regarding solutions to remedy marine problems.</td><td></td><td></td><td></td></tr>
</table>

Source: own elaboration.

Table 12. Development of the didactic proposal: sessions 13, 14, 15 and 16.

<table>
<tr><th colspan="3">SESSIONS 13, 14, 15 AND 16. AWARENESS CAMPAIGN</th></tr>
<tr><td rowspan="2">Curricular linkage</td><td colspan="2">Basic knowledges: Block A. Scientific culture. 2. Life on our planet.
- The kingdoms of nature from a general perspective.
- Ecosystems as a place where factors intervene.
- Relationship between human beings and ecosystems.
Block C. Companies and territories. 4. Ecosocial awareness.
- Climate change. Introduction to the causes and consequences of climate change.
- Ecosocial responsibility.</td></tr>
<tr><td>Specific competences: 5 y 6</td><td>Evaluation criteria: 5.1, 5.2, 5.3 y 6.1</td></tr>
<tr><td rowspan="3">Didactic objectives: 2, 5, 7 and 8</td><td>Timing: 4 sessions of 45 minutes each.</td><td>Spaces: Classroom, except where specified.</td></tr>
<tr><td colspan="2">Groupings: Groups of 4 except where specified.</td></tr>
<tr><td colspan="2">Materials: specified in each activity.</td></tr>
<tr><td colspan="3">Description of activities</td></tr>
<tr><td>Activity 1. Session 13. Recapitulation (10')</td><td colspan="2">Together, the entire class is reminded of what we learned in the oceanografic and what we have learned throughout the sessions.</td></tr>
<tr><td>Activity 2. Session 13. Thoughts (35')</td><td colspan="2">In this activity, groups are formed, which will be 4 people. They must agree and write the dialogue of which the awareness campaign is going to be, they must decide the place where they are going to record and the material they must prepare in order to think about the materials they need.</td></tr>
<tr><td>Activity 1. Session 14. Delivery (5')</td><td colspan="2">The dialogue is given to the teacher to correct it and thus have it ready to start recording the next day.</td></tr>
<tr><td>Activity 2. Session 14. Preparation (3')</td><td colspan="2">Each group, with the materials brought from home, must get to work and prepare the materials they need for the recording. Free materials, from cardboard, rubber foam, to whatever you want to use.</td></tr>
<tr><td>Activity 3. Session 14. Sharing (5')</td><td colspan="2">All groups briefly explain what they are going to do, how they are going to distribute the roles and the place where they will record, to be clear.</td></tr>
<tr><td>Activity 1. Session 15. Organization (5')</td><td colspan="2">The places where they will record are organized and each group goes to their chosen place, taking into account that they cannot disturb any other class.</td></tr>
<tr><td>Activity 2. Session 15. Recording (30')</td><td colspan="2">Each group records the campaign they have managed during the previous sessions. The materials they need are a video camera, the script and the materials they have previously prepared.</td></tr>
<tr><td>Activity 3. Session 15. Assembly (10')</td><td colspan="2">With the computers, each group edits or assembles the videos if necessary to have it ready; if they do not finish it in this time, they will have to finish it at home.</td></tr>
<tr><td>Activity 1. Session 16. Sending (5')</td><td colspan="2">All groups must send the videos to the teacher to be able to view them.</td></tr>
<tr><td>Activity 2. Viewing (35')</td><td colspan="2">The videos are watched one by one, the groups must evaluate each other and even themselves. The way to evaluate yourself is individually, on a piece of paper write as objectively as possible the strengths, weaknesses and a numerical grade based on this to the group in general.</td></tr>
<tr><td>Activity 3. Distribution of notes (5')</td><td colspan="2">The notes that have been previously written on a piece of paper are distributed so that each group knows their weaknesses and strengths. The way they are distributed is by giving each group the papers anonymously.</td></tr>
</table>

continued on following page

Table 12. Continued

SESSIONS 13, 14, 15 AND 16. AWARENESS CAMPAIGN			
Evaluation			
Indicators	**Achieved**	**Not achieved**	**In process (observations)**
To know Sustainable Development Goal 14.			
To recognize the problems of marine pollution.			
To promote environmental responsibility.			
To promote sustainable habits with the marine environment.			

Source: own elaboration.

At the same time, considering the theme of the work in question, Tables 13 and 14 propose the implementation of two complementary activities with the objective of reinforcing and enriching the learning of the students, connecting with their interests and applying the contents of the curriculum worked on in contexts outside the classroom.

Table 13. Development of the didactic proposal: complementary activity 1.

COMPLEMENTARY ACTIVITY 1: OCEANOGRAFIC VISIT		
Curricular linkage	Basic knowledges: Block A. Scientific culture. 2. Life on our planet. - The kingdoms of nature from a general perspective. - Ecosystems as a place where factors intervene. - Relationship between human beings and ecosystems.	
	Specific competences: 5	**Evaluation criteria:** 5.1
Didactic objectives: 1, 3, 4 and 7	**Timing:** complete day,	**Spaces:** Oceanografic Valencia
	Groupings: According to Oceanografic instructions.	
	Materials: -.	

continued on following page

Table 13. Continued

COMPLEMENTARY ACTIVITY 1: OCEANOGRAFIC VISIT	
Description of activity	
Activity 1: Visit	The visit is made to the Oceanografic of Valencia, they will be with the educators who work there for a duration of 1 hour and 15 during the visit they will take a tour of the most relevant facilities, they will walk through the lagoon of the facility getting to know the biodiversity, and they will enter the aquarium . They will also travel through the Mediterranean learning about the environments and how to protect animals from human threats. They will go to the poles, where they will meet the penguins and their ecosystems and discover how their environment is changing. Finally, they will take a trip for change., learning how the current problem can be changed. During their visit, the students must take photographs to later use them in the awareness campaign activity.

Source: own elaboration.

Table 14. Development of the didactic proposal: complementary activity 2.

COMPLEMENTARY ACTIVITY 2: GARBAGE COLLECTION		
Curricular linkage	Basic knowledges: Block C. Companies and territories. 4. Ecosocial awareness. - Climate change. Introduction to the causes and consequences of climate change. - Ecosocial responsibility	
	Specific competences: 6	**Evaluation criteria:** 6.1
Didactic objectives: 9, 10 and 11	**Timing:** One hour.	**Spaces:** Beach.
	Groupings: in couples.	
	Materials: gloves, garbage bags, hat, sunscreen.	
Description of activity		
Activity 1: Collection	Taking advantage of the visit to the Oceanografic, when it is over, we change course towards the beach, there we separate into pairs and each pair is given a garbage bag and a pair of gloves for each one, the activity consists of collecting the garbage that If we find it lying around, firstly, we will realize how much trash there is and secondly it will help us to classify it later in class by learning the correct place for each waste found.	

Source: own elaboration.

4.6. Proposal Evaluation

In accordance with the provisions of Decree 81/2022, of July 12, which establishes the organization and curriculum of Primary Education in the autonomous community of Castilla-La Mancha, the evaluation of the proposal will be formative, global and continuous, as indicated in the decree.

During the development and execution of the proposal, the competences acquired by the students will be evaluated, as well as the adequacy of the activities in terms of time, workload and usefulness for the acquisition of the expected knowledge.

The evaluation will be formative by means of constant data collection. The final result of the implementation will be collected as well as individually to each student.

The evaluation method will be largely through observation, but students will also be evaluated through their productions throughout the sessions. Each session includes an evaluation to verify that the objectives worked on during the development of the session have been achieved. Table 15 also includes a final evaluation to check the effectiveness of the implementation of the sessions. In this way, it will be possible to verify whether the objectives have been met, the appropriateness of the activities, among other aspects.

Table 15. Evaluation of the proposal design.

INDICATORS	YES	NO
The proposed objectives are met.		
The spaces are adequate.		
The activities are appropriate to the age of the students.		
The classroom climate allows the proposal to be carried out.		
The materials are appropriate for the sessions.		
Cooperative work is promoted.		
The timing is met in each session.		
The activities awaken the interest of the students.		
Communication and interaction between equals is enhanced.		
Involvement on the part of the students is shown.		

Source: own elaboration.

5. CONCLUSION

Today's society is increasingly aware of the problems and challenges arising from the environmental and climate crisis we are facing and the need to adopt a change in the practices developed so far in order to alleviate and reverse this situation. This panorama implies a challenge for all educational systems and it is up to them to provide new generations with an adequate education that will make it possible to amend the current state of the planet. The educational field, from its daily action, stands as a catalyzing agent that trains new generations with an eco-social awareness that will contribute to make the transition to a more sustainable future and will sow the necessary changes in today's society. To this end, it is necessary to address the sustainability-related issues described in the SDGs and provide students with knowledge, values, skills and attitudes that empower them as agents of change.

The purpose of this work was to address the inclusion of the SDGs in the educational curriculum of the Primary Education stage, specifically focused on the promotion of SDG 14, directly involved in promoting and encouraging attitudes, behaviors and habits aimed at conserving and sustainably using the oceans, seas and the resources they provide, through the design and planning of a didactic proposal aimed at the 4th grade of Primary Education. Its main objective is based on sensitizing, raising awareness and actively involving children in the conservation of natural resources, including the seas and oceans, as well as the fauna of their ecosystems. Life below water is of vital importance to human survival because of the ecosystem resources it provides, as well as its biological diversity. Therefore, the choice of the SDG chosen responds to the current emergency situation in which these environments find themselves.

Throughout this work, several aspects relevant to the teaching of issues related to sustainable development have been addressed. First, a theoretical reference framework was developed that addressed the evolution and conceptualization of sustainability, education for sustainable development and its current implications. Next, the different SDGs and the current action plan for sustainable development under the 2030 Agenda were discussed, with special emphasis on SDG 14, which focuses on life below water. On the other hand, the importance of addressing the SDGs from childhood was highlighted, justifying the role and contributions of the educational field for sustainable development and, specifically, in the classrooms of the primary education stage.

In relation to the approach and elaboration of the didactic proposal "Learning to take care of the environment", the didactic design has been based on active and participatory pedagogical approaches that seek to generate significant learning in students. In order to achieve this, a series of diverse and attractive strategies, resources and activities have been proposed to capture the students' interest and facilitate a deeper understanding of the topics covered. In addition, the proposal has been integrated in a coherent manner with current legislation, identifying points of connection with the school curriculum. In this sense, one of the fundamental aims of the proposal was not only to raise environmental awareness and literacy, but also to sensitize, motivate and involve students in taking action through activities that encourage critical reflection and active participation in environmental care and preservation initiatives, highlighting the importance of individual and collective actions for the conservation of the marine environment.

At this point it is important to highlight the importance of continuing to work not only on the approaches of SDG 14, but also on the claims set out in the rest of the SDGs in a comprehensive manner and involving all educational stages as a whole.

The intentional treatment of the SDGs is a powerful tool to generate correct student learning, where the promotion of sustainable values and attitudes and of environmental awareness and consciousness are crucial to turn students into citizens who are inclined and favorably predisposed to the effective conservation of the planet.

In conclusion, the didactic proposal presented in this work represents a significant effort to integrate environmental sustainability, focusing on marine life, in order to incorporate it into the classroom. This type of educational initiatives and actions are crucial to form informed, responsible and committed citizens, capable of facing the environmental challenges of the present and the future, in line with the global objectives established by the 2030 Agenda for sustainable development.

ACKNOWLEDGMENT

The researcher Antonio Morcillo-Martínez is beneficiary of a predoctoral contract from the Ministry of Universities of Spain (reference: FPU21/06754).

REFERENCES

Aibar, E. A. (2023). *Programa educativo de conservación del medio ambiente para mejorar la formación de conciencia ambiental en los estudiantes de quinto y sexto grado de educación primaria de la institución educativa N°1152 - Dos de Mayo, Cercado de Lima, periodo 2019* [Tesis doctoral, Universidad Nacional Federico Villarreal]. Repositorio Institucional UNFV. https://hdl.handle.net/20.500.13084/7237

Alacid, S. (2021). La sostenibilidad dentro del aula de Educación Infantil. *Journal of Research in Humanities and Social Science*, 9(10), 74–83. https://www.questjournals.org/jrhss/papers/vol9-issue10/Ser-5/K09107483.pdf

Alvarado-Herrera, A., Bigne, E., Aldas-Manzano, J., & Curras-Pérez, R. (2017). Una escala para medir las percepciones de los consumidores sobre la responsabilidad social corporativa siguiendo el paradigma del desarrollo sostenible. *Journal of Business Ethics*, 140, 243–262. DOI: 10.1007/s10551-015-2654-9

Ardoin, N. M., & Bowers, A. W. (2020). Early childhood environmental education: A systematic review of the research literature. *Educational Research Review*, 31, 100353. DOI: 10.1016/j.edurev.2020.100353 PMID: 34173434

Barnosky, A. D., & Hadly, E. A. (2016). *Tipping point for planet Earth: How close are we to the edge?* Editorial Macmillan.

Boeve-de Pauw, J., Gericke, N., Olsson, D., & Berglund, T. (2015). The effectiveness of education for sustainable development. *Sustainability (Basel)*, 7(11), 15693–15717. DOI: 10.3390/su71115693

Canto, E. M. (2022). *El rol del maestro en la educación para la sostenibilidad ambiental* [Tesis de maestría, Universidad de Otavalo]. Repositorio Digital Universidad de Otavalo. https://repositorio.uotavalo.edu.ec/handle/52000/708

Castro, A., & Renés, P. (11 al 14 de noviembre de 2018). *La Educación Infantil y el desarrollo sostenible: pequeños pasos, grandes pisadas*. XXXVII Seminario Interuniversitario de Teoría de la Educación. Educación en la Sociedad de Conocimiento y el Desarrollo Sostenible. Universidad de la Laguna, San Cristóbal de La Laguna, España. https://riull.ull.es/xmlui/handle/915/11681

Cernev, T., & Fenner, R. (2020). The importance of achieving foundational Sustainable Development Goals in reducing global risk. *Futures*, 115, 102492. DOI: 10.1016/j.futures.2019.102492

Chams, N., & García-Blandón, J. (2019). On the importance of sustainable human resource management for the adoption of sustainable development goals. *Resources, Conservation and Recycling*, 141, 109–122. DOI: 10.1016/j.resconrec.2018.10.006

Cicin-Sain, B. (2015). Conserve and sustainably use the oceans, seas and marine resources for sustainable development. *UN Chronicle, 51*(4), 32-33. Decreto 81/2022, de 12 de julio, por el que se establece la ordenación y el currículo de Educación Primaria en la comunidad autónoma de Castilla-La Mancha. *Diario Oficial de Castilla-La Mancha, 134*, de 14 de julio de 2022.DOI: 10.18356/c5857454-fr

Estrada, E. G., Mamani, H. J., & Huaypar, K. J. (2020). Eficacia del programa Cuidemos el ambiente en el desarrollo de la conciencia ambiental de estudiantes de educación primaria en Madre de Dios, Perú. *Ciencia Amazónica (Iquitos)*, 8(1), 85–98. DOI: 10.22386/ca.v8i1.282

Eyring, V., Gillett, N. P., Achuta Rao, K. M., Barimalala, R., Barreiro Parrillo, M., Bellouin, N., Cassou, C., Durack, P. J., Kosaka, Y., McGregor, S., Min, S., Morgenstern, O., & Sun, Y. (2021). Human influence on the climate system. In Masson-Delmotte, V., Zhai, P., Pirani, A., S.L. Connors, S.L., Péan, C., Berger, S., Caud, N., Chen, Y., Goldfarb, L., Gomis, M.I., Huang, M., Leitzell, K., Lonnoy, E., Matthews, J.B.R., Maycock, T.K., Waterfield, T., Yelekçi, O., Yu, R., & Zhou, B. (Eds.), *Climate change 2021: The physical science basis. Contribution of working group I to the sixth assessment report of the intergovernmental panel on climate change* (pp. 423–552). Cambridge University Press. DOI: 10.1017/9781009157896.005

Gil, C. G. (2018). Objetivos de Desarrollo Sostenible (ODS): una revisión crítica. *Papeles de relaciones ecosociales y cambio global, 140*, 107-118.

Hernando, M. J., & Dikam, F. P. (2016). De los objetivos de desarrollo del milenio a los objetivos de desarrollo sostenible. *Libre pensamiento*, (88), 60-69.

Hopkins, C. (2012). Twenty Years of Education for Sustainable Development. *Journal of Education for Sustainable Development*, 6(1), 1–4. DOI: 10.1177/097340821100600101

Juárez-Pulido, M., Rasskin-Gutman, I., & Mendo-Lázaro, S. (2019). El Aprendizaje Cooperativo, una metodología activa para la educación del siglo XXI: una revisión bibliográfica. *Revista Prisma Social*, (26), 200–210. https://revistaprismasocial.es/article/view/2693

Lomloe, U., & De La Democracia, L. E. (2020). Ley Orgánica 3/2020, de 29 de diciembre, por la que se modifica la Ley Orgánica 2/2006, de 3 de mayo, de Educación. *BOE*, 340, 122868–122953.

López, I. (2015). Sobre el desarrollo sostenible y la sostenibilidad: conceptualización y crítica. Barataria. *Revista Castellano-Manchega de Ciencias Sociales*, (20), 111-128. https://www.redalyc.org/articulo.oa?id=322142550007

Mabratu, D. (1998). Sustainability and Sustainable Development: Historical and conceptual review. *Environmental Impact Assessment Review*, 18(6), 493–520. DOI: 10.1016/S0195-9255(98)00019-5

Martín, C., & Navarro, J. I. (2015). *Psicología evolutiva en educación infantil y primaria*. Ediciones Pirámide.

Mastrángelo, M. E., Pérez-Harguindeguy, N., Enrico, L., Bennett, E., Lavorel, S., Cumming, G. S., Abeygunawardane, D., Amarilla, L. D., Burkhard, B., Egoh, B. N., Frishkoff, L., Galetto, L., Huber, S., Karp, D. S., Ke, A., Kowaljow, E., Kronenburg-García, Á., Locatelli, B., Martín-López, B., & Zoeller, K. (2019). Key knowledge gaps to achieve global sustainability goals. *Nature Sustainability*, 2(12), 1115–1121. DOI: 10.1038/s41893-019-0412-1

Murga-Menoyo, M. Á., & Novo, M. (2017). Sostenibilidad, desarrollo glocal y ciudadanía planetaria: referentes de una Pedagogía para el desarrollo sostenible. *Teoría de la Educación: Revista Interuniversitaria, 29*(1), 55-78. https://digital.casalini.it/10.14201/teoredu20172915578

Negrín, M. Á., & Marrero, J. J. (2021). La nueva Ley de Educación (LOMLOE) ante los Objetivos de Desarrollo Sostenible de la Agenda 2030 y el reto de la COVID-19. *Avances En Supervisión Educativa*, (35). Advance online publication. DOI: 10.23824/ase.v0i35.709

O'Flaherty, J., & Liddy, M. (2018). The impact of development education and education for sustainable development interventions: A synthesis of the research. *Environmental Education Research*, 24(7), 1031–1049. DOI: 10.1080/13504622.2017.1392484

Olawumi, T. O., & Chan, D. W. M. (2018). A scientometric review of global research on sustainability and sustainable development. *Journal of Cleaner Production*, 183, 231–250. DOI: 10.1016/j.jclepro.2018.02.162

Organización de las Naciones Unidas [ONU] (2015). *Objetivo 14: Conservar y utilizar sosteniblemente los océanos, los mares y los recursos marinos*.https://www.un.org/sustainabledevelopment/es/oceans/

Organización de las Naciones Unidas [ONU] (2023). *Acción por el clima.*https://www.un.org/es/climatechange

Organización de las Naciones Unidas [ONU] (2024). *Objetivos y metas de desarrollo sostenible.*https://www.un.org/sustainabledevelopment/es/sustainable-development-goals/

Pascual, J., & Navío, A. (2018). Concepciones sobre innovación educativa. ¿Qué significa para los docentes en Chile? Profesorado. *Revista de Currículum y Formación del Profesorado*, 22(4), 71–90. DOI: 10.30827/profesorado.v22i4.8395

Pauw, J., Gericke, N., Olsson, D., & Berglund, T. (2015). The Effectiveness of Education por Sustainable Development. *Sustainability (Basel)*, 7(11), 15693–15717. DOI: 10.3390/su71115693

Puig, J. M., & Martín, X. (2015). Para un currículum de educación en valores. *Folios*, 41(41), 7–22. DOI: 10.17227/01234870.41folios7.22

Pujolas, P. (2004). *Aprender juntos, alumnos diferentes. Los equipos de aprendizaje cooperativo en el aula*. Editorial Octaedro.

Pulido-Capurro, V., & Olivera-Carhuaz, E. (2018). Aportes pedagógicos a la educación ambiental: una perspectiva teórica. *Revista de investigaciones Altoandinas, 20*(3), 333-346. DOI: 10.18271/ria.2018.397

Ramos, A., & Torralba, A. (2020). Uso y potencial del Programa LIFE para la Educación Ambiental en educación formal, no-formal e informal, y especialmente en Educación Primaria. *Revista Eureka sobre Enseñanza y Divulgación de las Ciencias, 17*(3), 3501. https://digibuo.uniovi.es/dspace/handle/10651/54889

Rieckmann, M. (2018). Learning to transform the world: Key competencies in Education for Sustainable Development. In Leicht, A., Heiss, J., & Byun, W. J. (Eds.), *Issues and trends in education for sustainable development* (pp. 39–59). UNESCO publishing.

Rigal, R. (2006). *Educación motriz y educación psicomotriz en preescolar y primaria*. Inde Publicaciones.

Ruggerio, C. A. (2021). Sustainability and sustainable development: A review of principles and definitions. *The Science of the Total Environment*, 786, 147481. DOI: 10.1016/j.scitotenv.2021.147481 PMID: 33965820

Sachs, J. D., Kroll, C., Lafortune, G., Fuller, G., & Woelm, F. (2022). *Sustainable development report 2022*. Cambridge University Press., DOI: 10.1017/9781009210058

Sanahuja, J. A., & Tezanos, S. (2017). Del milenio a la sostenibilidad: Retos y perspectivas de la Agenda 2030 para el desarrollo sostenible. *Política y Sociedad (Madrid)*, 54(2), 521–543. DOI: 10.5209/POSO.51926

Sánchez, M. C., Campillo, J. M., & Vivas, V. (2021). *La formación del profesorado en didáctica de las ciencias sociales en el ámbito Iberoamericano*. EDITUM- Ediciones de la Universidad de Murcia., DOI: 10.6018/editum.2919

Sheehy, B., & Farneti, F. (2021). Corporate social responsibility, sustainability, sustainable development and corporate sustainability: What is the difference, and does it matter? *Sustainability (Basel)*, 13(11), 5965. DOI: 10.3390/su13115965

Suárez-Eiroa, B., Fernández, E., Méndez-Martínez, G., & Soto-Oñate, D. (2019). Operational principles of circular economy for sustainable development: Linking theory and practice. *Journal of Cleaner Production*, 214, 952–961. DOI: 10.1016/j.jclepro.2018.12.271

Tsalis, T. A., Malamateniou, K. E., Koulouriotis, D., & Nikolaou, I. E. (2020). New challenges for corporate sustainability reporting: United Nations' 2030 Agenda for sustainable development and the sustainable development goals. *Corporate Social Responsibility and Environmental Management*, 27(4), 1617–1629. DOI: 10.1002/csr.1910

United Nations Educational, Scientific and Cultural Organization [UNESCO] (2012). ESD—*Building a better, fairer world for the 21st century*. https://unesdoc.unesco.org/ark:/48223/pf0000216673

United Nations Educational, Scientific and Cultural Organization [UNESCO] (2020). *Education for sustainable development. A roadmap*. UNESCO.

United Nations General Assembly. (2015). *Transforming Our World: the 2030 Agenda for Sustainable Development*. https://sdgs.un.org/2030agenda

Velásquez, N. (2017). *Efecto del programa "Cuidando el medio ambiente desde mi aula", para mejorar la responsabilidad ambiental en los estudiantes del primer grado de educación primaria de la IE N° 88320, San Bartolo, Santa, Áncash, Perú, 2014*. [Tesis doctoral, Universidad Nacional del Santa Chimbote - Perú]. Repositorio Institucional Digital UNS. https://repositorio.uns.edu.pe/handle/20.500.14278/3096

Vilches, A., Gil, D., & Cañal, P. (2010). Educación para la sostenibilidad y educación ambiental. *Investigación en la Escuela*, (71), 5–15. DOI: 10.12795/IE.2010.i71.01

World Commission on Environment and Development [WCED] (1987). *Our common future. The Brundtland Report*. Oxford University Press.

KEY TERMS AND DEFINITIONS

Ecosocial Awareness: it involves the recognition and understanding of the connections between ecological systems and the well-being of human societies, showing how human actions have a direct impact on the planet and the quality of life of all the species that inhabit it. It involves being aware of how environmental issues impact communities and how social actions and policies can affect the environment. This awareness emphasizes the importance of integrating ecological sustainability with social justice to create a balanced, fair and equitable approach to addressing environmental challenges.

Environmental Commitment: it refers to the personal and moral responsibility of a person to contribute positively to the health of the planet and the preservation of the natural world, with the main objective being the construction of a cohesive and responsible society in terms of sustainability. It involves the dedication and willingness of individuals to participate and actively engage in behaviors, habits and actions that support the protection, preservation and care of the environment through the approach of initiatives such as the adoption of ecological practices, promotion of environmental policies, reduction of personal ecological footprints or the educational treatment of environmental issues, among others.

Environmental Consciousness: it supposes understanding the impact and implications of human activities on the natural environment, which means recognizing the various current environmental problems, understanding the causes and consequences that arise because of human influence, promoting attitudes that prioritize the health of the environment. planet and environmental sustainability, as well as promoting and making visible the set of policies, practices and initiatives aimed at the protection and preservation of nature at different levels.

Environmental Education: an interdisciplinary and cross-sectional pedagogical approach to the educational curriculum based on the systematic and intentional treatment and incorporation of environmental issues aimed at promoting the construction of knowledges, values, habits and pro-environmental practices with the purpose of promoting ecological awareness and care for the environment in the training of committed and responsible citizens.

Environmental Literacy: a concept that encompasses the set of practices, attitudes and values that stimulate a person's ability to critically recognize their belonging and dependence on the environment, being able to recognize the human impact on the environment and the interconnection of natural systems, as well as develop the skills that allow the adoption of behaviors and habits that support environmental sustainability. It involves the acquisition of knowledge, the understanding of concepts and the awareness of people regarding the current causes and problems that affect the environment, together with the reflective ability to apply this knowledge to make informed decisions and undertake responsible actions to protect nature.

Environmental Sensitization: a process aimed at increasing people's awareness and concern about environmental issues by encouraging and promoting a deeper understanding of current problems and implications. This process aims to cultivate a sense of responsibility and a more conscious and supportive attitude towards the environment, generating a significant impact on the emotional sphere of individuals that leads to the implementation of proactive behaviors and attitudes that contribute to balance. ecological and sustainability.

Sustainability: a model that seeks to guarantee the preservation and care of the environment and the promotion of social welfare through a balanced approach to economic development that takes into account the needs of the present, seeking to prevent the excessive exploitation of resources from threatening the survival of future generations and humanity.

APPENDIX

Table 16. Curricular foundation in relation to environmental sustainability and SDG 14

Area	Cycle	Course
Natural, Social and Cultural Environment Knowledge	Second	4th Primary Education
Curricular foundation		
Basic knowledges		

continued on following page

Table 16. Continued

Block A. Scientific culture 2. Life on our planet. - The kingdoms of nature from a general and integrated perspective from the study and analysis of the characteristics of different ecosystems. - Characteristics of animals that allow their classification and differentiation into subgroups related to their adaptive capacity to the environment: obtaining energy, relationship with the environment and perpetuation of the species. - Ecosystems as a place where biotic and abiotic factors intervene, maintaining a balance between the different elements and resources. Importance of biodiversity. Ecosystems of Castilla-La Mancha. - Ecosystem functions and services. - Relationship between human beings and ecosystems to cover the needs of society. Examples of adequate and inadequate uses of the natural resources of our planet and their consequences. Block C. Societies and territories 4. Ecosocial awareness - Climate change. Introduction to the causes and consequences of climate change, and its impact on the Earth's landscapes. Mitigation and adaptation measures. - Ecosocial responsibility. Eco-dependence and interdependence between people, societies and the natural environment. - Transformation and degradation of natural ecosystems by human action. Conservation and protection of nature. Animal abuse and its prevention. - Sustainable lifestyles. Responsible consumption and production, balanced and sustainable food, efficient use of water and energy, safe, healthy and sustainable mobility, and waste prevention and management.
Specific competences
5. Identify the characteristics of the different elements or systems of the natural, social and cultural environment, analyzing their organization and properties and establishing relationships between them, to recognize the value of cultural and natural heritage, conserve it, improve it and take action for its responsible use, especially that of our autonomous community. 6. Identify the causes and consequences of human intervention in the environment, from the social, economic, cultural, technological and environmental points of view, to improve the ability to deal with problems, seek solutions and act individually and cooperatively in their resolution, and to implement sustainable lifestyles and consistent with respect, care and protection of people and the planet.
Evaluation criteria
5.1 Identify the characteristics, organization, and properties of the elements of the natural, social, and cultural environment through inquiry and using appropriate tools and processes. 5.2 Identify simple connections between different elements of the natural, social, and cultural environment showing understanding of the relationships that are established. 5.3 Protect the natural and cultural heritage and value it as a common good, adopting respectful behaviors for its enjoyment and proposing actions for its conservation and improvement. 6.1 Identify ecosocial problems, propose possible solutions and put into practice sustainable lifestyles, recognizing respectful behaviors of care, co-responsibility and protection of the environment and sustainable use of natural resources, and expressing the positive and negative changes caused in the environment by human action.

Source: Own elaboration according to the provisions of Decree 81/2021 of July 12, which establishes the organization and curriculum of Primary Education in the autonomous community of Castilla-La Mancha.

Chapter 7
The Study of Relevant Social Issues in the Teaching of Geographic Space in Primary Education

Carmen Escribano Muñoz
https://orcid.org/0000-0001-9372-5331
International University of La Rioja, Spain

ABSTRACT

Primary education plays a fundamental role in shaping future citizens who are critical and engaged with their environment. In this context, the study of geographic space could become an essential tool for understanding the social and environmental realities of the world around us. When reviewing teaching practices, they are predominantly framed within a traditional approach that often focuses on memorization and abstract content. This makes it difficult for students to understand the relevance of geographic space in their daily lives and in the current society. To achieve these objectives, various didactic strategies are proposed throughout this chapter. These strategies are framed within a pedagogical approach that promotes knowledge construction through experience and critical reflection, enabling students to learn to understand social problems, investigating their causes and consequences, and participate in the search for solutions. This contributes to the construction of a more just, equitable, and sustainable society.

DOI: 10.4018/979-8-3693-7185-5.ch007

Copyright © 2025, IGI Global Scientific Publishing. Copying or distributing in print or electronic forms without written permission of IGI Global is prohibited.

1. GEOGRAPHICAL SPACE IN PRIMARY EDUCATION: IMPLICATIONS FOR THE TEACHING OF SOCIAL SCIENCES

Throughout history, philosophers and scientists have reflected on the nature of space, trying to understand what it is and how it relates to time and matter. Although we have made great progress in our knowledge, the concept of space remains a complex topic that has given rise to various interpretations. Beyond its physical dimension, space is a social setting where interactions occur, meanings are constructed and identities are defined (Santos, 2000, Souto, 2011). In line with this idea, Comes (2000) points out that space and time are not objective realities, but are mental constructions based on the representations that we make about reality, that is, we give meanings to these concepts based on our experiences, and This perception varies according to our culture and social context. As we interact with the environment, we create mental maps and time frames that allow us to orient ourselves and understand the world around us. Therefore, the concept of space transcends mere physical location, It is a complex cognitive construct that structures our understanding of the world and our interactions with it (Lefebvre, 2000, Massey, 2011).

The geographical space, in the words of Santos (1995):

[...] must be considered as an inseparable whole in which, on the one hand, a certain combination of geographical objects, natural objects and social objects participate, and on the other, the life that fills and animates it, that is, society in motion. The content (society) is not independent of the form (geographical objects), and each form contains a fraction of content. Space, therefore, is a set of forms, each of which contains fractions of society in motion. (p. 127)

In the educational field, the development and acquisition of spatial skills is essential, since they lay the foundations for meaningful learning, as it has a direct impact on people's daily lives. However, these spatial competencies are not only essential for functioning in daily life, but are also essential for addressing contemporary social challenges and forging a citizenry more committed to sustainable development (Granados-Sánchez, 2021). In the area of Social Sciences teaching, therefore, geography teaching is required that considers the experiences and perceptions of students, and in which classrooms are transformed into a space for reflection, discussion and research (Souto, 2018, Cordero & Svarzman, 2007, To achieve this, we must turn to theories of cognitive development, which give us a deep understanding of how children construct their representations of space. By knowing the cognitive processes involved in the acquisition of spatial notions, more effective learning activities can be designed and adapted to the needs of each student.

1.1. Cognitive Development and Spatial Representation

The domain of the space is one of the most important facts in the child development as it lets him/ her be involved in her environment and understand the reality where he/she lives.

Mastery of space is one of the most important aspects of a child's evolutionary development, as it enables the child to function effectively in their environment and grasp the reality in which they live. The physical, social, and cultural environment provides various circumstances and phenomena with which the child interacts, allowing them to internalize spatial concepts (Aranda-Hernando, 2016). Developmental psychology has devoted considerable attention to studying how children construct their understanding of space. In this regard, many authors have sought to establish an evolutionary model for the acquisition of spatial notions. Particularly noteworthy is the work of Piaget (1947), in collaboration with Inhelder, who proposed a theory on the development of spatial concepts—an evolutionary model that links cognitive growth to the progressive understanding of spatial notions. For example, Piaget (1978) suggested that children pass through different cognitive stages, during which their perception of space becomes increasingly abstract and complex. In the preoperational stage, children have an egocentric understanding of space, while in the concrete operational stage, they begin to coordinate different perspectives and establish more complex spatial relationships. It is during the formal operational stage that children develop an understanding of unfamiliar or abstract space, which introduces the concept of geographical space (Alonso Arenal et al., 2010, Trepat and Comes, 2002).

In Table 1, based on Piaget's theory, a guide is provided for teachers to design meaningful learning situations. By understanding children's cognitive processes, teachers can act as facilitators, guiding students in exploring the world and constructing their own knowledge.

Table 1. Notion of space and activities to develop.

From 5 to 8-years-old	**Preoperacional** - The child begins to master their environment and is capable of imagining living conditions different from those around them. - Their thinking is intuitive and egocentric - They have only a concrete idea of space and define things by their use.
	- Activities should involve concrete experiences and intuitive observations of their surroundings, as this is what interests them. - Teach them to find points of reference (buildings, trees, etc.). - Encourage observation and description of natural and human elements in the environment, as describing is, above all, organizing space. - Observation and description of different types of landscapes: natural, rural, urban. - The child's perception of the city: Younger children describe the city as a place where people live, referring to buildings and isolated elements that catch their attention (trees, swings, fountains, etc.). - Organize visits and itineraries.
From 9 to 11-years-old	**Concrete operations** - Starting around age 10, children undergo a rapid transformation and develop more objective thinking. They discover cause-and-effect relationships more through intuition than through reflective processes. Special interests begin to emerge, as they better understand what they read, have vivid imaginations, and progressively develop localization skills. - Their ability to observe more objectively will focus on studying the local environment. - The environment is no longer seen as a global reality but becomes an object of analysis. - These direct and analytical observations provide them with the skills to begin analyzing, classifying, and understanding the interdependence between objects.
	- Studying the local environment helps them acquire a method for understanding natural phenomena and human life. Starting from familiar places like a plaza or a museum, we can ask them to locate themselves on a map and find alternative routes. - Introduce the concept of "cardinal points" through a game like a treasure hunt in an open space at school, which helps them orient themselves using a map. - Identification and localization in different cartographic representations. - Ask appropriate questions to gather information from observing different types of landscapes, using various tools, and recording clear data.

Source: Own elaboration from Alonso Arenal et al. (2010), Trepat and Comes (2002)

Hannoun (1977), building on the theoretical foundations of Piaget and Inhelder (1947), proposes a didactic approach for teaching spatial understanding, noting that children go through progressive developmental stages in which they assimilate, perceive, and situate objects in space. According to the author, in the progressive comprehension of space, children pass through three essential stages: first, they experience space (distances, routes, the physical possibility or impossibility of reaching something), then they gradually become able to perceive space without having to experience it biologically (they begin to discover "over there"), and finally, approaching adolescence, they start to conceive of mathematical and abstract space. Hannoun (1977) called these three stages "lived space," "perceived space," and "conceived space."

According to these concepts, the ability to understand space and time results from the person's maturational development. However, we must also consider that these constructions arise from the experiences the child has in contact with their immediate reality. [...] We all think, feel, and act in spatial terms. We move every day through a concrete space, and the news we read in the newspaper or watch on television refers to specific spaces [...] We choose routes to get to a place, orient ourselves, construct explanatory theories about the world we live in, and act based on spatial schemas we build in our minds (Comes, 2000, p.127).

Vygotsky (1978) emphasized that the development of spatial notions is a deeply social and cultural process. According to him, children construct their representations of space through interaction with others. Initially, direct experiences with the physical environment are fundamental. However, it is through social interaction, especially with adults, that a qualitative leap in spatial understanding occurs. In this context, language acts as an essential tool for internalizing abstract spatial concepts. Through conversations, explanations, and the use of maps and models, children acquire more precise geographical vocabulary and develop increasingly complex spatial orientation and representation skills. From this pedagogical perspective, the way we conceive of space and time is influenced by certain social and cultural factors. In this sense, cultural norms, social conventions, and shared temporal structures influence the construction of a child's mental representations. For example, different cultures may have varying perspectives on certain social norms, the concept of time, or the organization of geographic space.

While Vygotsky highlights the importance of social interaction and cultural mediation in cognitive development, Bruner (1966) adds an evolutionary dimension and delves into how children represent space throughout their development, thus complementing Vygotsky's theory. Bruner suggests that children represent space in different ways as they develop, passing through three stages that are detailed below in Table 2.

Table 2. Stages of Children's Spatial Representation

Enactive representation	Children learn about space through direct interaction with objects. By exploring their surroundings, they build a basic understanding of spatial relationships between themselves and the objects around them.
Iconic representation	As children grow, they begin to use mental images to represent space, allowing them to remember places and distances without being physically present.
Symbolic representation	Finally, children develop the ability to use symbols and language to represent space abstractly. Maps, diagrams, and spatial terms are examples of this representation, enabling a more complex and flexible understanding of space

Source: Own elaboration from Bruner (2018)

By integrating Bruner's (2018) proposal on the development of symbolic representation in children with a socially focused geography centered on relevant issues, we can enhance the development of key competencies in primary students, such as the ability to think critically, communicate effectively, and actively participate in solving social problems. In this regard, one of the main goals for teachers today is to motivate students and encourage reflection and social and critical thinking (Pagès, 1998). School geography, by connecting spatial processes with social issues, becomes a fundamental tool for shaping future citizens who are critical and engaged with the society they are part of. By incorporating Bruner's proposal on symbolic representation from a social and critical perspective, students can analyze and evaluate complex information, identify patterns, and propose solutions to local and global problems, thus contributing to the construction of a more just and sustainable society (Souto, 2011, Comes, 2000, Batllori, 2011).

1.2. School Geography and its Implication in the Current Context

In the Social Sciences, geographic space refers to the area and places where human activities take place. Therefore, in addition to being considered a physical space, it is also a social and cultural space, as it is impossible to discuss sociocultural phenomena without properly locating them in the space and time in which they occur (Santos, 2000). Traditionally, geography teaching has focused on memorizing names, locating places, and describing physical characteristics. However, the sociocritical approach proposes to critically analyze the relationships between society and space, fostering social awareness in students that enables them to understand the causes and consequences of territorial transformations (Balaguer Mora, 2018, Souto, 2011). This didactic approach represents an improvement over the traditional conception of geography teaching and learning, as it is immersed in social processes.

From this perspective, school geography should not remain anchored in the geography of location and description but must move towards explaining social phenomena—a role that, until now, has been assigned solely to the teaching of time and history (Pagès, 2009). From a critical education standpoint, geography contributes to shaping critical citizens capable of analyzing social inequalities, territorial conflicts, and environmental issues, encouraging participation in transforming their surroundings (Ferrer & Vivancos, 2022, Balaguer Mora, 2018, Comes, 2000, Batllori, 2011).

Thus, school geography has become a fundamental discipline for understanding the world we live in. In a globalized context marked by challenges such as climate change, social inequality, and migration, geography provides tools to critically analyze and address these challenges. By integrating the study of socially relevant issues, geography enables students to understand the interconnections between social and spatial processes, thereby developing complex and critical thinking (Souto, 2018, Santisteban, 2019).

1.2.1. The Current School Geography, Implication, Challenges and Opportunities

School geography has undergone a significant transformation in recent decades, evolving into a much more dynamic and critical approach. This evolution is based on the need to understand the complex interactions between humans and the environment. Geography, therefore, helps us understand the world around us and make critical decisions. It studies how places change over time, how nature and society interact, and why inequalities exist between different regions. School geography, from a sociocritical perspective, connects knowledge with students' everyday lives, enabling them to understand global challenges such as climate change and sustainability (Souto, 2018, Ferrer and Vivancos, 2022). By developing skills like critical thinking and problem-solving, students gain the tools they need to make informed decisions and actively participate in building a more just and sustainable future.

For example, by analyzing climate patterns and their variations over time, students can understand the causes and consequences of climate change and evaluate different proposals to mitigate its effects. Similarly, by studying the global distribution of population and natural resources, they can comprehend the roots of social inequalities and mass migrations. From this paradigm, school geography, by providing the tools to understand space and the relationships between humans and their environment, fosters critical thinking that allows us to understand and address socially relevant issues in the classroom (Ferrer & Vivancos, 2022).

2. SOCIAL RELEVANT ISSUES AND EDUCATION FOR DEMOCRATIC CITIZENSHIP

Social issues, a phenomenon intrinsically linked to the construction of our societies, have been referred to by various terms over time. From the "issues" raised by Stenhouse (1987) to the "social controversy topics" addressed in the French context, they have been analyzed from multiple perspectives. López Facal and Santidrián (2011) define them as "socially relevant problems," meaning conflicts inherent to the life of societies.

Regarding the conceptual meaning of socially relevant problems, López (2010) states:

"They are those that generate conflict, opposing opinions, debate, social confrontation. Some are tragic and unexpected, such as a terrorist attack, an ecological disaster, or a human catastrophe like the Haiti earthquake... although they become hot topics around an event that draws media attention and brings them into focus, such as an outbreak of xenophobia revealing social fracture around immigration, an especially acute famine, or the exodus of tens of thousands of people fleeing war or a planned genocide." (p. 1-2).

We can say that socially relevant problems are those that directly affect people's lives and communities, generating debates, mobilizing society, and demanding solutions. These problems can be diverse in nature: economic, social, political, environmental, cultural, etc., and are often interconnected and rooted in the historical and geographical context in which they occur (Santisteban, 2019).

In the educational context, the inclusion of social issues in curricula, supported by authors such as Legardez and Simonneaux (2006), Oller (1999), and Pagès (1998), contributes to the development of essential citizenship skills for life in society. Addressing real-world problems allows students to develop skills in analysis, evaluation, and informed decision-making, as proposed by authors like Delors (1996) in his report on the four pillars of education.

This approach also advocates for problem-based learning or controversy-based learning so that students can interpret reality, propose solutions, and develop critical attitudes toward the consequences of scientific advances. It is essential for children to acquire an awareness, participate, and make informed decisions regarding the major challenges we face today (López Facal and Santidrián, 2011).

Santisteban (2019), in his article, delves into a social science teaching methodology that starts from social problems or controversial topics. This approach, rooted in Dewey's (1938) proposals, seeks to connect learning with real life and focuses on developing skills that foster critical thinking. From this perspective, addressing relevant social issues provides basic knowledge, methodological strategies, and interpretative frameworks that enable the understanding of the complex nature of reality (Ocampo & Valencia, 2019).

But why is it so important to address social issues in the classroom? First, because it allows us to connect theory with practice, linking the knowledge acquired in the classroom with the reality students live. Second, it promotes the development of key competencies for citizenship, such as empathy, tolerance, and social commitment. Third, it equips students with tools to critically analyze information, identify biases, and construct solid arguments (Ocampo & Valencia, 2019, Díaz Moreno & Felices la Fuente, 2017, Santisteban, 2019, Pagès & Santisteban, 2011).

In this sense, addressing social problems in the classroom can become a meaningful learning experience where students research, debate, propose solutions, and reflect on the implications of their actions. A teaching approach that advocates for the study of real social problems recognizes various opportunities for selecting, organizing, and structuring the content to be taught, while also expanding the teaching materials available for the classroom.

> *"[...] this contributes to emphasizing the idea that what should be taught is not predetermined, but rather that the constantly changing reality can and should be the source of knowledge and the formation of citizens committed to the ongoing transformation and reflection on society." (Ocampo and Valencia, 2019, p. 63)*

López Facal and Santidrián (2011) state:

> *"School life is not isolated from what happens outside the classroom, so conflicts sometimes break into them. A terrorist attack, a natural disaster, a political decision, or an accident caused by human error mobilizes different sectors of society that demand accountability and propose various alternatives to address the issue. Sometimes these conflicts are local: disputes over urban planning decisions and land use are often the most frequent. Others are significant for the entire autonomous community or the whole of Spain. For example, decisions on water transfers, forest fire management, forestry policy, the construction of major infrastructure, social policy including education, healthcare, pensions, and immigration policy, nuclear energy and the placement or rejection of such facilities, or responses to any emergency or catastrophe. Lastly, they can be of a global or international scope, such as military interventions in other countries or actions to respond to disasters in distant places, etc." (p.12)*

The educational potential of addressing these socially relevant issues suggests analyzing them in class, as they are complex and important topics that can spark students' interest and encourage a deep analysis of social reality. Therefore, it is worth diverting slightly from the curriculum and focusing on these topics (López Facal & Santidrián, 2011).

From a curricular perspective, a curriculum based on socially relevant issues directly leads to critical literacy for students, which should be action-oriented, meaning a commitment and engagement with these issues (Santisteban et al., 2016). Teaching social content through real-world problems facilitates the acquisition of competencies that are equally necessary in our society, which predispose us to critically engage with the vast amount of information we receive daily from various sources in numerous formats (Castellví et al., 2020). Today, given the importance of visual information and its fleeting nature, it is imperative that students learn to select it and recognize its level of validity. The presentation of socially relevant problems in the classroom, with the intent to analyze them and propose solutions, inherently involves questioning the established social order and the power and domination relations that preside over our world. Therefore, thinking and reflecting on these aspects, as well as on alternatives to this system, can lead to social change (Santisteban et al., 2016).

One of the educational goals of addressing current controversial issues is to learn to argue and support opinions with reliable sources of information, verifiable data, and logical reasoning (López Facal & Santidrián, 2011, p. 15).

Returning to the question: why is it so important to address social issues in the classroom? We noted that studying them fosters the development of key competencies for democratic citizenship. In this regard, authors like Andreotti (2006), Cox et al. (2017), Ortega and Pagès (2017), Escribano and Tosello (2020), and Gómez Saldivia (2022), among others, argue that education for citizenship should promote critical skills so that individuals can analyze the contradictions of social reality and, therefore, perceive, denounce, and act against social inequalities and injustices. From this perspective, educating for citizenship means teaching students to acquire a civic culture and learn democratic values. This implies creating a classroom climate where students can develop their personality, create emotional bonds with their peers and people in their immediate environment, work together, and learn to commit to common projects or find solutions to complex situations.

Education for democratic and equal citizenship requires the formation of critical and autonomous citizens, capable of exercising responsible, participatory, and creative citizenship, prepared for collective decision-making and resolving social conflicts. In this process, the criticality of any discourse involves the ability to identify power relations, making visible all the protagonists of historical and social developments, constructing plural gender identities in inclusive educational contexts, and recognizing alterity and difference as inherent components of true global, inclusive, and democratic citizenship (Ortega & Pagès, 2017, p. 102).

From this approach, teaching social sciences centered on socially relevant problems can be a valuable opportunity for citizenship education, as it fosters critical civic awareness. These problems—whether from the past, present, or future—can help develop skills such as recognizing, examining, analyzing, evaluating, and appreciating diverse perspectives on a specific issue (Pagès, 2019).

3. HOW TO INTEGRATE GEOGRAPHIC SPACE AND SOCIALLY RELEVANT ISSUES IN PRIMARY EDUCATION?

The proposal to integrate geographic space with social issues in primary education is essential for fostering meaningful and relevant learning. By connecting geographic content with social realities, students develop thinking skills (García, 2019), analyze information, and evaluate different perspectives (Rivera, 2016). This, in turn, encourages social awareness and a willingness to participate in the search for solutions, contributing to a holistic education that transcends cognitive aspects. This proposal aligns with the ideas of David Sobel (2004), who advocates for an education that connects students with their environment and prepares them to be active and engaged citizens. The goal of working with socially relevant issues is to promote effective communication and constructive dialogue. By sharing our opinions and listening to others, we learn to defend our viewpoints reasonably and appreciate different perspectives. This process allows us to grow intellectually and develop essential social skills for life in society.

Below, in Table 3, we present some strategies and practical examples to achieve this, which are developed in the following section.

Table 3. Strategies to join social relevant issue in the study of geographical space.

Starting from the familiar aspects	
	The closest environment
	Local problems
Using active methodologies	
	Research projects
	Collaborative work
	Learning based on problems
	Learning and service
	Field trips
	Role playing
	Case study
Incorporating the socio-emotional dimension	
	Inequalities
	Conflcits
	Sostenibility
Making used of a varied of didactics materials	
	Maps
	Images
	Videos
	Statistics
	Primary sources
Fostering critical thinking	
	Challenging questions
	Discussion and debate
	Encouragement of curiosity
	Constructive feedback

continued on following page

Table 3. Continued

Use of digital technology	
	Geographic Information System (GIS)
	Virtual Reality
	Virtual Learning environment

Source: Own elaboration

3.1. Starting from the Familiar: Connecting Geography with Students' Everyday Lives

When starting the study of geography with primary students, it is essential to establish a direct connection between geographic concepts and their immediate surroundings. This not only facilitates the understanding of content but also sparks their curiosity and allows them to see the relevance of geography in their daily lives. To achieve this, it is vital to step outside the classroom and explore the environment, use authentic materials, and encourage critical analysis and research. How can we do this?

3.1.1. The Immediate Environment as a Starting Point

By using the immediate environment as a starting point, we follow this premise and provide our students with the opportunity to connect with their surroundings and understand their place in the world (Sobel, 2004).

Personal Maps: Invite students to create maps of their route to school, indicating the places they visit, the modes of transportation they use, and the obstacles they encounter. By creating maps of their route to school, students are constructing a mental representation of their immediate environment, inviting them to explore their own world and develop a sense of belonging to the community. In addition to reflecting physical elements on the map, they can also investigate changes in the environment over time, using old photographs or stories from older people.

Detailed Observation: Organize outings to the nearby environment to observe changes in the landscape over time (e.g., the construction of a new building, the disappearance of a park). Detailed observation of changes in the landscape fosters a sense of connection with time and place, as by identifying changes, students become witnesses to the transformation of their surroundings. It would be interesting for them to complete a field journal to record their observations and reflections on the changes they observe, they can also use cameras or mobile phones to document the changes and create a visual archive. A classroom discussion can also be held to analyze the changes observed and their consequences for the community.

Neighbourhood Comparison: If possible, compare the school neighbourhood with other neighbourhoods in the city, identifying similarities and differences in terms of services, infrastructure, and quality of life. By comparing different neighbourhoods, students develop an awareness of diversity and social inequality by establishing criteria for comparing neighbourhoods, such as the availability of public services, green spaces, transportation, etc. This can help them reflect on the causes of these inequalities and discuss possible solutions. Why not expand the educational proposal by suggesting concrete actions to improve neighbourhoods, such as clean-up campaigns, the creation of urban gardens, or the organization of local events?

3.1.2. Local Problems: Identifying and Seeking Solutions

Collaborative Research: Divide students into groups to investigate a specific local problem. They can gather data, interview experts, and seek information online. It is important to guide students in selecting a specific problem that is complex enough to require in-depth research but manageable for the group. For example, the pollution of a nearby river.

Proposing Solutions: Once problems are identified, encourage students to propose creative and realistic solutions. They can create posters, write letters to authorities, or organize a awareness campaign. It is important to use creative techniques like brainstorming to generate a wide range of solutions, thereby fostering divergent thinking and imagination. From the brainstorming session and the debate to select the most relevant proposal, a detailed plan can be developed to implement the selected proposal, including a timeline, budget, and task assignments.

Connection with Subjects: Relate local problems to other areas of the curriculum, such as natural sciences (studying water quality and analyzing environmental impacts), social sciences (analyzing historical causes of a problem, comparing the local problem with similar situations in other parts of the world), or mathematics (calculating the economic impact of a solution).

3.1.3. Examples of Practical Activities

"Citizen Debate": Organize a debate on a local problem, inviting different viewpoints and fostering respectful dialogue.

"Investigation on River Pollution": Students can analyze water quality, identify sources of pollution, and propose solutions.

"Analysis of the Impact of Tourism on a Coastal Area": Students can investigate the effects of tourism on the environment and geographic space, and propose measures to promote more sustainable tourism.

3.2. Using Active Methodologies: Turning Students into Knowledge Explorers

Active methodologies are a set of pedagogical strategies that aim to engage students actively in their own learning process. Instead of being passive recipients of information, students become the protagonists of their learning, constructing their own knowledge through experience, exploration, and collaboration.

3.2.1. Research Projects

Students work independently or in groups to investigate a specific geographic topic, following the steps of the scientific method. For instance, students might explore the issue of water scarcity in their community. They can collect data on water consumption, identify the causes of the scarcity, and propose solutions such as creating a school garden or implementing water-saving measures.

3.2.2. Collaborative Work

Divide students into groups to analyze a map of their city and identify green spaces, industrial zones, and residential neighborhoods. Each group then creates a concept map that shows the relationships between these elements and how they affect people's lives.

3.2.3. Problem-Based Learning

Students solve real and meaningful problems. For example, the problem of pollution in a nearby river could be presented to the students. They work in teams to investigate the causes of the pollution and can be tasked with proposing solutions, subsequently creating a community awareness campaign.

3.2.4. Learning and Service

Students learn while solving real problems in their environment, contributing to the well-being of others. For instance, they may collaborate with a local organization to clean a nearby park. Before the activity, they research the importance of green spaces and the benefits of caring for the environment.

3.2.5. Field Trips

Field trips are a teaching strategy conducted outside the classroom, allowing students to learn directly and experientially in a real environment. These activities enable students to observe, explore, and analyze phenomena, concepts, and processes that are difficult to understand solely through books or theoretical explanations. For example, a visit to a natural science museum could be organized to learn about soil formation and ecosystems, and then relate what they learned to environmental problems in their community.

3.2.6. Role-Playing

Simulated scenarios are created for students to experience and learn. For instance, a simulation of a municipal meeting could be held to discuss a project for constructing a new shopping center in a green space. Students assume different roles (neighbours, businesspeople, politicians) and debate the pros and cons of the project.

3.2.7. Case Studies

Real situations are analyzed to apply acquired knowledge. One activity could focus on analyzing a real case, such as a natural disaster (e.g., an earthquake) and its geographic and social consequences. Students research the causes of the disaster, prevention measures, and response actions.

The key is to combine these methodologies to create richer and more meaningful learning experiences. For example, a research project on air pollution could be initiated, and then students could be divided into groups to work on different aspects of the problem (causes, consequences, solutions). Field trips could be organized to an air quality monitoring station or a factory to observe production processes and their environmental impacts. Finally, students could create a role-play to simulate a municipal meeting where different proposals for reducing air pollution are discussed.

3.3. Incorporating the Socioemotional Dimension: Exploring the Social Connections in Our Environment

By incorporating the socio-spatial dimension, we allow students to understand how these issues manifest in space and how they affect people living in different places.

We can address problems such as inequalities, conflicts, and sustainability. Below, in Table 4, we present some activities that exemplify the treatment of these concepts in the primary classroom.

Table 4. Activities to address the concepts of inequalities, conflicts and sustainability.

Inequalities	- Analyze satellite images to compare different neighborhoods or cities and observe differences in terms of infrastructure, vegetation, and population density. - Study cases of inequality around the world in different countries, such as the inequality between developed and developing countries, or the inequality within a single country.
Conflicts	- Study local conflicts: Investigate conflicts that have occurred or are currently happening in the community, such as disputes over the use of a park or the construction of a new road. - Analyze historical conflicts: Study historical conflicts that have impacted the current configuration of territory, such as wars, migrations, or colonization. - Simulate conflicts and seek solutions: Create simulated scenarios where students assume different roles and negotiate solutions to a conflict.
Sustainability	- The water cycle: Explore how water is a limited resource and how our actions can affect its quality and availability. - Waste management: Analyze the problem of garbage and propose solutions for reducing, reusing, and recycling. - Renewable energy: Investigate different sources of renewable energy and their benefits for the environment. - Biodiversity: Study the importance of biological diversity and the threats it faces.

Source: Own elaboration

3.4. Making used of varied didactics resources: Knowing the environment through resources

As we have noted throughout this work, the teaching of geography in primary education has undergone significant evolution in recent decades, largely due to the use of a wide variety of teaching resources. As Ferrer and Vivancos (2022) point out, the integration of maps, images, videos, statistics, and primary sources in the classroom creates a dynamic and enriching learning environment that promotes a deep understanding of geographic and historical concepts.

3.4.1. Maps: A Window to Space Beyond Representation

Maps are a fundamental tool in the teaching of geography. They allow students to visualize spatial relationships, identify distribution patterns, understand geographic processes, and develop their orientation skills. By working with maps, children learn to interpret symbols, scales, and projections, which enables them to relate graphical representation to reality. Furthermore, interactive maps and digital platforms offer new possibilities for virtually exploring the world and conducting more complex spatial analyses. Practical examples for classroom incorporation include using maps to analyze the distribution of basic services in different neighborhoods of the city (schools, hospitals, parks) and questioning whether differences exist between neighborhoods and what these inequalities are due to. Students can also analyze and investigate current or historical territorial conflicts through maps and reflect on the causes of these conflicts and how they affect local and global populations.

3.4.2. Images and Videos: Beyond Words

Images and videos complement the information provided by maps, adding a visual and contextual dimension that facilitates the understanding of concepts. Historical photographs, for example, allow students to imagine how places and people were in the past, while documentary videos provide a dynamic perspective on geographic and historical phenomena. By analyzing images and videos, students develop observation, interpretation, and comparison skills. Practical examples for classroom activities include analyzing representations of different countries or cultures in movies, advertisements, and textbooks, reflecting on what stereotypes are reinforced and how these representations influence our perception of the world. Another activity could involve analyzing images of landscapes before and after human activity (for example, deforestation or urbanization) and asking what the environmental and social consequences of these changes are based on their visualizations.

3.4.3. Statistics: Data That Speaks

Statistics are an essential tool for quantifying and analyzing geographic and historical information. By working with statistical data, students learn to interpret graphs, tables, and diagrams, identify trends, and make inferences. Additionally, statistics allow students to develop critical thinking and problem-solving skills by confronting situations that require analyzing and evaluating complex information (García, 2019). Through statistics, students can analyze data on income, education, and access to basic services in different countries or regions and pose questions to the class about what conclusions can be drawn from this data and what factors influence these inequalities. Another activity may focus on analyzing data about temperature, sea level, and greenhouse gas emissions so that students can understand the impacts of climate change in different regions of the world.

3.4.4. Primary Sources: Connecting With the Past

Primary sources, such as historical documents, letters, diaries, old photographs, archaeological objects, etc., allow students to establish direct contact with the past and construct their own interpretations. By analyzing these sources, children learn to contextualize historical events, appreciate different perspectives, and develop critical and reflective thinking. Examples of activities could include seeking testimonies from people affected by armed conflicts, migrations, or natural disasters and interpreting how these individuals experience these events, thus constructing meanings and narratives from these experiences. Additionally, an analysis of historical documents that reflect social struggles and movements for justice can be carried out, reflecting on what can be learned from these experiences.

3.5. Fostering Critical Thinking: Building Our Own Judgment

The development of critical thinking requires the implementation of a variety of teaching strategies that promote higher cognitive processes such as inquiry, analysis, and evaluation. From formulating open-ended questions that stimulate reflection to using authentic materials that encourage the contextualization of knowledge, these strategies invite students to go beyond memorization and actively and meaningfully construct knowledge (García, 2019).

Below, in Table 5, we present various socially relevant problems combined with strategies that promote critical thinking and some examples we consider suitable.

Table 5. Strategies to Foster Critical Thinking Based on Socially Relevant Issues in the Study of Geographic Space.

Socially Relevant Problem	Strategies to Foster Critical Thinking and Examples
Inequality and Social Exclusion	- Pose open-ended questions: Why do you think there are neighborhoods with such different quality of life? How does a city's history influence the current distribution of wealth? - Encourage discussion and debate: Organize a debate on the public policies necessary to reduce social inequality. -Use authentic materials: Analyze reports from international organizations on inequality and poverty worldwide. -Propose problems to solve: Design a project to create a map of inequality in your community and propose solutions.
Environment and Climate Change	- Use challenging questions: How do you think climate change will affect future generations? What are the economic and social implications of transitioning to a low-carbon economy? -Foster curiosity: Investigate the causes and consequences of global warming and its effects on different ecosystems. -Model critical thinking: Analyze a graph showing the increase in global temperatures and explain how that conclusion was reached.
Globalization and Migration	- Evaluate different perspectives: Compare the perspectives of countries of origin and destination regarding migration. -Use active learning strategies: Create a role-playing scenario where students assume the roles of migrants and citizens of a destination country. -Provide constructive feedback: Evaluate an essay on the impacts of globalization on local culture, offering suggestions for improving the argumentation
Conflicts and Peace	- Encourage discussion and debate: Organize a debate on the causes of armed conflicts and possible solutions. -Use authentic materials: Analyze news reports on current conflicts and discuss the different perspectives involved. -Propose problems to solve: Design a peace plan for a fictional or real conflict.

Source: Own elaboration

3.6. Use of Digital Technology: New Ways to Learn Geography

The incorporation of digital technology in teaching geographic space can significantly enrich the learning process and foster critical thinking (García, 2019). Below, we illustrate how to integrate Geographic Information Systems (GIS), Virtual Reality (VR), and Virtual Learning Environments (VLE) in this context:

3.6.1. Geographic Information Systems (GIS)

A Geographic Information System (GIS) is a tool that allows us to work with geographic information and contains data associated with each location. With the help of this system, spatial analysis of social problems can be performed, and simulations can be created, as it enables us to visualize geographic data related to issues such as inequality, poverty, climate change, etc. For example, maps can be created to show population distribution, pollution levels, and simulations of future scenarios such as the impact of climate change on a specific population. An example of an activity could be to create a map showing the distribution of wealth and poverty in a specific city.

GIS facilitates the creation of models to simulate future scenarios and evaluate different options. Their ability to integrate demographic, environmental, and economic data provides a holistic view of spatial phenomena. In the field of urban planning, GIS is used to identify growth areas, analyze mobility patterns, and evaluate the impact of public policies, thus contributing to more informed and efficient decision-making

3.6.2. Virtual Reality (VR)

Virtual reality is a technology that allows us to create a simulated environment generated by computer, enabling students to explore remote or historical places in an immersive way, such as a tropical rainforest, an ancient city, or a conflict zone. Virtual Reality can simulate real situations, such as a natural disaster or a social conflict, so that students can firsthand experience the consequences of their actions. An example of an activity could be to simulate a coastal flooding event to understand the impacts of rising sea levels.

Virtual reality has emerged as a promising tool to transform education, especially in disciplines that require a deep spatial understanding. Meta-analytic studies such as those by Di Natale et al. (2020) and Villena Taranilla et al. (2022) have solidified this trend, providing empirical evidence of the benefits of immersive learning. Di Natale et al. offer a general overview of the field, while Villena Taranilla et al. delve deeper into the effects on primary education. By integrating virtual reality into the Social Sciences, we can foster the development of geographic thinking, a fundamental aspect for understanding complex phenomena such as urbanization, as proposed by Brooks (2017). According to Brooks, geographic thinking involves analyzing spatial patterns, understanding relationships between geographic phenomena, and evaluating the consequences of human decisions. Our research aligns with this theoretical framework by exploring how a virtual reality experience can help secondary school students visualize and analyze urbanization processes, thus developing critical spatial skills and a deeper understanding of the social and environmental impacts of urbanization.

Virtual Learning Environments facilitate interaction among students from different locations while allowing teachers to monitor student progress and adapt activities to their individual needs. An example of an activity could be to create a virtual forum where students from different countries can discuss the causes and consequences of a conflict.

4. FINAL REFLECTIONS, LIMITATIONS AND FORESIGHT

The teaching of geography in primary education is at a crossroads marked by the need to adapt to an increasingly complex and globalized world (Pagès, 2019). On one hand, technological advancements and growing awareness of environmental and social issues offer new tools and perspectives for addressing this discipline (Ferrer and Vivancos, 2022). At the same time, geography faces the challenge of maintaining its relevance in an increasingly content-heavy school curriculum and adapting to new teaching and learning methods that consider the diverse learning styles of students. It is undeniable that geography plays a fundamental role in the development of critical and engaged citizens. By studying landscapes, ecosystems, and societies, students develop essential skills to understand the world around them and make informed decisions. However, for geography to be truly effective, it is necessary to overcome some obstacles (Balaguer Mora, 2018).

One of the main challenges lies in updating geography curricula. Traditional content, often focused on memorizing names and places, must be replaced with a more dynamic and problematizing approach (Santisteban, 2019). Global issues such as climate change, social inequality, and urbanization provide opportunities for students to apply their geographic knowledge to real situations and develop innovative solutions (Balaguer Mora, 2018). Another important challenge is teacher training. It is essential that teachers are trained to use active methodologies, digital resources, and interdisciplinary approaches in teaching geography. Furthermore, it is crucial that teachers feel confident and motivated to address complex and relevant topics, such as sustainable development or global inequality, among others (Ferrer and Vivancos, 2022).

Despite these challenges, teaching geography also presents great opportunities. By connecting geographic content with students' daily lives, skills such as critical thinking, problem-solving, and creativity are fostered. Additionally, geography can contribute to developing civic awareness and a sense of belonging to both local and global communities (Rivera, 2016, Escribano and Tosello, 2020).

We can say that, currently, the teaching of geography in primary education is at a turning point. The demand for critical citizens who are aware of global challenges calls for a profound renewal of this discipline. While significant progress has been made in incorporating innovative teaching resources and more problematizing approaches, challenges such as curriculum updating, teacher training, and the availability of suitable materials still persist. Nevertheless, the opportunities are immense. By connecting geography with students' daily lives and addressing relevant social issues, we can promote the development of key skills such as critical thinking, problem-solving, and global citizenship (García, 2019). Geography, by enabling students to understand the relationships between humans and their environment, provides them with tools to critically analyze the environmental, social, and economic issues affecting our world.

To maximize these opportunities, it is necessary to implement a series of innovative teaching strategies. Among them, we have highlighted project-based learning, which allows students to research topics of interest and apply their knowledge practically, the use of digital technologies, which facilitate access to updated information and the execution of collaborative activities, and field trips, which enable connections between theory and reality. Collaboration with other areas of knowledge, such as social sciences, natural sciences, and languages, enriches the understanding of geographic phenomena. In this regard, we believe that initial and ongoing teacher training is crucial to ensure the successful implementation of these strategies. Teachers must be trained to use active methodologies, integrate digital technology into their lessons, and address complex topics appropriately. Additionally, it is necessary to promote research in geography didactics to generate new knowledge and experiences that can be shared and applied in the classroom (Pagès, 2009).

On the other hand, one of the points not addressed in this chapter is the evaluation of learning, as it must also adapt to these new approaches. It is essential to develop assessment tools that allow for the evaluation of not only basic knowledge but also the skills, attitudes, and values promoted through geography teaching. Formative assessment, based on observation and continuous feedback, can be a valuable tool to support students in their learning process (Sanmartí, 2018).

As limitations, we can say that the implementation of critical geography in classrooms may encounter obstacles such as the lack of time in the curriculum due to the overload of content in the current curriculum, which often limits the time devoted to geography. This hinders the exploration of complex topics and the execution of practical activities that require more teaching time. Furthermore, the pressure to achieve certain curricular objectives may lead teachers to prioritize more traditional and less problematizing content. Additionally, some teachers may resist change due to insecurity in handling more active methodologies or a preference for more traditional methods. The scarcity of adequate material resources or unequal access to technological resources may also create gaps among students. Furthermore, assessing learning in this area can be complex, thus requiring the development of evaluation instruments that allow for the assessment of critical thinking skills, problem-solving ability, participation in debates, and project execution. This necessitates greater training for teachers in designing more flexible and contextualized evaluation instruments and rubrics.

However, despite these challenges, we perceive a promising outlook for the future of geography teaching and the study of geographic space. The rapid advancement of digital technologies, such as virtual reality and geolocation, will provide new ways to explore the world and understand complex geographic phenomena. This will allow students to immerse themselves in virtual environments, conduct simulations, and visualize data more intuitively. At the same time, the growing awareness of environmental and social issues, such as climate change and inequalities, will drive an increasing demand for informed and engaged citizens (Balaguer Mora, 2018). In this sense, school geography, by providing adequate didactic tools to understand the interactions between society and the environment, will position itself as a key discipline to address all these challenges.

The commitment of teachers and researchers is also fundamental. Continuous training, experience sharing, and research in geography didactics are essential to develop new strategies and materials that respond to the needs of 21st-century students and the future. Additionally, interdisciplinary collaboration will enrich the teaching of geography, connecting it with other areas of knowledge such as history, social sciences, languages, and mathematics.

In this context, we can envision a more interdisciplinary, participatory, and reality-connected critical geography. Research projects, classroom debates, and field trips should become regular tools to foster active and meaningful learning for our students. We must transform school geography into a discipline that not only transmits basic knowledge but also develops skills and competencies such as critical thinking, problem-solving, and creativity, preparing students to face the challenges of an increasingly complex and globalized world.

REFERENCES

Alonso Arenal, S., González Alonso, S., González Alonso, A. P., & González Alonso, M. (2010). *Didáctica de las Ciencias Sociales para la Educación Primaria*. Pirámide.

Andreotti, V. (2006). Soft versus critical global citizenship *education. Policy & Practice: A Development.Education Review*, 3, 40–51.

Aranda Hernando, A. (2016). *Didáctica de las Ciencias Sociales en Educación Infantil*. Síntesis.

Balaguer Mora, P. A. (2018). Geografía crítica y pensamiento crítico. *Actualidades Pedagógicas*, (72), 73–95. DOI: 10.19052/ap.5232

Binimelis, J., Usher, J., & Bagoly-Simó, P. (2023). How is Geography taught at Primary Education centers in Spain? Results of a state-wide study. *Boletín de la Asociación de Geógrafos Españoles*, 98.

Brooks, C. (2017). International differences in thinking geographically, and why "the local" matters. In Brooks, C., Butt, G., Fargher, M. (eds). *The Power of Geographical Thinking. International Perspectives on Geographical Education* (pp-169-179). Springer

Bruner, J. S. (2018). *Desarrollo cognitivo y educación*. Ediciones Morata.

Canal, M., Costa, D. & Santisteban, Al (2012). El alumnado ante problemas sociales relevantes: ¿Cómo los interpreta? ¿Cómo piensa la participación? En N. de Alba Fernández, F. F. García Pérez & A. Santisteban Fernández (Eds.), *Educar para la participación ciudadana en la enseñanza de las Ciencias Sociales,*.Vol. I (pp. 527-535). Díada Editora, S.L.

Castellví, J., Diez-Bedmar, M.-C., & Santisteban, A. (2020). Pre-Service Teachers' Critical Digital Literacy Skills and Attitudes to Address Social Problems. *Social Sciences (Basel, Switzerland)*, 9(8), 134–145. DOI: 10.3390/socsci9080134

Cordero, S. & Svarzman, J. (2007). *Hacer geografía en la escuela. Reflexiones y aportes para el trabajo en el aula.* Novedades educativas.

Cox, C., Magendzo, A., Muñoz, G., & Redón, S. (2015). ¿Cómo debiera ser la formación ciudadana en la escuela? *Revista Docencia*, 58, 4–17.

Delors, J. (1996). de la publicación: La Educación Encierra un Tesoro. Informe a la UNESCO de la Comisión Internacional sobre la Educación para el Siglo XXI. *Laurus*, 14(26), 136–167.

Dewey, J. (1938). The philosophy of the arts. *John Dewey: The Later Works*, 13, 357–368.

Di Natale, A. F., Repetto, C., Riva, G., & Villani, D. (2020). Immersive virtual reality in K-12 and higher education: A 10-year systematic review of empirical research. *British Journal of Educational Technology*, 51(6), 2006–2033. DOI: 10.1111/bjet.13030

Díaz Moreno, N., & Felices de la Fuente, M. D. M. (2017). Problemas sociales relevantes en el aula de primaria: La "cartografía de la controversia" como método. Reidics. *Revista de Investigación en Didáctica de las Ciencias Sociales.*, 1, 24–38.

Escribano, C. & Tosello, J. (2020). ¿Qué sabe el profesorado español de educación infantil sobre la ciudadanía global? Una aproximación desde la teoría fundamentada.

Ferrer, L. A., & Vivancos, A. E. (2022). *Didáctica de geografía e historia en educación primaria*. Editorial Síntesis.

García, D. (2019). La construcción de pensamiento crítico en el aula de geografía escolar. *GeoGraphos: Revista Digital para Estudiantes de Geografía y Ciencias Sociales*, 10(118), 179–205.

Gómez- Saldivia. P. A. (2022). Formar al profesorado en educación para la ciudadanía a partir de problemas socialmente relevantes. En Bel-Martínez, J.C, Colomer-Rubio, J.C. y de Alba, N. *Repensar el currículum de Ciencias Sociales: prácticas educativas para una ciudadanía crítica* (pp. 323-331). Tirant Humanidades.

Granados-Sánchez, J. (2021). Conceptos poderosos para la construcción de una ciudadanía sostenible. In Granados Sánchez, J., & Medir Huerta, R. M. (Eds.), *Enseñar y aprender geografía para un mundo sostenible* (pp. 177–199). Octaedro.

Lefebvre, H. (2000). La production de l'espace. *Anthropos: Studie z Oboru Anthropologie, Paleoethnologie, Paleontologie a Kvartérní Geologie.*

Legardez, A., & Simonneaux, L. (2006). *L'école à l'épreuve de l'actualité. Enseigner les questions vives*. ESF.

López Facal, R., & Santidrián Arias, V. M. (2011). Los "conflictos sociales candentes" en el aula. Iber:Didáctica de las Ciencias Sociales. *Geografía e Historia*, 69, 8–20.

Massey, D. (2011). *For space*. SAGE publications.

Oller, M. (1999). Trabajar problemas sociales en el aula, una alternativa a la transversalidad. En García Santa María, T. (Coord.). *Un currículum de ciencias sociales para el siglo XXI: qué contenidos y para qué* (pp. 123-131). Díada Editora, S.L.

Pagès, J. (1998). La formación del pensamiento social. En P. Benejam y J. Pagès, J. (Coords.). *Enseñar y aprender Ciencias Sociales, Geografía e Historia en la Educación Secundaria* (pp.152-164). Horsori.

Pagès, J. (2009). Enseñar y aprender ciencias sociales en el siglo XXI: reflexiones casi al final de una década. In Investigación en Educación, Pedagogía y Formación Docente, II Congreso Internacional. Libro (Vol. 2).

Pagès, J. (2019). La ciudadanía global y la enseñanza de las Ciencias Sociales: retos y posibilidades para el futuro. *Reidics:Revista de Investigación en didáctica de las ciencias sociales*. 5, 5-22.

Pagès, J., & Santisteban, A. (2011). *Les qüestions socialment vives i l'ensenyament de les ciències socials*. Universitat Autónoma Barcelona.

Rivera, J. A. S. (2016). La acción didáctica de las Ciencias Sociales y el desarrollo del pensamiento crítico. *Educación y Humanismo*, 18(31), 241–256. DOI: 10.17081/eduhum.18.31.1377

Sanmartí, N. (2018). *La finalidad principal de la evaluación es la regulación tanto de la enseñanza como del aprendizaje. 10 ideas clave*. Evaluar para Aprender. Graó.

Santos, M. (1995). *De la totalidad al lugar*. Editorial Oikos-Tau.

Santos, M. (2000). La naturaleza del espacio. *Ariel*.

Sobel, D. (2004). Place-based education: Connecting classroom and community. Nature and listening, 4(1), 1-7.

Souto, X.M. (2011). Identidades espaciales y territorios políticos. *Iber, didáctica de las ciencias sociales, geografía e historia,* 69, 55-63.

Souto, X. M. (2013). Investigación e innovación educativa: El caso de la Geografía escolar. *Scripta Nova. Revista Electrónica de Geografía y Ciencias Sociales*, 17, 1–36.

Souto, X. M. (2018). La geografía escolar: Deseos institucionales y vivencias de aula. *Boletín de la Asociación de Geógrafos.*, 2757(79), 1–31. DOI: 10.21138/bage.2757

Stenhouse, L. (1987). *Investigación y desarrollo del currículum*. Morata.

Trepat, C., & Comes, P. (2000). *El tiempo y el espacio en la didáctica de las ciencias sociales*. Graó.

Villena-Taranilla, R., Tirado-Olivares, S., Cózar-Gutiérrez, R., & González-Calero, J. A. (2022). Effects of virtual reality on learning outcomes in K-6 education: A meta-analysis. *Educational Research Review*, 35, 100434. DOI: 10.1016/j.edurev.2022.100434

Vygotsky, L. S. (1978). *Mind in society: The development of higher psychological processes*. Harvard University Press.

Chapter 8
Fostering Intercultural Skills Through Geography:
A Proposal for Future Primary Education Teachers

Cristina Honrubia-Montesinos
https://orcid.org/0000-0001-8564-4400
University of Castilla La Mancha, Spain

Lorena Alleres Otero
https://orcid.org/0009-0004-0541-9318
University of A Coruña, Spain

ABSTRACT

This chapter explores the importance of intercultural competence in today's education, emphasizing its role in promoting peaceful coexistence and social development in a multifaceted world. It analyzes pedagogical strategies that empower future educators to foster empathy, respect, and understanding of different cultures. Additionally, it identifies challenges in integrating interculturality into the educational curriculum, such as prejudice and lack of exposure to diversity. To address these obstacles, practical activities are proposed that integrate geography teaching, allowing students to contextualize their learning and build an inclusive environment. Ultimately, the purpose of the chapter is to help teachers become more aware, enabling them to act as agents of change in their communities

DOI: 10.4018/979-8-3693-7185-5.ch008

Copyright © 2025, IGI Global Scientific Publishing. Copying or distributing in print or electronic forms without written permission of IGI Global is prohibited.

1. INTRODUCTION

The present chapter highlights the significance of intercultural competence in today's world. Intercultural competence refers to the ability to engage and communicate respectfully and effectively with individuals from diverse cultural backgrounds, acknowledging and appreciating the variety of their contexts and perspectives (Normatova, 2024). This competence has become a key skill for professional success, personal development, and harmonious coexistence in multicultural societies.

In an increasingly interconnected environment, diversity has become widespread, reaching schools, workplaces, and markets, and is now a key factor for peaceful coexistence and global progress. Rather than being an obstacle, cultural variety serves as a valuable source of knowledge and experience (Esho et al., 2024). This essay examines the significance of interculturality in today's context, presenting the advantages and the challenges it entails. Specifically, it explores how these cultural interactions impact different areas, such as the economy, education, and interpersonal relationships, providing a broad view of its relevance.

Interculturality plays a vital role in building inclusive and equitable societies. As diverse cultural identities coexist, addressing the barriers that lead to division becomes imperative. Recognizing the unique and valuable contributions of each culture enhances collective well-being. In this way, a strong and cohesive social fabric can be woven, reducing tensions and fostering the development of fairer communities, where diversity is viewed not as a threat but as an asset (Eden et al., 2024).

Moreover, cultural pluralism contributes to economic growth and prosperity. In the business world, the ability to understand and collaborate effectively with individuals from different cultural backgrounds has become an essential skill. Such proficiency drives innovation and creativity while simultaneously opening doors to new markets and fostering stronger international business relationships (Chitondo & Chanda, 2023). Companies that promote cultural diversity within their teams tend to be more competitive globally, as they can approach problems from multiple perspectives and propose more comprehensive and creative solutions.

The ability to adapt and collaborate in diverse environments, known as intercultural competence, is just as crucial as technical skills in today's workplace. It serves as a catalyst for economic growth, facilitating more seamless and productive interactions between cultures (Ricardo et al., 2024). Additionally, it creates an inclusive workplace that encourages the development of innovative ideas and facilitates conflict resolution, which is essential in a globalized landscape.

Another area where interculturality shines is education. Educational institutions, from early childhood to higher education, shape individuals' perceptions of other cultures (Zalli, 2024).

In this context, the emphasis is on diversifying the curriculum to reflect the cultural richness of society while also creating an inclusive environment that promotes interaction and mutual learning among students (Malacarne, 2023).

This interaction occurs not only in the classroom but also in extracurricular activities, collaborative projects, and international experiences that enrich the educational process. AGIntegrating interculturality into educational processes helps cultivate more tolerant and empathetic citizens, who are equipped to actively participate in a pluralistic and connected society (Portera, 2023). Furthermore, it fosters a mindset of lifelong learning, as individuals who embrace cultural diversity are more likely to continue exploring and learning about different identities, positively impacting their teaching and personal growth (Maddux et al., 2021).

Emphasizing the development of these skills in higher education helps equip future educators to create classrooms that are rich in educational content as well as emotionally and socially engaging, thereby shaping a more inclusive society. Teachers with intercultural competence can transform teaching and learning dynamics, promoting respect and understanding in their classrooms, which contributes to the preparation of future generations (Foster, 2024).

However, despite the undeniable benefits of interculturality, there are also significant challenges. Resistance to accepting and understanding other cultures can stem from deeply ingrained prejudices, stereotypes, or a lack of exposure to diversity. To overcome these barriers, educators need to design and implement activities that foster open-mindedness and respect for differences. Prior work on critical thinking will also be necessary, as it helps students reflect on their limitations and appreciate the complexities of identity (Sikorska, 2024).

As mentioned, due to the variety of races, teachers must be able to engage with all students to create a sense of belonging among them (Graham et al., 2022). By organizing a classroom environment that celebrates diversity, future educators can help students feel connected to their peers, thereby enhancing their overall educational experience (Soloman, 2023).

Given the need for interculturality, teachers bear the primary responsibility for integrating it into the curriculum. Therefore, this essay aims to propose a series of activities designed to assist primary education teachers in incorporating interculturality into their classrooms. Through these activities, the goal is to educate students about cultural diversity, break down barriers, build bridges between cultures, and create opportunities for participation in projects (Amsler et al., 2020).

2. THEORETICAL REVIEW

Interculturality, rooted in respect, dialogue, and interaction among people, serves as a cornerstone for creating inclusive societies and enhancing mutual understanding (Freire, 1970; Bennett, 1993).

To achieve a society that embodies these values, García (2017) argues that schools are the most effective context for promoting equity, reflecting global diversity. In this regard, education should facilitate the active and reflective participation of students, contributing to the development of critical and culturally competent citizens (García, 2017). Moreover, geography emerges as a particularly suitable subject for developing intercultural skills (Pickles, 2012).

Exploring the intersection between cultural diversity and geography reveals a unique perspective that contributes comprehensively to understanding interculturality (Crang, 2013; Maffi, 2005). Considering these aspects, geography education provides a foundation for understanding spatial distribution and serves as a valuable tool for developing the complementary skills necessary for meaningful learning.

First, geography develops critical thinking by enabling students to analyze stereotypes, narratives, and political, economic, and environmental decisions (Tadaki et al., 2012; Sandahl, 2023; Bladh, 2020). This lays the groundwork for a deeper understanding of the inequalities and challenges faced by various countries (Bergua et al., 2024). In this context, authors such as Sauer (1974) and Harvey (2005) place geography at the heart of understanding cultural diversity (Mon, 2021).

Secondly, this discipline cultivates cultural empathy, as studying the ways in which different cultures inhabit and adapt their territories reveals the challenges these communities have faced or continue to face (Oswin, 2020; Lambert & Morgan, 2010). This perspective leads to a deeper respect for cultural differences and collective memory, opening discussions about the importance of preserving cultural heritage and passing it on to future generations (Crang, 2013). It provides students with a richer understanding that landscapes are not merely physical configurations but also bearers of stories, music, symbols, and meanings that represent community identities (Darby, 2020).

Geographical analyses of these spaces allow for an understanding of how cultures have shaped and been shaped by their environments, as well as the impact of historical events that have left their mark on the cultural landscape. This connects directly with power dynamics, as the distribution of space is rarely neutral and often results from policies, economic interests, and conflicts (Anderson, 2023). Such spatial inequalities can manifest in urban segregation, the marginalization of certain groups, and the appropriation of territories by dominant actors. This illustrates how power, in its various forms, shapes access to resources and affects communities' ability to maintain their identity and culture (Lacerda, 2023).

From a broader perspective on learning, it is proposed to leverage skills such as spatial thinking, which helps learners visualize and understand complex spatial relationships an essential competency for making informed decisions (Downs & Liben, 1991). Alongside developing the ability to interpret geospatial data, students can engage with essential technologies such as geographic information systems (GIS) and digital mapping (Pérez-del Hoyo et al., 2020).

As previously mentioned, these skills prepare students to tackle real-world challenges and contribute to sustainable and equitable solutions (Nelson, 2022). In other words, geography becomes a catalyst for developing key analytical skills (Parkinson, 2020).

Various methodological resources are suggested for teaching interculturality through geography to successfully apply this knowledge and meet the stated objectives.

One of these resources is metacognition, which refers to the ability to reflect on one's own learning process. This practice is essential for future educators to develop a conscious and effective teaching approach. According to Fox and Riconscente (2008), metacognition encompasses self-regulation, allowing students to evaluate and adjust their learning strategies based on their experiences.

This approach is particularly relevant in geography education, as future teachers can gain valuable insights by reflecting on how their own classroom experiences shaped their understanding of cultural and geographical diversity. Metacognition enables them to identify which methods and strategies were most effective for them and which could be beneficial in their future classrooms. Furthermore, integrating metacognitive practices in activities such as case studies or projects addressing global issues encourages collaborative skills by teaching students to negotiate and seek solutions (Fox and Riconscente, 2008).

Moreover, case studies provide a critical analysis opportunity. By investigating historical events that have impacted cultural diversity in their region, students can reflect on how geographical and political decisions have shaped cultural identities (Baskarada, 2014). This methodology allows them to connect theory and practice, making learning more meaningful.

Furthermore, this experiential learning enriches geographical and cultural understanding by directly engaging students with various cultures. It enables students to learn about identities while experiencing the richness of diversity in a real-world context (Nussbaum, 2019).

Project-based learning (PBL) is another experiential methodology that engages students in hands-on and collaborative projects to investigate intercultural themes. This approach encourages reflection on cultural diversity, deepening their understanding of its significance within geography (Thomas, 2000).

Finally, debates and forums, as previously mentioned, stimulate discussion and critical analysis of intercultural topics. Through these platforms, students cultivate communication skills and develop empathy for different perspectives, which is essential in our globalized world (Beames et al., 2023).

Together, these methodologies strengthen geography education and contribute to the development of educators who can effectively and thoughtfully address interculturality.

3. OBJECTIVES

The primary objective is to develop a project aimed at enhancing intercultural skills among undergraduate students in the Primary Education degree program, specifically within the course titled "Social Sciences and its Didactics: Geography." This initiative will focus on equipping future teachers with the necessary tools to foster understanding, empathy, and respect for diverse cultures through the lens of geography and social sciences.

4. PROPOSAL

4.1 Context

The course titled "Social Sciences I: Geography and Its Didactics" is offered in the Bachelor's degree program for Primary Education, specifically for 2nd-year students, and carries a total of 6 credits. The course is divided into two sections: physical geography and human geography content.

This structure not only provides a comprehensive understanding of geographical principles but also emphasizes the pedagogical approaches necessary for effectively teaching these concepts in a primary education setting.

4.2 Methodology

The methodology is grounded in critical reflection, which encourages students to actively question and analyze the content they encounter. This approach fosters an open and dynamic learning environment where students feel empowered to express their thoughts, challenge assumptions, and explore diverse perspectives. By engaging in critical thinking, students can develop a deeper understanding of the material and its implications in real-world contexts.

Furthermore, the methodology incorporates collaborative learning, which plays a vital role in enhancing the educational experience. Through group discussions, students share their findings, insights, and personal experiences related to the topics at hand. This collaborative approach not only promotes dialogue but also allows students to learn from one another, as they consider different viewpoints and ideas.

In this interactive setting, students can build on each other's contributions, leading to richer discussions and a more comprehensive grasp of the subject matter. The exchange of ideas fosters a sense of community and teamwork, making the learning process more engaging and meaningful.

Overall, the combination of critical reflection and collaborative learning enriches the educational experience, equipping students with essential skills for analyzing complex issues, communicating effectively, and working together toward shared goals.

The main learning outcomes of this course are the next ones:

Competences

- Know the school curriculum for social sciences and relate it to the social sciences syllabus of the degree program.
- Integrate historical and geographical studies from an instructional and cultural perspective.
- Promote the democratic education of citizenship and the practice of critical social thinking.
- Develop and evaluate curriculum content using appropriate teaching resources and promote the corresponding competencies in students.
- Students will demonstrate knowledge and understanding in a field of study based on general secondary education, often at a level supported by advanced textbooks, while also including some aspects from the forefront of their field of study.
- Students will know how to apply their knowledge to their work or vocation in a professional manner, demonstrating competencies through the formulation and defense of arguments and the resolution of problems within their field of study.
- Mastery of Information and Communication Technologies (ICT).
- Proficient oral and written communication.
- Ability to identify, classify, and develop types of learning activities for teaching History, Geography, and other Social Sciences.

Outcomes

- Be able to adapt to social, economic, and cultural changes and apply them to the knowledge of social sciences.

- Be able to explain the process of conceptualizing geographical space for children aged 6 to 12.
- Be able to explain and apply the techniques and methods specific to History (temporal categories, representations of cycles and stages, use of historical sources, employment of specific vocabulary, formulation of hypotheses, etc.), Geography (orientation and measurement of space, graphic and cartographic representation, direct observation, landscape analysis, etc.), and other social sciences (quantitative and qualitative methods, single case studies, etc.).
- Be able to identify the content, techniques, methods, and evaluation criteria for Geography and History in Primary Education curriculum.
- Be able to integrate historical and geographical studies from an instructional and cultural perspective.
- Be able to incorporate new technologies, both computer-based and audiovisual, into the teaching of History, Geography, and other Social Sciences.

4.3. Activities

Table 1. Analysis of personal experience in the educational system

Analysis of personal experience in the educational system
Outcomes Be able to identify the content, techniques, methods, and evaluation criteria for Geography and History in Primary Education curriculum
Contents The current curriculum of Social Sciences
Competences Know the school curriculum for social sciences and relate it to the social sciences syllabus of the degree program. Proficient oral and written communication

continued on following page

Table 1. Continued

Description of the activity In this activity you have the opportunity to remember and reflect on your own experience as social studies students, in Primary and Secondary. You have to answer a series of questions in an essay, regarding how you remember social classes and what differences you think there are with respect to how history and geography are currently taught in Primary and Secondary and how you think they should be taught. Development of the activity Read the text and answer the questionnaire. Surely over the years you have had a series of experiences in relation to the educational system, some pleasant, others unpleasant, good moments, bad moments..., of which you still have memories. In the same way, you have met teachers who, for some positive or negative reason, deserve special consideration. Think about some of those situations and tell them, describing the most significant and relevant details according to the script. • What memory do you have about Geography and History? • Has your experience as a student influenced the interest you currently have in Geography and History? • What happened in classes? What was the usual sequence of work? • What were the problems, difficulties, etc., in your opinion, that most worried the teachers and you as a student? • Do you think that teacher training in relation to the teaching of Social Sciences has changed compared to when you were students? • Do you currently think the same skills and content are required?
• Assessment criteria To use critical thinking to compare their experiences with the current curriculum of Social Sciences

Source: Own elaboration.

Table 2. Exploring the Local and the Global

Activity: Exploring the Local and the Global
Outcomes • Be able to adapt to social, economic, and cultural changes and apply them to the knowledge of social sciences. • Be able to integrate historical and geographical studies from an instructional and cultural perspective
Contents
• Competences Integrate historical and geographical studies from an instructional and cultural perspective. Proficient oral and written communication

continued on following page

Table 2. Continued

Description of the activity In this activity, we will highlight both local and global aspects to appreciate the differences and similarities between villages, as these elements enrich us culturally and socially. To achieve this, we will select a village from Castilla La Mancha and another from a different country and continent. The goal is for students to understand how different contexts can share characteristics while also being unique. Information to Collect For each of the selected villages, we need to gather essential information on the following aspects: Location: Indicate the continent, country, and region where the village is located. Physical Geography: Physical factors: Analyze the type of soils present in the area, the climate (whether it's warm, temperate, cold, etc.), the relief (mountains, valleys, rivers), as well as the flora and fauna found there. Describe the essential characteristics of the physical geography of the place, such as the presence of natural resources and soil quality for agriculture. Population Geography: Indicate the number of inhabitants in the village. If data is available, also specify whether the population is aging or if there are many births, which could reflect the social dynamics of the area. Cultural Geography: Explore the most notable cultural aspects, such as the language spoken, predominant religions, local traditions, typical clothing, and the gastronomy that characterizes the region. This will help students understand the cultural identity of each village. Economic Geography: Identify the economic activities carried out in the village. This can include information about specific crops (such as cereals, fruits, vegetables) or livestock (types of animals raised, dairy production, etc.). It may also address the importance of the local economy in the regional and national context. Steps of the Activity Selection of Villages: Each student will select a village from Castilla La Mancha and another from a different country and continent. It's important to share these choices in a collaborative mural to avoid repetition. Adapting the Information: The collected information should be adapted to make it understandable for the target audience (e.g., children of different ages). This may involve simplifying the language or using visual examples. Creating the Mural: A mural will be created that includes all the information. Students will have total freedom to be creative, using colors, images, and drawings to represent the selected villages. It's also suggested to include words in the foreign village's language to showcase linguistic diversity. Informal Presentation: An informal presentation of the murals will take place. There's no need to prepare a formal speech; classmates will be able to observe and comment on the displayed work. Exhibition of Creative Murals: The most original or creative murals will have the opportunity to be exhibited in the education faculty hall, allowing a broader audience to appreciate the students' efforts and creativity. Planning of the Activity The activity will be carried out during class over three weeks. About 30 to 45 minutes will be assigned in each session to work on gathering information and creating the murals. However, if the teacher observes that students are not working or if they finish ahead of schedule, the deadline may be moved up. Activity:

continued on following page

Table 2. Continued

• Assessment criteria Information gathered on physical geography, population, culture, and economy is comprehensive and well-researched. Information is directly related to the chosen villages and enhances understanding of their characteristics. The mural is colorful, engaging, and visually represents the key aspects of each village effectively. Use of unique materials, designs, or concepts that enhance the presentation and capture the audience's attention. Clear demonstration of understanding the cultural identity of each village, including traditions, languages, and gastronomy. Ability to articulate similarities and differences between the two villages, enriching the discussion around cultural diversity. Active participation in group discussions, sharing of responsibilities, and collaboration during the research and mural creation process. Willingness to contribute ideas and support peers, fostering a positive group dynamic. Ability to convey information clearly and confidently during the informal presentation, engaging the audience effectively. Incorporation of foreign language elements in a meaningful way, enhancing the linguistic diversity of the presentation.

Source: Own elaboration

Table 3. Analyzing a Population Pyramid

Analyzing a Population Pyramid
Outcomes • Be able to adapt to social, economic, and cultural changes and apply them to the knowledge of social sciences. • Be able to integrate historical and geographical studies from an instructional and cultural perspective
Contents Population Geography
• Competences Integrate historical and geographical studies from an instructional and cultural perspective. Proficient oral and written communication

continued on following page

Table 3. Continued

Description of the activity In this activity, we will focus on selecting a population pyramid from a specific place, which can be either a city or a country. The chosen location should be relevant in today's context due to its political, social, or economic situation. Steps to Follow Selection of the Population Pyramid: Choose a specific city or country that has been highlighted in recent news or studies due to significant issues such as migration, aging populations, economic challenges, or political changes. This relevance will help engage the students and provide real-world context. Justifying Its Importance: Explain why this population pyramid is significant. Discuss the current events or trends affecting the population, such as economic instability, changes in immigration laws, or health crises. This context will help students understand the implications of demographic changes. Explaining the Population Pyramid: Describe the shape of the population pyramid. Is it a broad base with a narrow top (indicating high birth rates and a younger population), or is it inverted (indicating an aging population)? Provide visual representations to make it easier for students to grasp these concepts. Analyzing the Shape: Discuss the specific characteristics of the population pyramid. For example: Expansive Pyramid: High birth and death rates, common in developing countries. Constrictive Pyramid: Lower birth rates and a higher proportion of older individuals, often seen in developed countries. Stationary Pyramid: Stable population, with relatively equal numbers across age groups. Explaining Historical and Social Context: Look for historical, social, political, or economic explanations for the observed shape of the pyramid. For instance: Historical events like wars, famines, or pandemics that may have impacted population growth. Social policies, such as family planning initiatives or immigration policies. Economic factors like job availability or educational opportunities that influence birth rates. Adapting Information for 6th Grade Students. After gathering and analyzing this information, the next step is to present it to 6th-grade elementary school students. Here are some tips for adapting the content: Simplify Language: Use clear and straightforward language that is age-appropriate. Avoid jargon and complex terminology. Use Visual Aids: Incorporate images of the population pyramid, graphs, and charts. Visual representations will help students better understand the concepts. Incorporate Videos: Find short educational videos that explain population pyramids or related topics. This can help reinforce the information in an engaging way. Utilize Technology: Consider using interactive apps or online tools that allow students to explore demographic data and create their own population pyramids. This hands-on approach can enhance learning. Engage with Questions: Encourage students to ask questions or share their thoughts about the population pyramid and its implications. This can foster a deeper understanding and make the lesson more interactive. By completing this activity, students will gain a better understanding of how demographic factors influence society. They will learn to analyze and interpret population pyramids and understand their relevance in today's world. This will not only enrich their knowledge but also develop critical thinking skills as they relate historical and social contexts to current events.
• Assessment criteria The selected city or country is current and significant due to relevant political, social, or economic issues. Clear explanation of why the location was chosen. Comprehensive analysis of the population pyramid, including shape, implications, and demographic trends. Inclusion of historical, social, and economic contexts that influence the pyramid's shape. Information is presented in a clear, age-appropriate manner suitable for 6th-grade students. Jargon is avoided, and complex concepts are broken down effectively. Effective incorporation of visual elements (graphs, charts, images) that enhance understanding. Visuals are relevant and clearly labeled. Encouragement of questions and active participation from peers during the presentation. Facilitation of discussions that deepen understanding. Effective use of technology (apps, videos, interactive activities) to engage students and reinforce learning. Clear explanation of the population pyramid's shape and its demographic implications. Students demonstrate understanding of different types of pyramids (expansive, constrictive, stationary).

Source: Own elaboration.

Table 4. Analyzing a Population Pyramid

Analyzing a Population Pyramid
Outcomes • Be able to adapt to social, economic, and cultural changes and apply them to the knowledge of social sciences. • Be able to integrate historical and geographical studies from an instructional and cultural perspective
Contents Curriculum of Social Sciences
• Competences Integrate historical and geographical studies from an instructional and cultural perspective. Proficient oral and written communication
Description of the activity As you know, the space of the students, as well as the evolutionary development, will determine how to develop geography. So, in this portfolio activity we are going to have to design 4 activities for your students. The following criteria must be met: • One of the centers is located in a rural environment, while the other is located in an urban environment. • One of the groups is 1st grade, while the other is 6th grade, in both centers. • You must explain the procedure of the activity, its objectives, the necessary materials, the grouping, the basic knowledge, the key and specific competencies that are going to be developed, as well as the evaluation criteria. • It is recommended that these activities promote active student learning and a procedural nature. • You have complete freedom when designing these activities, in case you want to involve both centers or separately. Also, it gives it a more realistic character if you design or include the materials that you are going to use, for example, design of Google Earth projects, National Geographic, APPANGEA... • Maximum: 6 sheets, minimum: 3 sheets.
• Assessment criteria Discuss what geography is and why it's important. Take students on a guided walk around the school grounds to identify local geographical features (trees, hills, rivers). Active engagement in the activity. Ability to grasp and explain key geographical concepts. Quality and originality of work produced Teamwork and cooperation during group activities. Clarity and effectiveness in sharing findings.

Source: Own elaboration.

5. KEY WORDS

The following keywords are fundamental for understanding the approach to intercultural education within the context of geography and teacher training. Each term addresses a crucial aspect that contributes to the development of intercultural competencies and the preparation of educators in an increasingly diverse environment.

Interculturality refers to the dynamic interactions and exchanges between different cultures. It includes the importance of understanding, respecting, and valuing cultural diversity while promoting dialogue and collaboration among people from various backgrounds. By engaging with diverse cultural perspectives, individuals develop skills that facilitate effective communication and coexistence in multicultural settings, ultimately enriching their own experiences and contributing to social cohesion.

This understanding of interculturality is essential within the realm of **human geography,** a subfield that focuses on the relationships between people and their environments. Human geography examines how human culture, societies, and economies interact with the spatial aspects of the world, exploring topics such as migration, urbanization, cultural landscapes, and social practices. This field provides insights into the complexities of cultural interactions and the spatial dynamics that influence communities.

In **higher education's** context, the cultivation of intercultural competence becomes increasingly important. Higher education, which encompasses the level of education that occurs after secondary school—typically involving universities, colleges, and other institutions—focuses on specialized knowledge, critical thinking, and research. As students from diverse backgrounds come together in these settings, higher education aims to cultivate informed and engaged citizens who can contribute to society. The collaborative learning helps environment academic understanding and intercultural awareness.

Preparing future educators relies heavily on teacher training to promote intercultural understanding. This process includes pedagogical theory, practical teaching experience, and skill development in various subject areas, all of which are vital for creating inclusive classrooms. By prioritizing the development of intercultural competencies, teacher training ensures that educators are prepared to meet diverse learning needs and cultivate a sense of belonging for all students. Incorporating interculturality into teacher training programs equips educators to interact with students from various cultural backgrounds and establishes an atmosphere of mutual respect and understanding.

Finally, the concept of **critical thinking** serves as a pillar for navigating the complexities of intercultural education. This essential skill involves analyzing, evaluating, and synthesizing information and arguments while questioning assumptions and identifying biases. Critical thinking encourages individuals to think independently, consider multiple perspectives, and engage in reflective thought. Within intercultural education, students can tackle complex cultural issues, challenge stereotypes, and develop a nuanced understanding of diversity, further enhancing their ability to thrive in a multicultural world.

6. LIMITATION AND FUTURE RESEARCH

The main limitation of this chapter is that it presents a proposal rather than empirical research. As a result, it cannot assert with confidence that the proposed methods effectively promote the acquisition of intercultural skills among future educators. Although the theoretical framework and rationale for integrating intercultural competencies into teacher training are clearly articulated, the absence of empirical evidence leaves the actual impact of these strategies untested. While the theoretical frameworks are sound, they need practical support to validate their assumptions and ensure that the proposed strategies achieve the desired intercultural objectives in diverse educational contexts.

Another methodological limitation is the potential lack of diversity in the sample, which future research should address seriously. To understand how outcomes may differ significantly, it is necessary to include participants from a variety of cultural, socioeconomic, and geographic backgrounds. Including a diverse sample would enhance the representativeness of the results and help identify how various cultural realities affect the effectiveness of the proposed pedagogical strategies. Without this diversity, the conclusions drawn risk being limited and not applicable to all educational settings. A lack of diversity could limit the applicability and generalization of findings to different educational settings. Moreover, the duration of the interventions could influence the results; short-term studies might not adequately capture the long-term effects of intercultural training on educators. Future research should consider a longer follow-up period to assess whether initial benefits are sustained or evolve over time.

To address these limitations, future research could take the form of a quasi-experimental study involving both an experimental group and a control group. The proposed design allows researchers to thoroughly assess the effectiveness of the intercultural training interventions. By comparing outcomes between the two groups, researchers could draw more definitive conclusions about the promotion of intercultural skills and competencies. This approach would enable the measurement of the immediate impact of the interventions and facilitate observation of how participants' diverse experiences shape their intercultural development. Additionally, the use of control groups would strengthen the internal validity of the study, providing a clear reference point for evaluating the success of the implemented strategies.

A suitable instrument for measuring these skills could be the Assessment of Intercultural Competence (AIC), which provides a structured approach to evaluating various dimensions of intercultural effectiveness. Utilizing such an established tool would lend credibility to the findings and allow for comparisons with other studies in the field. Furthermore, the AIC could be complemented by other instruments that measure teachers' self-efficacy in intercultural contexts, broadening the understanding of the impact of the proposed interventions on both teacher training and classroom practice.

Beyond quantitative measures, a mixed-methods approach may improve the research. By incorporating qualitative techniques such as interviews and focus groups, researchers can gather detailed insights into future teachers' perceptions and experiences related to intercultural appreciation. Qualitative data can reveal how educators perceive and interact with different cultures and identities, providing a deeper understanding of the nuances involved in intercultural competence. The combination of qualitative and quantitative approaches would provide researchers with the ability to capture statistically relevant data while also exploring the emotional and social dimensions of intercultural training, an important aspect often overlooked in purely quantitative studies.

A promising direction for future research involves examining the long-term effects of intercultural competencies on teaching practices. Investigating how the skills developed during training shape classroom interactions and teaching methods in practice may grant valuable insights. Such a study could clarify how intercultural competencies are applied in daily teaching and their role in fostering inclusive and equitable educational environments. Additionally, exploring the long-term effects would help determine whether intercultural competencies are consistently integrated into teaching practice or if they require ongoing reinforcement over time.

Exploring these future research directions could validate the proposed interventions and enrich the broader discourse on preparing educators to navigate and celebrate cultural diversity in increasingly globalized educational contexts. By systematically investigating these areas, we can equip future teachers with the intercultural skills necessary for creating inclusive and responsive learning environments. Implementing long-term intercultural training programs could transform how teachers approach diversity in their classrooms, promoting a culture of respect and appreciation that extends beyond the educational sphere.

In conclusion, interculturality serves as a foundational element for promoting fair and prosperous societies. By cultivating mutual respect and cultural understanding, it contributes to global economic growth and collective well-being. However, to ensure that diversity has a positive impact, challenges such as prejudice, and a lack of empathy need to be addressed. This study has presented a teaching strategy for incorporating interculturality into higher education through geography instruction, providing educators with practical guidance for integrating this approach into their classrooms. This strategy is based on the reconception of the subject of Geography, which is typically taught in a decontextualized manner, in order to view it as a catalyst for the intercultural education of future teachers.

Ultimately, interculturality contributes to achieving peaceful coexistence in today's interconnected world and serves as a significant factor in driving social progress and sustainable development. Prioritizing intercultural training in higher education ensures that future educators are well-prepared and motivated to create a culture of inclusion and respect in their classrooms. This empowerment allows them to become agents of change, shaping society toward a more just and equitable future.

REFERENCES

Amsler, S., Kerr, J., & Andreotti, V. (2020). Interculturality in teacher education in times of unprecedented global challenges. *Education and Society (Melbourne)*, 38(1), 13–37. DOI: 10.7459/es/38.1.02

Anderson, B. (2023). Forms and scenes of attachment: A cultural geography of promises. *Dialogues in Human Geography*, 13(3), 392–409. DOI: 10.1177/20438206221129205

Baskarada, S. (2014). Qualitative case study guidelines. *The Qualitative Report*, 19(40), 1–25.

Beames, S., Higgins, P., Nicol, R., & Smith, H. (2023). *Outdoor learning across the curriculum: Theory and guidelines for practice*. Taylor & Francis. DOI: 10.4324/9781003010890

Bennet, M. J. (1993). Toward ethnorelativism: A developmental model of intercultural sensitivity. In Paige, R. M. (Ed.), *Education for the intercultural experience* (pp. 21–71). Intercultural Press.

Bergua, S. B., Fields, D. L., & Gómez, N. B. (2024). *Geography and global change through interculturality and CLIL*. Universidad de Oviedo.

Bladh, G. (2020). GeoCapabilities, didaktical analysis and curriculum thinking – Furthering the dialogue between Didaktik and curriculum. *International Research in Geographical and Environmental Education*, 29(3), 206–220. DOI: 10.1080/10382046.2020.1749766

Chitondo, L., & Chanda, C. T. (2023). Effects of ethnic diversity on the development of a country: A case of Zambia. *International Journal of Current Business and Social Sciences*, 9(5), 01-13.

Crang, M. (2013). *Cultural geography*. Routledge. DOI: 10.4324/9780203714362

Darby, W. J. (2020). *Landscape and identity: Geographies of nation and class in England*. Routledge. DOI: 10.4324/9781003135852

Downs, R. M., & Liben, L. S. (1991). The development of expertise in geography: A cognitive-developmental approach to geographic education. *Annals of the Association of American Geographers*, 81(2), 304–327. DOI: 10.1111/j.1467-8306.1991.tb01692.x

Eden, C. A., Chisom, O. N., & Adeniyi, I. S.Chima Abimbola EdenOnyebuchi Nneamaka ChisomIdowu Sulaimon Adeniyi. (2024). Cultural competence in education: Strategies for fostering inclusivity and diversity awareness. *International Journal of Applied Research in Social Sciences*, 6(3), 383–392. DOI: 10.51594/ijarss.v6i3.895

Eden, C. A., Chisom, O. N., & Adeniyi, I. S.Chima Abimbola EdenOnyebuchi Nneamaka ChisomIdowu Sulaimon Adeniyi. (2024). Cultural competence in education: Strategies for fostering inclusivity and diversity awareness. *International Journal of Applied Research in Social Sciences*, 6(3), 383–392. DOI: 10.51594/ijarss.v6i3.895

Esho, E. O. (2024). The paradigm of peace enhancing peace culture. *Journal of Aggression, Conflict and Peace Research*, 16(1), 83–97. DOI: 10.1108/JACPR-12-2022-0765

Foster, K. J. (2024). New classroom teachers' perceptions of multicultural teacher preparation. *Educational Leadership & Policy Studies Dissertations*, 4. https://mavmatrix.uta.edu/edleadershippolicy_dissertations/4

Fox, E., & Riconscente, M. (2008). Metacognition and self-regulation in James, Piaget, and Vygotsky. *Educational Psychology Review*, 20(4), 373–389. DOI: 10.1007/s10648-008-9079-2

Freire, P. (1970). Pedagogy of the oppressed. *Continuum : an Interdisciplinary Journal on Continuity of Care*.

García, C. (2017). *La educación intercultural: Claves para una enseñanza inclusiva*. Editorial Universitaria.

Graham, S., Kogachi, K., & Morales-Chicas, J. (2022). Do I fit in: Race/ethnicity and feelings of belonging in school. *Educational Psychology Review*, 34(4), 2015–2042. DOI: 10.1007/s10648-022-09709-x

Harvey, D. (2005). The sociological and geographical imaginations. *International Journal of Politics Culture and Society*, 18(3-4), 211–255. DOI: 10.1007/s10767-006-9009-6

Lacerda, D. S. (2023). Investigating the political economy of the territory: The contradictory responses of organisations to spatial inequality. *Organization*, 30(5), 1074–1093. DOI: 10.1177/13505084211061239

Lambert, D., & Morgan, J. (2010). *Teaching geography 11–18: A conceptual approach*. Open University Press.

Maddux, W. W., Lu, J. G., Affinito, S. J., & Galinsky, A. D. (2021). Multicultural experiences: A systematic review and new theoretical framework. *The Academy of Management Annals*, 15(2), 345–376. DOI: 10.5465/annals.2019.0138

Maffi, L. (2005). Linguistic, cultural, and biological diversity. *Annual Review of Anthropology*, 34(1), 599–617. DOI: 10.1146/annurev.anthro.34.081804.120437

Malacarne, S. (2023). Mutual learning: A real way to cooperate? Evidence of more equal international development cooperation. *Italian Sociological Review*, 13(3), 347–369.

Mon, P. R. (2021). An introduction to the concept of landscape in geography. *The Asian Review of Civil Engineering*, 10(1), 20–25. DOI: 10.51983/tarce-2021.10.1.2941

Nelson, T. A., Goodchild, M. F., & Wright, D. J. (2022). Accelerating ethics, empathy, and equity in geographic information science. *Proceedings of the National Academy of Sciences of the United States of America*, 119(19), e2119967119. DOI: 10.1073/pnas.2119967119 PMID: 35507875

Normatova, A. (2024). Five ways to build intercultural competence and confidence in your classroom. *Educational Research in Universal Sciences*, 3(4), 512–514.

Normatova, A. (2024). Five ways to build intercultural competence and confidence in your classroom. *Educational Research in Universal Sciences*, 3(4 SPECIAL), 512-514.

Nussbaum, M. C. (2019). *The monarchy of fear: A philosopher looks at our political crisis*. Simon & Schuster.

Oswin, N. (2020). An other geography. *Dialogues in Human Geography*, 10(1), 9–18. DOI: 10.1177/2043820619890433

Parkinson, A. (2020). *Why study geography?* London Publishing Partnership.

Pérez-delHoyo, R., Mora, H., Martí-Ciriquián, P., Pertegal-Felices, M. L., & Mollá-Sirvent, R. (2020). Introducing innovative technologies in higher education: An experience in using geographic information systems for the teaching-learning process. *Computer Applications in Engineering Education*, 28(5), 1110–1127. DOI: 10.1002/cae.22287

Pickles, L. (2012). Geography and intercultural understanding. *Interaction*, 40(1), 9–11. https://search.informit.org/doi/10.3316/ielapa.038913886192132

Portera, A. (2023). Global versus intercultural citizenship education. *Prospects*, 53(3), 233–248. DOI: 10.1007/s11125-021-09577-3

Ricardo, C., Peñuela Epalza, M., Lombana Coy, J., Guevara Novoa, C. C., Ramírez, M., & Cano, J. (2024). *Transformaciones de la competencia intercultural. Prácticas de una comunidad de aprendizaje docente (CAD)*. Editorial Uninorte.

Sandahl, J. (2023). Engaging with scale and place: Geographical thinking on migration in middle-school. [ZGD]. *Zeitschrift für Geographiedidaktik*, 51(2), 55–69.

Sauer, C. O. (1974). The fourth dimension of geography. *Annals of the Association of American Geographers*, 64(2), 189–192. DOI: 10.1111/j.1467-8306.1974.tb00969.x

Sikorska, L. (2024). Different meanings of "diversity". *Linguistic, Literary, and Cultural Diversity in a Global Perspective*, 21, 8–26. DOI: 10.1075/fillm.21.01sik

Soloman, J. M. (2023). Embracing Diversity to Increase Belongingness and Foster a Culture of Success. *The Organizational Improvement Plan at Western University*, 321. https://ir.lib.uwo.ca/oip/321

Tadaki, M., Salmond, J., Le Heron, R., & Brierley, G. (2012). Nature, culture, and the work of physical geography. *Transactions of the Institute of British Geographers*, 37(4), 547–562. DOI: 10.1111/j.1475-5661.2011.00495.x

Thomas, J. W. (2000). *A review of research on project-based learning*. The Autodesk Foundation.

Zalli, E. (2024). Globalization and education: Exploring the exchange of ideas, values, and traditions in promoting cultural understanding and global citizenship. *Interdisciplinary Journal of Research and Development*, 11(1 S1), 55–55. DOI: 10.56345/ijrdv11n1s109

Chapter 9
Preparing Children for the Future With Environmental Education

Bedia Yıldırım
https://orcid.org/0000-0002-3167-4038
Aksaray University, Turkey

ABSTRACT

The natural environment, which has been damaged by human interventions, was initially able to tolerate human-induced negative impacts, but over time this destruction has reached the dimension of environmental disaster. Although steps were taken to address environmental problems from time to time, the measures taken were not sufficient and environmental problems continued to increase. Considering that the easiest, permanent and economical way to solve problems is through education, education should be given importance in the fight against environmental problems. The prerequisite for a livable environment is to raise an environmentally sensitive generation. The first target should therefore be the critically important early childhood period. In order to invest in our future, it is necessary to increase children's contact with nature, to ensure that they love and protect nature, and to develop their attitudes and skills to contribute to nature. So what are the duties of families and educators in this regard, what can we do for a more ecological, sustainable and livable environment?

DOI: 10.4018/979-8-3693-7185-5.ch009

Copyright © 2025, IGI Global Scientific Publishing. Copying or distributing in print or electronic forms without written permission of IGI Global is prohibited.

INTRODUCTION

Human beings and the environment they live in are inseparable. Within this integrity, it is considered a normal phenomenon for humans to interact with nature, use and change nature for their vital needs. However, the fact that humans use nature only for their own benefit without thinking about it leads to environmental problems and even environmental disasters that are difficult or even impossible to solve (Sadık, Çakan & Artut, 2011: 1067). Environmental problems are not only excessive and unplanned human intervention in the natural environment, but also destructive environmental formations that occur during the natural environment's own cyclical process, especially global-scale spatial changes (Özgen & Kahyaoğlu, 2011:137). Environmental problems are excessive and unplanned human interventions in the natural environment, as well as destructive environmental formations that occur during the natural environment's own cyclical process, especially global-scale spatial changes. (Özgen & Kahyaoğlu, 2011:137). Although various environmental problems have been experienced throughout human history, the intensification of economic activities for the needs of the increasing population has led the process to a dead end. In parallel with the rapid development of industry in the world, the environment, which is important for all living things, has started to be rapidly polluted, destroyed, in short, destroyed as a result of the consumption of natural resources to meet the need for raw materials, the increase in production wastes and unconscious damage to nature (Yalçınkaya, 2013: 417). Thus, environmental elements became unable to bear the burden brought about by the industrial revolution and unplanned urbanization. The modern era has been a period in which environmental problems have diversified and intensified. The effects of pollution on air, water and soil have become much more visible, nuclear hazards, biological threats, the risks of ozone depletion, the increase in waste load and the inadequacy of recycling processes, and global warming stand in front of humanity as important environmental problems (Kayaer, 2012:24- 27). For almost the entire 20th century, the world has acted with the belief that man, armed with science and technology, has the right to exploit nature unlimitedly to satisfy his desires, and it is in this context that the roots of environmental problems began to emerge (Ardoğan, 2012:89). As a result of the negative developments in the 1970s, leading figures in the fields of politics, education and science started to attach importance to the solution of environmental problems and although various steps were taken (Balkan Kıyıcı, 2009), human beings continued to live their lives without being aware of the damage they caused to the environment or without caring about the global scale of environmental pollution, which was thought to be insignificant (Ünal, Mançuhan & Sayar, 2001: 6).

One of the most important environmental problems encountered by humans is environmental pollution. Soil, water, air, noise, food, sea and ocean pollution, forest destruction, erosion, desertification, destruction of plant and animal species, nuclear wastes, ozone depletion, climate change are the most important environmental problems we face (Kırlıoğlu & Zeytin, 2015:754, Ertürk, 2017). These human-induced environmental problems such as consumption of natural resources, rapid population growth, industrialization, unplanned urbanization, air pollution, soil pollution are increasing day by day (Kaya, 2014; Tomar, 2009). Although geographical and economic phenomena such as industrialization, urbanization, tourism, population growth are shown as the cause of environmental problems, the main reason is the carelessness and inadequacies in environmental education, environmental sensitivity and environmental awareness, disorders and inaccuracies in environmental attitudes and behaviors (Atasoy, 2010: 31). Population growth is not an environmental problem but an important factor in the emergence of environmental problems. Destruction of forest areas to meet the needs of nutrition and shelter, unconscious hunting of land and water animals, destruction of natural areas, unplanned construction, depletion of natural resources and wastes generated as a result of human vital activities are problems that occur due to population growth (Gürbüz & Kışoğlu, 2011: 159). Of course, the environmental problems mentioned did not occur all at once. At first, the negative consequences of environmental pollution were not felt thanks to the self-renewal power of nature. However, the accumulation and diversification of pollution over time led to the formation of environmental problems (Yaman & Gül, 2018: 201). The deterioration in the ecological balance has caused many problems such as the increase in natural disasters, temperatures, desertification, storms, hurricanes, flood disasters and the drying up of lakes (Can, 2016: 59). Thus, environmental problems have ceased to be local, have become too big for nations to solve on their own, and have reached the point of destroying not only humanity but also all living species and world life (Kaypak, 2013: 20). However, over time, the idea that the environment is a trust for the future has become widespread and the importance given to the environment has increased globally (Zayimoğlu & Öztürk, 2015: 117). The need for raising young generations who will find solutions to the problems that are being experienced in our world, which is deteriorating more and more every day, and that are likely to be experienced in the future, has increased in parallel. Thus, it has become a necessity to raise environmentally conscious individuals to ensure that future generations live in a healthier environment (Yoleri, 2012: 100) In this context, it is seen as very important to gain environmental awareness and ecological literacy skills in early childhood, which form the basis of the characteristic traits, attitudes and behaviors in later years of life. Conclusion: What can we do to overcome the barriers to ecological literacy?

While children born in the years when the serious consequences of global climate problems were observed are expected to be more aware of environmental problems and more conscious and sensitive in their behaviors (Hedefalk, Almqvist, & Östman, 2015: 976), the fact that the situation is getting worse day by day has made it necessary to investigate the underlying factors. In this regard, the human perspective on nature is that it is an unlimited store of resources that can be taken and consumed in an unplanned manner. Considering that different factors such as the cultural perspective of the society in which one lives, the conditions of upbringing, moral understanding and legal rules, education and training processes cause this (Atasoy, 2005: 77), it is seen that these reasons are united on the human factor. This situation increases the importance of an understanding of education that will change the perspective of today's people towards nature and shape their values and attitudes in preventing environmental problems (Gökçe, Kaya, Aktay, & Özden, 2007: 454). In this direction, in order to create environmental awareness, first of all, philosophical awareness should be created. With the idea that "we are all tenants of the earth (Udall)", we should adopt the environment as our home. We should stop believing that humans are the "masters of the universe", think about the rights of all living and non-living beings and future generations, and direct environmental behaviors accordingly. In short, the way to a radical and sustainable change in the perspective on the environment and environmental problems is through education (Gül, 2013: 19). Because education is the most effective and permanent way to raise awareness about environmental problems. And the identification of this situation dates back to the 1960s. In 1964, W. Stapp's definition of education for the environment formed the basis for scientific studies after that date (Örnek, 1994) and since the 1972 Stockholm Conference (UNEP, 1972), environmental education has been recognized as an important tool in solving environmental problems. The definitions and principles of environmental education were put forward at the Conference on Environmental Education held in Tbilisi in 1977 (UNESCO, 1977). The first Conference on Environmental Education organized by the United Nations was the first meeting on this subject where the principles of environmental education were determined (Özdemir, 2007: 25) and its objectives were also included (Akçay, 2006). In the twenty-first century, education for the environment, international cooperation in solving problems and mutual solidarity between states are seen as prerequisites. Because without the adoption of common environmental policies among countries in the solution of global environmental problems and without the dissemination of environmental education in all countries, salvation will not be achieved in the long term (Atasoy, 2005: 120). Because environmental education is not an individual subject. Environmental education is education that aims to inform and raise awareness about the environment in which people are intertwined throughout their lives and cannot be isolated, the future of this environment and how to protect it (Dikmen, 1993: 22). Environmental education is a lifelong interdisciplinary approach that evaluates issues related to the environment, works towards solving problems that may occur, and acts with the aim of creating a social consciousness with knowledge, experience, skills, motivation, attitudes, personal and social responsibilities that will prevent the formation of new situations (Moseley, 2000: 24). It is the process of approaching the environment positively and becoming environmentally friendly by transforming it into a lifestyle (Gülersoy, Dülger, Dursun, Ay, & Duyal, 2020: 2361). At the same time, it is an indispensable tool for preventing environmental problems that could bring the end of the world (Shin, 2008). Environmental education means helping people learn about their environment, interacting with the environment, and using the skills and knowledge they need to protect or care for the environment (Türkoğlu, 2019).

The characteristics of environmental education can be listed as follows (Meredith, et al., 2000: 5) Environmental education is related to environmental issues and problems; - Environmental education is based on accurate and real information; - Environmental education has an interdisciplinary structure; - Environmental education is a lifelong learning process; - Environmental education is based on the needs and interests of children; - Environmental education utilizes the external environment as much as possible as a learning environment in environmental education.

The main goal of this education, which aims to raise individuals with environmental awareness, is to provide positive, permanent behavioral changes in individuals towards the environment (Erten, 2004: 3; Gülersoy, Dülger, Dursun, Ay, & Duyal, 2020: 2361). In addition to transferring ecological information, environmental education, which differs from other ecological education in terms of ensuring the development of attitudes towards the environment and the transformation of these attitudes into behaviors, also aims to gain positive, permanent, permanent, desired behavioral changes and to ensure the active participation of individuals in solving problems (Şimşekli, 2004: 84).

However, the aims of environmental education can be listed as follows (Lane, 2006: 10-11):

- To be aware: To help children gain awareness and sensitivity about the environment and its problems, develop perception and discrimination skills, process and generalize these perceptions, and use these new skills in various contexts.
- Knowledge: To help children gain a basic understanding of how the environment works, how people interact with it, how environmental problems are increasing and how they can be solved.

- Attitudes: To help children acquire environmental values and the willingness and motivation to participate in environmental protection and improvement activities.
- Skills: To help children acquire the skills to identify, research and contribute to the solution of environmental problems.
- Participation: To help children gain experience by taking an active role in solving environmental problems by using the knowledge and skills they have acquired about the environment.

There are basically 4 relationships during environmental education. At the core of these relationships is the purpose of individuals' participation in environmental education. These relationships as defined by Priest and Gass (1997); 1. Interpersonal relationships, which include issues such as good communication, cooperation, problem solving that people in a group establish with each other 2. Relationship with oneself, which includes psychological processes such as self-concept, self-confidence, self-efficacy related to how a person defines himself/herself. 3. Relations with the ecosystem, which refers to the interdependence with other living things in the ecological system 4. Ecistic relations, which refers to human interaction with natural resources. In other words, it is about human impact on the natural environment and how the environment affects human life.

Raising an environmentally literate citizen is at the center of environmental education (Roth 1992:10). As a matter of fact, by raising people with environmental literacy skills, both the emergence of new environmental problems will be reduced and it will be possible to solve existing problems (Kışoğlu, Yıldırım, Salman & Sülün; 2016).

Environmental literacy, which includes people's interaction with the environment, is to perceive and interpret the healthy functioning of environmental systems, which should be explained by observable behaviors as well as environmental perceptions, and to take appropriate measures to repair the factors that disrupt the natural functioning and make them sustainable (Roth, 1992: 34). It is an understanding of how people and societies mutually relate to natural systems and how to do so in a sustainable way (Orr, 1992: 142). The aim is to make society environmentally literate so that they can actively participate in dealing with all environmental problems. The aim of improving the level of environmental literacy is to encourage and motivate productive and responsible individuals to solve the environmental problems they face. In this direction, it is desired to strengthen the belief that they can contribute to the solution of environmental problems individually or as part of a group (Roth, 1992: 85).

The first step of environmental education is ecological literacy. Ecological literacy can be defined as exhibiting knowledge, skills, attitudes, values and understanding of nature, or understanding natural processes on earth (TEMA Foundation; 2015). Ecological literacy is an interdisciplinary field of study that aims to make individuals realize themselves, their place in nature and that they are a part of nature (Aydın et al., 2016: 1161), and it also provides knowledge about environmental problems such as climate change, pollution, change in living habits, overconsumption, invasive species. It supports the development of behaviors to conduct research on the solution of these problems and to take responsibility in solving problems (Lewinsohn et al., 2015: 154). Ecological literacy emphasizes that natural resources are not independent and unlimited from the ecosystem, but are natural, dynamic and interdependent. With this approach, it reveals the role of humans in both utilizing and protecting natural resources (Casper, Fernández-Giménez & Balgopal, 2020: 25). Ecological literacy education is the process of developing attitudes, value judgments, knowledge and skills for the protection of nature in individuals, ensuring the demonstration of nature-friendly behaviors, discovering ways to live in harmony with nature and seeing the results of these. In other words, an individual who receives ecological literacy education should be an individual who knows the principles of nature, can look critically at natural phenomena, develop ideas for solving problems in the environment, conduct research and inquiry in this direction, and communicate with nature correctly (TEMA, 2016). Therefore, from an early age, emphasis should be placed on developing ecological literacy to inform and increase the individual's attitude towards sustainability in the economic, ecological and social spheres (De Brito Miranda, Jófili & dos Carneiro-Leão, 2017: 192).

For sustainable life, every society has to equip children, who will constitute future generations, with attitudes, values, knowledge and necessary skills related to environmental protection and to educate future generations in the best way (Güler, 2010: 32). When it comes to education, the necessity of biological, psychological, social and cultural assistance in the education of human beings, who are biopsycho-sociocultural beings, makes education not only the work of theories. This makes it necessary and important to enlighten the family and parents about child education (Özdoğan, 2001: 3). This importance is especially evident in the early periods when there is no schooling yet. Because in this period, parents have a shaping effect both as the child's first teacher and as a role model. In this period, parental interaction and communication, attitudes and behaviors towards the child are very important. Because from an early age, children begin to understand the relationship between people and the world, how they change the environment, and how the weather changes the characteristics of a place (Kefi, Çeliköz, & Erişen, 2013: 300). Consciously or unconsciously, they have various knowledge about people, places and environmental problems. They are aware that different places have different climates, that different people live in different parts of the world, that the planet they live on is called Earth and its shape is round, that different kinds of living things live on the earth, and that environmental problems caused by humans in the world they live in (Catling, 2006: 59). Indeed, Berk (2015) emphasizes the first years of life as the period when children are most open to development in terms of developing cognitive, physical, sensory and social skills. In this period when skills are easily acquired and brain development is very rapid, children are more open to external stimuli and more flexible in changing behavior (Barrett, 2014: 3). Acquisitions, behaviors and attitudes acquired in this period greatly affect personality traits, value judgments and behavioral patterns (Berk, 2015). Since these years form the basis of the knowledge, skills and habits that children will acquire throughout their lives, they are considered to be a critical stage in which the foundation of approaches towards the environment is laid. Since environmental knowledge and environmental interests begin to take shape in early childhood, environmental awareness and environmental education gained in this period play an important role in the acquisition of environmental knowledge, attitudes and behaviors in the following years(Aral, Kandır & Yaşar, 2001; Basile, 2000; Elliot, 2003; Gülay, 2011; Hellden & Hellden, 2008; Onur, 2016; Smith, 2001; Taşkın & Şahin, 2008; Yoleri, 2012). Onur (2016) He states that there are two important reasons for starting environmental education at an early age. The first reason is that children develop a sense of responsibility, respect and positive attitude towards nature. The second reason is that children need healthy positive interaction with the natural environment and see the environment as a source of interest.

Today, children are born into a world with serious problems such as pollution, poverty and biodiversity loss (Hedefalk, Almqvıst, Ostman, 2014: 975) and grow up indoors where they interact less with nature (Talay, Aslan, & Belkayalı, 2010). However, children need to live in a healthy environment for their healthy growth and development. Because children are exposed to environmental hazards more than adults due to their immature organs, metabolism, natural curiosity and lack of knowledge. This exposure can occur at home, on the playground, at school and in the wider environment. Worldwide, three million children under the age of five are exposed to environmental risks every year (Çağlayan, 2015: 32). In short, it can be said that children are the individuals most affected by environmental pollution. As a matter of fact, the main environmental threats that are effective in the emergence of children's health problems are as follows: 1. Deprivation of Safe Water and Conditions: Although access to safe and potable water is a fundamental right for all people, biological contaminants and chemical pollutants can compromise water quality. This often results in the availability of water being more important than its quality. 2. Chemical Pollution and Radiation: Increased isolation, reduced ventilation and tobacco use 3. Destruction of Natural Resources: Currently, although there is enough food and drink in the world, about 149 million children are still malnourished. However, current population growth requires tripling food production within 50 years (http://www.cevrehekim.org.tr/sempbild.htm). Therefore, in order for children to be minimally affected by environmental problems, it should be aimed that they become more conscious, sensitive and well-educated individuals about the environment. It should not be forgotten that the environmental ethics that children acquire in relation to environmental protection will be effective until they reach adulthood and even for the rest of their lives (Anon, 1992). For this reason, it should be aimed that children become more conscious, sensitive and well-educated individuals about the environment. Through environmental education, children should learn their role in nature, the natural balance and the information necessary to live in harmony with nature (Geray, 1995: 665). Because the investment made to the individual at an early age is an investment in the future. For this reason, for the welfare, peace and happiness of future generations, it is necessary to provide suitable environmental conditions in every sense, while giving young children the awareness and motivation to protect and develop.

When it comes to children, the greatest responsibility undoubtedly falls on teachers. However, teachers' lack of knowledge about the environment and insufficient educational opportunities on the subject have been revealed in many studies as obstacles to environmental education (Pedretti & Nazir, 2014; Winther, Volk, & Shrock, 2002). In addition, in the process of environmental education, teachers generally have problems with the lack of materials, green areas, gardens, parks, negative attitudes of parents, financial problems arising in the excursion process, and crowded classrooms are the factors that cause teachers to have difficulties in environmental education. Apart from these, it is also known that they need guidance on how to conduct research, whom to consult, how to organize the educational environment, how to plan nature trips, and what content to hold meetings to raise awareness of parents (Okyay et al., 2021: 135).

Families' lack of attention to environmental education, lack of knowledge about what to do and how to do it, the impossibilities of urban and business life, prioritizing academic development and similar reasons are the main difficulties encountered in environmental education in children.

So, what should we pay attention to in order to raise an environmentally conscious generation?

As in every field of education, early education is essential in environmental education. Since preschool education aims to raise awareness of children at an early age in almost every subject, it is important to start at an early age when it comes to environmental education in order to raise children's awareness and to form habits of protecting and developing the environment early (Önder & Özkan, 2013: 46).

The first elements of acquiring contemporary environmental awareness begin in the family. It matures through different educational experiences (Türküm, 1998: 180). In short, environmental education should have an understanding of education that starts in the family, continues until school and aims to create environmental awareness in individuals after school (Akınoğlu & Sarı, 2009: 6). The importance of the family for children never diminishes, but with the start of school life, the dominant role model schema shifts from the family to teachers. Teachers become important models for children with the beginning of their educational life. When it comes to environmental education, in order for the education to achieve the desired results, the teacher who provides environmental education should not only have environmental knowledge, be sensitive to environmental problems and be an environmentally friendly individual, but also have the belief that he/she can implement the responsibility he/she has undertaken will contribute to the successful outcome of the education (Erkol & Erbasan, 2018: 812).

It is possible for individuals to exhibit environmentally friendly attitudes and behaviors, first of all, by getting to know nature from early childhood (Aini & Laily, 2010: 270). This is because studies show that children who have high interaction with nature are more sensitive about protecting nature (Karatekin & Aksoy, 2012: 1435).

As a prerequisite for sustainable environmental awareness, an emotional relationship with nature and living beings is needed through the careful use of environmental opportunities. Because man protects what he loves. Destruction of the outer world always leads to poverty of the inner world. Emotional connection with nature can lead to the development of the instinct to protect nature (Gebhard, 2013).

The educational environment should be planned so that children can interact directly with the natural environment and natural materials (Güler, 2010). As a matter of fact, Dewey also emphasized the importance of nature and natural materials in child development and education and emphasized that the child's close relationship with nature contributes to both the development and learning of the child (Davis, 1998).

It is thought that raising generations that are sensitive to environmental issues, have developed awareness and protect the environment will be realized through well-planned early childhood environmental education programs (Wells & Lekies, 2006; 2). Therefore, it would be beneficial to integrate environmental issues such as energy and water consumption, pollution problems, protection of animals and plants, and recycling into early childhood education programs (Schultz, 2002). In order to develop children's ideas and behaviors related to sustainability, practices such as reducing consumption, reusing, recycling, respecting the environment, reflecting and rethinking should also be included in environmental education programs (Okyay, Sayın, Demir & Özdemir, 2022: 2865). Basile and White (2000) also emphasize that a suitable environmental education program for early childhood should have a relationship with different disciplines, a content based on real life experiences and a communication carried out with the child. In summary, the following points should be taken into consideration when developing curricula to develop environmental awareness in children. - The environment is the totality of phenomena and processes, that is, the environment that encompasses everything except the individual. The environment is embodied in the dimensions of time, space and social relations, so the concept of sensitivity should be considered within this holism. - Environmental education should aim at the acquisition of rich knowledge; course topics should be determined to embody the concepts classified as diversity, variability, functionality, colors and aesthetics, which constitute the basic characteristics of the social or spatial environment. - In addition to the acquisition of rich knowledge, environmental education should aim to make children adopt contemporary and universal value judgments and to acquire attitudes and behaviors compatible with them. - The processing of the subjects should be programmed based on excursion and observation, and the acquired knowledge should be reinforced with practical, repetitive, visual and linguistic expression methods and should be used at every opportunity (Türksoy, 1991: 22).

Environmental education, which starts from the preschool period, should deal with the interaction of human and environment scientifically. It is inevitable to create a moral and value perspective towards the environment and at the same time to have its own rules. Along with these, there should be active participation that will enable children to live one-to-one (Mamedov, 1998: 26).

Because all kinds of changes in the natural environment affect children and fall within their field of observation. For this reason, it may not be enough to give children only the theoretical knowledge of these events. In addition, it should be ensured that they learn these situations by experiencing them and a ground should be prepared that will allow them to establish a cause-effect connection between them (Özkan-Kılıç, 2012: 41). In addition, it is useful to ensure permanent learning in the educational environment, to make teaching activities meaningful for students and to make them participate in the process, and to address more than one sensory organ of students in teaching activities where education is an important element (Ünlü, Üçışık, & Özey, 2002: 11).

Effective environmental education programs, methods and materials used in the program should arouse enthusiasm in children about environmental education (Onur, 2016, Akt., Güzelyurt & Özkan, 2019). For example, learning through play is very important in environmental education in early childhood. Because children are motivated by play by nature, focus on the process rather than the product, and are fully engaged in learning with mind, heart and body (Oltman, 2002: 30). Oyun, çocuklar için en kolay, en doğal ve en etkili öğrenme aracıdır (Tuğrul, 2007: 59). Bu nedenle erken dönemde çocuklara istendik davranışlar kazandırmak amacıyla doğayla iç içe olabilecekleri oyunlar, doğa yürüyüşleri, bitki ve hayvan bakımı uygulamaları yaptırmak onların doğa dostu, çevreye duyarlı bireyler olarak yetişmelerinde önemli rol oynayacaktır (Erten, 2004: 2361). In addition, activities such as talking, commenting and discussing about the environment may be preferred in environmental education as they will increase the permanence of learning (Kuzu, 2008: 337). In addition, children having the opportunity to share simple nature experiences with an adult who is interested in them or all the activities they will perform with their peers in nature will positively affect the social and emotional development of the child (Yayla-Ceylan & Ülker, 2014: 37-58).

Another important element in the process of environmental education is the materials used in the education year. In order to provide children with environmental awareness, it is very important to use educational materials that provide information about ecology and explain values related to personal behaviors (Kandır, Yaşar, İnal, Yazıcı, Uyanık, & Yazıcı, 2012: 23).

One of the important educational materials used in environmental education in preschool period is children's picture books (Gönen, Pektaş, Değirmenci, Bülbün, & Ünal, 2016 p249; Güzelyurt & Özkan, 2019: 20). While picture books, which are the first works that children encounter in their environment, support children's development, they also help them think critically and socialize (Gönen & Veziroğlu, 2013: 9). Illustrated children's books, which are one of the important educational materials used in environmental education practices, enable children to recognize the environment, gain environmental awareness, see environmental problems and develop solutions to existing problems. These books aim to be useful for children with their subjects and the sub-messages they contain for the purposes of environmental education. In line with this information, it is necessary to examine which topics the illustrated children's books suitable for the preschool period and prepared for environmental education cover and which objectives of environmental education they are prepared for (Güzelyurt & Özkan, 2019: 20). Because in environmental activities through stories, it is important to pay attention to issues such as what are the achievements and indicators related to the environment in the book, whether the story is suitable for children's developmental characteristics, whether the information provided by the book about the environment is real and accurate, whether the information in the pictures in the book is accurate, whether the text and picture relationship is compatible, whether the narrative language of the story is positive, whether the narratives in the book present a theme for the environment (Wells & Zeece, 2007, cited in Wells & Zeece, 2007, as cited in Sağsöz, 2018, p. 57). Sağsöz, 2018: 57). Reading children's picture books selected in accordance with these criteria by associating them with natural materials collected from nature such as acorns, stones and leaves is a very entertaining and nature-teaching activity for children (Cullinan, 1992, p.77). As can be seen, it is very important to make use of literary works to help children gain value awareness. Fairy tales are a literary genre that both children and adults love to read. For this reason, they can be easily and conveniently used at every stage of education and training from pre-school to university education in developing environmental awareness and teaching values (Kuzu, 2008: 327). In early childhood, fairy tales and story books on the environment can be used to support children's imagination and preparation for life (Türküm, 1998, p. 179). Thanks to fairy tales, children can learn some values at a young age and set themselves the goal of protecting and keeping them alive. Fairy tales can also be used to develop environmental awareness in children (Kuzu, 2008: 327).

Another important element is school gardens. School gardens should be more than areas where time is spent during breaks. In fact, they are opportunity areas where real life experiences can be experienced, especially in terms of environmental education. As a matter of fact, designing gardens as educational spaces that will help children develop positive attitudes towards the natural environment will both make educational activities fun and increase the permanence of the information learned. For this reason, it can be suggested to organize school gardens as natural habitats to include examples from both plant and animal habitats. With the miniature living spaces created in this way, children's contact with nature can be maximized so that they can adopt nature and strive to protect it.

As a result of the rapid development of technology, mobile devices, which are included in all areas of our lives, make our lives easier and enable us to communicate quickly when desired. The inclusion of mobile devices in all areas of life has brought about a transformation in education. Mobile devices used in the field of education have eliminated the limitation of education to the classroom and provided the opportunity to access information anytime and anywhere (Doğan & Seferoğlu, 2015). Educational technologies, which have become widespread in recent years, have brought a new perspective to education and brought computers, one of the most frequently used educational technologies, into the educational environment. Thus, computers have provided educators with the opportunity to prepare and use materials that can be used at different levels in education. As a matter of fact, computer animations prepared in computer environment have created multiple environments that increase students' comprehension of problems in environmental education, provide a better understanding of the connection points in ecosystems, and enable students to easily see how the problem at any stage will be reflected in the future (Aslan Efe, 2015: 136). Another technological product that can be used in environmental education is mobile devices. It is emphasized that mobile devices attract students' attention in the integration of environmental learning and are an unforgettable experience. It is observed that students are better organized in environmental education with mobile devices and that they attract students' attention better (Anderson et al., 2015). Nowadays, it is known that interactive games are more preferred. The successful design of the visual elements in the content is one of the main reasons why children turn to online games (Seyman & Gülaçtı, 2022: 33). For this reason, as in many areas of education, the use of online games in environmental education will make education fun and permanent.

Apart from these, Wilson (1996) emphasized the elements that should be considered in environmental education practices in early childhood as follows:

- Start with simple practices where children can feel safe.
- A warm and friendly environment should be created for children.
- It should be kept in mind that the pleasure to be derived from environmental education is as important as its content, and children should be provided with enjoyable and memorable experiences.
- Active participation of children should be ensured and all their emotions should be involved.
- Different cultures and perspectives should be introduced.
- Focus on the perfection and beauty of nature.
- Learning should be based on understanding concepts and analysis rather than rote and pattern-based learning.
- In environmental education activities, families as well as children should be involved in the process.

REFERENCES

Aini, M., & Laily, P. (2010). Preparedness of Malaysian pre-school educators for environmental education. *Pertanika Journal of Social Science & Humanities*, 18(2), 271–283.

Akçay, İ. (2006). Farklı ülkelerde okul öncesi öğrencilerine yönelik çevre eğitimi. [Environmental education for preschool students in different countries], (Master's Thesis), Uludağ University Institute of Social Sciences.

Akınoğlu, O., & Sarı, A. (2009). İlköğretim programlarında çevre eğitimi. [Environmental Education in Prımary School Curriculum]. *Marmara University Atatürk Education Faculty Journal of Educational Sciences*, 30(30), 5–29.

Anderson, M. P., Woessner, W. W., & Hunt, R. J. (2015). *Applied groundwater modeling: simulation of flow and advective transport*. Academic press.

Anon, S. (1992). *Çevre ve çocuk* [Environment and children]. Environment and Human Journal. Çevre Bakanlığı Yayınları.

Aral, N., Kandır, A. ve Can Yaşar, M. (2001). *Okul Öncesi Eğitim* [Preschool Education].Pre-School Education 1. Ya-Pa

Ardoğan, R. (2012). Tüketim, Nüfus Ve Çevre Sorunları: Orantısız Denklem [Consumption, Population and Environment Problems: disproportionate equation]. *Tarih Kültür ve Sanat Arastirmalari Dergisi*, 1(4), 81–106.

Aslan Efe, H. (2015). The Effect of Anımatıon Supported Environmental Education on Academic Achievement, Retention and Attitude Towards the Environment. *Journal of Computer and Education Research*, 3(5), 130–143. DOI: 10.18009/jcer.90852

Atasoy, E. (2005). *Çevre İçi Eğitim: İlköğretim Öğrencilerinin Çevresel Tutum ve Çevre Bilgisi Üzerine Bir Çalışma [Environmental Education: A Study on Environmental Attitude and Environmental Knowledge of Primary School Students]* (Doctoral dissertation, Bursa Uludag University Turkey).

Atasoy, E. (2005). *Çevre İçin Eğitim: İlköğretim Öğrencilerinin Çevresel Tutum ve Çevre Bilgisi Üzerine Bir Çalışma* [Education for the Envıronment:Primary Education Students' Environmental Attitudes And A Study On Environmental Knowledge]. (Doctoral dissertation) Uludag University.

Atasoy, E. (2010). Genel Coğrafya. [General Geography]. (2. Baskı). İstanbul: Ezgi Kitabevi Yayınları.

Aydın, M., Dündar, R. ve Korkut, Ş. (2016). Türkiye'de Ekolojik Okuryazarlık Eğitimine Yönelik Öğretmen Görüşleri. [Teacher Views Upon Ecological Literacy Education in Türkeye].*Abant İzzet Baysal University Journal of Faculty of Education,* Cilt 16, (USBES Özel Sayı II), 1160-1172.

Balkan Kıyıcı, F. (2009). Çevre eğitimi. [Environmental education].Sevinç, V. (Ed.), Eğitim fakülteleri için genel çevre bilimi içinde (s. 173- 183). Maya Akademi.

Barrett, P. M., Cooper, M., & Teoh, A. B. (2014). When time is of the essence: A rationale for earlier early intervention. *Journal of Psychological Abnormalities in Children*, 3(4), 1–9. DOI: 10.4172/2329-9525.1000133

Basile, C., & White, C. (2000). Respecting Living Thinks: Environmental Literacy for young Children. *Early Childhood Education Journal*, 28(1), 57–61. DOI: 10.1023/A:1009551705001

Basile, C. G. (2000). Environmental education as a catalyst for transfer of learning in young children. *The Journal of Environmental Education*, 32(1), 21–27. DOI: 10.1080/00958960009598668

Berk, L. E. (2015). *Child Development* (9th ed.). Illinois State University: Pearson.

Çağlayan, Ç. (2015). Çocuk ve Çevre. [Child and Environment].Turk J Public Health is, 1(2), 32-38.

Can, F. (2016). "Çevre Politikasının Ekonomik Araçları", [Economic Instruments of Envıronmental Policy]. Ömer Halisdemir *University* Faculty of Economics and Administrative Sciences Journal 9(3). 59-73.

Casper, A. M. A., Fernández-Giménez, M. E., & Balgopal, M. M. (2020). A tool for measuring ecological literacy: Coupled human-ecosystem interactions. *Journal of Agricultural Education and Extension*, 27(1), 21–34. DOI: 10.1080/1389224X.2020.1780139

Catling: (2006). What Five-Year-Olds Know of the World? -Geographical Understanding and Play in Young Children's Early Learning.' Geography, C/S. 91 (1): 55-74.

Cullinan, B. E. (1992). *Read To Me: Raising Kids Who Love To Read*. Scholastic.

Davis, J. (1998). Young Children, Environmental Education and Future. *Early Childhood Education Journal*, 26(2), 117–123. DOI: 10.1023/A:1022911631454

De Brito Miranda, A. C., Jófili, Z., & Dos Anjos Carneiro-Leão, A. M. (2017). Ecological literacy – preparing children for the twenty-first century. *Early Child Development and Care*, 187(2), 192–205. DOI: 10.1080/03004430.2016.1226353

Dikmen, S. (1993). *İlköğretim Kurumlarında Çevre İçin Eğitim* [Education for Environment in Primary Schools]. Çevre Eğitimi, Türkiye Çevre Vakfı Yayını.

Doğan, D., & Seferoğlu, S. S. (2015). Mobile devices and digital transformation in education. In Akkoyunlu, B., İşman, A., & Odabaşı, H. F. (Eds.), *Educational Technologies Readings* (pp. 541–545). TOJET.

Elliott, S. (2003). Patches of Green EPA Social Research Series Early childhood environmental education in Australia: scope, status and direction Environment Protection Authority 59–61 Goulburn Street, http://www.penrithcity. nsw.gov.au/uploadedFiles/patchesofgreen.pdf

Erkol, M. ve Erbasan, Ö. (2018). Öğretmenlerin çevre eğitimi öz-yeterliklerinin çeşitli değişkenler açısından incelenmesi. [Investigation of Teachers' Environmental Education Self-Efficacy in Terms of Different Variables]. *Erzincan University Journal of Education Faculty*, 20(3), 810-825. DOI: 10.17556/erziefd.446125

Erten, S. (2004) Çevre Eğitimi Ve Çevre Bilinci Nedir, Çevre Eğitimi Nasıl Olmalıdır? [What is Environmental Education and Environmental Awareness, How Should Environmental Education Be?]. Journal of Environment and Human, Publication of the Ministry of Environment and Forestry.. Sayı 65/66. 2006/25

Ertürk, R. (2017). İlkokul öğrencilerinin çevre sorunları ve çevre eğitimine yönelik algıları. [Environmental Problems Of The Primary School Students And Perceptions For Environmental Education]. *Inonu University Journal of the Faculty of Education*, 18(3), 12–24. DOI: 10.17679/inuefd.354142

Gebhard, U. (2013). *Kind und Natur. Die Bedeutung der Natur für die psychische Entwicklung*. Westdeutscher Vlg. DOI: 10.1007/978-3-658-01805-4

Geray, C. (1995). Çevre koruma bilinci ve duyarlılığı için halkın eğitimi. [Public education for environmental protection awareness and sensitivity]. *Yeni Türkiye Özel Sayısı*, 1(5), 664–669.

Gökçe, N., Kaya, E., & Aktay, S. ve Özden, M. (2007). İlköğretim Öğrencilerinin Çevreye Yönelik Tutumları. [Elementary Students Attitutes Towards Environment]. *Elementary Education Online*, 6(3), 452–468.

Gönen, M. ve Veziroğlu, M. (2013). Çocuk edebiyatının genel hedefleri. [General aims of children's literature].M. Gönen (Ed.), Çocuk Edebiyatı (ss. 1-11) içinde. Eğiten Kitap.

Gönen, M., Pektaş, D., Değirmenci, Y.G., Bülbün, B. & Ünal, Ş. (2016). Çevre konulu resimli çocuk kitaplarının resimleme ve içerik açısından incelenmesi. [Physical, Content and Content of Illustrated Children's Books Analyzing in Terms of Drawing Features].*Çocuk ve gençlik edebiyatında barış kültürü.*

Gül, F. (2013). İnsan-doğa ilişkisi bağlamında çevre sorunları ve felsefe. [Environmental Problems and Philosophy in The Context of Human-Nature Relationship]. *Pamukkale University Journal of Social Sciences Institute*, 14, 17–21. DOI: 10.5505/pausbed.2013.09609

Gülay, H. (2011). Ağaç yaş iken eğilir: Yaşamın ilk yıllarında çevre eğitiminin önemi. [You Cannot Teach an Old Dog New Tricks: The Importance of Environmental Education In The Early Years Of Life]. *TÜBAV Journal of Science*, 4(3), 240–245.

Güler, T. (2010). Çevre Eğitimi. [Environment Education].Berrin Akman, Gülden Uyanık Balat ve Güler, Tülin, (Ed.), *Okul Öncesi Dönemde Fen Eğitimi* (s.181-201). Pegem

Gülersoy, A. E., Dülger, İ., Dursun, E., & Ay, D. (2020). Nasıl bir çevre eğitimi? Çağdaş yaklaşımlar çerçevesinde bazı öneriler. [What Kind of Environmental Education? Some Suggestions in the Context of Contemporary Approaches]. *Turkish Studies (Elektronik)*, 15(5), 2357–2398. DOI: 10.7827/TurkishStudies.44074

Gürbüz, H., & Kışoğlu, M. (2011). Güncel çevre sorunları. [Current Environmental Problems]. H. Yazıcı & K. Arıbaş (Ed.), Günümüz dünya sorunları içinde (s. 153-190). Pegem Akademi.

Güzelyurt, T., & Özkan, Ö. (2019). Okul Öncesi Dönemde Çevre Eğitimi: Çocuk Kitaplarına Yönelik Bir İnceleme1 [Environmental Education in Pre-School Period: An Examination of Children's Books]. *Elementary Education Online*, 18(1), 20–30.

Hedefalk, M., Almqvist, J., & Östman, L. (2015). Education for sustainable development in early childhood education: A review of the research literature. *Environmental Education Research*, 21(7), 975–990. DOI: 10.1080/13504622.2014.971716

Helldén, G., & Helldén, S. (2008). Students' early experiences of biodiversity and education for a sustainable future. *NorDiNa: Nordic Studies in Science Education*, 4(2), 123–131. DOI: 10.5617/nordina.286

Kandır, A., Can Yaşar, M., İnal, G., Yazıcı, E., Uyanık, Ö., & Yazıcı, Z. (2012). *Etkinliklerle Bilim Eğitimi* [Science Education with Activities]. Efil.

Karatekin, K., & Aksoy, B. (2012, Winter). Sosyal bilgiler öğretmen adaylarının çevre okuryazarlık düzeylerinin çeşitli değişkenler açısından incelenmesi, [Examination of Teacher Candidates Of Social Studies' Environmental Literacy Level in Terms of Various Variables]. Turkish Studies- International Periodical For The Languages. *Literature and History of Turkish or Turkic*, 7(1), 1423–1438.

Kaya, M. F. (2014). Sosyal bilgiler öğretmen adaylarının çevre sorunlarına ilişkin algıları: Metafor analizi örneği. [Social Studies Teachers' Perceptions Related to Environmental Problems: A Sample Analysis of Metaphors]. *Turkish Studies*, 9(2), 917–931. DOI: 10.7827/TurkishStudies.6308

Kayaer, M. (2012), Hukuksal Araçlarla Çevrenin Korunması [Protection of the Environment Through Legal Instruments]. (Doctoral dissertation) Sakarya University SBE.

Kaypak, Ş. (2013b). Çevre sorunlarının çözümünde küresel çevre politikalarının önemi. [The Global Environmental Politics for the Solution of Environmental Problems]. *Journal of the Institute of Social Sciences of Muğla University*, 31, 17–34.

Kefi, S., Çeliköz, N. ve Erişen, Y. (2013). Okulöncesi Eğitim Öğretmenlerinin Temel Bilimsel Süreç Becerilerini Kullanım Düzeyleri [Preschool Teachers' Levels of Using the Basic Science Process Skills].Journal of Research in Education and Teaching, C/S. 2 (2): 300-319.

Kırlıoğlu, H., & Zeytin, M. (2015). "Belediyelerde Çevre Muhasebesi ve Raporlama: Marmara Bölgesinde Bulunan Belediyeler Üzerine Bir Durum Analiz", [Environmental Accountıng and Reporting in Municipalities: A Case Analysis on the Municipalities Located in the Marmara Region].ASOS JOURNAL (The Journal of Academic Social Science), Yıl: 3. *Sayı*, 10, 753–777.

Kışoğlu, M., Yıldırım, T., Salman, M., & Sülün, A. (2016). İlkokul ve ortaokullarda çevre eğitimi verecek olan öğretmen adaylarında çevre sorunlarına yönelik davranışların araştırılması. [Investigation of Behaviors Towards Environmental Problems in Teacher Candidates Who Will Give Environmental Education at Primary and Secondary Schools].Erzincan University Journal of Education Faculty, 18 (1), 299-318. DOI: DOI: 10.17556/jef.93507

Kuzu, T. (2008). Aytül Akal'ın masallarıyla çocukta çevre bilinci geliştirme. [Environmental Awareness in Children With Aytül Akal's Fairy Tales Development]. *The Journal of Selcuk University Social Sciences Institute*, 19, 327–339.

Lane F. J. (2006), Environmental Education Implementation in Wisconsin: Conceptualizations and Practices, Wisconsiv-Madison Üniversitesi, yayımlanmamış doktora tezi.

Lewinsohn, T. M., Attayde, J. L., Fonseca, C. R., Ganade, G., Jorge, L. R., Kollmann, J., & da Rocha, P. L. (2015). Ecological literacy and beyond: Problem-based learning for future professionals. *Ambio*, 44(2), 154–162. DOI: 10.1007/s13280-014-0539-2 PMID: 24973054

Mamedov, N. (1998). Environmental culture and education. Mamedov, N (Ed.). Environmental Education - Concept and Methodical in Approach. (pp.19-50). Izmir: Environmental Protection and Research Foundation Publications.

Meredith, J., Cantrell, D., Conner, M., Evener, B., Hunn, D., & Spector, P. (2000). Best Practices for Environmental Education: Guidelines for Success.

Moseley, C. (2000). Teaching for environmental literacy. *The Clearing House: A Journal of Educational Strategies, Issues and Ideas*, 74(1), 23–24. DOI: 10.1080/00098655.2000.11478633

Okyay, Ö., Demir, Z. G., Sayın, A., & Özdemir, K. (2021). Ekolojik okuryazarlık eğitiminin okul öncesi öğretmenlerinin ekolojik farkındalığı ve çevreye yönelik motivasyonlarına etkisi. [The effect of ecological literacy training on preschool teachers' ecological awareness and motivation towards the environment]. *Başkent University Journal of Education*, 8(1), 129–146.

Okyay, Ö., Sayın, A., Demir, Z. G., & Özdemir, K. (2022). Ekolojik Okuryazarlık Eğitiminin Okul Öncesi Çocukların Çevreye Yönelik Farkındalık ve Tutumlarına Etkisi. [The Impact of Ecological Literacy Training on Environmental Awareness and Attitudes of Preschool Childrens]. *Milli Eğitim Dergisi*, 51(236), 2851–2870. DOI: 10.37669/milliegitim.944271

Oltman, M. (2002). Natural Wonders: A Guide to Early Childhood for Environmental Educators.

Önder, A., Özkan, B. (2013). *Sürdürülebilir Çocuk Gelişimi.* [Sustainable Child Development]. *Environmental Education with Activities in Preschool.* Anı Yayıncılık

Onur, B. (2016). *Child environment nature: Environment and citizenship education.* İmge Bookstore.

Örnek, G. (1994). *Çevre eğitimi ve lise eğitim programlarındaki yeri.* Yayınlanmamış Yüksek Lisans Tezi, Hacettepe Üniversitesi Fen Bilimleri Enstitüsü.

Orr, D. W. (1992). *Ecological Literacy: Education and ahe Transition to a Postmodern World.* State University of New York Press.

Özdemir, O. (2007). A new environmental education perspective: "Education for sustainable development". *Education in Science*, 32(145), 23–38.

Özdoğan, B. (2001). Altı-on iki yaşlarındaki çocukların eğitimi ve okul başarıları. [Education and School Success of Children Aged Between Six and Twelve]. *Education in Science*, 26(120), 3–7.

Özgen, N., ve Kahyaoğlu, M. (2011). Farklı Fonksiyonel Özeliğe Sahip Yerleşim Ünitelerinde İkamet Eden İlköğretim Öğrencilerinin Çevre Sorunlarını Algılama ve Çözüm Önerileri: Fenomenografik Bir Araştırma [Perception of Primary School Students Residing in the Settlement Units With Diverse Functional Features on Environmental Problems and Solution Recommendations: A Phenomenographic Research]. *Electronic Journal of Social Sciences*, 10(38), 136–157.

Özkan-Kılıç, Ö. (2012). Okul Öncesi Dönem Resimli Öykü Kitaplarının Coğrafi Kavramları İçermesi Yönünden İncelenmesi, [Analyze whether picture story books published for preschoolers include geographic concepts]. (Master's Thesis), Çanakkale Onsekiz Mart University.

Pedretti, E., & Nazir, J. (2014). Tensions and opportunities: A baseline study of teachers' views of environmental education. *International Journal of Environmental and Science Education*, 9(3), 265–283.

Priest, S., & Gass, A. M. (1997). *Effective leadership in adventure programming*. Human Kinetics.

Roth, C. E. (1992). *Environmental literacy: Its roots, evolution and directions in the 1990s*. ERIC Clearinghouse for Science, Mathematics, and Environmental Education. https://files.eric.ed.gov/fulltext/ED348235.pdf

Sadık, F., Çakan, H., & Artut, K. (2011). Çocuk resimlerine yansıyan çevre sorunlarının sosyo-ekonomik farklılıklara göre analizi [Analysis of the Environmental Problems Pictures of Children from Different Socio-economical Level]. *Elementary Education Online*, 10(3), 1066–1080.

Sağsöz, G. (2018). Resimli çocuk kitaplarının barış eğitimine ilişkin bileşenler açısından incelenmesi. [Investigating illustrated children's books from the point of the components related to peace education]. (Doctoral dissertation), Ankara *University* Eğitim Bilimleri Enstitüsü

Schultz, P. W. (2002). In Dietz, T., & Stern, P. C. (Eds.), *Knowledge, information and household recycling: Examining the knowledge-deficit model of behavior change. New Tools for Environmental Protection- Education, Information, and Voluntary Measures* (pp. 67–82). National Academy Press.

Seyman, N., & Gülaçtı, İ. E. (2022). Mobile Application Designed to Raise Environmental Awareness in Children and the Results of the Application. [SSSJournal]. *Social Sciences Studies Journal*, 8(93), 32–40. DOI: 10.26449/sssj.3762

Shin, K. H. (2008). Development of environmental in the Korean kindergarten context. Unpublished Doctoral Thesis, University of Victoria, Canada.

Şimşekli, Y. (2004). Çevre bilincinin geliştirilmesine yönelik çevre eğitimi etkinliklerine ilköğretim okullarının duyarlılığı [Sensitivity of Elemantary Schools to the Environmental Education Activities For Increasing Environmental Knowledge]. *Journal of Faculty of Education Uludag University*, 17(1), 83–92.

Smith, A. (2001). Early childhood a wonderful time for science learning. *Australian Primary and Junior Science Journal*, 17(2), 18–20.

Talay, İ., Aslan, F., & Belkayalı, N. (2010). Okul Öncesi Eğitim Kurumlarında Doğa Dostu ve Çocuk Katılımı Temelli Dış Mekan Tasarım Yaklaşımları Bir Proje Önerisi. [Nature Friendly And Child Participation Based Outdoor Design Approaches in Preschool Institutions: A Project Proposal]. *Kastamonu Education Journal*, 18(1), 317–322.

Taşkın, Ö. ve Şahin B. (2008). Çevre kavramı ve altı yaş okul öncesi çocuklar. *Pamukkale University Journal of Education*, 1(23), 1–14.

TEMA (2011-2015) Türkiye Erozyonla Mücadele Vakfı Genel Müdürlüğü Arşivi,

Tomar, A. (2009). *Toprak ve su kirliliği ve su havzalarının korunması* [Soil and water pollution and protection of watersheds.]. TMMOB İzmir Kent Sempozyumu.

Tuğrul, B. (2007). Etkili okul öncesi eğitim programlarının özellikleri [Elektronik versiyon] [Characteristics of effective preschool education programs]. Symposium on preschool education (s.50-93). Antalya.

Türkoğlu, B. (2019). Opinions of preschool teachers and pre-service teachers on environmental education and environmental awareness for sustainable development in the preschool period. *Sustainability (Basel)*, 11(18), 4925. Advance online publication. DOI: 10.3390/su11184925

Türksoy, Ö. (1991). Çevre duyarlılığı eğitiminde bilişsel yaklaşım: Çocuk ve temel çevre özellikleri. [Cognitive approach in environmental sensitivity education: The child and basic environmental characteristics.]. *Education in Science*, 15(80).

Türküm, A. S. (1998). *Çağdaş toplumda çevre sorunları ve çevre bilinci* [Environmental problems and environmental awareness in contemporary society]. Çağdaş Yaşam Çağdaş İnsan. Anadolu Üniversitesi Açık Öğretim Fakültesi İlköğretim Öğretmenliği Lisans Tamamlama Programı.

Ünal, S., Mançuhan, E. ve Sayar, A. (2001). Çevre bilinci, bilgisi ve eğitimi. [Environmental awareness, knowledge and education]. Marmara Üniversitesi Teknik Eğitim Fakültesi Matbaası

UNEP. (1972). United Nations Conference on the Human Environment, United Nations: https://www.unep.org/Documents.Multilingual/Default.asp?DocumentID=97

UNESCO. (1977). Intergovernmental Conference on Environmental Education organized by Unesco in co-operation with UNEP, Final Report, Tbilisi (USSR).

Ünlü, M., Üçışık, S. ve Özey, R. (2002) . Coğrafya eğitim ve öğretiminde haritaların yeri ve önemi. [The Importance of othe Maps in Geography Education and Instruction].lnternational Journal of Geography and Geography Education, 5, 9-25.

Vakfı, T. E. M. A. (2016), https://www.tema.org.tr/Sayfalar/Calismalarimiz/ com

Wells, N. M., & Lekies, K. S. (2006). Nature and the life course: Pathways from childhood nature experiences to adult environmentalism. *Children, Youth and Environments*, 16(1), 1–24.

Wilson, R. A. (1996). *Starting Early: Environmental Education During The Early Childhood Years*. ERIC Clearinghouse for Science Mathematics and Environmental Education.

Winther, A. A., Volk, T. L., & Shrock, S. A. (2002). Teacher decision making in the 1st year of implementing an issues-based environmental education program: A qualitative study. *The Journal of Environmental Education*, 33(3), 27–33. DOI: 10.1080/00958960209600812

Yalçınkaya, E. (2013). İlköğretim 8. Sınıf öğrencilerine göre çevre sorunları: Nitel bir çalışma. [Environmental Issues to 8th Grades Elementary Education Students: A Qualitative Study]. *lnternational Journal of Geography and Geography Education*, 27, 416-439

Yaman, K., & Gül, M. (2018). Kuruluşundan Günümüze Avrupa Birliği'nin Çevre Politikası [Environmental Policy of the European Union since its foundation]. *Journal of Economy Business and Management*, 2(2), 198–217.

Yayla Ceylan, Ş. ve Ülker, P. (2014). Okul öncesi dönem çocuklarında çevre eğitimi neden önemlidir? [Why is environmental education important for preschool children?]. (Ed.: H. Gülay Ogelman). *Çocuk ve çevre. Küçük Çocuklar ve Çevre Eğitimi El kitabı*. Eğiten Kitap, s. 37-58. http//tez2.yok.gov.tr

Yoleri, S. (2012). Çocuk ve Çevre: Okul Öncesi Çocuklar Arasında Çevre Bilinci Oluşturma. [Children and the Environment: Creating Environmental Awareness Among Preschool Children]. *Buca Faculty of Education Journal*, 34, 100–111.

Zayimoğlu Öztürk, F., & Öztürk, T. (2015). "Öğretmen Adaylarının Çevre ve Çevre Eğitimi ile İlgili Görüşleri (Ordu Üniversitesi Örneği) " [Opinions of Pre-Service Teachers About Environment and Environmental Education Ordu University Sample]. Balıkesir University Journal of Social Sciences Institute, 18(33). 115-132

KEY TERMS AND DEFINITIONS

Early Childhood: A biological transition period that covers the first 8 years of life, has a significant impact on the child's future life, and is a period of rapid growth and development.

Ecological Literacy: The attitudes and skills to protect all the values of nature, to recognize the ecosystem and to live in harmony with the environment in a sustainable way by being aware of being a part of nature.

Environment: Is the living space that encompasses all living and non-living things, contains all living things and in which living things interact with each other throughout their lives.

Environmental Pollution: It is a threat element that leads to the unnatural deterioration of the environment and adversely affects the living and non-living elements living in it.

Environmental Education: Education that aims to acquire attitudes, knowledge and skills that should start in the family and continue throughout life for the protection of the environment and a sustainable world.

Environmental Education Program: A developmental program for the environment, based on the active participation of children in the learning process, including awareness, attitude, knowledge and skills related to the environment based on excursions and activities.

Family Involvement in Environmental Education: The process by which parents and other family members participate in their children's education program to increase awareness, sensitivity and sustainability.

Chapter 10
Learning on the Move:
Merging Geography With Road Safety Education

Cristina Honrubia-Montesinos
https://orcid.org/0000-0001-8564-4400
University of Castilla La Mancha, Spain

Laura Moreno Macías
Independent Researcher, Spain

Paula Tarancón Sánchez
https://orcid.org/0009-0002-0792-1710
Independent Researcher, Spain

ABSTRACT

Didactic itineraries serve as a valuable tool for teaching Social Sciences, particularly Geography. Therefore, a tour has been organized through the streets so that first-year students can learn about transportation systems and basic road safety rules. Thanks to the experience that provide this project, the benefits of didactic itineraries has been highlighted, as students acquire meaningful learning through them by directly engaging with their surroundings and classmates, which they find more engaging and motivating compared to traditional classroom lectures. Due to this, these itineraries are an ideal resource to incorporate into educational programs.

DOI: 10.4018/979-8-3693-7185-5.ch010

Copyright © 2025, IGI Global Scientific Publishing. Copying or distributing in print or electronic forms without written permission of IGI Global is prohibited.

1. INTRODUCTION

This paper presents a didactic experience carried out through the preparation and implementation of a didactic itinerary for teaching Geography in Primary Education. The relevance of the topic is analyzed based on the contributions of various authors mentioned in the theoretical foundation, which served as the basis for planning the experience. The didactic itinerary for the first year of Primary Education (hereinafter, PE) is described, along with the different preparation phases, its contextualization, and its foundation, which includes all the elements extracted from Decree 81/2022 that regulates Primary Education in Castilla-La Mancha and that are related to the didactic itinerary itself. In addition, to the activities carried out, the resources that have been used and the corresponding evaluation of the proposal are presented.

The justification for didactic itineraries lies in their effectiveness and multiple benefits in the teaching of Geography, as well as Social Sciences. Among these, it can be highlighted that students learn through direct contact with reality, breaking away from traditional methodologies and increasing student motivation.

For this reason, it is essential to consider this tool when working on geographic content, as it is not simply about taking a field trip, but rather achieving certain pre-established objectives through the outing. In other words, meaningful learning should take place during this session outside the classroom. However, the teacher's role and planning are crucial, as they must be familiar with the different phases of the itineraries: the preparation phase, the development phase, and the interpretation phase. Likewise, the education professional at this stage must master curricular elements, time management, as well as space and resources, to carry out a suitable educational proposal.

2. BACKGROUND

To begin with, it is important to highlight the significance of Geography in the teaching-learning process of spatial dimensions, as it allows the acquisition and understanding of information about space and the events that occur within it. Geography, therefore, is defined as the science that studies the relationships between society (human beings) and the environment. This discipline delves into how humans occupy territory and the outcomes of this occupation in that space (Martínez-Fernández et al., 2015). For students, acquiring geographic knowledge is essential as it fosters the development of skills that enable them to understand and analyze the territory through the location and distribution of relevant events. Furthermore, it equips students with the ability to obtain, interpret, evaluate, and apply this information across various everyday contexts (Sebastià and Tonda, 2011).

As a result, it is crucial to encourage geographical reflection and move away from the mere exercise of locating places on a map (Martínez-Fernández et al., 2015).

Despite the recognized importance of geographical reflection, scientific literature suggests that traditional teaching methods and reliance on textbooks still dominate Geography lessons in Primary Education (PE) (Binimelis et al., 2023; Morote, 2017). This reliance on outdated teaching methods contributes to student demotivation, as many perceive Geography as an extensive and unengaging subject (Morote, 2017). In response to this demotivation, didactic itineraries should be implemented as a strategy to bring learning to life. These itineraries, understood as outdoor activities carried out by teachers and students, take various forms such as fieldwork, field days, or excursions. However, it should be noted that this methodology is not new, as it has been advocated by several renowned educational theorists, including Comenius, Pestalozzi, and Rousseau, who sought to incorporate such activities into their teaching practices (Gómez et al., 2017). In fact, the use of didactic itineraries has been studied on an international level, with evidence supporting their effectiveness in educational contexts (Anđelkovi et al., 2018; Grubar et al., 2018; Marvell and Simm, 2018).

Historically, the Institución Libre de Enseñanza (ILE) was a pioneer of didactic itineraries in Spain, emerging in the late 19th century. According to Ortega (1998), Crespo (2012), and Liceras (2018), educators like Bartolomé Cossío, Torres Campos, and Giner de los Ríos advocated for the use of excursions to break away from teaching methods centered solely on theory. Their goal was for students to learn through practice and direct observation of their surroundings. This approach laid the foundation for the ILE's educational philosophy. In the 20th century, the prominence of didactic itineraries grew with pedagogues such as Dewey, Decroly, and Pestalozzi, who promoted the concept of "classroom-walks." More recent pedagogical renewal movements continue to support and develop this methodology (Liceras, 2018). Nevertheless, as Morote-Seguido (2019) notes, traditional Geography continues to dominate both educational practices and public opinion. This resistance to adopting innovative methods, such as didactic itineraries, has resulted in confusion or contradiction between classroom learning and students' real-life experiences regarding terms like "space" and "geographical space." To overcome this disconnect, it is crucial to design didactic itineraries that are both interdisciplinary and cross-cutting, and to abandon the perception of itineraries as mere "non-scholarly activities" (Morote-Seguido, 2017).

The benefits of implementing didactic itineraries have been thoroughly researched and documented. These itineraries align with the four pillars of education proposed by UNESCO: 1) Learning to be; 2) Learning to live together; 3) Learning to know; and 4) Learning to do. Beginning with "learning to be," scientific literature indicates that students acquire ethical and social values through this tool, as they are required to take responsibility for their surroundings. This promotes social interaction, as well as interaction with the landscape. With respect to "learning to live together," the cooperative nature of these activities fosters collaboration and mutual respect among students (de la Fuente, 2015). Moreover, students undeniably "learn to know" and "learn to do" through outdoor activities, which allow them to connect classroom learning with their real-world experiences, achieving meaningful and profound learning while simultaneously developing their thinking skills by interacting with their environment (Jerez et al., 2018).

In recent years, the evolution of didactic itineraries has been driven by the inclusion of new technologies, such as the Global Positioning System (GPS) and mobile applications compatible with Geographic Information Systems (GIS), including ArcGIS and QField. These tools enhance student motivation and engagement (García and Delgado, 2013; Fuller et al., 2006; Fombona and Pascual, 2016; Moreno et al., 2015). However, despite the growing availability of cartographic tools like Google Maps and Google Earth, their integration into Geography classes has been limited, with many educators still hesitant to use these technological resources.

Didactic itineraries, combined with the use of technological tools, are essential for addressing road safety education. This concept is embedded in current Spanish legislation, which includes the following objective: "n) To develop daily habits of active, autonomous, and healthy mobility, promoting road safety education and attitudes of respect that contribute to traffic accident prevention." According to the Royal Decree on Primary Education, road safety is a fundamental component of the Physical Education curriculum. In the first cycle, it is introduced as: "Rules of use: road safety education for pedestrians. Safe, healthy, and sustainable mobility." In the second cycle: "Rules of use: road safety education for bicycles. Safe, healthy, and sustainable mobility." Finally, in the third cycle: "Rules of use: road safety education through physical and sporting activity. Safe, healthy, and sustainable mobility."

Similarly, road safety is highlighted in the Natural, Social, and Cultural Environment curriculum, with key learnings such as: "Road safety. The city as a space for coexistence. Basic rules for walking and using transportation." "Road safety. The city as a space for coexistence. Traffic rules, road signs, and markings. Safe, healthy, and sustainable mobility for pedestrians or transportation users."

Specific competency:

To know and internalize basic rules for coexistence in the use of public spaces, especially as pedestrians or transportation users, becoming aware of the importance of safe, healthy, and sustainable mobility for both individuals and the planet.

After reviewing the importance of didactic itineraries and road safety education, supported by scientific evidence and educational legislation, the overarching goal of this chapter is to design an intervention proposal utilizing didactic itineraries for first-grade primary students to address road safety education. This proposal seeks to ensure that students not only acquire essential knowledge but also engage in practical, real-world experiences that solidify their understanding of road safety, environmental awareness, and responsible mobility. The intervention will also emphasize the use of technology to further enhance the learning experience, integrating tools like GPS, GIS apps, and digital maps to make the process more interactive, engaging, and relevant to students' everyday lives. Through this approach, it is expected that students will develop a deeper, more meaningful understanding of geography and road safety, fostering lifelong learning and responsible behavior.

3. OBJECTIVES

The general objective of this chapter is to design and implement a didactic itinerary for first-grade students that promotes the acquisition of road safety knowledge and skills, integrating geographical understanding, responsible mobility, and the use of new technologies for meaningful and experiential learning.

4. MAIN FOCUS ON THE CHAPTER

4.1 Justification

The justification for this proposal is rooted in the theoretical framework previously outlined, where the numerous benefits of implementing this methodological strategy are clearly demonstrated. Didactic itineraries have been shown to enhance students' learning experiences by promoting active engagement, fostering critical thinking, and making abstract concepts more tangible through direct interaction with their surroundings. These itineraries provide opportunities for students to apply theoretical knowledge in real-world contexts, which in turn facilitates more meaningful and lasting learning outcomes. This approach aligns with contemporary pedagogical trends that emphasize experiential and interdisciplinary learning, making it an ideal method for teaching subjects like Geography and road safety.

Furthermore, in considering the specific context of the group where this proposal will be carried out, it is important to note that these students have not previously been exposed to didactic itineraries or worked with content related to road safety education. This lack of prior experience underscores the relevance and necessity of introducing such an innovative approach, as it can fill gaps in their learning and provide them with essential knowledge and skills they have not yet developed. The absence of earlier exposure to road safety education also highlights an opportunity to address critical life skills that are necessary for ensuring the students' safety and fostering responsible behavior in public spaces. By integrating didactic itineraries into their learning process, students will benefit from a dynamic, hands-on educational experience that is both engaging and educationally enriching.

This specific group of students stands to gain significantly from the introduction of didactic itineraries, as these strategies can help bridge the gap between theoretical classroom-based instruction and practical, real-life application. Moreover, the use of this innovative methodology has the potential to stimulate greater interest and motivation among the students, providing them with a refreshing departure from the traditional teaching methods they may be accustomed to.

4.2. Context

The didactic itinerary is specifically designed for first-grade students in the subject area of Natural, Social, and Cultural Knowledge, focusing on the theme of "Transportation." This topic is particularly relevant as it encompasses a wide range of essential concepts that help students understand the various means of transport, their functions, and their significance in everyday life. By exploring this theme, students will be able to connect their learning to their daily experiences, fostering a deeper understanding of how transportation affects their communities and environments.

The class is composed of 15 students, including 5 boys and 10 girls. All students are performing at a similar level in the area of Natural and Social Knowledge, which allows for a cohesive learning environment where they can work together effectively. However, it is important to note that one student in the class has a different linguistic background; she is Moroccan and does not speak Spanish as her mother tongue. This linguistic difference presents both challenges and opportunities for inclusive teaching practices.

To support this student, various visual aids and codes were utilized during the presentation of information related to the topic. These visual tools—such as images, diagrams, and pictograms—are crucial for enhancing comprehension and engagement, especially for students who may struggle with language barriers. However, while these adaptations were beneficial, they were somewhat limited in scope. This highlights the need for further consideration of differentiated instruction techniques to fully support all learners in the classroom, particularly those who come from diverse linguistic and cultural backgrounds.

Overall, the didactic itinerary aims to create an inclusive and engaging learning experience for all students, emphasizing collaboration and interactive learning. By exploring the theme of transportation, the students will not only acquire knowledge about different modes of transport but also develop essential skills such as critical thinking, communication, and teamwork. This approach ensures that all students, regardless of their linguistic abilities, can participate actively in the learning process and benefit from the rich educational experiences that didactic itineraries offer.

4.3. Methodology

The operational and participatory methodology has been intentionally selected to guide the development of this proposal. This approach is particularly effective as it enables students to engage actively in their learning process, fostering an environment where they can reflect on their prior knowledge and connect it to new concepts. In this setting, the role of the teacher shifts from a traditional lecturer to that of a facilitator—acting as a guide, companion, and resource generator who supports students in their exploration and understanding of the subject matter.

The didactic itinerary is structured around the three phases outlined by Liceras (2018), ensuring a comprehensive and organized approach to the learning experience:

Preparation Phase

In this initial phase, careful planning is essential. The selection of the outing day is made, and informative notes are prepared for the parents or legal guardians of the children. These notes serve to keep families informed about the itinerary, including the various stops planned along the route, the activities that will take place prior to the outing, and the materials that students are required to bring. This communication not only helps to prepare the students but also engages the families in the educational process, creating a supportive learning community.

Implementation Phase

The implementation phase is the heart of the itinerary, lasting two hours, during which students actively participate in various activities. These activities take place on the streets of the town and at the different pre-planned stops, allowing students to experience learning in a real-world context. This hands-on engagement helps to solidify their understanding of the transportation theme while also promoting skills such as teamwork and critical thinking. By navigating their local environment, students gain valuable insights into how transportation systems operate and their impact on daily life.

Interpretation Phase

Following the outing, the interpretation phase takes place back in the classroom, where students engage in activities designed to reinforce and consolidate the knowledge gained during their excursion. In line with Mínguez's (2010) recommendations, small groups of 4 or 5 students are formed to reflect on their experiences, discussing what they learned and what aspects of the outing they enjoyed the most. Although Mínguez suggests beginning this reflection process individually before discussing it in groups, the decision was made to facilitate group discussions from the outset, given the young age of the students. This adaptation allows them to collaboratively think through their experiences and express their ideas in a supportive environment.

To capture their reflections, the students contribute their thoughts to a poster, which will eventually serve as a page in a larger book authored by the entire class. This collective project not only reinforces their learning but also instills a sense of ownership and pride in their contributions.

Flexibility in Grouping

It is important to highlight the flexibility in grouping strategies throughout the activities. The itinerary includes a variety of group dynamics—individual activities, such as completing worksheets; partner activities like "Who is Who?" which promote peer interaction; and larger group activities that encourage collaboration and teamwork among all students. This diverse approach caters to different learning styles and needs, ensuring that all students are engaged and able to contribute meaningfully to their learning experience.

Overall, this comprehensive methodology not only enhances the educational experience but also promotes a sense of community and collaboration among the students, equipping them with the knowledge and skills necessary for their continued learning in Natural, Social, and Cultural Knowledge.

4.4. Legislative Foundation

The project has been designed taking into account the current legislation. In this regard, the Royal Decree, which establishes the organization and minimum teachings of Primary Education, as well as the decree from the autonomous community of Castilla La Mancha, has been considered. Thus, the general specific competencies, and basic knowledge and assessment criteria related to the theme addressed throughout the itinerary—transportation—can be enumerated

Table 1. Assessment Criteria

Key competences	Language competence
Specific Competencies	1. Use digital devices and resources safely, responsibly and efficiently, to search for information, communicate and work individually, in teams and on networks, and to rework and create digital content in accordance with the digital needs of the educational context.
Descriptors	Locate, select and contrast, with due accompaniment, simple information from two or more sources, evaluating their reliability and usefulness based on reading objectives, and integrates and transforms it into knowledge to communicate it by adopting a point of view creative, critical and personal to the so respectful of intellectual property.
	Interprets and transmits the most important elements of some scientific, mathematical and relevant to processes, reasoning, technological information in a clear and truthful manner, using the appropriate scientific terminology, in different formats (drawings, diagrams, graphs, symbols...) and taking advantage of critically, ethically and responsibly digital culture to share and build new knowledge.
	Perform guided searches on the Internet and use simple strategies for the digital treatment of information (keywords, selection of information relevant, data organization...) with a critical attitude on the contents obtained.
	Create, integrate and rework digital content in different formats (text, table, image, audio, video, computer program...) through the use of different digital tools to express ideas, feelings and knowledge, respecting intellectual property and copyright of the content you reuse.
	Participate in school activities or projects through the use of virtual tools or platforms to build new knowledge.
	Work cooperatively, and share data and content in environments securely restricted and supervised digital an open and responsible attitude towards its use.
	Know the risks and adopt, with the guidance of the teacher, preventive measures when using digital technologies to protect devices, personal data, health and the environment, and begins with the adoption of habits of critical, safe, healthy and sustainable use of these technologies.
	Starts developing digital solutions simple and sustainable (reuse of materials technological, block computer programming, robotics educational...) to solve specific problems or challenges proposed in a creative way, requesting help in case necessary.
	Experiment creatively with different media and supports, and various plastic, visual, audiovisual, sound or physical, to elaborate artistic and cultural proposals
Evaluation criteria	1.1 Use digital devices and resources safely and according to the needs of the context educational.
Key competences	Language competence
Specific Competencies	2. Pose questions and provide answers to simple scientific inquiries, using different techniques, instruments, and models characteristic of scientific thinking, to interpret and explain facts and phenomena occurring in the natural, social, and cultural environment.

continued on following page

Table 1. Continued

Descriptors	Express facts, concepts, thoughts, opinions or feelings orally, written, signed or multimodal, with clarity and adaptation to different everyday contexts of their personal, social and educational environment, and participates in communicative interactions with a cooperative attitude and respectful, both to exchange information and create knowledge to build personal ties
	Perform guided searches on the Internet and use Simple strategies for the digital treatment of information (keywords, selection of information relevant, data organization...) with a critical attitude on the contents obtained.
Evaluation criteria	2.1 Show curiosity about nearby objects, facts and phenomena, asking questions and making predictions.
	2.2 Search for simple information from different safe and reliable sources in a guided manner, using it in research related to the natural, social and cultural environment.
	2.3 Participate in scheduled or guided experiments, when the research requires it, using simple techniques of inquiry, safely using instruments and recording observations clearly.
Key competences	Competence in linguistic communication
Specific Competencies	9. Participate in the environment and social life in an effective and constructive way from respect for democratic values, human and childhood rights and the principles and values of the Spanish Constitution and the European Union, valuing the role of the State and its institutions in the maintenance of peace and comprehensive citizen security, to generate respectful and equitable interactions and promote the peaceful and dialogued resolution of conflicts.
Descriptors	
Evaluation criteria	Know and internalize basic rules for coexistence in the use of public spaces, especially as pedestrians or as users of means of transportation, becoming aware of the importance of mobility safe, healthy and sustainable for both people and the planet
Key competences	Competence in linguistic communication
Specific Competencies	6. Communicate and represent, individually and collectively, mathematical concepts, procedures and results, using oral, written, graphic, multimodal language and appropriate terminology, to give meaning and permanence to mathematical ideas
Descriptors	Use, in a guided way, some methods inductive and deductive reasoning mathematician in known situations, and selects and use some strategies to solve problems reflecting on the solutions obtained.
	Use scientific thinking to understand and explain some of the phenomena that occur at your around, trusting in knowledge as a driving force development, using tools and tools appropriate, asking questions and making simple guided experiments.
	Interprets and transmits the most relevant elements of some scientific, mathematical and relevant to processes, reasonings, technological information in a clear and truthful manner, using the appropriate scientific terminology, in different formats (drawings, diagrams, graphs, symbols...) and taking advantage of critically, ethically and responsibly digital culture to share and build new knowledge.

continued on following page

Table 1. Continued

	Participate in actions founded scientifically to promote health and preserve the environment and living beings, applying principles of ethics and safety and practicing responsible consumption
Evaluation criteria	6.1 Show sustainable lifestyles and value the importance of respect, care, co-responsibility and protection of the elements and beings of the planet, identifying the relationship of people's lives with their actions about the elements and resources of the environment such as soil and water.
Basic knowledge	Knowledge of the Natural, Social and Cultural Environment: B. Technology and Digitalization - Devices and resources of the digital learning environment according to the needs of educational context. C. Societies and territories. 3. Civic literacy. - Road safety. The city as a space for coexistence. Basic rules when traveling as pedestrians or as users of means of transportation. Mathematics Area: A. Number sense. 1. Counting. - Various counting and systematic counting strategies in everyday life situations in quantities up to 999.

Source: Own elaboration.

4.4 Objectives

The objectives for this educational stage encompass both academic and practical life skills that aim to foster comprehensive learning. These objectives include:

h) Understanding Key Concepts in Natural and Social Sciences: One of the fundamental goals is to familiarize students with essential topics in Natural Sciences and Social Sciences, alongside key aspects of Geography and History. This objective also extends to the appreciation of diverse cultural expressions and the understanding of how these elements influence societies and environments.

n) Promoting Healthy and Autonomous Mobility Habits: Another important goal is to encourage the development of everyday habits that support active, independent, and healthy mobility. Through this, students will not only cultivate personal autonomy but also learn the importance of road safety education. They will adopt attitudes of respect toward others that directly contribute to reducing traffic accidents, fostering a more conscientious and responsible approach to transportation in their daily lives.

The specific didactic objectives for this particular learning unit have been carefully designed to address the theme of transportation and road safety in an engaging and practical way. These objectives include:

Identeying and Naming Vehicles: Students will learn to recognize different vehicles, such as cars, buses or bicycles. To associate the name with the corresponding vehicle, they must have a clear understanding of the characteristics of each vehicle.

This objective also aids in language development because students will use new vocabulary.

Recognizing Different Types of Transport: Students will learn to identify various modes of transportation, ranging from land, air, and water vehicles, to distinguish between them in terms of function and context.

Distinguishing between Passenger Transport (individual and collective) and Goods transport: This objective is related to the second objective but it focusses on helping students understand the differences between the transportation of people and the goods. It is essential for recognizing how and where different transport systems operate.

Understanding Where Different Transports Operate: This involves teaching students to associate specific transport modes with their corresponding environments—whether they travel on roads, in the air, or across water—and to appreciate the infrastructure that supports each.

Learning About Common Traffic Signs: The students will become familiar with basic traffic signs they are likely to encounter in their daily routines, ensuring they can recognize and understand their significance.

Respecting Key Rules of Road Safety: Students will be introduced to fundamental road safety rules, fostering an attitude of responsibility toward following them both as pedestrians and future drivers.

Discovering the Concept of a Heliport: This objective adds a unique element to the learning experience by introducing students to the concept of a heliport, expanding their understanding of transport-related infrastructure beyond common forms like roads and railways.

Visiting a Parking Lot: A hands-on visit to a parking lot will give students real-world exposure to how transport systems function, emphasizing the importance of organization and safety in public spaces.

Active Participation in Activities: Students will be encouraged to fully engage in all proposed activities, ensuring a dynamic and interactive learning process.

Collaborating with Peers: In tasks that require teamwork, students will practice collaboration, learning to work effectively with others and developing social and cooperative skills.

4.5. Description or Development of the Proposal

This educational proposal is designed around a well-structured learning journey, combining classroom-based teaching with an experiential field trip. The proposal comprises three key phases: pre-trip classroom activities, the actual educational field trip, and post-trip reflection and activities. This approach ensures a well-rounded learning experience that integrates theory, observation, and practical application.

4.5.1. Temporal Sequencing and Spatial Organization

The learning process begins with a set of preparatory classroom sessions, followed by the educational field trip, and concludes with reflective activities.

The pre-trip phase included four classroom sessions dedicated to preparing students for the concepts they would encounter during the field trip. This preparation is essential to ensure students are equipped with the necessary background knowledge to maximize the benefits of the trip.

During the field trip, students made three main stops:

The first stop was at the town's heliport, offering a unique opportunity for students to observe a lesser-known part of transport infrastructure. The second stop was at Rosaleda Park, where students could reflect on how transport connects different parts of the town and affects public spaces and the final stop took place at a parking lot, providing students with a practical demonstration of how transport infrastructure operates in a day-to-day context.

Throughout the journey, the group made several brief stops to discuss traffic signs and observe different modes of transport encountered along the way, such as cars, vans, motorcycles, bicycles, tractors, and ambulances. These short discussions allowed students to make real-time connections between the theoretical knowledge acquired in the classroom and practical observations in their environment.

The final part of the sequence involved a post-trip phase back in the classroom. Here, students worked in groups to create presentations or visual representations that reflected what they had learned throughout the itinerary. This phase allowed students to consolidate their understanding and share their learning experiences with their peers.

4.5.2. Activities in the Didactic Itinerary

As previously mentioned, and in accordance with Liceras (2018), the didactic itinerary follows a three-phase model: preparation, execution, and reflection. Each phase plays a crucial role in reinforcing the students' learning experience. The activities are designed to gradually build up the students' understanding, from theoretical explanations to practical applications and finally to reflective discussions.

4.5.2.1. Pre-Trip Phase in the Classroom

The pre-trip phase comprises a series of preparatory activities aimed at familiarizing students with the key concepts and ideas they would explore in the field trip. These activities include:

Table 2. Means of transport

<table>
<tr><th colspan="2">TITTLE OF THE ACTIVITY
We are learning means of transport.</th></tr>
<tr><td colspan="2">Didactic objectives
Identying and Naming Vehicles.
Distinguishing between Passenger Transport (individual and collective) and Goods transport.
Recognizing Different Types of Transport
Active Participation in Activities</td></tr>
<tr><td colspan="2">Basic knowledge
Knowledge of the Natural, Social and Cultural Environment:
B. Technology and Digitalization
- Devices and resources of the digital learning enviromnet according to the needs of educatiomal context.
C. Societies and territories.
3. Civic literacy.
- Road safety. The city as a space for coexistence. Basic rules when traveling as pedestrians or as users of means of transportation</td></tr>
<tr><td>Key competencies
• Personal, social and learning to learn competetence
• Language competence
• Digital competence</td><td>Evaluation criteria
Show curiosity about nearby objects, facts and phenomena, asking questions and making predictions</td></tr>
<tr><td colspan="2">Description of the activity
Different Modes of Transport. In this introductory lesson, students were first introduced to the concept of transportation as a means of movement. The teacher explained that there are passenger transports, which can be either individual (like cars and motorcycles) or collective (like buses and trains), as well as goods transports (e.g., trucks, cargo ships). To reinforce this concept, the teacher asked students to name different types of transport they knew, creating an interactive discussion as a brain storming that helped link their prior knowledge to new information. After the discussion, we will use the digital resources to complete an activity that involves classifying different types of vehicles. This activity was created using Wordwall and here is the link to access it: https://wordwall.net/resource/79055884 . Finally the students will complete an individual worksheet to solidify their understanding of different transport types (see Appendix A for details).</td></tr>
<tr><td colspan="2">Assessment's instruments
Observation
Questions
Checklist rubric</td></tr>
<tr><td>Individualized educational inclusion measures
Images of different modes of transportation: Include photos or illustrations of the types of transportation mentioned in the lesson (cars, motorcycles, trains, trucks, boats, etc.). Images can help make the visual connection with Spanish vocabulary and facilitate comprehension.
Adaptation of individual worksheet with simplified vocabulary</td><td>Resources
Images or videos of different vehicles
Wordwall
Individual worksheet</td></tr>
</table>

Source: Own elaboration.

Table 3. Plane landing

<table>
<tr><th colspan="2">TITTLE OF THE ACTIVITY
The plane is landing</th></tr>
<tr><td colspan="2">Didactic objectives
• Understanding Where Different Transports Operate
• Active Participation in Activities.</td></tr>
<tr><td colspan="2">Basic knowledge
• Means of transport</td></tr>
<tr><td>Key competencies
• Personal, social and learning to learn competetence
• Language competence</td><td>Evaluation criteria
• Show curiosity about nearby objects, facts and phenomena, asking questions and making predictions</td></tr>
<tr><td colspan="2">Description of the activity
Where Do Transports Go?. In this session, students learned to distinguish between land, air, and water transports. The teacher explained how land-based transports travel along roads and railways, water transports sail on rivers and seas, and air transports fly in the sky. To engage the class, the teacher asked whether a series of vehicles were land, water, or air-based, and how they moved. Students participated actively by answering questions and volunteering to classify the vehicles. Afterward, they completed an individual worksheet that reinforced the lesson. This worksheet will be included in the project diary that will include all the exercises, worksheets and notes taken during each of the sessions (Visualize the Appendix)</td></tr>
<tr><td colspan="2">Assesssment's instruments
Observation
• Questions
• checklist rubric</td></tr>
<tr><td>Individualized educational inclusion measures
Include photos to help her to understand where different transports operate.
Adaptation of the individual worksheet with visual aid and simplified vocabulary.</td><td>Resources
• Images or videos of different vehicles
• Project diary (Individual worksheet)</td></tr>
</table>

Source: Own elaboration.

Table 4. Structuring the mind

<table>
<tr><th colspan="2">TITTLE OF THE ACTIVITY
Structuring our minds</th></tr>
<tr><td colspan="2">Didactic objectives
• Understanding Where Different Transports Operate.
• Active Participation in Activities</td></tr>
<tr><td colspan="2">Basic knowledge

Knowledge of the Natural, Social and Cultural Environment:
B. Technology and Digitalization
- Devices and resources of the digital learning enviromnet according to the needs of educatiomal context.
C. Societies and territories.
3. Civic literacy.
- Road safety. The city as a space for coexistence. Basic rules when traveling as pedestrians or as users of means of transportation</td></tr>
<tr><td>Key competencies
• Language competence
• Personal, social and learning to learn competetence
• Digital competence</td><td>Evaluation criteria
• Show curiosity about nearby objects, facts and phenomena, asking questions and making predictions.</td></tr>
<tr><td colspan="2">Description of the activity
What Do Transports Need? In this session, students explored the infrastructure and resources required by different types transports. Using two Genially presentations (interactive digital presentations), the class examined the various structures needed for transport, such as roads, railways, and ports, as well as the energy sources and materials each type of transport relies on (e.g., gasoline for cars, electricity for trains). Students participated by taking turns to answering questions, with each student receiving help from a classmate if needed. This activity ensured continuous engagement, as everyone had to remain attentive and involved. After completing the Genially presentations, a "Who's Who?" activity involving modes of transportation was carried out. Students were placed in pairs and tasked with identifying the mode of transport based on a series of guided questions displayed on the digital board. Depending on the responses received, conclusions were drawn, allowing students to guess the transportation type they had in mind.</td></tr>
<tr><td colspan="2">Assesssment's instruments
Observation
• Questions
• checklist rubric</td></tr>
<tr><td>Individualized educational inclusion measures

Include photos to help her to understand where different transports operate.
Adaptation of the individual worksheet with visual aid and simplified vocabulary.</td><td>Resources
• Individual worksheets
• Laptops
• Genially presentations</td></tr>
</table>

Source: Own elaboration.

Table 5. Life of the vehicle

<table>
<tr><td colspan="2">TITTLE OF THE ACTIVITY
The life of the vehicles</td></tr>
<tr><td colspan="2">Didactic objectives
• Understanding Where Different Transports Operate.
• Learning About Common Traffic Signs.
• Respecting Key Rules of Road Safety
• Active Participation in Activities.</td></tr>
<tr><td colspan="2">Basic knowledge
Knowledge of the Natural, Social and Cultural Environment:
C. Societies and territories.
3. Civic literacy.
- Road safety. The city as a space for coexistence. Basic rules when traveling as pedestrians or as users of means of transportation</td></tr>
<tr><td>Key competencies
• Personal, social and learning to learn competetence
• Language competence
• Civic competence</td><td>Evaluation criteria
• Show curiosity about nearby objects, facts and phenomena, asking questions and making predictions.</td></tr>
<tr><td colspan="2">Description of the activity
A custom-made story was projected, featuring a family of characters: "Mama Caravan," "Papa Bus," and "Blue and Orange, the Little Cars," who introduced various traffic signs. As each sign appeared, students were asked if they recognized it, and several were familiar with their meanings. Afterward, a worksheet was distributed for students to individually color the traffic signs from the story in the story and write their meanings underneath. This activity aimed to reinforce their understanding of traffic signs and basic road safety rules in preparation for the upcoming outing.</td></tr>
<tr><td colspan="2">Assesssment's instruments
Observation
• Questions
• Checklist rubric</td></tr>
<tr><td>Individualized educational inclusion measures

Include photos/ images of the different characters of the story with their names to help to understand the different types of vehicles and clarify the storyline .This visual help enhance student´s reading comprehension.
Adaptation of the individual worksheet with visual aid and simplified vocabulary.</td><td>Resources
• Individual worksheets
• Laptops
• The story</td></tr>
</table>

Source: Own elaboration.

4.5.2.2 Second Phase: Field Trip

Table 6. Trafiquin

TITTLE OF THE ACTIVITY **We learn with Trafiquín**	
Didactic objectives • Understanding Where Different Transports Operate. • Identying and Naming Vehicles. • Learning About Common Traffic Signs • Respecting Key Rules of Road Safety • Discovering the Concept of a Heliport • Visiting a Parking Lot • Active Participation in Activities	
Basic knowledge Mathematics Area: A. Number sense. 1. Counting. - Various counting and systematic counting strategies in everyday life situations in quantities up to 999. Knowledge of the Natural, Social and Cultural Environment: C. Societies and territories. 3. Civic literacy. - Road safety. The city as a space for coexistence. Basic rules when traveling as pedestrians or as users of means of transportation.	
Key competencies • Personal, social and learning to learn competetence • Language competence • Mathematical Competence in science technology and engineering. • Civic competence.	Evaluation criteria • Show curiosity about nearby objects, facts and phenomena, asking questions and making predictions.

continued on following page

Table 6. Continued

<table>
<tr><th colspan="2">TITTLE OF THE ACTIVITY
We learn with Trafiquín</th></tr>
<tr><td colspan="2">Description of the activity
At the school entrance, students were introduced to a character named "Trafiquín". It was explained that Trafiquín wanted to walk through the town's streets but did not know any traffic signs or rules, and needed the students' help. When asked if they were willing to assist, the students eagerly responded with enthusiasm. Each student was then given a necklace featuring a drawing of Trafiquín, labeled as his helper. The character was displayed prominently, holding a stop sign in one hand and a yield sign in the other.

First Stop: Heliport. Upon arriving at the heliport, the students were asked to form a circle around the "H" marked on the ground. It was explained that a heliport is where helicopters take off and land. Several photographs of different helicopters and various heliports were shown. Using a drawing of a helicopter, the different parts were pointed out (cockpit, tail, landing skids, tail rotor, and main rotor). This led to a team game where groups of 4 or 5 students raised a banner with either the letter A or B, depending on what they thought was the correct answert. Questions like, "Where do helicopters take off and land? A: Airport or B: Heliport?" were asked, and the students discussed it within their team before raising B. Finally, a drawing of a helicopter at the heliport was given to color at home.

Second Stop: Rosaleda Park. The second stop was at a large green area in the town, the biggest park available. There, students were asked to form small groups of about 4 and complete a worksheet on road safety. They were asked to select the correct option based on the image provided. Additionally, they had to draw one of the traffic signs they had seen along the route and write what it indicated.

Third Stop: Parking Lot. The final stop of the outing took place at the parking lot, where it was explained that this is the place where vehicles are parked, and that different types of parking exist depending on the mode of transport. To enhance understanding, photographs of parking areas for motorcycles, electric scooters, bicycles, caravans, buses, trucks, cars, etc., were shown. After observing these images, different transportation modes were presented, and students were asked whether they could be parked in the lot where they were located. Finally, students worked in small groups to complete a brief worksheet listing the vehicles parked there. They counted the vehicles and represented the data in a bar graph, integrating mathematical skills into the activity. This activity was conducted while seated on benches at the Senior Center across the street, ensuring the group didn´t obstruct traffic. On the way back to school, a bicycle parking area was spotted, which was also pointed out and discussed</td></tr>
<tr><td colspan="2">Assesssment's instruments
Observation
• Questions
• checklist rubric</td></tr>
<tr><td>Individualized educational inclusion measures

To engange students who learn often better through visual representation, additional visual aids such as colorful images of Trafiquín and the various vehicles were shown. Incorporate tactile materials for example 3D models of vehicles or traffic signs.
Adaptation of the individual worksheet with visual aid and simplified vocabulary.</td><td>Resources

• Diary Project (with the individual worksheet)
• Necklace of Trafiquín helper
• Images of vehicles
• Notebooks and color pencils to draw.</td></tr>
</table>

Source: Own elaboration.

4.5.2.3. Third Phase (Post-Trip)

To conclude the itinerary, a joint reflection was conducted in the classroom through a game.

Table 7. Snowball game activity

<table>
<tr><th colspan="2">TITTLE OF THE ACTIVITY
Snowball game</th></tr>
<tr><td colspan="2">Didactic objectives
• Understanding Where Different Transports Operate.
• Identying and Naming Vehicles.
• Learning About Common Traffic Signs
• Respecting Key Rules of Road Safety
• Active Participation in Activities
• Understanding Where Different Transports Operate</td></tr>
<tr><td colspan="2">Basic knowledge
Knowledge of the Natural, Social and Cultural Environment:
C. Societies and territories.
3. Civic literacy.
- Road safety. The city as a space for coexistence. Basic rules when traveling as pedestrians or as users of means of transportation.</td></tr>
<tr><td>Key competencies
• Personal, social and learning to learn competetence
• Language competence
• Civic competence</td><td>Evaluation criteria
• Show curiosity about nearby objects, facts and phenomena, asking questions and making predictions.</td></tr>
<tr><td colspan="2">Description of the activity
Snowball Game. Students were gathered to the hallway and asked to form a circle. Small teams were created, and each team was given a sheet of paper. A ball was held, which was thrown to one student to share their thoughts. The student had to say aloud something they learned the previous day during the outing, so that the rest of their classmates could write it down on the sheet. When one student finished, they passed the ball to a classmate who had not yet spoken, allowing everyone to participate and relive the experience by recalling in detail everything they had seen. Finally, the activity was repeated, but this time they shared what they liked the most and wrote that down on the sheet as well.</td></tr>
<tr><td colspan="2">Assesssment's instruments
Observation
• Questions
• checklist rubric</td></tr>
<tr><td>Individualized educational inclusion measures

Assign different roles, such as note-taker or presenter to students encourage to their participation. Use or provide images of some difficult concepts to clarify information and improve our student's' speaking skills.</td><td>Resources
• Diary Project
• Sheet of paper
• Pencil</td></tr>
</table>

Source: Own elaboration.

Table 8. Becoming Trafiquins assistant

<table>
<tr><th colspan="2">TITTLE OF THE ACTIVITY
We became Trafiquín's assistants!</th></tr>
<tr><td colspan="2">Didactic objectives
• Active Participation in Activities</td></tr>
<tr><td colspan="2">Basic knowledge
Knowledge of the Natural, Social and Cultural Environment:
C. Societies and territories.
3. Civic literacy.
- Road safety. The city as a space for coexistence. Basic rules when traveling as pedestrians or as users of means of transportation.</td></tr>
<tr><td>Key competencies
• Personal, social and learning to learn competetence.
• Language competence.
• Civic competence.</td><td>Evaluation criteria
• Show curiosity about nearby objects, facts and phenomena, asking questions and making predictions.</td></tr>
<tr><td colspan="2">Description of the activity
ouring Sisante with Trafiquín . Each team had to capture on a large piece of cardboard that I gave them what they had written the day before on the sheet of paper. They could decorate it however they wanted. Finally, I made the front and back covers and by joining the cards from the 3 groups we formed a book in which we could capture our entire adventure through the streets of our town. In the last pages I included a photograph of everyone and the thumb prints of each of us with our names below to remember who we lived with this fun experience in which we were Trafiquín's assistants.</td></tr>
<tr><td colspan="2">Assesssment's instruments
Observation
• Questions
• checklist rubric</td></tr>
<tr><td>Individualized educational inclusion measures

Assign different roles, such as writer or decorator to students encourage to their participation. Use or provide images of some difficult concepts to clarify information and improve our student's speaking skills.</td><td>Resources
• Sheet of paper (written in the last activiy)
• Decorations
• Cardboard
• Photos</td></tr>
</table>

Source: Own elaboration.

CONCLUSION

After implementing this didactic experience, several conclusions were drawn from the process, deepening the understanding of the value and effectiveness of didactic itineraries as an educational resource.

To begin with, the primary objectives set for this proposal were successfully met. Throughout the preparation and execution phases, significant time was invested in thoroughly researching the structure, phases, benefits, and significance of didactic itineraries. This research provided a solid foundation for the learning experience.

Additionally, an educational outing through the streets of Sisante was organized with a 1st-grade class, focusing on road safety education. This was aligned with Decree 81/2022 for the subject area of Knowledge of the Natural, Social, and Cultural Environment in the first cycle of Primary Education. Aligning the experience with the official curriculum ensured that the proposal was not only pedagogically sound but also fully compliant with the educational framework, serving as a reference throughout the process. This adherence to the curriculum is essential, as it ensures that the activities contribute to the holistic development of students within the mandated learning goals.

The specific didactic objectives set for the students were also achieved. Activities designed prior to the outing were crucial in preparing the students for the experience. During these sessions, students were introduced to various modes of transportation, learning to recognize them and understand where and how each operates. They were also introduced to common traffic signs, a key aspect of road safety education.

On the day of the outing, students engaged in a rich, immersive learning experience. They had the opportunity to explore real-world locations such as the heliport and parking lot and actively identify different types of transport and traffic signs studied in class. This hands-on experience effectively bridged the gap between theoretical knowledge and practical application, reinforcing their learning in an engaging and meaningful way.

In the post-activity phase, students consolidated their learning through group work, creating a book that captured the key concepts and activities covered in the sessions. This group-based reflection allowed students to review the content, while also fostering teamwork, creativity, and critical thinking. The reflection process encouraged collaboration and helped solidify the lessons learned during the outing.

This process emphasized the educational value of didactic itineraries. As Liceras (2018) highlights, didactic itineraries enhance student learning by providing context and real-life experiences that engage students beyond the classroom. Throughout this experience, students demonstrated increased motivation and enthusiasm while learning outside the traditional classroom environment. Their engagement with their surroundings helped reinforce the lessons covered in the classroom.

The reflection phase, as noted by Mínguez (2010), is a critical component of any educational experience, offering students the opportunity to internalize the lessons learned. In this case, the creation of a reflective book allowed students to review and express their understanding of the key concepts. The level of detail students recalled from the trip and their ability to articulate what they found most interesting showed active engagement and successful consolidation of learning.

The students' ability to retain the material and add personal reflections on the most exciting or impactful aspects of the outing demonstrated a high level of engagement. This aligns with the goal of fostering students who not only learn but also actively engage with the content, making meaningful connections between classroom learning and the world around them.

In conclusion, this experience has highlighted the value of didactic itineraries as a dynamic and interactive educational resource. These experiences offer a stimulating way for students to learn, fostering motivation, curiosity, and a deeper connection to the material. As Hernández et al. (2012) emphasize, didactic itineraries serve as lessons outside the classroom, significantly enhancing student engagement and retention of knowledge. By incorporating more of these itineraries into teaching plans, educators can ensure that students not only learn theoretical aspects of a subject but also experience it in the real world, catering to diverse learning styles and providing a well-rounded educational experience

REFERENCES

Aguilar, O., Rodríguez, E., Armiñana, R., & Padilla, A. (2023). Un itinerario didáctico, buenas prácticas para la formación de profesores de Geografía. *Revista Conrado, 19*(91), 320-331. 1990-8644-rc-19-91-320.pdf (sld.cu)

Alcántara, J., Rubio, S., & Mora, M. (2014). Google Earth como herramienta para formadores en la preparación de itinerarios didácticos. [Trabajo Académico, Universidad de Córdoba]. Universidad de Córdoba. https://helvia.uco.es/bitstream/handle/10396/15443/geografia6.pdf?sequence=1

Benejam, P. (2003). Los objetivos de las salidas. *Didáctica de las Ciencias Sociales, Geografía e Historia* (7-12). Íber. Los objetivos de las salidas (1).pdf

Centro para la Educación y Desarrollo de la Excelencia (CEDEC). (s. f.). *Banco de rúbricas y otros documentos*. Recuperado de Banco de rúbricas y otros documentos | Cedec (intef.es)

Chicano, C. (2015). Itinerario didáctico: la Catedral de Granada y su entorno. [Trabajo de Fin de Grado, Universidad de Granada]. Universidad de Granada. ITINERARIO DIDÁCTICO (ugr.es)

Claval, P. (2020). El papel del trabajo de campo en la geografía: de las epistemologías de la curiosidad a las del deseo. [Trabajo de campo, Universidad Nacional Autónoma de México]. Instituto de geografía. 2448-7279-igeo-103-e60283.pdf (scielo.org.mx)

Crespo, J. M., Gómez, M. L., Cruz, L. A. (2018). Una aproximación a los Parques Nacionales y sus paisajes a través de itinerarios didácticos. *Revista Espacio, Tiempo y Forma, serie VI*(11), 121-140. Vista de Una aproximación a los Parques Nacionales y sus paisajes a través de itinerarios didácticos (uned.es)

Decreto 81/2022, de 12 de julio, por el que se establece la ordenación y el currículo de Educación Primaria en la comunidad autónoma de Castilla-La Mancha.

García, A. (2004) El itinerario geográfico como recurso didáctico para la valoración del paisaje. *Didáctica Geográfica, 2ª época, 6*(1), 79-95. https://didacticageografica.age-geografia.es

Gómez, A., Corrochano, D., Parra, G. (2017). Itinerarios didáctico-naturales en educación primaria: el noroeste de Zamora. *Didáctica Geográfica*, (18), 111-131. Vista de ITINERARIOS DIDÁCTICO-NATURALES EN EDUCACIÓN PRIMARIA: EL NOROESTE DE ZAMORA (age-geografia.es)

Guerra, F. J. (2022). Medios para la interpretación del patrimonio. Planificación y gestión. *Servicios de Educación y Estudios Ambientales* (144-218). S.L. (SEEDA). capitulo_muestra_guia_digital_IP.pdf (juntadeandalucia.es)

Hernández, A. M., Martínez, C. M., Torres, I. C., & Hernández, L. A. (2012). La excursión integradora en la enseñanza de aprendizaje de la carrera Biología-Geografía. [Redalyc.La excursión integradora en la enseñanza aprendizaje de la carrera Biología-Geografía]. *Revista Ciencias Holguín*, XVIII(2), 1–12.

Jerez, O., Serrano de la Cruz, M. A., & García, J. L. (2018). Propuesta metodológica para la planificación de un itinerario didáctico en un paisaje protegido. En A. García (Ed.), *Contribución didáctica al aprendizaje de la geografía* (717-733). https://www.researchgate.net/publication/340037981_Propuesta_metodologica_para_la_planificacion_de_un_itinerario_didactico_en_un_Paisaje_Protegio

Liceras, A. (2018) Los itinerarios didácticos en la enseñanza de la geografía. Reflexiones y propuestas acerca de su eficacia en educación. *Revista UNES. Universidad, Escuela y Sociedad*, (5). https://www.google.es/Los itinerarios didácticos en la enseñanza de la geografía. Reflexiones y propuestas acerca de su eficacia en educación

Liceras, A. (2022) Aportaciones de Giner de los Ríos y la ILE a la enseñanza del paisaje. Las excursiones. *Revista de Investigación en Didáctica de las Ciencias Sociales,* (10). Vista de Aportaciones de Giner de los Ríos y la ILE a la enseñanza del paisaje. Las excursiones (unex.es)

López, J. A., Mora, M., Arrebola, J. C., Medina, S. (2017). Itinerarios didácticos interdisciplinares en el Grado de Educación Primaria: una propuesta en la ciudad de Córdoba. [Trabajo Académico, Universidad de Córdoba]. Didáctica de las Ciencias. 68_-_Itinerarios_didacticos_interdisciplinares_en_el_Grado_de_Educacion_Primaria.pdf (uab.cat)

Marrón, M. J. (2013). Tratamiento interdisciplinar de la geografía, la historia y el arte con estudiantes del Grado de Magisterio. Una propuesta de enseñanza activa a través de un itinerario didáctico. [Trabajo Académico, Universidad Complutense de Madrid]. Innovación en la enseñanza de la geografía ante los desafíos sociales y territoriales. Tratamiento interdisciplinar de la Geografía, la Historia y el Arte con estudiantes del Grado de Magisterio. Una propuesta de enseñanza activa a través de un itinerario didáctico (dpz.es)

Martínez, R., & Tonda, E. M. (2014). Nuevas perspectivas conceptuales y metodológicas para la educación geográfica. *Grupo de Didáctica de la Geografía de la Asociación de Geógrafos Españoles* pp. 227-229. INTERIOR VOLUMEN I.indd (age-geografia.es)

Mínguez, Mª del C. (2010). El paisaje como objeto de estudio de la Geografía. Un Itinerario Didáctico en el Marco de la Semana de la Ciencia de la Comunidad de Madrid. *Revista Didáctica Geográfica* (11), 37-62. Vista de EL PAISAJE COMO OBJETO DE ESTUDIO DE LA GEOGRAFÍA. UN ITINERARIO DIDÁCTICO EN EL MARCO DE LA SEMANA DE LA CIENCIA DE LA COMUNIDAD DE MADRID. (age-geografia.es)

Molina, M. P. (2018). *La Educación Patrimonial en la Didáctica de las Ciencias Sociales en Primaria*. [Trabajo Académico, Universidad de Córdoba]. Contextos Educativos. bibybape,+12_MolinaTorres.pdf

Moreno, L. (2024). Recorriendo Nuestras Calles. [Storyjumper]. "Recorriendo nuestras calles" - Read stories online. Create books for kids | StoryJumper

Morón, H. (2016). *El itinerario didáctico como herramienta para la activación patrimonial: una experiencia docente desde el parque María Luisa*. [Trabajo académico, Universidad de Sevilla]. EL ITINERARIO DIDÁCTICO COMO HERRAMIENTA PARA LA ACTIVACIÓN PATRIMONIAL: UNA EXPERIENCIA DOCENTE DESDE EL PARQUE MARÍA LUISA (us.es)

Núñez Romero, V. (2017) Las salidas escolares como recurso motivador para los niños [Trabajo Fin de Grado, Universidad de Valladolid]. Universidad de Valladolid. Las salidas escolares como recurso motivador para los niños

Olave, D. (2005). El itinerario didáctico: Una propuesta metodológica para el análisis geo-histórico local. [Redalyc.El itinerario didáctico: una propuesta metodológica para el análisis geo-histórico local]. *Revista Geoenseñanza*, 10(2), 197–208.

Pulgarín, R. (1998). La excursión escolar como estrategia didáctica en la enseñanza de la geografía. [Trabajo académico, Universidad de Antioquia]. La Gaceta Didáctica nº 2. La_salida_de_campo_como_estrategia_didactica-libre.pdf (d1wqtxts1xzle7.cloudfront.net)

Rubio, M. (2023). Los itinerarios didácticos virtuales como herramienta pedagógica complementaria a los itinerarios didácticos reales en geografía e historia en 1º de ESO. [Trabajo de Fin de Máster, Universidad Católica de Murcia]. UCAM. https://*Rubio_García_Marina.pdf (ucam.edu)

Sánchez, E. (2010) Innovación e interdisciplinariedad en los itinerarios didácticos de ciencias sociales. *Revista Electrónica de Didácticas Específicas,* (3). https://www.google.es/Innovación e interdisciplinariedad en los itinerarios didácticos de ciencias sociales

Sánchez Ogallar, A. (2010). El trabajo de campo y las excursiones. *Parte II: Estrategias y Recursos Didácticos*, 159-184. A S Ogallar Itinerarios.pdf

Sousa, S. A., García, D., & Souto, X. M. (2016). Educación geográfica y las salidas de campo como estrategia didáctica: un estudio comparativo desde el Geoforo Iberoamericano. *Revista Bibliográfica de Geografía y Ciencias Sociales, vol. XXI*(1.155), 1-19. Educación Geográfica y las salidas de campo como estrategia didáctica (uv.es)

Vilarrasa, A. (2003). Salir del aula. Reapropiarse del contexto. *Íber*, (36), 13-25.

KEY TERMS AND DEFINITIONS

Active Methodology: Active methodology describes teaching strategies that involve learners in the educational process through participation and engagement. Rather than simply receiving information, students are encouraged to take part in hands-on activities, collaborative projects, and problem-solving tasks. This approach aims to boost motivation, enhance understanding, and develop skills applicable to real-life situations.

Geography: Geography is the discipline that examines the Earth's physical features, environments, and the interactions between humans and their surroundings. It looks at both natural elements, like landforms and climates, as well as human factors, such as cultures and economies. This field helps to understand the spatial distribution of various phenomena and how human activities influence and are influenced by the environment.

Primary Education: Primary education refers to the initial phase of formal schooling, usually targeting children from ages 5 to 11 or 12, depending on the region. The focus is on building essential skills in literacy, numeracy, and social understanding while promoting overall personal development. This stage of education lays the groundwork for further learning and aims to foster a well-rounded growth experience for young learners.

Road Safety: Road safety encompasses the strategies and practices designed to reduce the risk of accidents and injuries on the road. It involves following traffic laws, using safety equipment like seat belts and helmets, and ensuring vehicles are well-maintained. Public awareness initiatives and infrastructure improvements, such as signage and safe crossings, also play a crucial role in creating a safer environment for all road users.

Transport System: A transport system refers to the organized network and methods that usually is used to move people or good. This transport system change depends on the vehicle so is essential to understand where the different vehicles operate because the infrastructure (like roads, railways or airports) also change. In addition, an effective transport system is crucial for economic development that influence on our actual society.

APPENDIX

Project Cover Page

First Session Worksheets

Second Session Worksheets

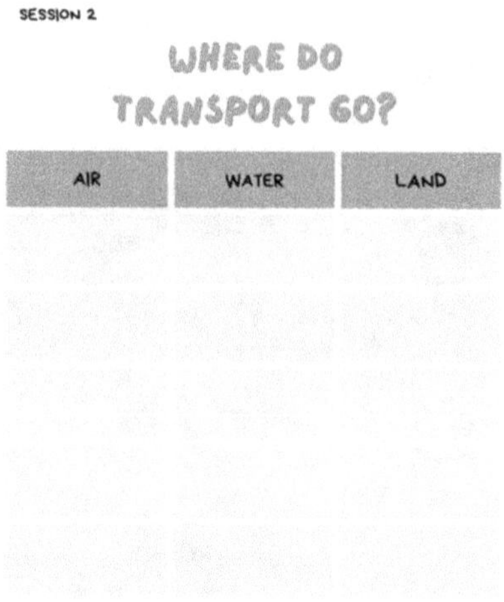

Third Session Worksheet

Section 2
Interdisciplinarity, Culture, and Geography

The social needs and requirements of the present requires multidisciplinary consideration that includes educational and pedagogical approaches involving other related disciplines that contribute to generate progress in these issues in a holistic way. For this reason, the second part of the book devotes efforts to present links with the contributions made by language and literature, the arts or the impact of technological advances.

Chapter 11
The Revival of Lucía Baquedano's Work in Youth Literature

Jesús Guzmán Mora
https://orcid.org/0000-0001-5687-0169
University of Castilla-La Mancha, Spain

ABSTRACT

Recovering Lucía Baquedano's work is essential for enriching juvenile literature and highlighting overlooked Spanish women authors. Cinco panes de cebada, Baquedano's novel, reflects rural Spanish life during the late Franco era and the country's modernization. The protagonist, Muriel, an urban teacher in a rural setting, undergoes a journey of adaptation and identity-building as she immerses herself in her new community. Over time, Muriel defies the stereotype of the traditional rural teacher by prioritizing her students' learning and well-being. The novel's confessional tone suggests an autobiographical quality and serves as a potential model for today's educators. The article ultimately argues for the renewed appreciation of works like this in educational contexts, as they address timeless themes such as teacher vocation and the transformative power of education for disadvantaged communities.

1. INTRODUCTION

Spanish Children's and Youth Literature (CYL) was made possible, for much of the last century, thanks to a substantial group of women writers, including, among others, Elena Fortún (1886-1952), Carmen Kurtz (1911-1999), María Luisa Gefaell (1918-1978), Lucía Baquedano (1938), and Pilar Mateos (1942). Unfortunately, all of them have fallen into an unjustified obscurity, which today makes it difficult to

DOI: 10.4018/979-8-3693-7185-5.ch011

Copyright © 2025, IGI Global Scientific Publishing. Copying or distributing in print or electronic forms without written permission of IGI Global is prohibited.

include their work in the school reading canon or for readers to enjoy them, due to the lack of reprints or the inability to find their books on the shelves of public libraries in our country. Thus, one of the challenges facing academic research is the recovery of these authors and others from their circle to rigorously re-evaluate their texts and bring them back to the literary stage.

With this premise, we conceive this work as part of the ongoing effort to recover the works of these writers, a project we have undertaken at other times (Guzmán Mora & Sevilla-Vallejo, 2019). On this occasion, we find it fitting to position Lucía Baquedano's career within the literary system, as she is one of the most awarded authors in Spanish CYL during the second half of the 20th century. To this end, we structure our study into three sections: first, we approach the role of women writers in the field of Spanish CYL. Second, we focus on Lucía Baquedano, particularly her novel *Cinco panes de cebada* (*Five Barley Loaves*), and third, we present the conclusions of our study. According to our hypothesis, the recovery of these women writers is not about fulfilling any gender quota; rather, it is essential due to the significance of their works in the history of Spanish children's and youth literature.

We discover nothing new when we point out that Children's and Youth Literature has always been a fertile ground for women writers, which also prevents us from separating the authors' careers from their personal lives. The case of Elena Fortún, whom we cited at the beginning of our study, is one of the most interesting examples of recovery and revitalization in recent years, thanks to the efforts of scholars like Nuria Capdevilla-Argüelles (2005) and María Jesús Fraga Fernández-Cuevas (2019), as well as contemporary writers like María Folguera. Forgotten by critics and readers alike, and only rediscovered through the iconic character Celia in the adaptation of this imaginative young girl's adventures in a Spanish Television series in the early 1990s, Fortún now occupies a prominent place in one of the flagship collections ("Biblioteca Elena Fortún") of the publisher Editorial Renacimiento. The reissue of her works with introductory studies and the recovery of an unpublished manuscript, *Oculto sendero* (*Hidden Path*), written during the early stages of the author's exile and published in 2016, have once again placed María de la Encarnación Gertrudis Jacoba Aragoneses y de Urquijo—Elena Fortún's real name—back into the Spanish literary scene (Díaz Fernández & Mérida Jiménez, 2021).

As we indicated at the beginning of the chapter, the recovery of Spanish women authors in Children's and Youth Literature does not stem from any type of parity approach to the literary canon that would lack rigor. In fact, the influence they have had on some contemporary authors is anything but trivial: Maruja Torres confessed that if there has been a "fairy godmother" in her life, or "a piñata that suddenly burst open in front of me and spilled out its gifts, it was Carmen Kurtz" (1999, p. 22). This influence was so significant that it inspired one of the characters in her novel *Mientras vivimos* (*While We Live*), which won the Planeta Prize in 2000. Returning to Fortún, the best example is the aforementioned María Folguera. A novelist, playwright, and former director of the Teatro Circo Price in Madrid, Folguera has approached the figure of Elena Fortún through the theatrical diptych *Sendero Fortún*, for which she directed *Celia en la revolución* (*Celia in the Revolution*), with a script by Alba Quintas, and *Elena Fortún*, with her own script and direction, along with the novel *Hermana. (Placer)* (*Sister. Enjoyment*) (Britland, 2024). The author confirms this in various interviews she has given over the years.

Elena Fortún is one of my most important and essential influences. I started reading her when I was young, and later I discovered the Celia books she wrote in her maturity, which are more tragic and possess a realism that I find incredibly interesting. Then these new editions of her unpublished work arrived. So it's natural that my writing reflects some of that learning, but I haven't done it consciously, nor is it an exercise in homage to her style. It's just that I'm very interested in her realistic perspective, which is also infused with fable and poetry. I also admire her love for the Spanish language. She reflects a lot on the lyricism of simple, popular Spanish, and I identify with that deeply (López, 2021).[1]

It has been remarked that Children's and Youth Literature provides women writers with the economic stability that adult literature does not offer. Likewise, although this genre, as Pedro C. Cerrillo indicates, is "first and foremost literature" (2007, p. 44), it is imperative to remember that until not too long ago, it was treated as second-rate literature or a gateway to adult literature. This false belief has left a void (or, we could say, a space emptied) by the men themselves that the women writers, in tune with their "own marginality" (Vázquez, 2007), would have occupied (Pérez Alonso, 2015). However, it is curious how those who were sidelined during their time are once again marginalized when it comes to including them in various literary histories. Thus, it is at least interesting to note the explanation some scholars provide when pointing out how these authors are not mentioned in recent studies on the subject:

Regarding the period we aim to study, it is important to highlight that abundant studies on post-war Spanish literature have appeared in recent years. Almost all of them note the significance of incorporating women novelists into the literary scene of the time, in overviews, monographs, or encyclopedias. As García Padrino states, it is curious, on the other hand, that the participation of many of these women writers in the world of children's and youth literature remains silenced. Perhaps this silence is a way to "ennoble" these authors literarily, since engaging in children's literature often implies falling into a "feminine" or devalued literature, as devalued as the romance novel (Díaz-Plaja Taboada, 2011, pp. 65-66).

2. LUCÍA BAQUEDANO

Earlier, we presented a group of authors that, while small, is nonetheless unjustly limited, serving as a brief sample of the names of women who have occupied a space within the cultural and literary field of 20th-century Spain. The shift in publishing dynamics and the relevance that Children's and Youth Literature has gained within academic studies and reading habits seem to condemn these writers to a second marginality. Our aim is to rescue these authors, whose stories are often reduced to a single page or, at times, sadly, a footnote within manuals on the subject.

Lucía Baquedano Azcona, born in 1938 in Pamplona and a professional secretary, is one of the most significant authors of 20th-century Spanish Children's and Youth Literature and has a long publication history, of which we will enumerate a significant portion without aiming for exhaustiveness: *Cinco panes de cebada* (*Five Barley Loaves*, 1981), *Fantasmas de día* (*Ghosts by Day*, 1987), *De la Tierra a Harley* (*From Earth to Harley*, 1988), *Nosotros, los otros y los demás* (*We, the Others and the Rest*, 1991), *La casa de los diablos* (*The House of Devils*, 1992), *¡Pobre Antonieta!* (*Poor Antonieta!*, 1994), *Me llamo Pipe* (*My Name is Pipe*, 1995), *El pueblo sombrío* (*The Gloomy Village*, 2002), *El trébol de esmeraldas* (*The Emerald Clover*, 2005), *Un fantasma para Ángela* (*A Ghost for Ángela*, 2011), *Las aventuras de Tachín* (*The Adventures of Tachín*, 2014), and *Las vacaciones de Tachín* (*Tachín's Holidays*, 2016) (Bellido, 1998).

Despite her long career, there are not many academic works dedicated to Lucía Baquedano's oeuvre. A bibliographic search yields brief biographical notes that underscore the necessity of this work and the need to continue researching the writer. Nonetheless, approaches to her life and work, such as that by Teresa Bellido (1998), allow us to track down an author who, as the scholar rightly points out, has received various awards such as the Gran Angular for *Cinco panes de cebada*, the Barco de Vapor for *Fantasmas de día*, which was included in the IBBY Honor List and in the Annual Bulletin of Recommended Books by the Jugendbibliothek (Youth Library) in Munich. She also received, among others, the prize from the Spanish Catholic Commission for Childhood for *La casa de los diablos*.

Her work, in line with Bellido's study, is primarily set in rural contexts and features children, teenagers, and young people as main characters, while also incorporating fantastical elements such as ghosts or Martians. Although it is true that the author opts for realistic settings, we must consider that her texts, at times, diverge from the issues faced by the real-life counterparts of her protagonists, who are more concerned, as we will see below, with romantic matters.

3. CINCO PANES DE CEBADA (FIVE BARLEY LOAVES)

As the author herself has confessed, her first work, *Cinco panes de cebada* (*Five Barley Loaves*), was a text she wrote "around 1960, with no specific purpose, just because I liked writing" (Unattributed, 2010). It has already been noted that it was awarded the Gran Angular in 1979 and that it would be published by the SM publishing house in 1981, making it the 17th book in the "Gran Angular" collection and the first title in Spanish to appear in it (Guijarro Arribas, 2021). The story follows Muriel, a young teacher whose first assignment is Beirechea, a fictional village to which she arrives with little enthusiasm for her profession. All her suspicions about the difficulties she will face are confirmed by the protagonist as soon as she arrives at the place, as she has to confront her character and personality with those of the rural inhabitants. These, as Muriel will soon notice, are more concerned with farming tasks, for which they regard knowledge as useless in their lives. Thus, thanks to her role as a teacher both inside and outside the school and the alliance she establishes with other local figures, such as the priest, she begins to shift her initial desire to leave into an awareness of the influence she can exert among her new neighbors. Additionally, and in line with other novels by the author, the romantic element will play an essential role in the development and conclusion of the plot.

Cinco panes de cebada (*Five Barley Loaves*) does not escape the scarcity of academic studies on Lucía Baquedano noted earlier. This work, which can be considered her best or at least the one with the greatest journey, is only mentioned in a few manuals on Children's and Youth Literature that provide a historical overview of this genre in our country. Thus, Jaime García Padrino states the following:

> *The recovery of a reality surpassed by the passage of time, laden with memories or personal experiences, inspired Cinco panes de cebada (1981) by Lucía Baquedano, a young adult novel in which the central protagonist, a young teacher in a small village in Navarre, undergoes a challenging process of personal maturation thanks to her experiences and relationships with other characters, among whom the local priest stands out. This process is well described but not without passages laden with sentimentality. (2018, p. 256)*

As can be seen, this reference, along with others we have not included, has an encyclopedic nature and does not reflect the results of specific research studies on the work and/or the author. This is particularly surprising because this fiction, besides being Baquedano's first and having received a prize of significant importance in the Children's and Youth Literature (CYL) of the time, such as the Gran Angular, has been recently reissued by SM and reached, as of August 2021, its fifty-third edition. Thus, we take advantage of this gap to offer a study on it that focuses on two aspects we consider of special interest: the treatment of rural space and the exploration of teacher identity through the character of Muriel.

3.1. The Rural Space

Recent years have led to discussions about the existence of the neorural novel in Spain (Mora, 2018), and in some cases, this discovery seems to overlook the deep roots of the village context that have accompanied Spanish narrative since ancient times. Although its resurgence as a counterpoint to the almost total presence of urban space is evident, along with the widespread idea of empty Spain (Molino, 2016), as if the depopulation of rural areas were an ongoing process rather than one that has already been completed (Cortizo Álvarez, 2020), this is not a new phenomenon in our literature. The greatest example of this issue is likely Miguel Delibes, who "when he wrote *El camino* [...] was labeled reactionary, and forty years later was publicly recognized as an ecologist" (Vázquez Fernández, 2007, p. 126). Thus, *Cinco panes de cebada* could be considered one of the roots of this type of text within contemporary CYL narrative.

In no case do we consider this work to fall within the postulates of ecocriticism (Glotfelty, 1996) since the author's intention is not to show, through her text, the negative consequences of man's relationship with nature. In fact, we could even venture to say that there exists a certain anthropomorphic tendency in the way it is presented, as we will see below, in observing how human behavior is shaped through its relationship with the environment, rather than how the environment influences humans. In fact, in a statement from the author, when asked about the preponderance of the rural environment over the city in her texts, she responds as follows: "I believe that villages lend themselves more to adventure than cities do. Children can run freely in them, the doors of houses are always open, and everyone knows all the neighbors. On the other hand, my parents were from a village, and I myself have spent time in one of them" (Unattributed, 1992).

As we can see, the presence of this context in Lucía Baquedano's creations responds to the need to construct bonds between the characters, for which her own experience seemed to indicate that it would be easier to do so by portraying life in the countryside. Thus, it would be complicated for the work to be used instrumentally, as it does not address ecological issues, a fact that some studies consider "one of those invisible subjects of the curriculum with an interdisciplinary character that can be worked alongside school subjects" (Quiles Cabrera & Sánchez García, 2017, p. 68).

Sylvia, Muriel's sister, at the beginning of the novel, tells us about Beirechea that "Not even the name of the village appears in the Encyclopedia. It must be a dump" (2021, p. 7). This presents us with a place devoid of identity in the text, a village that the author insists on naming, but for the characters—especially for Sylvia, who later expresses her desire for Muriel to return to the city—it could be any rural space. Just a bit earlier, at the start of the novel, the protagonist herself reflected a similar thought: "I was only twenty-one then, and perhaps that's why I felt so disappointed when I learned that my destination was a village" (2021, p. 7). This place is presented to the protagonist not as a reward for a job well done but as a punishment, despite having fulfilled her obligations in the competitive exams: "If I did everything so well that I even earned the congratulations of the board, why were they now giving me a school in such a small village? How many students would I have? Maybe nine?" (2021, p. 7).

Muriel's expectations, upon arriving in the village, are fulfilled as she observes the school. We have previously mentioned "empty Spain," a term that, although currently trendy, is nothing new. The young teacher goes to a place that, while not abandoned, was already suffering the effects of the migration of people from the countryside to the city. Remember that the novel, despite having been published in the early 1980s, was written by Lucía Baquedano almost two decades earlier. As we said, the state of the school reflects what Muriel expected to find and shows her the reality she must face:

I was in a plain, and the first impression I had upon entering it I will never forget. In the very center of the classroom, there were two beautiful mice eagerly eating something [...]. I took a quick glance around the classroom and was dismayed. The walls were the saddest and most faded shade of blue, and the light bulb, too small, was trapped inside a glass ball, also blue, casting a dim glow. The ceiling was a poem of drip stains. I began to imagine a school populated by blue children, with lessons accompanied by the click, clack of raindrops on the floor. (2021, p. 12)

This image contrasts with what the protagonist observes one morning upon waking up. The relationship she establishes with the space initially leads her to be disappointed with the condition of the school or the house where she is supposed to stay, and, of course, with the environment in which she must live. The image presented in this passage of the village is just a glimpse of the progressive change she will experience in relation to her conception of space.

But how! Was this the same village as the day before? Now it seemed much more beautiful to me. The mountains, which the day before had seemed threatening, were now a beautiful green-blue color. The gray and sad houses now showed their balconies filled with pots of flowers of all colors. The well-kept gardens were full of fresh vegetables. And many, many trees laden with pears and apples. Farther away, the other fields made me smile with embarrassment (2021, p. 30).

Muriel not only observes the village through its places. She arrives in Beirechea with a suitcase full of prejudices based on her stereotypical idea of rural people. This concept of stereotype, as Lippmann suggested, should be understood as "the images in our minds that mediate our relationship with reality. These are crystallized representations, preexisting cultural schemas through which one filters the reality of the environment" (Amossy & Herschberg Pierrot, 2010, pp. 31-32). Her fixed ideas about others lead her to be surprised, for example, by the presence of a priest who rides a motorcycle and with whom she establishes a relationship that goes beyond formal interaction, and who "dressed now in a green chasuble, looked like a completely normal priest" (2021, p. 48). But it will be this character who provides Muriel with the necessary coordinates to get to know the people who inhabit Beirechea. Although she does not abandon a moralizing tone, in which the voice of Lucía Baquedano is present through the intradiegetic narrator (Muriel) warning her of the necessity of marrying one of the young men in the village or how she will one day do so and leave, her voice also serves to guide her during her first moments in the place:

No, daughter; I'm not laughing. On the contrary, because everything you're telling me seems very serious. It's true that you've ended up in a place with closed-minded, shy people, as you say. I know that for you, a young girl who is not used to villages, it would be much easier if we were all open, communicative, and friendly, right? But never try to understand men, Muriel. Love them. And when you have learned to love them, you will see that it doesn't matter if you don't understand them. (2021, p. 80)

3.2. The Identity of the Teacher

Those of us who work in a Faculty of Education frequently hear that our students are characterized by the vocation they exhibit towards their profession. This, observable during school visits during the practical periods of future teachers, convinces us, in a significant number of cases, of the importance of this impulse among our students, to the point of elevating it to the level of an axiom. Therefore, it is not surprising that one can speak of the existence of a "teacher identity" within the "professional identity" of individuals who dedicate themselves to teaching. That has been defined as:

the way in which the teacher is included in the educational activity is to discuss the feeling of belonging or not belonging to this profession. The teacher seeks to respond differently to the demands posed by their educational practice, and thus re-signifies it, reconstructing their imaginary around the profession of educating and, with it, reconstructing their identity, which allows the teacher to locate themselves in time and history, with a sense of permanence and prolongation. It is the possibility of integrating the past with the present, embodied in the uniqueness of oneself, in what one shares with others, in what one does not share, present in the ongoing negotiation with society, with students, with the institution, with the collective, and with oneself. (Gómez Gómez, 2005, p. 13, cited in Serrano Pastor et al., 2015, p. 126)

If the teacher's identity influences the way the profession is practiced, it is important to highlight how this is approached by adding the parameters of rural and feminine contexts. In the first case, if it is generally impossible to separate the school from its context due to the impact it has on the school and the synergies that are generated in both directions, this is even more true when the school is located in the countryside. As studied (Montoya *et al.*, 2022, pp. 15-16) through the collection of a series of testimonies from rural teachers, they "recognize each other as significant subjects in the professional history, bearers of knowledge and co-actors in constructing their own experience. Likewise, they recognize the context, the rurality, as a school in itself [...]. Rural teachers are aware of their impact on the territories and communities."

For the second case, we take into account the relationship between the teaching profession and women beyond the academic sphere. The association of formal education in early childhood with the task assigned to women has its starting point in the Enlightenment and its impact on Spanish educational legislation since the Moyano Law of 1857 (Sonlleva Velasco *et al.*, 2019). Thus, we find a very defined identity profile based on the characteristics of the profession, the place where it is practiced, and the gender of the person carrying it out. These characteristics seem to center around the teaching vocation, something that José Cardona Andújar has pointed out when referring to this novel:

> *A great deal has been written from the perspective of pedagogy about the concept and impact of the vocational dimension of the teaching function, particularly exploring its educational variable, which is the most desirable, creative, and noble aspect of it. Literature, being less methodical but more aesthetic and with greater linguistic rigor (it's disheartening to read some pedagogy manuals), offers beautiful narratives that reflect on and invite us to think about what vocation is, how that calling, that gift received, emerges, manifests, and consolidates in the everyday reality of the profession; and it does so by delving into the generous commitment of the true teacher, that gardener of souls, as Ramón y Cajal would say. (2016, p. 270)*

Five Barley Breads does not provide any temporal reference that situates it in a specific period, although it cannot be denied that its plot unfolds during the present of Lucía Baquedano herself. Here we should consider not so much the date of its publication, 1981, or the date of the award, 1979, but rather the date of its writing, that is, the 1960s. Muriel is a prototype of a rural teacher from the period we can identify with the later Franco regime and the onset of certain modernizing trends in the country. Sonsoles San Román Gago makes a brief reference to this work in her study on rural teachers during Francoism, noting that "there are writers who, without being teachers, have written in the first person identifying the essay as autobiography" (2015, p. 362). We do not believe that Baquedano is using fiction for this purpose beyond the moral Catholic vision expressed through the character of the priest and Muriel's behavior towards Javier. However, we do observe how the use of intradiegetic narration gives the text a confessional, autobiographical aspect. In fact, from the Theory of Literature, according to the worlds proposed by Tomás Albaladejo (1992), which take reality as a reference and distinguish three types (I, model of the world of the true, II, verisimilar fictional model, and III, implausible fictional model), we would need to classify Baquedano's text within the first model (the world of the true).

If we assume that Baquedano wants to reflect how the life of a newly graduated teacher in a rural environment and in her time was, we must also take into account that she is not allowing her novel to be dominated by the rural teacher depicted in studies such as those by Sanz Simón and Sonlleva Velasco (2020). These abound with narratives about the sobriety of rural teachers and their concern for transmitting the same values that national Catholicism had proposed for women in their daily lives (Richmond, 2004).

In fact, the rural context and the chosen profession, beyond the possibilities for adventure that the former offers—a point the author herself acknowledged in a previously cited interview—allow her to portray the character in development, which can cautiously relate her to the Bildungsroman genre. At least three characteristics of this type of text can be observed in Muriel: she is a young protagonist; the world around her, in this case, the rural setting, generates conflict within her; and lastly, the ending is not destructive; in fact, it is happy (López Gallego, 2013).

Thus, during her early days in Beirechea, Muriel observes how education has been relegated to a secondary status in the village. The mayor neglects everything related to the school, and for the locals, the school does not contribute anything to their daily tasks related to farming. Here, the author highlights a reality that can only be understood from the socio-historical framework of the novel: rural Spain faced serious problems concerning the education of children. Documents from the time, such as that presented by Sánchez Fernández on the current situation of preschool education, conclude that "the situation of children living in marginalized areas—rural and suburban—should be viewed with special sensitivity through the adoption of specific programs that counteract the disadvantaged socio-cultural situation in which they find themselves" (1982, p. 89). Muriel soon understands that her mission is not to turn the rural space into an urban one but to make those who inhabit it understand that formal education would impact their lives. Hence her reflection and the response she receives:

What could I do for Juanita Jotajuria to understand that even though "after all, she would stay in the village," it was necessary to learn, that culture would help her do anything better, any job she would have to perform in life? Well, I don't see how knowing subjects and predicates is necessary for milking cows, she calmly replied. (2021, p. 30)

The lack of resources in the school falls within the aforementioned effort to counter the conditions in which rural schools found themselves. The school's upkeep, undertaken, unsurprisingly, by Muriel herself along with Ana María, a young woman her age, contrasts with the apathy of the mayor, who deems it unnecessary to paint the walls—serving as a clear example of this. At the same time, the scarcity of academic materials is mitigated by the teacher, who makes an intentional selection of texts that she plans to purchase with her own money. As can be seen, this choice is no trivial matter:

I went up to my room without stopping in the kitchen and grabbed a notebook. I wrote down:

All of Elena Fortún's books.

All of Joanna Spyri's books.

Some by the Countess of Ségur.

All of Jules Verne's books.

Some by Karl May.

Look for some classics of world literature adapted for children. (2021, p. 47)

Muriel's construction of her identity as a rural teacher is gradual throughout the novel, grounded in the commitment she builds with the community and her students. Although there are various examples reinforcing this idea, her sister's visit to the village to offer her a job in the city early in the story clearly shows the person Muriel becomes over time:

"I want to be a village teacher," I repeated. "I want my kids to be able to study and have culture. Only then will they know how to choose their destiny. Some will leave, I know, and others will stay [...] but they'll be happier than they are now because, having made the choice, they will love their work; they will have had two options and chosen the one that most appeals to them, you understand?" (2021, p. 62)

Muriel's identity, however, is not constructed by her actions alone—it also finds reflection in others. At the start of this section, we mentioned that her purpose is not merely to transmit her own beliefs, though she certainly holds them. Rather, her greatest mission is to instill the value of knowledge as a catalyst for change among the villagers and to reinforce the role of school as a vehicle for social mobility. Her students come to understand this, as does Josetxo, one of the schoolchildren's parents:

Finally, almost as if by chance, he took notice of the books. He said it wasn't a bad idea, that books always educate, that if he had more time... that he didn't understand much, not being a learned man...

"There's always a moment to spare, and it would help José a lot. He'd probably get more interested if he saw you reading, if he saw that books are something serious, something for grown-ups. Don't you think?"

I saw he was pleased, hesitating a little, unsure of what to do.

"Well... if you'd be so kind..."

"What subject would you like?"

"Well, look, what can I say... For me, history..."

"Perfect! Take something with you and pass on your enthusiasm to the boy."

He left very satisfied, carrying *Episodios Nacionales* by Pérez Galdós under his arm. I liked, too, that he didn't mind being seen around the village with a book (2021, p. 101).

4. CONCLUSION

Lucía Baquedano's work, along with those of her fellow female authors, deserves to be rediscovered, republished, and appreciated in our time. This is not due to any signs of decline in today's publishing industry, which is thriving and probably producing more books than in previous decades. Ironically, this very fact hastens the risk of forgetting those authors who were once widely read and awarded but, with few exceptions—like Elena Fortún, who has been the subject of a comprehensive literary revival, or the occasional reprints of works by Carmen Kurtz, María Luisa Gefaell, and Baquedano herself—seem to be vanishing from the shelves.

Cinco panes de cebada stands as an interesting example within Young Adult Literature, portraying the rural Spanish life in the 1960's. Though the settings and characters may appear somewhat archetypal and stereotypical, they provide a faithful representation of the space-time context portrayed through Beirechea, a village that could just as easily be any small town in Navarra. Characters like the priest, Javier, and Muriel subtly reflect the social changes taking place in Spain at the time, changes that Baquedano skillfully captures.

Muriel is not the typical rural teacher of the dictatorship era. In fact, her identity as a rural teacher is progressively shaped throughout the novel. Initially reluctant to embrace this role, she ultimately finds herself fitting into it—not by following the conventional expectations of the dictatorship, but by placing the student and knowledge at the core of her mission. This alignment even positions her, within the town's social structure, on par with the priest.

The confessional tone of the novel might lead readers to interpret it as autobiographical, had Baquedano actually worked as a teacher. Instead, she drew from her experiences as a secretary and writer, using first-person narration to present the story as something authentic, lending itself as an *exemplum* for current and future teachers. Although the conditions in rural Spain have changed, the text achieves a certain timelessness when considering elements such as vocation, the value of knowledge in students' lives, and the evolution of a teacher's identity over their career. This identity, often idealized by education students, is depicted here as a long journey that does not always lead to expected outcomes, shaped by the teaching environment and the students themselves.

The history of Spanish Young Adult Literature remains incomplete if authors like Lucía Baquedano are confined to mere mentions in academic manuals, if their works are found only in second-hand bookstores, and if their texts are not incorporated into classrooms across all educational levels. Like her fellow authors, Baquedano aimed her work at a wide range of young readers, spanning childhood to adolescence. In a literary sense, Baquedano, much like María Folguera with Elena Fortún, serves as a valuable role model for future writers.

This study aims to avoid any kind of advocacy that promotes the recovery of Lucía Baquedano's work and that of other authors solely because they are women writers. The fact that literary history has sometimes failed to do justice to the role these authors played in shaping readers and the entire literary system—evidenced by the awards they received and their editorial success—is not a sufficient reason for their inclusion as a mere quota requirement. Rigor must be the foundation of philological and didactic studies, which is the primary reason for studying and recovering the work of these writers. *Cinco panes de cebada* aligns with these criteria due to its timeless themes, the significance of its author, and the novel's place within her broader body of work, as well as the need to fairly and accurately rewrite the history of Spanish Youth Literature (CYL) in the 20th century. Texts like this one stand out due to the unique circumstances of their creation and publication, as well as the author's personal history and relationship with literature.

Through this analysis, we explore the potential of *Cinco panes de cebada.* The rural school within Spain's "emptied" regions faces a perspective often detached from its reality, viewed from an urban distance. The lack of resources needed to enable students to complete their studies under conditions similar to those in the city requires, at the higher levels, relocation to more populated areas or even the decision to abandon their education altogether. Novels that explore the identity of teachers—one of the most complex professional identities—provide insight into the concerns and aspirations of those committed to educating future generations. There is a growing need within educational faculties to offer spaces that showcase such texts through Literature Didactics and extracurricular activities, as an essential resource for those entering a profession with many facets.

Rediscovering and recovering Baquedano's work represents both a philological and educational need, best approached from three directions: rewriting the history of Spanish Youth Literature from the past century, reissuing texts that current readers and scholars cannot easily access and incorporating this and other authors into the canon of Youth Literature and classroom literature.

REFERENCES

Albaladejo, T. (1992). *Semántica de la narración: la ficción realista*. Taurus.

Amossy, R., & Herschberg Pierrot, A. (2010). *Estereotipos y clichés*. Eudeba.

Baquedano, L. (2021). *Cinco panes de cebada*. SM.

Bellido, T. (1998). Lucía Baquedano, vida y obra. *Revista TK, 5*, 29-36. https://www.asnabi.com/revista/tk5/05bellido.pdf

Britland, J. (2024). El feminismo transtemporal y transnacional: mujeres que se unen y rebelan por la escritura y el *performance* en Elena Fortún y *Hermana (Placer)*, de María Folguera. En Á. Martín Pérez, S. Sevilla-Vallejo & J. Guzmán Mora (Eds.), *La dramaturgia femenina reclama su sitio* (pp. 345-362). Renacimiento.

Capdevilla-Argüelles, N. (2005). Elena Fortún (1885-1952) y *Celia*. El *Bildungsroman* truncado de una escritora moderna. *Lectora: revista de dones i textualitat, 11*, 263-282. https://revistes.ub.edu/index.php/lectora/article/view/7127/9019

Cardona Andújar, J. (2016). *Saber docente desde la pedagogía poética*. Universidad Nacional de Educación a Distancia.

Cortizo Álvarez, T. (2020). La España vacía: De sintagma a geografísmo. *Polígonos. Revista de geografía*, 32, 159–178. https://revpubli.unileon.es/ojs/index.php/poligonos/article/view/6411/4933

del Molino, S. (2016). *La España vacía*. Turner.

Díaz Fernández, E., & Mérida Jiménez, R. (2021). María de la Encarnación Gertrudis Jacoba Aragoneses y de Urquijo. *Estudios Hispánicos*, 29, 55–63. DOI: 10.19195/2084-2546.29.6

Díaz-Plaja Taboada, A. (2011). *Escrito y leído en femenino: novelas para niñas*. Ediciones de la Universidad de Castilla-La Mancha.

Fraga Fernández-Cuevas, M.ª J. (2019). "Lo que Celia dice" en otros países: versiones de las novelas de Elena Fortún. *CLIJ: Cuadernos de literatura infantil y juvenil, 32*(290), 24-33.

García Padrino, J. (2018). *Historia crítica de la Literatura Infantil y Juvenil en la España actual (1939-2015)*. Marcial Pons.

Glotfelty, C. (1996). Introduction: Literary Studies in an Age of Environmetal Crisis. In Glotfelty, C., & Fromm, H. (Eds.), *The Ecocriticism Reader: Landmarks in Literary Ecology* (pp. xv–xxxvii). University of Georgia Press.

Guijarro Arribas, D. (2021). De la circulation internationale à la globalisation des normes editoriales: Le cas des livres pour « adolescents » en France et en Espagne. *Reseaux (London)*, 226-227, 185–207. https://shs.cairn.info/revue-reseaux-2021-2-page-185?lang=fr&contenu=resume

Guzmán Mora, J., & Sevilla-Vallejo, S. (2019). Carmen Kurtz en el cine infantil de la década de 1970: intervenciones y adaptaciones. En M. Marcos Ramos (Coord.), *El cine como reflejo de la historia, de la literatura y del arte en la filmografía hispano-brasileña* (pp. 301-309). Universidad de Salamanca, Centro de Estudios Brasileños.

López, C. (2021). María Folguera: "Venimos de una tradición en la que lo femenino se construía desde el sacrificio". *Eldiario.es*, 23 de mayo de 2021. https://www.eldiario.es/cultura/libros/maria-folguera-venimos-tradicion-femenino-construia-sacrificio_128_7950455.html

López Gallego, M. (2013). Bildungsroman. Historias para crecer. *Tejuelo, 18*, 62-65. https://dehesa.unex.es/bitstream/10662/4521/1/1988-8430_18_62.pdf

Montoya, G., Valencia, L., Vargas, L., García, J., Franco, J., & Calderón, H. (2022). Ruralidad, educación rural e identidad profesional de maestras y maestros rurales. *Praxis & Saber*, 13(34), e13323. DOI: 10.19053/22160159.v13.n34.2022.13323

Mora, V. L. (2018). Líneas de fuga neorrurales de la literatura española contemporánea. *Tropelías.Revista de Teoría de la Literatura y Literatura Comparada*, 4(4), 198–221. DOI: 10.26754/ojs_tropelias/tropelias.201843071

Pérez Alonso, J. E. (2015). Las que cuentan. Narradoras en la literatura infantil y juvenil española. *Espéculo.Revista de Estudios Literarios*, 55, 11–28.

Quiles Cabrera, M.ª del, & Sánchez García, M.ª R. (2017). Educación medioambiental a través de la LIJ: Claves para la formación del profesorado en los grados de magisterio. *RIFOP: Revista interuniversitaria de formación del profesorado: continuación de la antigua Revista de Escuelas Normales, 31*(90), 65-77.

Richmond, K. (2004). *Las mujeres en el fascismo español: la Sección Femenina de la Falange, 1934-1959.* Alianza.

San Román Gago, S. (2015). Madres sociales de la aldea: La maestra española durante el nacional-catolicismo. *Educação e Pesquisa*, 41(2), 359–373. DOI: 10.1590/s1517-97022015041806

Sánchez Fernández, M. P. (1982). La educación preescolar en España en el momento actual (I). *Vida escolar, 220-221*, 78-89. http://hdl.handle.net/11162/79094

Sanz Simón, C., & Sonlleva Velasco, M. (2020). La maestra franquista en el imaginario del alumnado femenino. Un estudio a través de testimonios orales. *Revista Diversidade e Educaçao, 8*, 213-237. https://repositorio.furg.br/handle/123456789/10820

Serrano Pastor, F. J., Martínez España, A., & Montesinos Navarro, A. M. (2015). Construcción de la identidad docente de los futuros maestros de Educación Infantil y Educación Primaria: entre la vocación y la profesionalización. In Mirete Ruiz, A. B., & Nortes Martínez-Artero, R. (Eds.), *Investigación e innovación: una constante necesaria en la formación del profesorado* (pp. 125–136). Servicio de Publicaciones de la Universidad de Murcia.

Sonlleva Velasco, M., Martínez Scott, S., & Barba Martín, R. (2019). Las marcas del género en las trayectorias biográficas de las maestras de educación infantil. Analizando nuestra historia para desmontar el patriarcado. *Cabás, 22*, 47-60. https://doi.org/DOI: 10.35072/CABAS.2019.17.63.006

Torres, M. (1999). Mi hada madrina. *Cuadernos de estudio y cultura, 11*, 21-24. https://acec-web.org/PDF/CUAD/11.pdf

Unattributed (1992). Entrevista a Lucía Baquedano. *Peonza, 22*, 23-27. https://www.cervantesvirtual.com/obra-visor/peonza-revista-de-literatura-infantil-y-juvenil--60/html/0272fb60-82b2-11df-acc7-002185ce6064_21.htm

Unattributed (2010). Lucía Baquedano: "Después de los padres, es el maestro quien debe encauzar bien a un niño". *Nuestro Tiempo, 660.* https://nuestrotiempo.unav.edu/es/grandes-temas/despiece/lucia-baquedano

Vázquez Fernández, M.ª I. (2007). *Miguel Delibes, el camino de sus héroes*. Pliegos.

ENDNOTE

1 All the quotations have been translated into English by the chapter's author.

Chapter 12
Elena Quiroga:
A Writer to Rediscover

María Isabel Gil-García
https://orcid.org/0009-0006-0522-3977
University of Castilla-La Mancha, Spain

ABSTRACT

The aim of this paper is to revalue the Spanish writer Elena Quiroga, who received some recognition in her literary beginnings, winning prizes such as the Nadal, allowing her to insert herself into the narrative panorama of the 1950s. However, her figure was forgotten in Spain after her death. This writer stands out for her constant willingness to experiment in her narrative forms, which leads us to speak of a personal narrative universe. The great richness and variety of her work is the main reason why she deserves critical attention. For this reason, after studying the historical and literary panorama in which Elena Quiroga is situated, a corpus of three works has been selected, two short novels and a novel, to illustrate the evolution of her work by comparing them. These works are Trayecto uno, Plácida, la joven and Tristura.

DOI: 10.4018/979-8-3693-7185-5.ch012

Copyright © 2025, IGI Global Scientific Publishing. Copying or distributing in print or electronic forms without written permission of IGI Global is prohibited.

1. INTRODUCTION

The objective of this work is to recover the writing of Elena Quiroga. Although her literary beginnings were recognised with significant awards such as the Nadal Prize, and her work was integrated into the narrative landscape of the 1950s, the figure of the writer was mostly forgotten after her death.

Her works were acknowledged outside Spain, with a considerable amount of research on her novels as opposed to the few works that can be found in Spanish. The only monograph that exists on her work is the product of the North American Phyllis Zatlin, the researcher who has been most dedicated to Quiroga.

Possible causes for the lack of attention she received have been studied, including the situation of women writers during Franco's regime and censorship, the presence of constant experimentation that could have led to her work misunderstanding, her life on the periphery of literary circles, and the early interruption of her production for a decade until her death.

In recent years, interest in the writer has grown in Spain, leading to the publication of critical editions of her work, such as *La enferma* (2013) – previously there were only critical editions of *Escribo tu nombre*, by Phyllis Zatlin (1992), and *La última corrida*, by Andrés Amorós (1994)–) with an extensive introduction by Gregorio Torres Nebrera. Likewise, articles have been published in scientific journals, theses, and renowned critics such as Soldevila, Sobejano, Martínez Cachero and Darío Villanueva have pointed out the unfair and scant critical attention paid to her work.

Since she is considered worthy of further attention, a corpus of two short novels and a novel —*Trayecto uno*, *Plácida, la joven*, and *Tristura*— was selected for an individual study in the first instance, which reflected the versatility, richness, and flexibility of her narrative forms. From this first approach it emerged that although the narrative structure is different the focus is common to all three works: the female perspective.

Therefore, the main aspects obtained from this study will be presented here, followed by a comparison of them, which will justify Elena Quiroga's constant desire for renewal, translating it into a significant evolution.

2. ELENA QUIROGA DE ABARCA

Elena Quiroga de Abarca was born in Santander on 26th October 1921. She was the daughter of the Count of San Martín de Quiroga, from Viloira. She spent in Galicia most part of her early childhood, alternating stays between the Galician town and the Cantabrian capital with her mother's family. At the age of two, she lost her mother and was raised under the influence of her grandmother, something that is important to highlight due to the presence that orphanhood will have in several of her works (Torres, 2013).

She spent time in cities such as Bilbao, where she studied at a religious school, Rome, where she continued her studies during the Civil War, and Barcelona. At the age of twenty-one, she settled with her father in La Coruña, until in 1950 she married the illustrious historian Dalmiro de la Válgoma. Then she moved to Madrid, where he would remain for much of his life, until he died on October 3, 1995 (Galdona, 2018).

As many authors have pointed out, her beloved Galicia is very present in most of her narrative world, highlighting elements of the Galician landscape, language, customs and idiosyncrasy, thereby showing the limited vital conditions of women in the Galician rural world. Although to a lesser extent, Madrid will also have a presence in some of her works.

3. ELENA QUIROGA IN THE NARRATIVE OF THE SECOND HALF OF THE TWENTIETH CENTURY

Elena Quiroga's novelistic trajectory -whose publications began in 1949 with *La soledad sonora* and end in 1973 with *Presente profundo*-, covers an important part of the second half of the 20th century. Therefore, it is necessary to know the panorama and the narrative currents of those years. Moreover, given the versatility of this writer's work – which began in realism fields and progressively moved towards formal experimentation–the opinions of critics differ regarding her immersion in one literary current or another.

Soldevila (1980), considers that Quiroga belongs to the “second post-war promotion”, which includes writers born between 1912 and 1922, who began to publish years later than their historical generation. This will justify they have a new attitude towards the previous tradition from their beginnings.

Sobejano (1970), in a first classification of the contemporary Spanish novel, includes it within “Existential novel. Other conflictive novelists”. In a later edition of 2006, following Eugenio de Nora, he establishes the classification in two subgroups, within those born between 1905 and 1920, where he includes Quiroga with those who approach a narrative that could be called realistic, such as Cela, Laforet and Delibes. However, given the later development of her novels, it could be considered that Quiroga is also close to the second subgroup, with authors who, in a markedly realistic line, show an impulse of formal renovation or tend to pose an intellectual or moral problem that dominates the story (Sobejano, 2006). Thus, Martínez Cachero refers that after her realist-naturalist attempts, she changed course, personally exploring modes and resources (1997).

This diversity of opinions already demonstrates the versatility and evolution of his work, which makes it difficult to classify it in a single current.

3.1 The Short Novel Situation

During the second half of the twentieth century, the genre of the novel dominated the Spanish literary scene, relegating the short novel to second place, despite the interest it generated among certain authors, including Elena Quiroga. According to Martínez Arnaldos (1996), studies on this genre were concentrated on formal aspects and specific authors such as Clarín, Pardo Bazán and Unamuno, without considering its historical evolution. Arnaldos justifies the need to analyse the short novel within its social and publishing context. He remarks that the rise of this genre in Spain took place in the first third of the 20th century, with publication in literary magazines (Arnaldos, 1996). Pujante (2010) extends this period to the mid-twentieth century, highlighting the importance of the collection *La novela del sábado* (1953-1955), which brought back authors of short novels and attracted many young women writers. This collection, successful among the reading public, published works with a wide variety of themes that combined tradition and cosmopolitanism. Quiroga participated in this collection with *Trayecto uno* and *La otra ciudad*, becoming the beginning of her career in the short novel. However, this glimmering of the genre was brief and did not manage to consolidate over time (Pujante, 2010).

3.2 The Nadal Prize and Female Novelists

Martínez Cachero (1997) reflects on women novelists and the importance of prizes in the rise of their writing. During the post-war years, there was a significant incorporation of women into the cultivation of the Spanish novel. This fact was recognised and became more evident when prizes began to be awarded to women writers. First, Carmen Laforet received the Nadal Prize in 1945, and she was followed by Elena Quiroga, Dolores Medio, Carmen Martín Gaite and Ana María Matute, among others.

In this respect, an issue of the *Boletín de la Biblioteca Menéndez Pelayo 2023* has recently been published. Its aim is to study the relationship between this incorporation of women into the literary field and literary prizes as a means of making them visible (Sotelo and Ripol, 2023). This research shows that, although on many occasions the Nadal Prize means the consecration of some of the women writers, as in the case of Carmen Laforet's *Nada*, on other occasions it means their silence, as their work never receives the attention of the critics. Thus, in the case of Elena Quiroga, her second work was the winner of the Nadal Prize in 1950, but her later work has not been fairly valued.

Despite this, there is no doubt that Carmen Laforet's triumph with her first novel, as a young and unknown girl, was a boost to the self-esteem of many women who wanted to be writers and who, as a result, were encouraged to submit their writing for literary prizes. I was six years later that *Viento del norte* was awarded (Cabello, 2023). Carmen Laforet herself recognised the importance of this fact:

> *Al oír el nombre de Elena [Quiroga], pensé en todos los nombres de mujer que el Nadal ha dado a conocer, aunque hasta hoy no haya decidido entregarse a ninguno en estos años, como si hubiera quedado asustado de la experiencia corrida al caer la primera vez sobre mí [...] El tema de las mujeres escritoras me es, desde luego, muy grato porque lo siento y lo conozco, pero aquí sólo quiero ceñirme al hecho de que el Nadal 1950 lleve un nombre de mujer, y que yo quizá por eso, tengo un presentimiento de buena suerte para la bella novela que, sin duda, es Viento del Norte. (Laforet, 1951, quoted in Cabello 2023, pp. 26-27)*

As we know today, *Viento del norte* did receive attention thanks to the Nadal and so did Quiroga, who managed to enter the novelistic panorama. However, many of her other works did not achieve the same success and her presence has faded over the years.

Although her long period without publication –since 1973– made her a practically forgotten writer in the 1980s, she received another important recognition, which made her the second woman to become a member of the Real Academia Española in 1983, occupying chair 'a', replacing Zunzunegui (Torres, 2013). On 8 April 1984 she delivered her speech, *Presencia y ausencia de Álvaro Cunqueiro*, which defended the figure and work of the Galician author, and served as an example to illustrate the love for the Galician landscape and figures that Quiroga felt so strongly about and captured in her work (Torres, 2013). This would be the author's last published work.

3.3 Reception of Elena Quiroga in the Franco Regime

Elena Quiroga received a great reception abroad, even with translations of her works into Finnish, German, Swedish, Italian, French and English, being valued by both literary critics and the reading public (Brent, 1959). Outside Spain, critics offered a more objective view of her work, emphasising her experimentation and originality. Kristen R. Barney (1993) refers to her 'silent revolution', especially for works such as *La enferma* and *Presente profundo*, which did not initially receive the critical attention they deserved. Although her themes, such as female sexuality, divorce and the moral degeneration of post-war Spain were considered daring in Francoist Spain, censorship did not always detect the subversive nature of her work, possibly due to her modernity and the narrative techniques she employed (Barney, 1993; Legott, 2014). These techniques included the female or child's perspective, which could have been seen as inoffensive, or the presence of autobiographical elements. Quiroga also confessed to having contacts within the censor system, which may have facilitated the publication of some of her works. Moreover, the fact that she was a woman sometimes made criticism of her works less severe, as censors did not consider works written by women to be dangerous (Legott, 2014).

4. THREE INTERCONNECTED NARRATIVES

We will now analyse three of her works, two short novels and a novel: *Trayecto uno* (1953), *Plácida, la joven* (1956) and *Tristura* (1960). It is convenient to justify the choice of these three works from her entire narrative career. In the study of her whole novelistic trajectory, we have mainly followed the North American Phyllis Zatlin (1977, 1992), the researcher who has been most devoted to the study of Quiroga.

As it has already been mentioned, Quiroga's narrative is characterised by a permanent evolution, in which the author experimented with different narrative procedures. Thus, these three works represent three different modes of narration: the first is presented as an objectivist narrative with multiple perspectives, the second as an epistolary narrative, and the third as a double homodiegetic narrative. In fact, despite being constituted as three different modes of narration, they share a common characteristic in terms of the focus of their narration: the female perspective. Moreover, all of them share a few common themes and techniques.

To compare these three works, a previous work has been carried out, in which the particular analysis of each of them is specified, in relation to its genre and narrative modes. Although we are not going to focus into each one in depth in this study, it is necessary to contextualise them to correctly understand the common aspects they share.

4.1 *Trayecto Uno*

The first of the selected works is *Trayecto uno*, a short novel published in the collection *La Novela del Sábado* in 1953. Quiroga would later include it in her own collection *Plácida, la joven y otras narraciones* (1956), where the work would occupy second place in order of appearance.

Trayecto uno is a short novel that will be particularly different to the rest of her production for several reasons. Firstly, this work is the first of those she had published to date to be set in Madrid, as critics such as Vidal Ortuño (2006) and Torres Nebrera (2013) have pointed out. Subsequently, other works such as *Algo pasa en la calle, La otra ciudad* and *La careta* would also be set in the capital. Secondly, this work reflects a different technique to the two previous ones, that of the objective narration that Zatlin (1992) would link to Cela's *La colmena* and which is close to the social novel (Torres, 2013).

With a similar tone to the next two works to be analysed, this short novel narrates the bus journey of colourful people on 15 January 1953. In this story we see fragments of the lives of very different people, from girls leaving school to wealthy ladies and gentlemen, a bricklayer, a nanny with a sick baby, a young couple and, above all, Anuncia, the watchful girl through whose eyes we can see the Madrid landscape at a Thursday sunset in which she discovered poetry. In this brief journey on the Cartagena-Moncloa line, the reader will end with the circular perception of time that unites its passengers and driver.

This work stands out to its approach to the social novel –with the spatiotemporal reduction and the collective protagonist (Vidal, 2006)– and for the formal renovation of its narrative procedures and intertextual elements. In terms of narrative procedures, we can underline the oscillating narrator (Vidal, 2006) and the stream of consciousness or interior monologue in some of the characters (*Trayecto uno,* p. 133):

Esta mañana... Estaba en clase, simplemente, y mordía el revés del lápiz, mientras el profesor explicaba Filología. Palabras, palabras... Solas, desmenuzadas, buscando el origen. Le empezó a zumbar el corazón, la cabeza despejadísima, como una campana de cristal. Palabras sueltas, rostros atentos, el aire casi gozo. Respiró hondamente. «Estoy subyugada...» «Aquello» se acercaba, la entraba como aire por una espita, luego a chorros: orejas, lengua poseídos... Se puso de pie.

–¿Qué le pasa a usted, señorita?

volvió a sentar, casi automáticamente. ¿Qué le pasaba? Se estaba volviendo loca. Sí. Sí. Atención... «Me estoy volviendo loca»

[...]

«Me voy. Me voy. No puedo más. A la calle, a cualquier sitio. Que me dé el aire en la cara. No tener que fingir, no tener que estar atenta. Salir...»

Another technique that could be emphasized is the way of activating memories through the mechanism of involuntary memory (Vidal, 2006), which leads us to think of Proust:

Anuncia apretó en su mano la violeta. Y entonces al triturarla, despidió un olor suave, insistente. Ella conocía aquel olor, más intenso y al mismo tiempo de aire y de frescura. Se vio en el otero, a la salida de Palencia, yermo el cerro, termo la tierra en torno. (Trayecto uno, p.131)

The intertextuality in this work is also remarkable, as it contains numerous examples of Hispanic poets and authors –Pedro Salinas, Juan Ramón Jiménez, Alberti, Rainer María Rilke, Gregorio Marañón– which demonstrate Quiroga's interest and literary knowledge not only in narrative but also in poetry, and in the theme of reading, which is also repeated in *Tristura* (1960) and *Escribo tu nombre* (1965).

4.2 *Plácida, La Joven*

Plácida, la joven was published in 1956, entitling the author's seventh book. This short novel returns to the Galician world of *Viento del norte* (1950) and *La sangre* (1952) and to Catholic existentialism (Torres, 2013).

This story had received little attention from critics until some works have been published about it on the last years. Elena Olazagasti (1988) considered this work to be a synthesis of the whole Quiroga's narrative, emphasising elements that had already been used in previous works and announcing others that would not be seen until the following years. Other critics such as Torres Nebrera (2013) and Vidal Ortuño (2022) have also pointed out the formal perfection of this short novel and have dedicated interesting comments on it.

Fernández Almagro (quoted in Olazagasti, 1988) was the first to point out the similarities of this work with previous ones, focusing on three typical factors of Quiroga: lyricism, psychologism, and Galician emotion (p.135). Articulated under the epistolary structure, these elements will be central to the book.

The technique used in the short novel to present and introduce the narrative is the epistolary technique. Although the typical epistolary structure is not followed (Zatlin, 1977), the framework of the letter becomes a central and essential element for the narrative to develop and makes possible that the reader receives the information in the manner it does (Marks, 1980; Olazagasti, 1988; Vidal, 2022).

The reading starts with the following situation: an unnamed narrator is writing a letter to her absent husband while she is on holiday in her Galician hometown. She is narrating the story of Plácida, a young woman who has just died in childbirth. As Olazagasti (1988) points out, using the words of Tadea –the protagonist of *Tristura* and *Escribo tu nombre*–, a letter is more than a simple transmission of information, it is 'a communication from soul to soul ('comunicación de alma a alma') (p.141). The letter that the narrator writes to her husband will serve not only to tell Plácida's story, but also to unburden herself of her own guilt, as well as leading her to immerse herself in reflections and memories.

Although the book begins with a heartfelt expression of sadness at the young woman's death, the husband –and with him the reader–, learns shortly afterwards that the two women did not know each other personally (*Plácida, la joven*, pp. 9-11):

> *...Que ayer estaba triste porque Plácida ha muerto. Aún hoy, mientras te escribo, me pesa el corazón. [...] que el corazón no duele en alegría, no pesa o se aligera, llegas a prescindir de él, o a no importarte que se te despedace en la tristeza. Duele, te digo. Oprime. Es aquí, justo a la izquierda del pecho [...] Y dirás: ¿Quién es Plácida? Repasas velozmente en los recuerdos comunes, espigas, hojeas esta carta. En balde. No la conocías. Tampoco yo.*

The author of the letter needs to resort to the perspectives of the neighbours to complete the young woman's story. Therefore, the narration is carried out by a homodiegetic-witness narrator, who addresses to a narratee to tell the facts that come to her through other characters (Olazagasti, 1988; Vidal, 2022; Zatlin, 1992).

With the basis on the epistolary structure -which allows the development towards confession or intimate diary- the author develops a series of narrative procedures, some of them innovative, and other ones already used previously. As Phylis Zatlin (1977) pointed out in her study of Quiroga, *Plácida, la joven* is characterised by a highly subjective technique, with a lyrical and sensorial style and a narrative that does not follow a chronological order. At the beginning of this section, it was mentioned that Almagro summed up the work in lyricism, psychologism, and the emotion of Galicia. The theme of orphanhood (Olazagasti, 1988), and the link between woman and land (Torres, 2013) must be added to this. Furthermore Olazagasti (1988) pointed the work as a synthesis of Quiroga's narrative not only in terms of themes but also in terms of technique. The final fragment of the story is a good example of this synthesis of procedures (*Plácida, la joven,* pp. 71-72):

> *Y el día será en aquellas costas -quien sabe- brumoso, de menuda lluvia. Y él pensará en todo lo sucedido y, sin darse cuenta lo verá entre bruma y lluvia también. Pensará: "Alá xa andan a vendimar", y no sabe que está tan bajo el sol y la tierra tan buena, tan alegre. No habrá nadie que le cuente, cuando regrese, que hubo, en el día de hoy, un misterioso gozo de la tierra y del aire, impropios del otoño. Nadie para decirle que, en verdad, no parecía posible que, en la hora dorada y verde de una tarde tan dulce, una mujer que se llamaba Plácida muriera.*

4.3 *Tristura*

Tristura was published in 1960, with the initial subtitle of «Secreto de la infancia. Novela de una niña», which was later deleted. This was the first part of the *Grandes soledades* trilogy which was in fact never finished.

If *Tristura* and *Escribo tu nombre* are the two works that have received most critical attention, the author also acknowledged that the Tadea novels were her favourites in an interview with Phyllis Zatlin (1977). The first has been selected for this work because it is the one that initiates this new set of formal and temporal experimentations, and because in it the theme of childhood and innocence of the girl dominate the perspective of the novel in a representative way. Both works have autobiographical elements (Zatlin, 1977).

This novel is an example of a Bildungsroman, in which the two realities that Tadea has lived are presented: the freedom and freshness of her home in Galicia and the stifling, dark atmosphere of her grandmother's house in Santander.

In the Spanish literary scene of the 1950s the theme of childhood took the central position in a significant number of novels, this was particularly noteworthy in novels written by women (Ciplijauskaité, 1988). Torres (2001) finds the explanation for this event in the Civil War and the post-war period, which marked the collective unconscious and coincided with the childhood and adolescence of these authors.

Since the piece has no plot – it tells of the child's reaction to daily events–, the most important aspect of Tristura is the careful psychological development of Tadea's character, as Zatlin points (1993). If Quiroga had already presented solitary characters resulting from a traumatic childhood in several of her works, she gives here the most complete development of such childhood experiences (Zatlin, 1977). The Galician name that gives the novel its title reflects the feeling that devastates the girl during these two years of her troubled childhood.

Structurally, this work can be considered the most complex of the three works studied in this chapter. The novel is narrated from the perspective of Tadea, a homodiegetic narrator, written simultaneously on two levels: the implied narrator of the adult woman, who recalls the crucial and trivial episodes that shaped her personality, and presents them from the girl's perspective (Marks, 1980; Zatlin, 1977). The existence of the adult narrator is not directly expressed in the text, but her presence can be glimpsed in some sequences, as descriptions and syntheses that could only come from a perspective more mature than that of a child (Zatlin, 1977, 1993). The following fragment serves as an example:

Veíamos el alto de Miranda cuando dimos la vuelta. El viento nos empujaba por la espalda, camino de casa. No era fuerte viento, sino fresco, ya no se veía el cielo rosado en bandas estriadas, sino gris muy transparente. (Tristura, p. 189)

Despite that fact, Zatlin (1977, 1993) notes that all episodes and characters are shown through Tadea's conscience and perspective. Because of her naïveté there are real life situations that she does not understand which are left to the reader`s empathy.

In addition to the complexity of the structure, there are many formal techniques and innovations. Time becomes a crucial structural factor (Torres, 2013). As the author herself acknowledged in an interview, the time of childhood in Tristura 'is not relived, it is lived. It is a profound present [...]' (Quiroga, quoted in Uribarri, 1969, p.27). This explains why the adult narrator re-actualises the past in the present (Torres, 2013), because it is the voice of the child narrator that we mainly read.

To bring the slow tempo of childhood to writing, Quiroga makes use of a whole series of resources, which are the ones that most attract the reader's attention in this novel and which will achieve their immersion in the deep present. First, it is noticeable that the two tenses of the novel are typographically marked in different ways, using italics to present the memories in the house in Galicia –which are usually characterised by his conversations with Tina–, and round letters to narrate the present in the house in Santander. These memories are always associated with the situation Tadea is witnessing in present time.

5. EVOLUTION AND COMPARISON: CHILDHOOD, YOUTH AND MATURITY

The reason for choosing these three works has already been justified. In this section we will comment the aspects that connect them after the individual analysis of each one.

5.1 The Narration of Female "Self"

Biruté Ciplijauakaité (1988) studies the modes of creation of the new women's writing, –women writing consciously as women–, which does not necessarily imply a feminist approach. In this study, she reviews some European women authors who conform to the first-person mode of narration in search of this new type of writing. Among the works selected there are some novels by Elena Quiroga. For this reason, it has been considered convenient and appropriate to mention certain fundamental issues in her work, which have been related to the three narratives we are concerned with.

In the 19th century women had to copy the patterns established by male writers to prove that they were capable of writing. In the 20th century the situation had improved somewhat and they could go in search of an original expression that is due only to their intimate personality. Even if this first stage had been overcome, women's relationship to themselves remained undefined, so that the self-questioning continues (Ciplijauskaité, 1988).

This author postulates that in the new women's writing it is no longer of interest to tell reality in the extradiegetic or 'objective' mode, but to express the interior as 'immediately as possible' (Ciplijauskaité, 1988, p. 18). In *Trayecto uno*, Quiroga does a mixture of both. Given the interest in getting close to the people, she sometimes uses this more objective perspective, but she soon abandons it when she focuses her attention on the characters, especially the girl with the fringe and Anuncia. Both reveal their deepest concerns in the piece, rebelling against what is 'proper', in both cases thanks to their reading experience: the girl against her parents' prohibition and Anuncia in her way of looking at the world.

In *Plácida, la joven*, and in *Tristura*, she will adopt this feminine perspective, in a more intimate way, centred throughout the story on a female 'self'. The former still creates a distance between her own personality and the person she represents, with the technique that Ciplijauskaité applies to *La enferma*, but which could well fit Plácida: creating a frame with the female narrative voice in the first person to then focus the interest on a different, non-speaking character (1988, p. 16).

In the process of this new writing, a constant effort of self-awareness can be observed in the writers, which requires an appropriate language. For this reason, the recourse to the first person serves as the most appropriate mode for psychological enquiry (Ciplijauskaité, 1988) –hence, although *Trayecto uno* encompasses several characters, it uses free indirect speech to allow these characters expose themselves–. Thus, this 'liberated' language serves two purposes: 'it expresses the reaction to the social repression of the past times and leads towards self-knowledge' (Ciplijauskaité, 1988, p.17).

The author also theorizes of the awakening of consciousness in a girl child, which we see expressed in *Tristura* through Tadea's perspective. In this type of novel, the central focus is on the critique of the education that girls receive, which tends to silence the conscience instead of helping to awaken it. The outside world is so important that it marks the girl's life for the rest of her life. She also refers to the tendency in Bildungsroman written by women to replace rebellion with concentration, which implies a less linear narrative, where oppression leads to introspection and oddities in behaviour (Ciplijauskait, 1988). Thus, we see how Tadea is silenced throughout the novel, only to end with an unusual awakening: «Tristura... Era exacto. Y roencia también [...] El corazón roído por la muerte, ahora sabía qué era muerte... Y ansia, ansia, ansia.» (*Tristura,* pp. 242, 243).

Following Ciplijauskaité (1988) to instil these modes of feminine perception, the female authors renounce to the rules of language composition, a different lexicon is introduced and the use of syntax is modified. These non-norms are the ones that predominate in Quiroga in a recurrent and different way in her production. That is why it has been necessary to comment on them in these three pieces, given that each of them uses a different way of narrating the female experience.

There has been an evolution of the 'self' concept, whereby it is not only of interest to tell a story, but also to manage to speak as women, to analyse oneself, to raise questions and to discover unknown aspects that had not been expressed until then (Ciplijauskaité 1988). In *Plácida, la joven,* some authors (Olazagasti, 1988; Vidal, 2022) pointed out the coincidence of author-narrator. Telling the story of Plácida served the narrator to confess to her husband in a letter, which allowed her to express herself and to unburden herself emotionally when she needed to. In the letter the narrator questions herself, discovering her own guilt. On occasions the letter is not entirely intimate, in the fragments in which she addresses her husband, as well as evidencing the need to justify her emotions: «te lo aseguro» (p. 10), «(Tú me has dicho que las corzas lloran para morir)» (p. 19), «¿Habla sola Eduarda cuando tú y yo no estamos?» (p. 29).

The evolution of the female first-person novel was developing from social realism to psychological reality (Ciplijauskaité, 1988). These three books could respond to this evolution if *Trayecto uno* was narrated by a homodiegetic narrator. As we have already pointed out, Quiroga started from the grounds of the social realism that had a place in the panorama of the time. Despite this, she already introduced the intention of focusing attention on a female character who would have the strongest voice in the work through the interior monologue. In *Plácida la joven*, she takes an intermediate step in introspection, as the consciousness of what is written appears to be regulated. It is in *Tristura* that the most introspective relate is produced in the consciousness of a woman, who in this case is a child.

We can consider that, although Quiroga declares in an interview that she does not conceive of a difference between masculine and feminine modes of writing (Esquiza, 1976), her writings show some aspects that do have to do with this particular mode of expression.

5.2 Female Characters

Perhaps it is because of this desire for self-discovery and intimate psychological exploration that the choice of the protagonists lies in the fact that they are women. Whether or not the authors and narrators coincide in the works analysed, it is undeniable that the perspective or focus adopted in them is that of a woman. Although each novel is particular and therefore its characters are particularly defined by the context of the work in which they are set, there are similarities.

The character most emphasised in *Trayecto uno* is Anuncia, which is why we could say that she is the protagonist of the work. Anuncia stands out for her sensitivity, her view of the world seen through the eyes of poetry, which awakens in her a feeling of tenderness and solidarity towards people, being observant, analytical and rebellious in her way. The unnamed character of the girl with the fringe stands out for her difference from the other girls. Her discovery of reading as a moment of intimacy shows she is also rebellious against her parents' impositions on what she should and should not read. Villanueva (2019) points she is an antecedent of Tadea, so similarities have been seen in the conflictive family relationships and restrictive education (Torres, 2013). Furthermore, in *Tristura* she also discovers poetry and in *Escribo tu nombre* she immerses herself in more significant readings (Vidal, 2006). Both girls begin their poetic experience by memorising poems.

Although the focus of the narrative in *Plácida, la joven* is the woman who gives the short novel its title, the character we get to know in greater depth is the narrator. The sensitivity seen in Anuncia is heightened in the author of the letter, who is capable of enormous suffering and pain at the death of a woman she does not know. This narrator is now an adult woman, but her concerns and reflections resemble those of Anuncia, whose age could coincide with that of Plácida. Plácida and Anuncia, despite being at two similar stages in life, they have in practise completely different lives. Anuncia lives in Madrid and, although it is mentioned that she does not have great economic power, –her place of residence, and perhaps some other information unknown to the reader–, it allows her to go to university. On the other hand, Plácida, who is anchored in a Galician village and without family, distances from the modern, urban life that Anuncia observes through the windows of the bus and is forced to work and get married. She ends dying in childbirth of a daughter who may repeat her mother's footsteps.

Coming from the same place as Plácida - a Galician village - but from a wealthy family, –like the narrator of the second work analysed–, we have the character of Tadea. Tadea is the character to whom we have the most direct access. Thanks to the way of narrating and the resources used by the author, we see the girl's immediate reactions, her thoughts and her memories. In *Tristura* there is a character who is specially closed to Tadea. She is Julia, the old woman who empathises with the girl and comfort her when she is sad. Because of her social position, poorer than her relatives, the woman is also misunderstood in the house, eating with the servants instead of with the rest of the family.

The fundamental difference in these characters, apart from their social class, is the time of life in which they are. Tadea is a child, Anuncia and Plácida are young women, the narrator of the letter is an adult woman and Julia is an old woman. Although the way they understand the world differs due to the difference in maturity of each of them, they all interpret it with an exacerbated sensitivity –or so it is perceived from the reader's point of view– which opposes the lack of humanity of the rest of the characters. Sometimes this emotiveness makes them sad, sometimes because of guilt, as it is the case of the narrator of *Plácida, la joven* (1975, pp.10-11):

Aún hoy, ya ves, te lo aseguro, miro por esta ventana del despacho hacia el patio de entrada con su cerdo gigante, y todo lo percibo a través de esta suerte de entristecido estar: porque el aire parece que lleva luz distinta, y la tierra huele apagada, conteniéndose, y en el estanque el agua de la fuente chorrea amortiguada, queda: avergonzado todo porque Plácida ha muerto.

Anuncia, on the other hand, feels human solidarity and empathy (p.123):

«"La quiero..." Sintió que una pena enorme le atravesaba el corazón. "¡Ayúdala, oh Dios!" Y se quedó tranquila, infinitamente, dulcemente tranquila. Se había acercado a la mujer que sufría». What makes Julia suffer is hunger –«una roencia, una tristura...» (p. 86)–, and in the case of Tadea, it is prohibition and absence what causes sadness: «aquel dolor sin nombre, desprendiéndose de mí, sintiéndose desprenderse, un adiós largo, enorme, voceado en la noche helada de la estación desierta, perdida [...] (p. 242)».

5.2.1 Existential Awakening in Search of an Identity

The search for an identity in the women writers mentioned in the previous section is represented by the female characters themselves in the works discussed here. Zatlin relates this search to the existential awakening experienced by the characters in Quiroga's works. For these characters, it will be important that the 'self' they find is authentic, and to this end they will oppose closed systems, whether secular or religious, that seek to deny the individual the right to develop that self (1977). In the case of the three books studied here, we find that these systems could be capitalism, precarious work and the Catholic religion understood in a prohibitive way.

This existential awakening is what makes the characters reflect on the world around them, finding they do not fit in it and, sometimes, rebelling against it. In doing so, Quiroga exposes themes such as freedom and loneliness in her works, which relates to her own personal disillusionment with institutionalised religion and her emphasis on existential philosophy (Zatlin, 1977).

Anuncia discovers her identity in poetry and in the love to people, when it seems that she had been looking for it: «Todo había empezado por ahí», «Y "aquello" estalló dentro de ella, explotó, la destrozó y la dejó limpia», «¿Es que antes no veía? » (*Trayecto uno*, p. 133-134). The narrator of the letter seeks her identity through the figure of Placida, who serves as a pretext to discover her human side and to remember, in turn, her childhood with her sister Eduarda, reconstructing the dramatic story of her parents. Torres (2001) pointed out that the adult Tadea in *Tristura* undertook a journey into the past because there she could find the origin of a sought-after identity. The adult narrator finds in the reconstruction of her past the incidents that marked her personality, so that is why the feelings that cause sadness and anxiety are highlighted -mainly the repression of the house in Santander and the absence of her loved ones- but also those that have generated affection, and are always associated with the same images -either of her house in Galicia or of Julia.

5.2.2 Orphanage

In the construction of characters' psychology, Quiroga shows the importance of childhood in human development, making reconstructions of this part of life. It is significant and perhaps binding that in the account of the childhood of these characters it is discovered that many of them are orphans –as is the case of the ones we analyse here–, at least of one parent.

First, in the memory of Cristo del Otero, Anuncia mentions her father, who disappeared in the Spanish Civil War. The case of *Plácida, la joven*, is especially particular, because there are four orphans: Eduarda, the narrator, Plácida, and her newborn daughter. Tadea's maternal orphanage is discovered through the words of the characters, constituting the reason why the aunt is in charge of Tadea's care. Moreover, all of them lost their mother in the first part of their childhood, almost always in childbirth. It seems to be a pattern that encloses a cyclical structure in the way they enter the world, which marks their lives from birth (Olazagasti, 1988).

Torres (2013) points out that Tadea shares the fate of early orphans, who forge their personality without a mother figure and learn to use rebellion as a defence. In her case, in the face of the discovery of the existence of evil and injustice.

For her part, Eduarda creates an armour after her father's second marriage to her own friend, following the death of her mother. When her friend –now her father's wife– dies, she pretends to raise her sister –the narrator of the letter– as a father figure. Perhaps losing her mother and feeling the disappointment of her father makes her wanting to instil in her sister the tools of defence that she has had to learn, and that her father would not give her. Thus, the narrator of the letter grows up in an already broken family environment, with the coldness of Eduarda's upbringing and her father's mild displays of affection.

In worse living conditions, Placida must learn to survive alone in the world, working precariously throughout her life and travelling from house to house to find a roof over her head, but never finding a home. Not even when a glimmer of hope appears in her womb is she able to escape this fate, which condemns her -and with her, her daughter's life- to an early death.

Orphanhood could be directly related to the search for a lost or snatched identity. After a life of absences and facing the world alone, they need to return to the beginning to find themselves. Many of Quiroga's works feature orphan characters, in addition to the three mentioned here, *Viento del norte, Algo pasa en la calle, La careta, La enferma* and *Presente profundo*. For this reason, Zatlin considers that Quiroga's great sensitivity in developing orphan characters has personal roots (1992).

5.3 Social Commitment and Paratexts

Analysing paratexts is an interesting task in Quiroga's books because all of them are dedicated to real people (Vidal, 2022):

- In Trayecto uno: A mis compañeros de autobús Cartagena-Princesa-Moncloa (por Ferraz)
- In Plácida, la joven: A una mujer – y en ella a todas las trabajadoras mujeres del campo gallego – que murió cerca de mí, en octubre pasado, sin que hubiéramos cruzado palabra… Era muy joven. Se llamaba Plácida
- In Tristura: A María Luisa Caturla. A mi marido.

In *Trayecto uno* it was already pointed out that one of Quiroga's aims was to get closer to the people of the street (Vidal, 2006) and, for this reason, she chooses the perspective of the heterodiegetic narrator, alternating perspectives to gain access to the different realities of the bus passengers. In doing so, she intends to show the reality of the different social classes in post-war Madrid denouncing the situation at the same time.

In *Plácida, la joven* there is a clear denunciation of the situation of Galician working-class women in the representation of Plácida's figure. There are many references to the difference in life between the narrator and her neighbours, as well as between women and men in the village.

As *Tristura* is her most personal and autobiographical book, she dedicates it to her husband and to María Luisa Caturla, a friend of hers. The situation she is narrating is very much inspired by her own experiences, so that she cannot be objective in her judgement. If the action is presented from the perspective of Tadea - a child - criticism cannot appear explicitly, as she is still naïve to verbalise certain judgements. Therefore, it is up to the reader to decide whether to condemn the situations to which she is subjected.

Thus, although she has expressed that she did not intend to represent the status of women in Spanish society (Quiroga quoted in Zatlin, 1977), this information is in her novels. Whether she intended to it or not, Quiroga shows different life possibilities in the women of her books, in which social class plays a fundamental role. They are all united by a feeling of incomprehension and anguish in the face of the society that surrounds them, whether for one reason or another. And for this reason, the portrayal of almost all her characters is marked by the context in which they find themselves. Hence, she does not directly blame the women of the village for accepting Placida's death, but presents them as part of the rural environment, where the priority work to be continued, despite a loss: 'Han aceptado el tiempo mientras yo me rebelo. Porque pienso. Ellas no tienen tiempo para pensar' (*Plácida, la joven,* p. 63).

The origin of this social commitment and this empathy for humanity that she shares with the protagonists of her works of fiction, could have its origins in the Spanish Civil War. The author has stated that the war marked her personality, accentuating her sensitivity towards other people: «Desde siempre ha predominado en mí un sentimiento de humanidad que todo aquello acrecentó» (Quiroga quoted in Esquiza, 1974, pp. 11). This could be the origin of her desire to create sensitive characters, who do not ignore the world around them, even though it is a hostile place, which she criticises both explicitly and implicitly.

5.4 Time Conception, Memory and Circularity

Aspects of the handling of time in the three works have been discussed and it is important to relate them here. Marks (1981) pointed out that Quiroga was one of the first Spanish post-war writers to experiment with time. Indeed, only two of her works follow a chronological order that is not altered at any point. Some have a linear chronology, but are interrupted by flashbacks or projections into the future; others take place in a brief period of real time, in which subjective time appears, allowing the psychology of the characters to be developed. Finally, in some novels the chronological events are very limited in time and occur during a single evening, but the past is evoked so frequently that it merges with the present, creating a simultaneous narrative on two temporal planes.

Trayecto uno traces the brief period of a bus journey, which, as Baquero Goyanes (1951) pointed out, coincides with the moment of its reading. Although it follows a linear time, there are flashbacks that are obtained by focusing on the different characters.

Plácida, la joven, belongs to the group in which the absolute present becomes an inherent part of the narrative. Since the narrator must present the events she recounts in a letter, she will try to do so logically, following a chronological order (Marks, 1981). However, analepsis is also observed, shifting Plácida's story to narrate key moments in the narrator's family history.

For its part, *Tristura* belongs to the group in which external chronology is subjectively interrupted by analepsis (Marks, 1981). The most significant in the play is the 'deep present' (Quiroga, quoted Uribarri, 1969, p.27) that serves as a structure, as it was already noted.

It is interesting to note that in all three works the procedure of activating memory is used, giving rise to retrospections, mainly through the senses and mental association. If the example of the smell of the violet in *Trayecto uno* has already been mentioned, there are also examples in which it is sight which makes the association, such as the driver who remembers his daughter when he sees her resemblance in Anuncia. In *Tristura*, Tadea constantly makes associations as a result of the words she hears from adults, although the most interesting fragment is that of the storm, with the sound of the wind and the rain:

– Nadie vendrá a buscarme. La tormenta…
– Aún está lejos.

... terneros brincaban y entraban atravesando los macizos, espantados, escapándole a Eustaquio.

– Huelen tormenta.

La Mora ladraba, ladraba, exasperada, loca, corría en dirección del trueno, con la grupa erizada, el pelo de la cabeza tieso, duro. Leontina se reía [...]
Tormentas en el campo no tenía que ver con tormentas en los altos de la ciudad. No se oía a los animales. La Diana atada con la cadena gorda a su caseta, se habría metido dentro, sobre la paja del suelo. Se oía al cielo, nada más.

In *Plácida, la joven*, because it is a letter, which requires planning, it is more difficult to find the involuntary association. Despite this, there are some examples in which memory is activated by association of what is observed through the window of the office in which she is writing (*Plácida, la joven,* p.35):

Sé -fijo sé- que al abrir la puerta del cuarto, sin encender la luz, veré el balcón abierto y a ti, acodado en el balcón, mirando hacia afuera. (Al fondo, temblorosas como las luces de un gran trasatlántico, el pueblo de Bayona se estira. Luces azuladas, amarillentas. Ha desaparecido el contorno, el paisaje: sólo las luces temblando en la noche, suspendidas en lo oscuro.) Y el cielo está fabulosamente estrellado y tembloroso también.

After this fragment of very detailed and lyrical description of the Galician landscape she returns to the present to address the narratee: «Perdida estrella de nuestro cielo de verano. Ayer noche no estuve en el balcón porque tú no estabas [...] No te he dicho que antes de llegar al correo nos cruzamos con una mujeruca [...]» (p. 36).

Circularity is another aspect linked to the way time is treated. In *Trayecto uno*, it is represented by the route of the line, constant, embodied above all in the bus driver, Domingo, and the money collector, Blas el Largo. The circularity is most evident in Domingo, who involuntarily recalls that his daughter was run over and killed by a bus like the one he is forced to drive day after day. The time jumps in *Tristura* are circular (Marks, 1981), as the reader, taken by the adult narrator to the 'fictitious present' of the child Tadea, moves to the past - to her memories of Galicia, but also to the more recent past in the house in Santander - to return to that present. This is the case in all the analepsis, which we can call 'repetitive' (Garrido, 1993, p. 169), since they underline the opposition between past and present. In *Plácida, la joven,* the idea of circularity is marked by the early orphanhood of the women -the deaths of their mothers occurring in up to four cases-, the ultimate representation of which is seen in Plácida and her daughter (Olazagasti, 1988).

Through these works we can see the evolution that Quiroga has carried out with the handling of temporality, introducing some procedures in *Trayecto uno* and *Plácida, la joven*, and developing them with total perfection in Tristura, achieving, as Marks (1981) points out, total control of the internal narrative rhythm. Indeed, 'chronic time' (Garrido, 1993, p. 158) is not so important in his novels -especially in those studied here- and psychological time is what really dominates the presentation of the plot. Quiroga's particular way of 'organising and feeling' these two types of time constitutes the figurative time of the works (Garrido, 1993, p. 162), being the main point of interest for this study.

5.5 Spaces

The time that structures the works is linked to a space, which is important so that memory can be activated and evocations can be made. Since the senses appear on a spatio-temporal expression, the chronotope is necessary as a form of sensory knowledge (Garrido, 1992).

Space as a concept is something essential for the human being and for the artist, as it allows them to project their basic concerns and obsessions onto it, and to express his feelings through the dimensions or objects of space. Consequently, space is never indifferent to the novel character (Garrido, 1992). This is demonstrated in the different works with their own words. In *Trayecto uno*, contemplating the streets of Madrid and its shops leads Anuncia to consider her economic possibilities. In *Plácida, la joven*, the landscape that the narrator observes through her window changes her mood: «reencontrar la tristeza que creí momentáneamente ida al asomarme a la mañana apacible, limpia, soleada y verde» (p. 46). In *Tristura*, it is worth recalling the example from chapter XVIII, in which Tadea notices the closed doors of the house as opposed to the open gates of the estate, among other examples (p. 147):

> *MIENTRAS estudiábamos, cerrada la puerta de nuestro cuarto. Mientras dormíamos, cerrada la puerta. Para jugar, al fondo de los plátanos, cerradas las cancillas de la huerta, sin traspasar las aberturas del seto. Cerrada la puerta grande del jardín, rematando el muro. Cerrada la puerta del baño, con Suzanne dentro [...] Cerrado el camisón sobre nuestros cuerpos al desnudarnos.*

Spaces even become symbols: of asphyxia - the house -, of serenity - Julia's room -, of freedom - the pazo -, of playing - the banana trees - in *Tristura*; of community - the bus -, of wealth and modernity - the shops on Gran Vía -, of poverty - the streets of the periphery -, in *Trayecto uno*; and of intimacy - the office -, of meeting - the living room -, of poverty and prosperity - the old and the future house of Plácida -, in *Plácida, la joven*.

Galicia and Madrid are the two main places in which Quiroga's works are set. *Trayecto uno* is set in Madrid, *Plácida, la joven* in Galicia, and so there the memories evoked by *Tristura*, although the main action takes place in Santander. They will be especially important because of the symbolism associated with them, and because of the different ways of life of the characters in them.

5.6 The Reader's Role

Because of the wealth of narrative procedures, poetic resources, intertextual references, symbols, and themes, Quiroga's works demand an active reader, albeit to different degrees.

This aspect was pointed out by Vidal Ortuño (2006) regarding the complicated syntax of *Trayecto uno*. Here attention must also be paid to focalisation and dialogue, since the narrator does not always introduce the discourse of the characters –direct and free indirect discourse being very frequent.

In *Plácida, la joven* the reader will necessarily have to read in depth to differentiate the two converging narratives - the story of Plácida and that of the narrator's conscience - as well as to understand the symbolism that leads Plácida to be equated with the land, and the determinism that operates in the Galician countryside.

Perhaps the greatest reading effort is made in *Tristura*. Throughout the reading, it must be understood that the narration takes place on two levels, that of Tadea as an adult and that of Tadea as a child, which are not explicitly expressed, and in which two tenses are also mixed. The formal complexity will require a deeper reading: there are reiterated features such as the elision of *dicendi* verbs and hyphens in the dialogues, the long parentheses that imply a regression to the recent past -as opposed to the italics that represent the past in Galicia-, the intertexts, the nominal phrasing (Vidal, 2006) and the break in syntax in the introspections, where the free indirect discourse and the interior monologue prevail.

Furthermore, it has already been mentioned that she carries out a social critique that is not always explicit. Only the reader who has the will to understand and analyse all the meanings hidden in her works will be able to unravel the true and profound quality of her writing.

We could here show three examples of three different types of descriptions made in each piece, which demonstrate the complexity and variety of her descriptions' syntax. The first one is a description of the shops and advertisements that Anuncia observes through the windows of the bus (*Trayecto uno,* pp. 98, 100):

Carteles en colores: "Visite usted Sevilla." La Giralda. Plaza Mayor de Salamanca. Pórtico de la Gloria. Palmeras. Cactus. Playa. Un sol amarillo sobre el papel: "Invierno en Málaga." Un barco sobre lamar. Zocos. Viajar...[...] «... el autobús contorneaba la Cibeles. Luz. Agua. Agua delgada, curva, luminada. Una isleta de luz.

The second one is a symbolic and lyric description of nature made by the writer of the letter in *Plácida, la joven* (p. 66):

Miran al sol. Es bueno tener sol mientras siga fresca la tierra allá, sol y aquel airecillo oloroso, que huele a cebada en palleiro, a maíz recogido, a hierba tierna, a la uva agria, al dulzón excremento de los bueyes, al pelo de las cabras, a una mar cercana que desde el valle no se ve. Bueno que el aire baje desde la casita con balsámico olor de árboles y cruce la aldea y brice la tierra recién apisonada ayer.

The last example corresponds to a description made by Tadea, with her child view (*Tristura,* p.148): «Bajar despacio, a pulso, la ventana de guillotina: prado de Piano, sendero a Cueto, prado de Piano. Vacas. Al fondo, en alto, mar».

6. CONCLUSION

After having delved into the historical-literary context, writing conditions, general literary trajectory, and brief biography of Elena Quiroga, it has been observed that there is an assorted and disparate opinion of the critics in the insertion of her work in one or another literary current, as well as a devaluation of her writing at the beginning of her career. After studying and analysing three of her works in detail, a series of conclusions can be drawn, based both on personal views and on the bibliographical sources used for this purpose.

Throughout her career she experimented mainly with the genre of the novel, although she devoted special attention to the short novel in a period when it was barely present on the literary scene. The author started out in realist sphere and gradually moved towards existentialism and the representation of inner reality with a renewal of narrative forms. The selected works, which represent three different moments in her production (1953, 1956 and 1960), have served as a clear example of this evolution.

Firstly, it has been observed that, despite their differences, she chooses one or several women as her main focus, although not always in the same narrative mode. In *Trayecto uno* -a story close to realism- the heterodiegetic narrator accesses to different perspectives, although his focus is Anuncia. In *Plácida, la joven* - an existentialist epistolary story - the homodiegetic narrator - an unnamed woman - uses her handwriting to give voice to another woman and to herself. In *Tristura* -an experimental Bildungsroman- different narrative levels access to the point of view of a child, always using the homodiegetic narrator. Thanks to the point of view adopted in the three books, the author achieves a very precise portrait of the experiences in the life of a woman, from her childhood, through her youth, to her maturity. In addition, the experiences of old age are also portrayed through secondary characters. The presentation of these experiences shows both the outside world and the inner reaction of the characters, which often leads to existential reflection. The level of closeness to the characters also increases from one work to the next, in a similar way to the development of the narration of a female 'self' in women writers.

Although her works are not described as feminist, as she did not intend to be, perhaps involuntarily, the author used techniques, procedures and dealt with themes that have been studied as typical of women's writing, especially the narration of the feminine self and the self-exploration of an identity -which her characters and the writer herself would realise.

It has also been observed that the use of experimental resources is already present in the first work analysed here, distancing from what would be expected in a narrative described as realistic. In *Plácida, la joven*, Quiroga's particular way of writing an epistolary story does not conform to tradition either. Among all, the most experimental of the works analysed is *Tristura*, in which he culminates in the use of the procedures that will allow him to present the inner world of a young girl in an unusual way.

Also noteworthy is the reiteration of the theme of the orphanhood of her characters, an element that considerably marks their lives and coincides with the author's personal circumstances.

Her literary skills are unquestionable in the spatio-temporal dimension. The space of her works, divided between the north of Spain - Galicia, Santander - and Madrid, personification of the different lifestyles and social classes of the time, which give rise to a social criticism - sometimes explicit and sometimes implicit - makes clear the personal significance that Quiroga associates with these places. Likewise, playing with time becomes a fundamental part of her works, making truly experimental contributions in which the reader's effort is necessary. On both planes, the sensory component and memory are very much present.

Quiroga has had a 'deep, abundant, inexhaustible friendship' with books, which has served him as a source of 'thought and feeling, a stimulus for a willing imagination, nourishment and enrichment'[1]. This may justify her personal taste for introducing the theme of reading into her female characters, especially in childhood and adolescence. It is not surprising, therefore, that the author's literary mastery is reflected in the works with frequent intertextual references and a deep lyricism, which converge very successfully with her narrative modes.

With all these motifs, extracted from the study of *Trayecto uno*, *Plácida, la joven* and *Tristura*, we attempt to illustrate the evolution and varied richness of her work, which, starting from common places, never loses the will to innovate in her narrative production. It is possible that she did not consciously seek to vindicate women, but the reading of the experiences of her characters manages to generate emotion in today's reader - especially in the female reader, who understands the intimacy of these experiences - and perhaps, if she had received greater attention in her time, she could have been a point of reference for other women. It is for this reason that we have sought to recover the figure and work of Elena Quiroga, with the aim of showing in this work that her silencing and forgetting is unfair, as her writing deserves attention and study by both the reading public and literary critics.

REFERENCES

Amorós, A. (1994). Introducción. En Elena Quiroga, *La última corrida* (pp. 7-29). Editorial Castalia.

Baquero Goyanes, M. (1951). *Problemas de la novela contemporánea*. Ateneo.

Barney, K. R. (1993). The Quiet Revolution of Elena Quiroga. *Revista Hispánica Moderna, 46*(1), 103–116. https://www.jstor.org/stable/30205865

Brent, A. (1959). The novels of Elena Quiroga. *Hispania*, 42(2), 210–213. DOI: 10.2307/335895

Cabello, A. (2023). Una habitación propia: El Premio Nadal y la escritora profesional en España. *Boletin de la Biblioteca de Menendez Pelayo*, 99(3), 21–47. DOI: 10.55422/bbmp.950

Ciplijauskaité, B. (1988). *La novela femenina contemporánea (1970-1985). Hacia una tipología de la narración en primera persona.* Anthropos.

Equiza, P. (1976). Elena Quiroga y la novela española de posguerra. *Mundo Hispánico*, 336, 10–12.

Galdona, R. I. (2018). Real Academia de la Historia. *Elena Quiroga de Abarca.* https://dbe.rah.es/biografias/10566/elena-quiroga-de-abarca

Garrido, A. (1992). *El texto narrativo*. Editorial Síntesis.

Legott, S. (2014). Negotiating censorship in the postwar spanish novel divorce and civil marriage in Elena Quiroga's "Algo pasa en la calle". *Revista de escritoras ibéricas, 2,* pp. 121-144. DOI: 10.5944/rei.vol.2.2014.12949

Marks, M. A. (1980). Elena Quiroga's Yo Voice and the Schism between Reality and Illusion. *Anales de La Narrativa Española Contemporánea*, 5, 39–55. https://www.jstor.org/stable/27741025

Marks, M. A. (1981). Time in the Novels of Elena Quiroga. *Hispania*, 64(3), 376–381. https://www.jstor.org/stable/342041. DOI: 10.2307/342041

Martínez Arnaldos, M. (1996). Deslinde teórico de la novela corta. *Monteagudo. Revista de Literatura Española, Hispanoamericana y Teoría de la Literatura, 1*, pp. 47–66. https://revistas.um.es/monteagudo/article/view/76881

Martínez Cachero, J. M. ª. (1979). *Historia de la novela española entre 1936 y 1975*. Castalia.

Olazagasti, E. (1988). "Plácida, la joven", síntesis de la narrativa quiroguiana. *Anales de La Literatura Española Contemporánea*, 13(1/2), 135–146. https://www.jstor.org/stable/27741844

Pujante, C. (2010). Notas para una recreación de la teoría sobre la novela corta española del siglo XX: A propósito de las escritoras en La Novela del Sábado (1953-1955). *Castilla: Estudios de Literatura, 1,* pp. 38-59. https://uvadoc.uva.es/handle/10324/12182

Quiroga, E. (1985). *Plácida, la joven y otras narraciones*. Plaza y Janés.

Quiroga, E. (2013). *La enferma*. Cátedra.

Quiroga, E. (2022). *Tristura*. Bamba.

Sobejano, G. (1975). *Novela española de nuestro tiempo*. Prensa Española.

Sobejano, G. (2006). *Novela española de nuestro tiempo. (En busca del tiempo perdido)*. Marenostrum Comunicación.

Soldevila, I. (1980). *La novela desde 1936*. Alhambra.

Sotelo, M., & Ripoll, B. (2023). El Premio Nadal y las narradoras de posguerra. *Boletin de la Biblioteca de Menendez Pelayo*, 99(3). Advance online publication. DOI: 10.55422/bbmp.958

Torres, B. (2001). *Estudio de los modos narrativos en el discurso autobiográfico de* Tristura, *de Elena Quiroga*. Estudios y ensayos 50. Universidad de Málaga.

Torres Nebrera, G. (2013). Introducción. En Elena Quiroga, *La enferma* (p. 9-91). Cátedra.

Uribarri, R. (1969). *Elena Quiroga habla para el*. Diario de Navarra. Diario de Navarra.

Vidal, J. M. (2006). *Trayecto uno,* novela corta de Elena Quiroga. *Montearabí*, 36, 63–76.

Vidal, J.M. (2022). *Plácida, la joven*, de Elena Quiroga. Entre la denuncia y la autoficción. *Hécula: revista de la Fundación Castillo-Puche, 9*, pp. 95-106.

Villanueva, D. (2019). *La trayectoria novelística de Elena Quiroga*. Volvoreta. Revista de Literatura.

Zatlin, P. (1977). *Elena Quiroga*. Twayne Publishers.

Zatlin, P. (1992). Introducción. En Elena Quiroga, *Escribo tu nombre*, (pp. 9-21). Austral.

ENDNOTE

1 Words of the writer in a letter to Rafael Lapesa, prior to her entrance speech to the Real Academia Española in 1988 (Quiroga quoted in Lapesa, 1988, p. 326).

Chapter 13
Literature That Dismantles Borders Hispanoamerican:
Short Stories and Interculturality in the Spanish Language and Literature Classroom

Laura Hernández González
https://orcid.org/0000-0002-9276-3974
University of Castilla-La Mancha, Spain

ABSTRACT

Due to the increasing presence of foreign students in classrooms, Spanish educational institutions face the challenge of developing an intercultural educational model to provide quality education to all students, regardless of their geographic and social background. Specifically, in the subject of Spanish Language and Literature, incorporating texts from the Hispano-American literary tradition can facilitate the academic and social integration of many Latin American students enrolled in Spanish schools. This article reviews the didactic potential of the work of Horacio Quiroga, considered one of the great masters of the Hispano-American short story, for the simultaneous development of both literary and intercultural competencies. The article analyzes the constructive, formal, linguistic, and thematic characteristics of Quiroga's stories, which not only offer an artistically enriching reading experience but also provide an opportunity for our migrant and native students to engage in intercultural dialogue.

DOI: 10.4018/979-8-3693-7185-5.ch013

Copyright © 2025, IGI Global Scientific Publishing. Copying or distributing in print or electronic forms without written permission of IGI Global is prohibited.

1. INTRODUCTION

1.1. Cultural Diversity in the Spanish Educational System

The Spanish educational system, particularly in public schools, faces the challenge of integrating an increasing number of foreign students (882,814 in 2022, according to OBERAXE), who currently represent nearly 11% of students in compulsory education. This trend is expected to increase in the coming decades due to phenomena such as globalization, climate change, armed conflicts, and poverty, all of which contribute to migration flows.

Clearly, the presence of students from various geographical and ethnic backgrounds, with different cultures and languages, requires changes in teaching practices and learning models to ensure that our educational system remains inclusive and provides quality education to all students, regardless of their geographical or social origins. Since the early 21st century, Spanish schools have been compelled to evolve from a model based on the cultural and linguistic homogeneity of students to the development of teaching-learning processes that adapt to a new multicultural and multilingual context.

Traditionally, the coexistence of individuals from diverse geographical origins, ethnicities, cultures, religions, or languages in the same social space has been a significant source of tension and conflict, as the most economically, politically, and socially powerful group often imposes its language, culture, and customs on others. This situation is reflected in the school system through an assimilationist educational model, in which students are convinced that to integrate and succeed academically and socially, they must abandon their cultural traits and adopt those of the dominant group within the educational institution. From this perspective, the migrant student feels "obligated" to change, while the rest of their classmates can continue expressing their cultural peculiarities since these are deemed acceptable and "correct" within the school community. As Martín Rojo (2004) states:

La ideología asimiladora da lugar, naturalmente, a modelos educativos destinados a la asimilación cultural de los estudiantes y a la sustitución lingüística (conocidos como modelos laissez faire), sin que consideren en ningún momento la posibilidad de introducir cambios que involucren a toda la comunidad escolar. (Martín Rojo, 2004, p. 36)

Clearly, the assimilationist model is based on the implicit or explicit belief in the superiority of one culture (that of the privileged and dominant group) over another (that of the minority group with less economic and social power).

However, since the mid-20th century, both the increase in ethnic, cultural, and linguistic diversity in European societies and the rise of certain movements fighting against racism and xenophobia have led to the implementation of a new model of coexistence, known as multiculturalism, both in society and education. The multicultural educational model does not require members of culturally and linguistically minority groups to abandon their cultural traits but allows them to express them freely, promoting an attitude of tolerance toward diversity. However, in both schools and society, there is still a rigid separation between culturally diverse individuals, so there is little socialization or cultural exchange between groups from different geographical backgrounds. The immediate consequence of the multicultural model, both in society and within school communities, is the ghettoization of minority groups with less socio-political and economic influence. As Baralo (2002) notes:

> *Las sociedades multiculturales incluyen diferentes grupos culturales, de diferentes nacionalidades y etnias, dentro de un mismo territorio, pero no necesariamente en contacto unos con otros. A menudo las diferencias se ven en estas sociedades como algo negativo y encuentran una justificación para la discriminación. Las minorías son toleradas de forma pasiva pero no son aceptadas ni valoradas. (Baralo, 2003, p. 152)*

In contrast to the assimilationist and multicultural models, interculturalism is based on valuing cultural differences as an opportunity for individual and social enrichment. In this way, cultural differences are accepted and respected (as in the multicultural model), but they are also seen as an opportunity for dialogue and exchange between different cultural communities.

In the educational sphere, intercultural education is a comprehensive model that affects all students, as it is based on the idea that both migrant and native students should base their coexistence not only on tolerance for cultural differences but also on a genuine desire to understand and learn about different cultural realities. As Besalú (2002) explains:

> *La educación intercultural es fundamentalmente una actitud y un comportamiento relativo al tipo de relaciones que se dan entre las culturas particulares que conviven en un ambiente determinado. Se propone atender las necesidades afectivas, cognitivas, sociales y culturales de los individuos y de los grupos posibilitando que cada cultura exprese su solución a los problemas comunes. Es un medio para promover la comunicación entre las personas y para favorecer las actitudes de apertura en un plano de igualdad. No se dirige, pues, sólo a las minorías étnicas o culturales, sino que es una educación para todos. La educación intercultural es una educación para y en la diversidad cultural y no una educación para los culturalmente diferentes. (Besalú, 2002, p. 71)*

In the future, globalization and increased migration will lead to increasingly diverse societies, so the educational system must prepare future generations to live together and understand cultural differences as opportunities for mutual enrichment. In this sense, intercultural educational models appear to be the only ones capable of preparing students to face the challenges of future societies, although this will require a transformation of educational practices that will affect not only students but also teachers and the organization of schools, as Martín Rojo (2004) pointed out two decades ago:

Para que la diversidad resulte un enriquecimiento mutuo, los centros de enseñanza han de abrir espacios para que esa diversidad se manifieste en condiciones de igualdad, y los profesores han de formarse para conocerla mejor y poder sensibilizar al alumnado. A diferencia de las políticas asimiladoras, en las que son los alumnos de origen extranjero, y solo ellos, los que deben cambiar, en las políticas integradoras se espera que tanto el centro escolar como todos los que participan en él se abran a una realidad diversa (...) [En los] modelos interculturales se favorece el conocimiento de formas y usos tanto propios como ajenos; (...) se destacan particularmente los puntos comunes, y las diferencias se contextualizan dentro de un marco en el que los factores sociohistóricos y económicos son fundamentals. (Martín Rojo, 2004, p. 40)

In this sense, educational legislation, schools, and teachers must strive to ensure that the diverse cultures of our students are present in classrooms, so all are equally valued and considered worthy of being known and studied by all students. This will promote the integration of migrant students in the classroom, as they will feel that their culture also has something to contribute to school knowledge and that it is not primitive, inferior, or something they must forget to integrate into the educational community. This has been explored by Carter-Thuillier and collaborators (2019) in their study of the inclusion of traditional Latin American games in Physical Education classes in Spain:

Los alumnos latinoamericanos valoran positivamente cuando los monitores utilizan dentro de sus clases juegos o prácticas corporales que se vinculan con su cultura de origen. Asimismo, esto puede entenderse como intento de incorporar la cultura de los estudiantes al aula y, por consiguiente, facilitar la integración social de dicho alumnado. (Carter-Thuillier, López-Pastor, Gallardo-Fuentes & Carter-Beltrán, 2019, p. 50)

In any case, the shift toward an intercultural educational model will be achieved through a comprehensive transformation of curricula and teaching-learning processes, not through a set of isolated actions added to a fundamentally monocultural model, as is currently dominant in Spanish schools. This was already noted in 1991 by Alegret:

se deberá pasar de los 'añadidos curriculares' a estrategias más globales e imaginativas que posibiliten transformar las acciones centradas en los programas y currículos, en acciones que impliquen a la escuela en su conjunto y donde las deficiencias debidas a la estructura social tengan prioridad frente al 'déficit cultural' individual o colectivo. Otra de las necesidades que habrá que plantearse será la toma de posición que cada centro educativo deberá adoptar como tal frente al tema del racismo. Es de este modo como se podrá cambiar la posición hasta ahora vigente de considerar los problemas racistas como problemas individuales que afectan o a los alumnos o a los profesores o a los padres, dando la espalda de este modo al núcleo del problema, cuál es su dimensión social global. (Alegret, 1991, p. 1)

Therefore, adopting an intercultural educational model involves a radical transformation of our educational system to ensure it provides quality education to all students, considering their cultural diversity.

1.1.1. Latin American Students in Spanish Classrooms

Currently, students from Latin America make up approximately 27% (INE, 2022) of the total foreign students in our country's schools. Latin American students have certain advantages over foreign students from other regions when it comes to integrating into the Spanish educational system. The first and most obvious is that, in most cases, their mother tongue is Spanish, the language of instruction in our schools, so they do not have to learn a new language while adapting to a new school and social environment.

At the same time, Spain's prolonged presence in its Latin American colonies led, to varying degrees depending on the territories, to the assimilation of indigenous cultures and languages by the Spanish. The result of this process is that the cultures of Latin American countries have many elements in common with Spanish culture (religion, festivals, folklore...), along with some traits inherited from pre-Columbian cultures and certain manifestations of other cultures that emerged in American societies during and after the colonial period (African, Portuguese, Italian, German, etc.).

Thus, the strong cultural and linguistic ties between Spain and its former colonies have created a long tradition of migration between the two regions. While in the late 19th and early 20th centuries, America was the destination for many Spaniards due to both economic and political reasons, since the late 20th century, Spain has become the destination for many Latin American migrants, particularly from Colombia, Venezuela, Ecuador, Peru, and Argentina (according to INE, 2022 data).

However, despite the apparent advantages for social and school integration, Hispano-American students face specific problems in successfully integrating into our educational system, which can lead, in the most severe cases, to their ghettoization. First, Hispano-American students confront racism and xenophobia, which are increasingly present in classrooms and are exacerbated if the student exhibits Indigenous ethnic traits. Indeed, the study by Andújar Llosa and his collaborators (2022) reveals that 25% of the population aged 15 to 29 is openly racist, although discrimination levels are lower for Hispano-Americans than for sub-Saharan and Moroccan populations.

In addition to suffering various forms of racial discrimination, foreign students are often concentrated in certain schools, which promotes their marginalization and exacerbates their curricular delays, as explained by Fernández, Valbuena, and Caro (2017):

Con respecto al entorno escolar, las manifestaciones de segregación o prejuicio se dan de diferentes maneras y en contextos diversos: entre los niños, especialmente en las actividades de ocio o recreo; en las actitudes de algunos padres con respecto a los niños de minorías étnicas consideradas como conflictivas; o bien en el contexto del propio centro a través de resoluciones que a veces resultan discriminatorias. Dos son los principales factores de rechazo percibidos por la población inmigrante en los centros escolares: en primer lugar, los prejuicios raciales. En segunda instancia, la pobreza, como factor de discriminación genérico. Esto es, el problema radica en la concentración de alumnos de un determinado perfil socioeconómico y educativo más que en el perfil étnico-cultural. De hecho, la falta de recursos económicos y humanos y la baja calidad de la educación en aquellos centros más desfavorecidos existían anteriormente a la llegada de inmigrantes. (Fernández, Valbuena & Caro, 2017, pp. 11-12)

Furthermore, racism and xenophobia sometimes lead to problems of bullying and a feeling of lack of belonging to the educational community on the part of the migrant student, as shown in the Report from the Spanish Observatory on Racism and Xenophobia based on the results of the 2018 PISA Report:

Según los resultados obtenidos, los estudiantes extranjeros muestran valores más negativos, y estadísticamente distintos del de los nativos, (...) indicando un menor sentimiento de pertenencia a la escuela y una mayor exposición a sufrir episodios de acoso escolar. En los nativos de padres inmigrantes, la exposición al acoso es similar a la de los extranjeros, sin embargo, el sentimiento de pertenencia es significativamente mayor que en los extranjeros, aunque se sigue situando significativamente por debajo del de los nativos. (Observatorio Español del Racismo y la Xenofobia, 2022, p. 34)

On the other hand, in some of their countries of origin, migrant students faced difficulties accessing formal education or the academic level they achieved was significantly lower than that of students their same age enrolled in the Spanish system (it is estimated that the average curricular lag with which a migrant student enters our educational system is one academic year). This leads to both the repetition rate and the early school dropout rate for foreign students exceeding those of native students by more than 50%. The curricular gap linked to migration, which is sometimes very difficult to address despite existing support measures, is, however, reduced in Hispano-American students: according to the OBERAXE report (2002), migrant students who speak Spanish at home narrow their gap with native students by 35%

in reading and scientific competence, a reduction that, however, is not observed in mathematical competence.

It is striking, moreover, that despite the advantages of having Spanish as a mother tongue, Latin American students sometimes have to face rejection from teachers and students for using a dialectal variety different from the Peninsular variety. Thus, Martín Rojo (2003), in his study in Madrid's educational centers, noted that teachers sometimes encourage students to eliminate the characteristic dialectal features of their variety and replace them with Castilian ones, considering Northern Peninsular Spanish as a desirable standard. Montes Pedrajo (2023) has recently studied the attitudes of native Spanish students towards different Latin American varieties, concluding that, perhaps due to sociocultural and economic factors, some are perceived more positively (the Rioplatense) than others (those from the Caribbean region). Similar conclusions can be drawn from Yraola's study (2014):

> *El país donde el mayor número de informantes considera que hablan más "correctamente" es España, con un porcentaje de 55,8%, seguido de Colombia (9,3%), Argentina (6,3%) y Chile (4,5%), justificándose la elección por la comprensión idiomática. En cuanto a la "incorrección", reina la incertidumbre o desconocimiento generalizados: el 44,6% de los informantes no saben o no responden, y se mencionan España, Cuba, México y Argentina, pero con porcentajes inferiores al 10% (…).*
>
> *En las preguntas de control acerca del agrado o desagrado por otras formas de hablar, en agrado, con un alto porcentaje (80%), España queda de nuevo en primer lugar, seguida de Chile, Colombia y Argentina; en desagrado, también España, seguida de Cuba y Estados Unidos, y los países centroamericanos junto con Paraguay y Uruguay, los grandes ausentes entre los españoles. (Yraola, 2014, pp. 631-633)*

Thus, Latin American students enrolled in Spanish classrooms (around 238,000 students) face the challenge of adapting to a new educational system (for which they must overcome certain curricular gaps, especially regarding mathematical competence) in a climate where, at times, they are still discriminated against due to racist and xenophobic prejudices, such as those that lead many of their peers and teachers to urge them to abandon their own dialectal traits.

1.2. Latin American Literature in Spanish Secondary Education Curricula

Latin American literature holds a testimonial presence within the curriculum of Spanish Language and Literature in Secondary Education and Bachillerato in Spain. Although current legislation is quite lax regarding literary education and does not prescribe specific works or authors, the contents that are evaluated in university entrance exams across different regions have shaped a kind of shared school canon among most textbook publishers. Due to the absence of Latin American literature in the university entrance exams of most autonomous communities, its presence in the Spanish Language and Literature manuals at all levels is minimal. For instance, in the 4th year of ESO, the study of Rubén Darío and some of the writers from the Boom (primarily García Márquez and Cortázar) is addressed, figures who reappear in some books aimed at the 2nd year of Bachillerato, as indicated by the study by Suárez Hernán and Bermúdez Martínez (2024), which includes various surveys of Spanish language and literature teachers in Spanish high schools:

> *La presencia de la literatura hispanoamericana se ve reducida a la inclusión de Rubén Darío en el Modernismo, como no podía ser de otra manera, y a algunos textos de Gabriel García Márquez y Julio Cortázar, principalmente. Reafirmamos así la idea de la limitación de la presencia en las aulas a unos pocos trazos y al fenómeno conocido como boom hispanoamericano. Los docentes muestran disconformidad con la limitación de la educación literaria a la llamada «literatura española». Los entrevistados plantean las carencias de los programas con respecto a la inclusión tanto de textos procedentes de literaturas periféricas peninsulares como de otras culturas que también conforman la cultura Española. (Suárez Hernán & Bermúdez Martínez, 2024, p. 61)*

This "school forgetfulness" of Latin American literature, which may seem inexplicable given its success among readers and the significance of studies on Latin American literature conducted at Spanish universities (for example, Barrera, 2019; Pedraza Jiménez, 2002) can be understood if we consider that the school literary canon in our country is largely determined by the romantic concept of national literature (literature as the expression of a nation's soul). This view, like others, prevents our students from studying literary works written in languages other than Spanish (whether or not they are Spanish), obstructing their access to the Latin American literary tradition.

This nationalist conception of literary tradition, characteristic of 19th-century Romanticism, hinders our students from obtaining a truthful perspective on the literary fact since literature is shaped through the dialogue between texts written in various times and languages. This intertextual connection is particularly relevant in the case of Spanish and Latin American literature, which, according to Carrión, cannot be conceived as independent traditions:

> *No puede entenderse la poesía de Pablo Neruda sin la de Lorca (y viceversa); ni la de ambos sin Juan Ramón Jiménez. No puede entenderse la novelística de Goytisolo sin la de Sarduy o Fuentes. No puede entenderse el ultraísmo sin el diálogo trasatlántico. No puede entenderse la narrativa hispanoamericana del siglo XXI sin las lecturas cruzadas. (2010, p. 249)*

Although the literary education of our students should not be restricted solely to the study of texts originally written in Spanish, if the established curricular tradition aims to prioritize the Hispanic canon over the universal, nationalist prejudices should be set aside, and those Latin American texts that are fundamental within a shared literary tradition, such as the Hispanic, should be integrated into the curriculum since their readers and writers never knew borders.

In addition to their intrinsic literary and cultural value, the inclusion of Latin American texts in the school canon can help consolidate an intercultural educational model since a significant number of Latin American students in our classrooms will see their culture represented in school content, at least in the literary field. As teachers, we must consider that the content addressed in the classroom, which is included within academic curricula, acquires added social value, as it becomes part of "high culture," which deserves to be preserved, taught, and learned by future generations. Thus, studying Latin American literature in Spanish educational institutions means that our students from Latin America see, at least in this area, their culture legitimized and valued equally to that of Spain within the school curriculum, both in their own eyes and in those of their peers. In this way, an intercultural school curriculum also facilitates intercultural social coexistence by teaching students to value, appreciate, and feel as their own cultural manifestations from various peoples and eras, and to identify the essentially human elements underlying all of them.

Moreover, by addressing Latin American texts in our subject, we will improve the integration of Latin American students in the classroom, as we will discuss content on which they have more knowledge than their native peers, allowing them to present themselves as "experts" in front of their classmates. This changes their usual role in the classroom since, often, due to the curricular mismatch mentioned earlier, students from Latin America struggle to keep pace with the class, which can lead to attitudes of demotivation, inhibition, or disruption.

In contrast, proposing activities, projects, or learning situations in which these students feel more competent than their peers will enhance their academic self-esteem and motivate them to continue engaging in our subject, even in areas not related to Latin American literature.

2. OBJECTIVES

The main objectives of our research are:

- To analyze the didactic potential of Horacio Quiroga's short stories in the subject of Spanish Language and Literature.

- To demonstrate how Latin American literature, specifically the work of Horacio Quiroga, can foster a literary education based on interculturality in the Spanish Language and Literature classroom.

3. METHODOLOGY

Our research is framed within the field of Literary Didactics and has an inductive nature, as through the didactic analysis of Quiroga's texts, we determine the extent to which they can serve for developing tasks, projects, and learning situations that contribute to an intercultural literary education for our students.

4. THE SHORT STORIES OF HORACIO QUIROGA

Horacio Quiroga (Salto, 1878 - Buenos Aires, 1937) was a Uruguayan writer who developed his work at the beginning of the 20th century. Although he also cultivated lyric poetry and theater, Quiroga is primarily recognized as one of the great masters of the Latin American short story, a narrative subgenre to which some of his most well-known works belong: *Cuentos de amor de locura y de muerte* (1917), *Cuentos de la selva* (1918), *El salvaje* (1920), *El desierto* (1924), or *Los desterrados* (1926).

Influenced by European narrative, Quiroga's work is particularly connected with the short stories of Edgar Allan Poe and Guy de Maupassant. However, in the case of *Cuentos de la selva*, critics have noted that the Uruguayan author sought inspiration in Rudyard Kipling's novel *The Jungle Book*. Quiroga's work is part of the Latin American Modernism movement, whose main representative is Rubén Darío, although many of his stories, which clearly aim to portray the reality of the Argentine region of Misiones, show the influence of earlier movements such as Realism and Naturalism.

In addition to his literary work, Quiroga is known for his famous *Decálogo del perfecto cuentista*, in which he gives ten guidelines to young writers for crafting a good story. However, it is often noted that many of Quiroga's stories do not follow the guidelines of the *Decálogo*, as, influenced by Modernism, some of the Uruguayan texts fall into a certain baroque style, with excessive use of epithets.

Horacio Quiroga is, in fact, a writer of contradictory and complex work who, according to some scholars of his figure, such as Abreu (2014), has always moved along borders: spatial borders, given his preference for setting his stories in the jungle of Misiones, a territory bordering several nations, but also between the jungle and civilization; artistic borders, as his texts have been linked to such varied movements as Modernism, Naturalism, Realism, regionalism, or fantastic literature; and stylistic borders, since his work combines detailed descriptions with broad ellipses or precise and abrupt endings with certain concessions to ambiguity.

The influence of Quiroga's work is fundamental in the subsequent development of the Latin American short story, as it helped establish it as one of the preferred genres for authors such as Borges, Rulfo, and Onetti, and of course, the writers of the Boom (especially figures like Cortázar, Miguel Ángel Asturias, and Carlos Fuentes).

4.1. Thematic Characteristics

Horacio Quiroga addresses a variety of issues in his more than one hundred sixty stories, but Milagros Ezquerro (1996) has established five basic thematic axes around which the Uruguayan's tales are grouped: I) Goces y dolores del amor; II) Los maleficios; III) La lucha del hombre contra la naturaleza; IV) Historias de animales; y V) Los hombres de la selva (Milagros Ezquerro, 1996, pp. 1381-1382).

As can be observed, all these issues are, in some way, a reflection of the tumultuous biography of Quiroga himself. Thus, his stories are influenced time and again by some of the most significant events of his life, such as his retreat to Misiones, the illness and death of his wife, or the harsh experience of loneliness tied to romantic suffering.

In any case, the Uruguayan's taste for mysterious and rare themes stands out, as well as his attraction to the strange; a facet in which critics have wanted to see the influence of Poe (Cadová, 2007). There is also a certain attraction to madness in his texts, seeking to approach man stripped of his humanity and subjected solely to the call of instinct. Death is, likewise, a constant in Quiroga's short stories, viewed from a fatalistic and resigned stance, perhaps due to the repeated misfortunes that marked the writer's life. The harshness of life in Misiones may have led Quiroga to conceive of death not as a transcendental issue but as an unavoidable occurrence in human existence.

The themes Quiroga addresses in his stories are quite appealing to adolescent readers, as juvenile novels focused on love, mystery and intrigue (often with supernatural elements) and adventures are, in fact, the most abundant and successful within the publishing market. Thus, we believe that many of Quiroga's tales, due to their exoticism, emotional intensity, or their ability to generate tension and intrigue, can motivate secondary school students towards reading.

Cuentos de la Selva, featuring animals, resemble fables and acquire a didactic tone that may be less appealing to adolescents, which is why this collection of stories will likely be read with more enjoyment in the final years of primary school or the first year of secondary school.

4.2. Stylistic and Constructive Characteristics

Quiroga's narrative is largely defined by its variety, so critics often find it challenging to assign the Uruguayan's work to a specific artistic trend. Indeed, Quiroga's style is multifaceted and changing, accommodating both the somewhat elaborate and sensory modernist prose of his literary beginnings as well as texts linked to Regionalism and Realism, which authentically reflect the ways of living and expressing oneself in Misiones. There are also stories that adopt avant-garde elements, in which, for example, ellipsis plays a central role, requiring the reader to take an active role in the story's creation.

However, critics have noted how, as Quiroga moved away from modernist forms, he evolved towards a more refined and direct prose, a predominant style in what the author called his "cuentos de monte," set in the jungle of Misiones, characterized by their density and marked drama. González Picado (1985, p. 13) highlights some of the most distinctive features of Quiroga's style: brevity; direct development from beginning to end without digressions; unity of action; surprising endings; a reduced number of characters; the connection between atmosphere and tone of narration; and the provision of truthful explanations for the fantastic (as occurs, for example, in the famous tale *El almohadón de plumas*).

Moreover, a fundamental characteristic of Quiroga's narrative is its symbolic nature, which allows his stories to offer various levels of meaning, as noted by Souto Alabarce (2003) and Zum Felde (1959):

> *Es el único escritor de América, y de los pocos fuera de ella, que ha traspasado las tres dimensiones normales de la narrativa; de ellas, dos, las exteriores, son las comunes: dimensiones planas, comprenden lo simplemente objetivo de la realidad, superadas por la tercera, de profundidad psicológica, de análisis subjetivo, de mundo interior, la de los novelistas más agudos, raramente intentada en nuestra literatura. Pero Quiroga opera en otro plano, el de lo subliminal, que estando "más allá" —como reza uno de sus títulos— sólo puede ser frecuentado por quien tiene su naturaleza. (1959, p. 408)*

Thus, the interpretation of certain texts by Quiroga, such as *A la deriva* (Arango, 1982) or *El almohadón de plumas* (Montenegro, 2006), has been the subject of debate among critics due to their significant density.

The internal construction of Quiroga's stories is, like his style, very varied. Alongside simple narratives that respect the Italian structure, such as the Cursiva, aimed at a child audience, we find stories where Quiroga takes advantage of the experimental possibilities offered by a short subgenre like the short story to construct circular narratives (*Los mensú*), stories that begin *in extrema res* (*El perro rabioso*), or those built around foreshadowing (*La gallina degollada*). In any case, as can be read in his theoretical texts on the short story subgenre, Quiroga, perhaps based on the narrative models of Maupassant and Poe, gives great importance to both the beginning of the story (which must capture the reader's attention) and the ending, aiming for it to be surprising, astonishing, and, at the same time, explanatory of all the elements developed throughout the story.

The stylistic and structural aspects of Horacio Quiroga's tales make them suitable reading for developing the literary competence of our students since, despite their sober and direct style that makes them seem simple, many possess a certain constructive complexity that will allow students to familiarize themselves with different narrative procedures.

5. DIDACTIC POTENTIAL OF QUIROGA'S WORK FOR INTERCULTURAL EDUCATION IN SPANISH LANGUAGE AND LITERATURE

Certain characteristics of the work of Horacio Quiroga make it a highly valuable resource for developing various intercultural educational approaches in the subject of Spanish Language and Literature.

5.1. Advantages of Short Stories for Classroom Reading

Currently, secondary school students, especially those with a lower degree of literary competence development, find it difficult to maintain concentration and sustained attention for the time necessary to develop extensive and autonomous reading of lengthy texts. These difficulties in maintaining attention among new generations can be explained by the fact that young people have grown up surrounded by new technologies that overstimulate them and often lead them to develop multiple tasks simultaneously while concentrating less on each one (McCrindle, 2014; Castro, Patera, and Fernández, 2020).

Thus, most cultural and entertainment products aimed at young people are characterized by their brevity or even their fragmented nature (consider, for example, videos on platforms like *TikTok* and *YouTube* or serialized novels shared on *Wattpad*).

Therefore, reading brief narratives, such as short stories, may be more accessible and motivating for students than using longer texts or excerpts that are sometimes presented out of context. The short story, as conceived by Quiroga, is a perfect literary artifact based on brevity and economy, guiding the reader, using only those elements strictly necessary, towards a surprising conclusion. In fact, one of the key constructive elements of Quiroga's short stories is their rhythm: the fact that each sentence and each paragraph smoothly lead, almost imperceptibly, to the next; culminating in the literary explosion of the conclusion.

Due to their structural and stylistic characteristics, Quiroga's stories are very easy to read: they are short and, in many cases, lack extensive descriptions or digressions, focusing solely on the essential elements of the story. Furthermore, Quiroga keeps the reader's attention through constant action, typically featuring stories that present a closed and impactful conclusion that, while surprising, satisfies the reader's horizon of expectations.

Thus, during a class session, it is easy to combine the intensive reading of a Quiroga short story with various activities and tasks aimed at ensuring critical and evaluative comprehension, through methodologies such as didactic questioning, debate, or dialogical construction of the story's meaning.

5.2. Adaptation of Quiroga's Short Stories to Various Levels of Reception

Horacio Quiroga wrote dozens of short stories, with varied lengths, styles, and structural characteristics, so within his work, we can easily find tales adapted to different tastes, interests, and levels of literary competence.

In this regard, it is worth highlighting Quiroga's *Cuentos de la selva* (1918), eight stories dedicated to his children, Eglé and Darío, which have become a classic text in Hispanic children's and young adult literature, although initially, the author conceived them as a minor work within his production. In fact, for generations, the entire collection or some of the narratives included in it have been published in richly illustrated volumes aimed at younger readers. Perhaps the success among younger readers of this collection of stories set in the jungle of Misiones can be explained by its inclusion of many of the characteristic elements of literature aimed at children and adolescents. The first and most evident is its subgenre, as the short story, due to its brevity and narrative condensation, allows for intensive and complete reading in a single session, whether by the child or a mediator, facilitating the literary reception process. Furthermore, given that fable is a narrative subgenre characteristic of children's literature featuring animals that embody human vices and virtues with a didactic and moralizing purpose, Quiroga's *Cuentos de la selva* easily relate to it since many of its stories feature personified animal protagonists, making it easier for child readers to identify with the characters:

Animal stories are far removed from the real world, and it is easy for children to identify with the protagonists, especially when they are puppies. These little animals think and act like children. Their adventures can influence their particular consciousness in a certain way. Their mistakes also serve as a lesson. By seeing the actions of animals and associating them with their qualities, children can intuitively know the results of their uncontrolled feelings as well as the benefits that the good ones bring them (Pan, 2021, p. 101).

On the other hand, although we cannot consider Quiroga's work as instrumental literature, unlike what occurs with fables, in some of the narratives included in this volume, there is a certain invitation to critical and moral reflection that sometimes borders on didacticism, through allegory or moralizing examples, as Fleming (1991, p. 57) suggests regarding stories such as *La abeja haragana* or *Las medias del flamenco*.

However, Quiroga never falls into simplification, as he writes from a deep respect for the child reader and, therefore, does not hesitate to present them with morally complex situations or scenes of great harshness, where evil, violence, or death are present. This is clarified by María Fernanda Maquieira, editor of *Cuentos de la selva*, in an interview published in *La Nación*:

> *Es un libro fundamental, que se sigue leyendo porque Quiroga ha construido un universo auténtico, fuerte y emocionante. Gusta a chicos y a grandes porque los hace entrar en una aventura increíble, pero además hay una forma de contar esas historias sin ornamentos, directa y visual, sin ñoñerías ni moralejas. Para su época, Quiroga fue un avanzado, un maestro del cuento. (Blanc, 13/08/2018)*

However, unlike what often happens in his stories for adults, the *Cuentos de la selva* have a happy ending, where poetic justice prevails and good triumphs over the forces of evil, as Pan (2021) explains:

The uniqueness of this work [*Cuentos de la selva*] can be appreciated, in which Quiroga shows his utmost tenderness because the stories are written for children. For this reason, although the process can be bloody and difficult, the fight between good and evil always ends with the triumph of the heroes. The end of *Cuentos de la selva* always presents scenes of harmonious coexistence between men and animals (Pan, 2021, p. 102).

From a formal perspective, *Cuentos de la selva* presents a simple and accessible style for child readers. In this collection of stories, Quiroga avoids extensive descriptions, uses relatively simple literary resources (mainly personifications of animals and other elements of nature, comparisons, and hyperboles), and shows a preference for dialogue and reflections of orality, connecting his work with popular literary forms aimed at a child audience. In fact, the desire to connect his work with folk narratives leads the Uruguayan author to resort, for example, to a relatively abundant use of onomatopoeia, a resource that, as Pan (2021, p. 99) suggests, helps immerse the child reader in the fictional environment and aids in understanding the story.

Thus, *Cuentos de la selva* is a work that, although highly valued by adult audiences, contains numerous elements that make it very attractive to child readers, allowing it to be used as a didactic resource in the last years of Primary education. As our students acquire literary competence, they can enrich their interpretation of *Cuentos de la selva*, leading to a much deeper understanding of these narratives or read other, more complex stories by Quiroga, such as those collected in *Anaconda* or *Cuentos de amor, de locura y de muerte*, which pose a greater degree of stylistic and structural difficulty and also address more problematic issues from an emotional and moral perspective, such as death, the exploitation of the weak, madness, or unrequited love.

5.3. Rioplatense Spanish in Quiroga's Work

Horacio Quiroga, quite innovatively for his time, does not hesitate to reflect the reality of the jungle in Misiones using the linguistic variety characteristic of this place: Rioplatense Spanish. Naturally, the Uruguayan writer constructs his work employing the grammatical constructions and lexical terms characteristic of this dialectal area, which, along with many other descriptive resources, manages to immerse the reader accustomed solely to the literary use of Peninsular Spanish in the reality of the jungle in Misiones.

The fact that our students can read literature in various varieties of Spanish, a circumstance they are not very accustomed to, is beneficial in many aspects. Firstly, Quiroga's stories showcase a variety of Spanish different from the Peninsular variety, used in a real communication context (literary communication) and in various fictional communicative contexts (those represented in each story), which allows students to come into contact with it naturally and necessitates activating various strategies linked to their communicative competence to unravel the meaning of the message.

Moreover, using Quiroga's texts, various activities can be proposed where students identify the linguistic features characteristic of Rioplatense Spanish and establish a comparison with Peninsular Spanish or with other varieties of Spanish with which they are familiar or about which they can obtain information. In any case, our students should observe a variety of Spanish different from their own in use, not, as is often seen in the manuals of our subject, reduced to a list of characteristic traits at each of the various levels of grammatical analysis.

In these types of activities, the Latin American students present in the classroom whose mother tongue is a variety of Spanish different from Peninsular Spanish can take on a certain prominence by presenting themselves to their peers as "experts" on the subject, a circumstance that will undoubtedly contribute to improving the academic self-esteem of students who, due to their condition as migrants, sometimes exhibit low levels of integration in the group and some curricular lag.

From a linguistic perspective, exploring the causes of the differences between Peninsular Spanish and Rioplatense can lead our students to analyze various historical events, such as the colonization process in America, to learn about some aspects of the history of the Spanish language that explain phenomena like *voseo* (form of address with implications for the morphology of pronouns and verbs) or *rehilamiento* (friction in the articulation of certain consonants), or to understand phenomena like interlinguistic borrowing, which arises when contacts occur between various languages.

Additionally, in his portrayal of Misiones, Quiroga's stories show the linguistic reality of a multilingual and border environment, where Rioplatense Spanish coexists with the indigenous language of this area (Guaraní), with Brazilian Portuguese, and also with other foreign languages, such as English. From a sociolinguistic perspective, and using some of the tales from *Cuentos de amor, de locura y de muerte* as a basis, it would be very interesting for high school students to reflect on the phenomenon of diglossia and how economic and political dominance relationships translate into the cultural and linguistic sphere. Thus, in the stories, there is an absolute primacy of European languages over indigenous ones that reproduces the relationship of dominance that, first, the colonizers and subsequently their creole descendants, have imposed on Native Americans.

Furthermore, by employing Rioplatense variety in his literary creations, Horacio Quiroga advocates for this variety: he positions it as a valid Spanish for literary expression and artistic and cultural creation. In this sense, the fact that our native students read literature written in a variety of Spanish different from Peninsular Spanish frees them from widespread linguistic prejudices that consider Castilian as the most correct or beautiful Spanish; that is, the only Spanish suitable for literary creation. Rioplatense Spanish, but also Guaraní and the multilingualism characteristic of the region of Misiones, are dignified through Horacio Quiroga's literature, as they become the raw material for a beautiful artistic product capable of producing aesthetic pleasure.

Simultaneously, Latin American students find in Quiroga's work a literary reference written in a variety different from Peninsular Spanish, contributing to their perception of their own way of speaking Spanish as fully valid and suitable for literary creation or academic use.

5.4. Literature as a Window to the World: Misiones in Quiroga's Work

Hispano-American literature helps enrich our students' worldview, distancing them from the ethnocentrism of most works that are common readings in the school context. In this way, one of the teachers interviewed by Suárez Hernán and Bermúdez Martínez (2024) emphasizes how Hispano-American literature broadens their students' horizons:

Y luego me encanta, también, aunque sean pequeños, que no les parezca que toda la literatura va sólo limitada al barrio, a la ciudad o a España; hay un mundo fuera y tenéis que ver nombres extranjeros y tenéis que ver de todo. Así que eso no sólo no me coarta sino..., y la literatura hispanoamericana también me resulta fascinante. (Suárez Hernán & Bermúdez Martínez, 2024, p. 157)

Much of Horacio Quiroga's short story production is set in a real space, Misiones, a jungle region in northwest Argentina where, around the course of the Paraná River, the borders of Paraguay, Brazil, and Argentina converge, but which sometimes takes on a symbolic dimension, representing a world in which man, in solitude, far from the refuge of civilization, must fight to survive in a hostile environment.

Thus, Quiroga's stories transport us to an environment completely removed from our daily lives, both historically and geographically, as well as in terms of nature, culture, and language. Each of his narratives brings the reader a piece of the reality of Misiones: thanks to the Uruguayan author, we learn about the lives of wealthy landowners, but also about exploited indigenous people, and, of course, about the animals (both those that accompany humans and the more savage ones that emerge from the depths of the jungle to try to exterminate them). Through reading, we come to understand their thoughts, empathize with their feelings, and ultimately share their existential framework, for, as Nussbaum (2010) argues, literature promotes narrative or empathetic imagination, understood as "la capacidad de pensar cómo sería estar en el lugar de otra persona, de interpretar con inteligencia el relato de esa persona y de entender los sentimientos, los deseos y las expectativas que podría tener esa persona" (Nussbaum, 2010, p. 132).

Quiroga's tales are filled with references to a culture that is geographically, culturally, and historically distant from that of our students: in terms of customs and social practices, the objects used by the characters, and also regarding the fauna and flora that surround them. Furthermore, far from providing a simplistic or Manichean view of the environment of Misiones, Quiroga consciously distances himself from *topica* such as *beatus ille*, *menosprecio de corte y alabanza de aldea*, or the *bon savage* in his vision of the jungle, portraying it in all its complexity and grandeur:

Está claro que, en semejante entorno, el hombre y la naturaleza no son iguales. Todos los elementos naturales tienen ahí dimensiones colosales, fuera de proporción con las fuerzas humanas: el calor, la humedad, las fiebres, los ríos, las crecientes, los tigres, las víboras, las hormigas carnívoras, la miel narcótica, las distancias, la soledad. Cada uno de esos elementos basta para dar la muerte al menor tropiezo, a la menor distracción, al menor error, o por simple fatalidad. La naturaleza es la figura polimorfa del destino, del implacable fatum, fuerza descomunal con la cual el hombre viene a medirse, a enfrentarse. El hombre, fascinado por esa naturaleza, está, indefectiblemente, habitado por la hubris, el orgullo de la criatura que intenta luchar contra las fuerzas sobrenaturales, pues la naturaleza, ahí, es una supernaturaleza. (Ezquerro, 1996, p. 1380)

Quiroga writes, in any case, with deep respect for nature, denouncing all those actions of man that threaten to destroy his environment (the overexploitation of resources, animal abuse, the extermination of certain species...), for he understands that human beings are the only animals capable of destroying their ecosystem and, with it, themselves. Thus, already in the early years of the 20th century, the literature of Horacio Quiroga advocates positions that today we would classify as ecological, always defending those characters who try to find a balance between progress and the conservation of natural resources for future generations (a clear example can be found in *El paso del Yabebirí*). Quiroga thus opposes the dominant ideology of his time, which viewed the jungle as an obstacle to the material progress of the Argentine nation. In the words of Wilde (2007):

> *Hacia fines del siglo XIX, la selva misionera era considerada por los intelectuales formadores de la idea de la nación argentina como un paisaje inhóspito y hostil. Constituía una frontera para la empresa civilizatoria, un espacio "virgen" que debía ser fecundado por la industria y una activa política de poblamiento. (Wilde, 2007, p. 88)*

In fact, despite the warnings of Quiroga and other thinkers with ecological awareness, the jungle of Misiones, which encompassed almost the entirety of the province in the mid-19th century, was reduced to just over 2,700,000 hectares in the 1950s.

For all these reasons, reading some of Quiroga's stories can lead to interdisciplinary projects with subjects such as Geography or Biology, analyzing the various species of fauna and flora from Misiones that appear in the narratives, as well as the different references the stories make to weather conditions that provide an idea of the specifics of the tropical climate, and then reflecting on the various threats the jungle ecosystem faces and discussing measures to preserve it for future generations.

Quiroga's narrative thus allows our students to approach a geographical, cultural, and historical reality that is completely different from their own, with customs and ways of life that are entirely foreign to them. This circumstance can serve as a starting point for reflecting on the Eurocentrism that cultural products we commonly consume often present and the lack of representation, both in literature and in film and other arts, of other ways of life, sometimes not so distant geographically and historically from our own as those presented in Quiroga's stories: those characteristic of rural environments, those of migrant populations from various countries, and those of our own grandparents...

In this way, the work of Horacio Quiroga offers our students the opportunity to learn about new life experiences and cultural realities, more specifically, those of the Hispano-American realm. In this sense, Quiroga's stories allow for the design of a project in which students can create a social and cultural overview of the region of Misiones in the early decades of the 20th century, observing both its peculiarities and the tensions it faced. From here, our students can expand their project by characterizing other Hispano-American regions, exploring their history, social and demographic characteristics, and their culture. For the execution of this project, Latin American students will be encouraged to act as guides for their peers, becoming leaders in research to help their native classmates become familiar with the cultural diversity of Hispano-America. In this regard, Núñez (2005) provides interesting guidelines for classroom work based on literary works of intercultural themes:

> *Desde el punto de vista didáctico, las pautas metodológicas deben procurar la integración de la cultura de los alumnos extranjeros en las dinámicas del aula. No bastan las visiones desde fuera, la mirada antropológica o etnográfica, hace falta la presencia real y constante, por medio del debate, el comentario, el diálogo, y eso puede hacerse muy bien a través de obras como las que hemos visto que pueden contribuir a hacer aflorar estereotipos y prejuicios para desmontarlos. La lectura de estos libros debe dar pie a que los alumnos se interroguen a sí mismos y unos a otros, que se pidan datos sobre las formas de vida en sus países, que pongan en cuestión los tópicos, que vean a las personas detrás de las marcas raciales o culturales. (Núñez, 2005, p. 618)*

5.5. Neither Civilization nor Barbarism: Intercultural Encounter and Reconciliation in the Work of Horacio Quiroga

Since the 19th century, much of Hispano-American literature has focused on the search for a political, social, cultural, and artistic identity that distances American peoples from their colonial past. Rejecting the oppression exercised by the metropolis and its civilizing ideal, literature claimed the cultural specificity of American peoples as well as a return to the natural state of man in a wild and liberating environment, embodied in the jungle. On the other hand, some intellectuals, like Sarmiento, argued that the new American states could only be built and progress from a "European" civilizing education that distanced them from the primitivism of indigenous cultures.

Quiroga's narrative breaks in many ways with the dichotomy between civilization and barbarism, adopting, from a deeply intellectual and critical perspective, a much more nuanced and less Manichean view of colonial and neocolonial issues.

Firstly, the jungle portrayed by Quiroga is by no means an idyllic *locus amoenus*; it is a dangerous place that, while fascinating, constantly threatens man by subjecting him to a continual struggle for survival. However, at the same time, the jungle is conceived as an ecological treasure that needs to be preserved and protected, to some extent, from neocolonial overexploitation. Ordóñez Díaz (2021) highlights in his study the complex vision of the jungle that Quiroga's narrative offers us:

La descripción de las heridas que el fuego y el machete abren en las entrañas del bosque (…), provocando un desarreglo duradero de los ecosistemas locales, pone al desnudo la violencia inherente al proceso colonizador. La llegada de los humanos a Misiones es presentada como un acontecimiento perturbador de efectos demoledores. Eso no significa, empero, que Quiroga fuese contrario a la causa del progreso (…). En la obra de Quiroga, por lo tanto, el impulso progresista que recurre a la ciencia y a la técnica como herramientas para modificar el entorno en función de las necesidades humanas coexiste con la conciencia plena de los daños ecológicos que la empresa modernizadora va dejando a su paso. (Ordóñez Díaz, 2021, p. 355)

On the other hand, in many of his animal-centered tales, Quiroga questions the animality of these beings in contrast to the supposed "humanity" of man. The Uruguayan writer presents individuals, whether animal or human, with behaviors, attitudes, and moral values not predefined by their condition. For example, *Historia de dos cachorros de coatí y dos cachorros de hombre*, included in *Cuentos de la selva*, showcases the humanity of the coatimundis, who are capable of sacrificing themselves to help children they have grown fond of; or *El paso del Yabebirí* narrates how the rays protect a man who has fought against fishing with dynamite. However, these same two stories use respective animals to embody evil: the snake and the tiger, portrayed as ruthless predators, governed solely by instinct.

Similarly, in the stories featuring human characters, Quiroga presents well-rounded characters with a certain psychological complexity, whose behavior is not determined by their ethnic origin or the social group to which they belong, but by their individual characteristics and moral sense. In this way, the Uruguayan writer avoids falling into generalizations and Manichaeism and encourages critical reflection from the reader, who ultimately judges his characters.

Thus, Quiroga's literature warns the reader against prejudice and stereotypes, urging them not to fall into overgeneralization but to judge each individual on their own merits—an essential attitude if we want to prevent racism and xenophobia and promote intercultural coexistence in our classrooms.

On the other hand, as previously noted, literature fosters empathetic imagination and allows our students to step into the shoes of men and women from cultures very different from their own. This development of empathetic capacity is a necessary first step toward intercultural coexistence. Hence, it is important for our students to access texts like those of Quiroga, which reflect diverse cultures. As Martínez-León and their collaborators (2017) state:

[La] identificación empática por parte del lector con personajes culturalmente diversos, que le conduce a reparar en la común condición humana, más allá de los rasgos o valores distintivos, a percibir el rico mundo interior de los otros y la complejidad de sus pensamientos y sentimientos. La activación de este tipo

de imaginación, a la que acompaña el desarrollo moral del lector, redunda en su capacidad de preocuparse por el bienestar y la fortuna de los personajes, participando de sus experiencias y emociones y comprendiendo sus particulares circunstancias. Se han destacado la narrativa y, en particular, las novelas como especialmente favorables a su desarrollo por el peso concedido a la descripción de situaciones y formas de vida que pueden ser distintas a la realidad del receptor, como también a los sentimientos y a la subjetividad de los personajes. Por tratarse, en otras palabras, de textos que presentan la realidad de los otros lejos de toda pretensión de objetividad u homogeneización. (Martínez-León et al., 2017, p. 160)

Moreover, although, as Bermúdez (2010) points out, many texts of Hispano-American literature are susceptible to an intercultural reading, Quiroga's work stands out for advocating reconciliation among the various ethnicities and cultures present in Misiones. In this way, the stories of the Uruguayan author value the "civilizing" contributions of Europeans in America, defend the incorporation of their scientific and technical advancements, and, above all, their political, legislative, cultural, and artistic imprint. However, this adoption of European science and culture is not presented as a reason to subjugate Americans or to undervalue their heritage, as Quiroga's narratives assert the extraordinary ecological heritage of America and also its significant human and cultural capital, advocating for its preservation and protection. This defense of what is American is expressed, according to Arévalo Viveros (2009), in Quiroga's tales' aim to "give voice" to the American jungle, since until then, literature had only reflected the reality of the white European man:

la valoración que debemos hacer de Quiroga debe apuntar a reconocer su enorme esfuerzo por inventar un lenguaje selvático de América, en una época en la que la mayoría de escritores aspiraban a simular de manera correcta las modas literarias venidas de Europa. (...) no es más que hacer de la selva casa de la palabra. Hacer de la narración, el lugar de los tigres y el agua. (Arévalo Viveros, 2009, p. 131)

Quiroga defends the jungle, the American landscape, and integrates it with the literary culture inherited from Europe, thereby breaking with the traditional dichotomy between civilization and barbarism, as Eduardo Lolo (2006) points out in reference to *Cuentos de la selva:*

uno de los temas básicos de los Cuentos de la selva para niños es una versión de la dicotomía antitética Civilización versus Barbarie muy distinta de las desarrolladas con anterioridad por Sarmiento y Martí: la alianza entre Civilización y Barbarie, expresada como una convivencia solidaria entre sus respectivos componentes positivos. La 'humanización' de los animales aporta la 'justificación' necesaria para el desarrollo de este tema; justificación que, analizada fuera de su contexto categórico, ha dado pie a que se interpretase a los animales como símbolos de los indios (…).

[Pero] los personajes humanos no alcanzan su entera dimensión sino cuando aparecen interactuando con los animales 'humanizados'; que es decir, al ser 'elevados' (según los postulados de la categoría literaria infantil) por el contacto directo con lo prístino de la vida. (Lolo, 2006, pp. 156-157)

Thus, culture in early 20th century America emerges in Quiroga's tales from multiplicity and unity, from an ideal conjunction between European values and culture and the substrate and specificity of what is American.

In this way, we can conclude that Quiroga, even in the early 20th century, conceived a model of intercultural coexistence for Misiones, in which two cultures would coexist harmoniously, enriching each other mutually. Therefore, the short stories of Horacio Quiroga are fully relevant in a society and educational system like the current ones, which must face the challenge of coexistence in a culturally diverse environment, and can serve as an impetus for our students to reflect on interculturality and the challenges it poses: overcoming ethnic and cultural prejudices and stereotypes, valuing both their own cultural elements and those of others, and fostering openness to dialogue and exchange... Similarly, many of the issues Quiroga reflected in his stories continue to be contentious today and can provoke debate in classrooms: power relations between different cultures and peoples, the conception of "progress" and "civilization" and their moral limits, the exploitation of natural resources, neocolonialism, etc.

6. CONCLUSION

Reading and studying literature are cultural and artistic experiences, but they can also provoke vital transformations in the reader's worldview, in how they perceive reality and recognize themselves and others. It is important to remember that literature is largely constructed through empathy when the reader shares the thoughts and feelings of characters entirely foreign to their reality, coming from diverse times and spaces. Reading is, therefore, recognizing what connects us to these strange beings, delving into that humanity that, despite our differences, we share.

For all these reasons, literature is an undeniably valuable resource for promoting intercultural education; even more so if, as in the case we propose, we address in the classroom the reading of texts that directly reflect the culture of part of our student body, as this fosters their social and cultural integration within the group.

Specifically, the stories of the Uruguayan Horacio Quiroga, in addition to possessing undeniable literary and artistic values, present themselves as a valuable educational resource for our students to develop their intercultural competence.

Ultimately, art and literature offer themselves as an end in itself, but also as a bridge to that culturally diverse society in which future generations will have to coexist.

REFERENCES

Abreu, C. (2014). Horacio Quiroga, a writer on the limits. *A contracorriente, 11* (2), 302-322.

Andújar Llosa, A., Sánchez Díaz, N., Pradillo Viñerta, Sh., & Sabín Galán, F. (2022). *Jóvenes y racismo. Estudio sobre las percepciones y actitudes racistas y xenófobas entre la población joven de España. Noviembre de 2022*. Centro Reina Sofía sobre Adolescencia y Juventud, Fundación Fad Juventud. DOI: 10.5281/zenodo.7268038

Arango, M. A. (1982). Sobre dos cuentos de Horacio Quiroga: Correlación en el tema de la muerte, el ambiente y la estructura narrativa en "A la deriva" y "El hombre muerto". *Thesaurus*, XXXVII(1), 153–161.

Arévalo Viveros, D. F. (2009). El cuento es la selva: lectura crítica-ecológica de *Los cuentos de la selva* de Horacio Quiroga. *452°F. Revista electrónica de teoría de la literatura y literatura comparada, 1*, 121-132

Baralo, M. (2003). Mestizaje e interculturalidad en la variación diatópica y su incidencia en español/ LE. En Pérez Gutiérrez, M. & Coloma Maestre, J. (eds.). *El español, lengua del mestizaje y la interculturalidad: actas del XIII Congreso Internacional de la Asociación para la Enseñanza del español como Lengua Extranjera (ASELE)*. ASELE, 152-164.

Barrera, T. (2019). *Historia de la Literatura Hispanoamericana III: siglo XX*. Cátedra.

Bermúdez, M. (2010). Literatura infantil y juvenil hispanoamericana y educación intercultural: perspectivas de lectura. In Herrera, J., Abril, J. M., & Perdomo, C. (Eds.), *Estudios sobre didácticas de las lenguas y sus literaturas* (pp. 126–138). Universidad de la Laguna.

Besalú, X. (2002). *Diversidad cultural y educación*. Síntesis.

Blanc, N. (13/08/2018). Cuentos de la selva: el centenario de un clásico salvaje. *La Nación*.

Cadová, R. (2007). La influencia de Edgard Allan Poe en Horacio Quiroga. *Études romanes de Brno, 37 (1)*, 149-157.

Carrión, J. (2010). Las estructuras y el viaje. (Hacia un nuevo hispanismo). In Ortega, J. (Ed.), *Nuevos hispanismos interdisciplinarios y trasatlánticos* (pp. 239–251). Iberoamericana/Vervuert. DOI: 10.31819/9783954871537-014

Carter-Thuillier, B., López-Pastor, V., Gallardo-Fuentes, F., & Carter-Beltran, J. (2019). Inmigración y deporte escolar: Un caso de estudiantes latinoamericanos en España. *Journal of Sport and Health Research*, 11(2), 41–52.

Castro, A., Patera, S. & Fernández, D. (2020). ¿Cómo aprenden las generaciones Z y Alpha desde la perspectiva docente? Implicaciones para desarrollar la competencia aprender a aprender. *Aula abierta, 49* (3), 279-285.

Ezquerro, M. (1996). Los temas y la escritura quiroguianos. En Baccino Ponce de León, N. & Lafforgue, J. (coords.). *Horacio Quiroga. Todos los cuentos.* ALLCA XX, 1379-1414.

Fernández, M., Valbuena, C., & Caro, R. (2017). *Evolución del racismo, la xenofobia y otras formas de intolerancia en España. Informe-encuesta 2017. Resumen.* Ministerio de Trabajo, Migraciones y Seguridad Social.

Fleming, L. (1991). Introducción. En Quiroga, H. *Cuentos.* Cátedra.

González Picado, J. (1985). Teoría y práctica del cuento en Hispanoamérica: El caso de Horacio Quiroga. *Filología y Lingüística*, XI(1), 5–14.

Instituto Nacional de Estadística (INE). (2022). *Población extranjera por comunidades y provincias, nacionalidad y sexo*. Recuperado de https://bit.ly/3BtJiEK

Lolo, E. (2006). *Cuentos de la selva* de Horacio Quiroga: La "otra" dicotomía civilizaciónVS barbarie. *Círculo: Revista de Cultura*, XXXV, 151–159.

Martín Rojo, L. (2004). Escuela y diversidad lingüística y cultural. In *Martín Rojo, L. (dir.). ¿Asimilar o integrar? Dilemas ante el multilingüismo en las aulas* (pp. 19–69). Ministerio de Educación y Ciencia.

Martínez-León, P., Ballester Roca, J., & Ibarra-Rius, N. (2017). Identidades fronterizas y pertenencias plurales: Diversidad e interculturalidad a través de la educación literaria. *Didáctica.Lengua y Literatura*, 29, 155–172. DOI: 10.5209/DIDA.57135

McCrindle, M. (2014). *The ABC of XYZ: understanding the global generations.* UNSW Press.

Montenegro, P. G. (2006). Símbolos inadvertidos en *El almohadón de plumas* de Horacio Quiroga. *El cuento en red: estudios sobre la ficción breve, 14* (2).

Montes Pedrajo, C. (2023). *Actitudes lingüísticas de estudiantes cántabros de bachillerato hacia las variedades americanas del español: diagnóstico e intervención* [TFG inédito]. Universidad de Cantabria.

Núñez, M. ª P. (2005). Literatura juvenil y educación intercultural: ¿estética vs. axiología? En Herrera, F. et alii (eds.): *Inmigración, interculturalidad y convivencia, IV*. Instituto de Estudios Ceutíes / Ciudad Autónoma de Ceuta, 609-618.

Observatorio Español del Racismo y la Xenofobia (OBERAXE). (2022). *Informe sobre la integración de los estudiantes extranjeros en el sistema educativo español. Resumen*. Ministerio de Inclusión, Seguridad Social y Migraciones.

Ordóñez Díaz, L. (2021). *Ríos que cantan, árboles que lloran: Imágenes de la selva en la narrativa hispanoamericana*. Universidad del Rosario. Universidad de los. *Andes (Salta)*.

Pan, Z. (2021). *Cuentos de la selva* of Horacio Quiroga: A regionalism adapted for children readers. *International Journal of Humanities and Social Science*, 11(7), 96–104.

Pedraza Jiménez, F. B. (2002). *Manual de historia de la literatura hispanoamericana*. Cénlit.

Quiroga, H. (1917). *Cuentos de amor, de locura y de muerte*. [(2019). Alma].

Quiroga, H. (1918). *Cuentos de la selva*. [(2011). Anaya].

Quiroga, H. (2005). *Cuentos* (de Leonor, F., Ed.). Cátedra.

Quiroga, H. (2008). *Anaconda y otros cuentos* (de Consuelo, L., & Bueno, J. A., Eds.). Edaf.

Souto Alabarce, A. (2003). Introducción. In Quiroga, H. (Ed.), *Más cuentos, Cuentos de la selva*. Porrúa. VII-XV.

Suárez Hernán, C. & Bermúdez Martínez, M. (2024). La presencia de la literatura hispanoamericana en Educación Secundaria Obligatoria y Bachillerato en España. Concepciones del profesorado. *América sin Nombre, 31,* 147-164. DOI: 10.14198/AMESN.24540

Wilde, G. (2007). De la depredación a la conservación: Génesis y evolución del discurso hegemónico sobre la selva misionera y sus habitantes. *Ambiente & Sociedade*, 10(1), 87–106. DOI: 10.1590/S1414-753X2007000100006

Yraola, A. (2014). Actitudes lingüísticas en España. In Chiquito, A. B., & Quesada Pacheco, M. Á. (Eds.), *Actitudes lingüísticas de los hispanohablantes hacia el idioma español y sus variantes. Bergen Language and Linguistic Studies (BeLLS), 5, 551-636*. Disponible., DOI: 10.15845/bells.v5i0.685

KEY TERMS AND DEFINITIONS

Civilization or Barbarism: A dichotomy around which American intellectuals addressed colonial and neocolonial issues since the 19th century. "Civilization" represents European, urban, Western culture, while "barbarism" is associated with the genuinely American: the jungle, rural life, indigenous peoples, or gauchos. However, Horacio Quiroga's narrative breaks with this antithesis and advocates for intercultural reconciliation.

Cuentos de la Selva: A volume of children's stories published by Horacio Quiroga in 1918. It comprises eight tales featuring animals from the jungle of Misiones. Despite their harshness, Quiroga's stories uphold high moral values and advocate for the protection of nature.

Horacio Quiroga: A Uruguayan writer from the early 20th century, considered one of the great masters of Hispano-American short stories. His narratives, due to their thematic, formal, and linguistic characteristics, allow students to simultaneously develop their literary and intercultural competencies.

Intercultural Competence: A set of knowledge, skills, and attitudes that enables our students to live harmoniously in a culturally diverse society, valuing cultural differences as opportunities for personal enrichment and developing proactive intercultural dialogue.

Intercultural Educational Model: An educational model that promotes the appreciation of cultural diversity, as well as dialogue among individuals and culturally diverse groups, to foster mutual enrichment.

Misiones: A jungle region located at the border between Argentina, Brazil, and Paraguay, along the upper Paraná River. Quiroga lived there for years, making the jungle an omnipresent motif in his narrative.

Rioplatense Spanish: A variety of Spanish spoken in Argentina and Uruguay. It is characterized by phenomena such as *voseo* (in some areas) and the *rehilamiento* of the "*ll*". This is the variety employed by Horacio Quiroga in his stories.

Chapter 14
Pre-Service Teachers' Conceptions on the Expressive and Writing Abilities of Current Conversational AI

Maria Navío-Inglés
https://orcid.org/0009-0006-9127-4876
University of Castilla-La Mancha, Spain

Sergio Tirado-Olivares
https://orcid.org/0000-0002-8557-5115
University of Castilla-La Mancha, Spain

Paula O'Connor-Jiménez
https://orcid.org/0009-0002-4817-2752
University of Castilla-La Mancha, Spain

ABSTRACT

In the post-truth era, where fake news abounds, advancements in artificial intelligence (AI), particularly conversational AI, are producing tools capable of generating increasingly human-like text. This mixed-methods study explores future teachers' perceptions about the ability of conversational AI to write argumentative texts. To this aim, a group of 87 pre-service teachers compared two anonymised texts: an AI-generated one and a text written by a fellow student. Results indicate that most participants believed the AI-generated text would outperform the human-written one in an exam, though opinions were divided on which text conveyed more emotions.

DOI: 10.4018/979-8-3693-7185-5.ch014

Copyright © 2025, IGI Global Scientific Publishing. Copying or distributing in print or electronic forms without written permission of IGI Global is prohibited.

Participants generally associated better writing and information richness with AI, while they linked subjectivity and emotivity to the human author. However, not all correctly identified the author of each text, highlighting the need for further preparation of future teachers on current AI abilities and limitations.

1. INTRODUCTION

Technology has become an integral part of our daily lives (Ghomi & Redecker, 2019). Its relevance is such that teachers are now required to develop their teacher digital competence to effectively integrate technology into their tasks and to help students acquire the digital competence that they will also need (Redecker, 2017). Among all existing technologies, artificial intelligence (hereinafter, AI) has gained attention during the last years (Pelletier et al., 2024), particularly with the recent advances in generative AI. This type of AI is characterized by generating new content based on the data it possesses (Lim et al., 2023; Pavlik, 2023) and has contributed to some chatbots being able to produce human-like text (Dale, 2021).

At the same time, we live in a world where information is instantly accessible to anyone and spreads rapidly. However, not all the sources we find are reliable, as post-truths and fake news are increasingly frequent realities that we have to deal with (Chinn et al, 2021; Moreno-Vera, 2023). In this context, students, regardless of their educational stage, need to strengthen their critical thinking skills to question, analyze and evaluate the impact of the information they find (Tirado-Olivares & Cózar-Gutiérrez, 2024).

As teachers are responsible for instructing students, they must help them develop the skills they need in the 21st century world. Consequently, both teachers and students have to be cognizant of the AI-based tools that can be used to generate loads of information instantly and that are available to everyone (Kasneci et al., 2023; Zhang & Tur, 2023). For this reason, the present study centers on exploring the ability of the generative AI tool ChatGPT to generate an argumentative text through knowledge and opinion that pre-service teachers have about its writing potential. To this aim, two groups of future teachers from different academic years followed a set of questions to analyze a text created by AI and another written by a student without knowing their authors of either. Thus, the objectives of the study (O.S.) are as follows:

O.S.1. To ascertain whether the students can identify a text generated by AI and what features they attribute to its expression.

O.S.2. To determine if there are any discernible differences based on the academic year of the students.

1.1. The Emergence of Technologies and the Digital Competence in the 21st Century

The technological advances that have occurred over recent decades have brought about a transformation in the functioning of society. These changes have continuously affected key sectors such as the economy, healthcare, and culture (Ait Baha et al., 2024; Kasneci et al., 2023; Lee et al., 2022). Digital technology has become an essential component of the way we interact, work, and learn, so now it is integral to daily activities ranging from communication with others to obtaining information on almost any topic (Carrión-Martínez et al., 2020). This requires the possession of skills that enable individuals to participate fully and appropriately in today's digital society (van Laar et al., 2017).

The educational field is not immune to this transformation, and the integration of technology into both classrooms and other activities related to teaching is now widespread. Evidence of this is seen in the development of initiatives aimed at digitizing schools in various countries, such as one-to-one laptop programs. This project, which has expanded across many regions since the 1990s, is based on providing each student with a laptop for use in the classroom and, in some cases, at home (Zheng et al., 2016). Moreover, various school curricula already include the acquisition of basic digital literacy that enables students, from an early age, to become proficient in the use of digital technologies (Organic Law 3/2020; OECD, 2016). In accordance with this, even universities are developing and implementing projects aimed at promoting teachers' digital competence. One such example is the TECSOCO project, among others, which focuses on developing and disseminating strategies and technological tools to enhance this competence in subjects related to Social Science teaching (Morote Seguido & Colomer Rubio, 2024). In recent years, educational institutions have therefore advocated for the implementation of new pedagogical methodologies that integrate digital tools to enhance teaching and for the promotion of technological proficiency, ultimately aiming to prepare citizens capable of facing the challenges of the 21st century.

In line with this, various reports and international institutions assign a prominent role to technology in both current and future educational landscapes (e.g., European Commission, 2020; OECD, 2020; Pelletier et al., 2023; Pelletier et al., 2024). Furthermore, they highlight not only the relevance of technology as an educational tool, but also its potential to transform the teaching-learning process, fostering an increasingly active, inclusive, and personalized approach. In this regard, the latest *EDUCAUSE Horizon Report* (Pelletier et al., 2024) identifies generative AI as a key technology. This follows the proliferation of tools based on it, such as ChatGPT, Copilot or Gemini. The report also emphasizes that the changes this type of AI could bring about in education still require a level of preparation and understanding that many people have yet to achieve.

One of the pillars of this transformation process is the development of digital competence in both students and educators. The interest in this has led to the establishment of various models and frameworks for developing this competence. Among them, the *Digital Competence Framework for Citizens* (DigComp) (Vuorikari et al., 2016), initially published by the European Commission in 2013, stands out. It defines digital competence as the "confident, critical and responsible use of, and engagement with, digital technologies for learning, at work, and for participation in society" (European Commission, 2019, p.10). To equip individuals with these abilities the role of teachers is crucial, as they are responsible for guiding students in developing the competences and skills they will need. Consequently, besides being digitally competent themselves, they must be prepared to effectively integrate current digital technology into their teaching tasks. For this reason, specific frameworks and guidelines have also been developed for this group, such as the *European Framework for the Digital Competence of Educators* (DigCompEdu) (Redecker, 2017) and the *Technological Pedagogical Content Knowledge* (TPACK) model, developed by Mishra and Koehler (2006). The first one is a guide created by the European Commission to assist educators in integrating digital technology into their teaching practice and defines six key areas of digital competence, ranging from the use of tools for teaching to facilitating the development of students' own digital skills (Redecker, 2017). On the other hand, the TPACK model offers a framework that establishes that effective teaching in technological environments requires knowing how to teach (pedagogical knowledge), what to teach (content knowledge), and how to use technology (technological knowledge) (Koehler et al., 2013; Mishra & Koehler, 2006).

Teacher digital competence is, therefore, a fundamental element for the success of technology-based educational initiatives (Martín Paragas et al., 2023). Teachers not only need to be familiar with digital tools but also need to be able to apply them effectively in the classroom. However, the low level of digital competence among in-service and pre-service teachers reveals a gap between the ideal digital competence and the actual digital competence level (e.g., Claro et al., 2024; Dias-Trindade et al., 2021; Tzafilkou et al., 2023), which needs to be addressed through continuous and specific training. This may have an impact on the education of students, who are currently exposed to digital environments and devices from an early age and are often not fully aware of the risks associated with the use of digital technologies (Chaudron et al., 2018). Thus, as derived from what the authors mentioned beforehand claim, mastery of technologies that are readily available and ongoing training to achieve it should not be limited to acquiring technical skills. It is also essential to promote a deep understanding of their functioning and possibilities, as well as a critical and reflective attitude towards them, both among teachers and students.

1.2. Conversational AI and its Role in the Educational Landscape

One of the technologies that is having a significant impact in today's society and is expected to have a great influence in the future, as pointed out by internationally recognized reports such as the *EDUCAUSE Horizon Report* (Pelletier et al., 2023; Pelletier et al., 2024), is AI. The term *artificial intelligence* was coined by John McCarthy in 1956 under the belief that machines could simulate human intelligence. Since then, numerous definitions have been proposed for this concept, evolving alongside this kind of technology itself. Rouhiainen (2018) highlights the difficulty of defining this concept and suggests that AI refers to technological systems capable of learning from data, making decisions, and performing tasks reproducing human intelligence. Similarly, Russell and Norvig (2021) conceive AI, from a broad perspective, as any technique that enables a computer to replicate and even surpass the human decision-making process to accomplish complex tasks autonomously.

Since the 1950s and the emergence of what are considered the first AI-based systems, AI has evolved considerably. To achieve this, the integration of machine learning and deep learning techniques has been crucial. Machine learning is a subfield of artificial intelligence that enables systems to automatically learn relationships and patterns and improve from experience (Bishop, 2006). In other words, it is based on the use of algorithms to analyze data, learn from it, and make decisions based on the information available. Deep learning, on the other hand, is a subfield of machine learning that involves using neural networks with many layers, which simulate the structure of the human brain, allowing for the advanced learning of data patterns and the execution of more complex operations (Janiesch et al., 2021). These techniques have enabled machines to learn from large amounts of data, enhance their ability to recognize patterns, and perform complex tasks. Following this evolution, AI now underlies a variety of technological systems and devices, ranging from facial recognition systems to virtual assistants (Akgun & Greenhow, 2022; Rapp et a., 2021). The level of sophistication it has achieved has made it possible to integrate AI into key sectors such as healthcare, business, and even education (Ait Baha et al., 2024; Kasneci et al., 2023; Lee et al., 2022; Zhang & Aslan, 2021).

One of the current applications of AI is conversational AI and chatbots. These are characterized by their ability to engage in a conversation with the user, simulating the interaction that could occur between two people (King, 2023; Shawar & Atwell, 2007). This technology is based on Natural Language Processing (NLP), which allows machines to learn, understand, and produce natural language (Hirschberg &

Manning, 2015). NLP uses computational techniques to analyze large amounts of textual or voice data. This has enabled the emergence of what are known as language models (LMs) and large language models (LLMs), depending on the amount of data they are trained on. They are designed to predict the probability of a word sequence in a language and, therefore, to understand, analyze, and generate text (Chang et al., 2024). In the latter group, one model has gained prominence recently due to being the underlying language model for tools like ChatGPT: the Generative Pretrained Transformer (GPT). This model is not limited to working with the information it is provided with but is characterized by its ability to produce new content and generate text that closely resembles what a human would produce too (Dale, 2021; Lund & Wang, 2023; Pavlik, 2023). To achieve this, it undergoes a supervised training phase followed by an autonomous training phase (Budzianowski & Vulić, 2019; Radford et al., 2018). Such is the capacity of today's conversational AI tools that studies that focus on analyzing the features of AI writing and even people's ability to distinguish between text produced by a human being and text produced by a chatbot have recently begun to emerge (e.g. Guo et al., 2023; Mitrović et al., 2023).

The progress described and the capability of conversational AI to process and generate language similarly to how a human would have led to its introduction in many fields, including education. The systematic review by Okonkwo and Ade-Ibijola (2021) shows that chatbots have already been used, for example, for assessment, research and development, administrative tasks, and especially as support for teaching and learning. However, recent LLMs and AI-based tools, such as the GPT model and ChatGPT respectively, have triggered new studies that explore their possibilities (e.g., Lo, 2023; Zhang & Tur, 2023). In line with this and specifically in the field of Social Sciences teaching, de Lázaro (2024) explores not only the potential of ChatGPT in assisting the development of learning situations to teach Geography, but also the student's perceptions of its possibilities and limitations. This increase in studies on this topic is due to the fact that further research is needed regarding the more advanced capabilities current generative AI has (Cong-Lem et al., 2024), which implies new usage possibilities, challenges, and limitations that must be explored. In this regard, some authors also acknowledge that teachers need to improve their digital and AI-related skills (Kasneci et al., 2023; Zhang & Tur, 2023).

1.3. Communication and Critical Thinking in the Technological and Post-Truth Era

The rise of technological tools and resources such as the Internet and mobile phones has profoundly transformed the way we communicate with other people and interact (Goodman-Deane et al., 2016; Rogers, 1986). The advent of digital platforms allows for the rapid spread of information on a global scale, making it accessible to millions of users in an instant. This mass dissemination of information, while beneficial in terms of connectivity and accessibility, also brings about challenges regarding the accuracy and reliability of the content being shared.

One of the main challenges of mass information dissemination is the reliability of the content that is published and accessed. As pointed out by Moreno-Vera (2023), post-truth is one of the main problems to face in today's society. The Oxford English Dictionary (n.d.) defines post-truth as "relating to or denoting circumstances in which objective facts are less influential in shaping political debate or public opinion than appeals to emotion and personal belief". The prefix "post", in this case, would allude to the idea of truth being eclipsed (McIntyre, 2018) and would refer to a context in which misinformation is widespread in the media (Barzilai & Chinn, 2020; Chinn et al., 2021). In this scenario, what is commonly known as *fake news* proliferates. Fake news is information that has been fabricated intentionally to mislead, mimicking the media content's form, but not its usual organization process that intends to guarantee accuracy and credibility (Buckingham, 2019; Lazer et al., 2018). However, this term does not refer to pieces of information that contain mistakes that are corrected and come from organizations that strive to ensure reporter's accuracy (Watson, 2018).

In the current reality described, critical thinking becomes an increasingly important skill that students must develop (Chinn et al., 2021; Kozyreva et al., 2023). The ability to analyze, evaluate, and synthesize information reflectively is crucial for navigating such a complex and ever-changing informational environment, especially considering that fake news continually affects issues relevant to society and daily life, such as politics and health among others (Lazer et al., 2018). As an intellectual process that involves other processes like conceptualizing, analyzing and evaluating information (Paul, 1992), critical thinking allows individuals to question the truthfulness of that information, identify biases, and assess the reliability of sources; thus, they do not just consume information but actively engage with it by analyzing and questioning it. The need for individuals to develop critical thinking skills is reflected in current educational curricula, which often advocate for its integration (e.g., Organic Law 3/2020), while also emphasizing the importance of media literacy. In line with this and in order to promote the abilities mentioned above, international initiatives such as Stanford University's 'Civic Online Reasoning' project underscore the importance of fostering lateral reading among students. This approach encourages students to cross-reference information from one source with other sources to verify its truthfulness. As such, lateral reading is proposed as an alternative to the traditional vertical reading, which relies on a single source. Initiatives like this are based on simulating activities that are actually carried out by specialists in the field of social sciences and are linked to analyzing, contrasting and interpreting sources.

As seen before, the relevance of developing critical and technological skills intensifies due to the capabilities of generative AI tools. We find ourselves in front of the possibility of anyone having at their disposal tools that can replicate human expression and that have millions of data and examples of language usage to do so (Dale, 2021; Pavlik, 2023) in order to produce and disseminate information, which might not be always accurate (Labadze et al., 2023), at high speed. With generative AI-mediated systems like ChatGPT, we face a complex technology that is being integrated with positive results into the current educational reality. At the same time, it is a tool capable of easily producing increasingly human-like discourse in an era marked by the mass dissemination of information that is not necessarily true, and interactions mediated by digital environments. Therefore, it is essential to evaluate the knowledge that future teachers possess about this technology, considering the projected role it will play in the future of education and the impact it may have on current and future society and that social sciences teachers cannot be oblivious to it. Consequently, the present mixed study aims to understand the perceptions of prospective teachers of two different academic years regarding the linguistic capabilities and limitations of the conversational AI ChatGPT. To do this, pre-service teachers were asked to analyze and compare two texts that had been previously anonymized: one generated by another student pursuing a degree in teaching, and another generated by ChatGPT (version 3.5).

2. MATERIALS AND METHODS

2.1 Design

Considering this evolving educational context, the current chapter seeks to understand the perceptions of prospective teachers regarding the linguistic capabilities and limitations of the conversational AI ChatGPT. To this aim, a mixed study was conducted through the analysis and comparison of two texts: one generated by a student pursuing a degree in teaching, and another generated by ChatGPT (version 3.5).

2.2 Participants

The perceptions of 87 students from two different academic levels are analyzed: 46 students from the first year of the Early Childhood Education Teacher Degree (hereinafter EC1) who were enrolled in the 'Knowledge of the Social and Cultural Environment' subject, and another 41 students from the third year of the same degree

(EC3) enrolled in the 'Didactics of the Natural, Social and Cultural Environment' subject.

In the context of this study, it is crucial to highlight the distinct educational backgrounds of the two research groups, EC1 and EC3. On the one hand, EC1 students were in their first month at university and, therefore, had practically no previous contact with aspects related to the didactics of this area. This lack of experience offers a unique perspective, as these students are essentially blank slates on the use of AI in the teaching-learning process, including its potentials and risks. On the other hand, students involve in the EC3 group already had a training background in the use of educational methodologies and tools and had already had their first contact in a real educational center. This prior exposure provides them with different lens through which they view and interact with the teaching-learning process.

The differences that could arise due to this unequal educational background justify that, when evaluating the results, these are analyzed in the two subgroups in isolation. Each subgroup, EC1 and EC3, should be evaluated independently to account for their unique experiences and learning trajectories. This is crucial in assessing their view of generative AI. But, at the same time, allows a comparative analysis between the two groups that can yield valuable insights into the impact of prior exposure to educational methodologies on their perceptions. This dichotomy in educational background underscores the importance of considering these factors when defining the objectives of this research. It not only informs the research design but also shapes the interpretation of the findings, ultimately contributing to a more nuanced understanding of the integration of IA in the teaching-learning process.

2.3 Procedure

This research necessitated the availability of pre-existing texts produced by both an AI and a student-in-training. Consequently, the foundation stone of this study is located in the work carried out by Tirado-Olivares et al. (2023). In that research, authors conducted a parallel analysis of two texts, each cantered around the significant societal repercussions of World War II. The texts utilized in that study had previously undergone evaluation by university students with similar characteristics of social context, educational interest, gender, and age. However, a key distinction in our research approach lies in the comparative analysis between two distinct academic courses, a factor not considered in the previous study. Given these circumstances, the texts from the Tirado-Olivares et al. (2023) study were considered ideal for fulfilling the objectives of our research.

The same procedure was effectuated for the analysis of the text in both research group. The text analysis process was conducted within a single class session, lasting approximately an hour. The two texts were introduced to the students without any indication of their respective authorship. This approach ensured that the students were unaware that one of the texts could potentially have been generated by an AI. This method of blind review allowed for an unbiased analysis of the content, focusing exclusively on the information included on the texts rather than preconceived notions about the source. This strategy aligns with our research objective of understanding the integration of AI-generated content in the teaching-learning process, particularly in terms of content creation and critical analysis.

Once the students had read both texts and were able to ask those questions they considered pertinent, they were asked four open-ended questions that would allow the collection of information for the present study. These questions were derived from Tirado-Olivares et al. (2023).

- Which text do you believe deserves a higher score in an exam? Why?
- Which one is able to convey more emotions and feelings? Why?
- Do you think a machine or AI can make a comment such as these ones?
- If I told you that one of these two texts was written by an AI, would you believe it? Which one do you think it would be?

These questions were carefully chosen to elicit students' perceptions of the texts, their emotional resonance, and their beliefs about the capabilities of AI in generating such content. The responses to these questions provide valuable insights about students' understanding and acceptance of AI-generated content in an educational context. As can be seen, both the third and last questions focus on whether participants are able to identify the ability of AI to generate critical texts.

2.4 Data Analysis

For data analysis, student responses were exported and coded into a digital database for further analysis using two different software programs.

For the quantitative data, the answers to each question were coded. First, if the students thought that the text written by humans deserved a higher score, it was coded as (1), if the text written by AI deserved a higher score, as (2), and when each text deserved the same score, as (0). The same coding was applied to which text was able to convey more emotions and feelings. For the third question, the coding was (1) whether they felt the AI was capable of writing similar text or (2) otherwise. For the last question, the coding was (1) whether the participants thought the first text was written by an AI, (2) whether they thought the second was, or (0) whether they still thought neither was. Once the questions were coded, it was possible to know the percentage of students' responses on each of these aspects and also to compare the two groups. For this second aspect, the inferential chi-square test was performed. To carry out all this descriptive and inferential analysis, the SPSS software (vs.29) was used.

Subsequently, the Atlas.ti software was employed for qualitative analysis. The latest version of this software (vs.24) was used, ensuring access to the most up-to-date features and capabilities. Thanks to this software, an open coding of the concepts was done, enabling the identification of the concepts most frequently highlighted by the students in their responses to the posed questions. This approach allowed us for a nuanced exploration of the students' perspectives, providing valuable insights into their understanding and interpretation of the texts. Two of the study investigators conducted the content analysis of the data, and any doubts or disagreements were resolved through discussion. In the results section, codes have been underlined for better identification.

3. RESULTS

This section has been divided into the four questions asked of students. When reporting the data, first the response percentages for each question are shown, followed by the inferential analysis between groups and reporting some of the students' statements that justify the reason for their response.

3.1 Question 1 - Text That Deserves a Better Grade

The vast majority of respondents felt that AI-generated text should score higher on an exam. In fact, on this question, there are no major differences among students depending on the university course they are in (X^2 (2) = 2.02; p = .365). Only six students from group EC1 and eight students from group EC3 considered that the student-generated text deserved a higher grade. This number of students corresponds to 13 and 20 percent of the total group respectively. All these data are visible in Table 1.

Table 1. Crosstabs table of consideration of a higher score on a test depending on the group to which the respondent belongs

			Question 1			
			Both Equal	**Student text**	**AI text**	**Total**
Group	EC1	Count	0	6	40	46
		% within group	0.0%	13.0%	87.0%	100%
	EC3	Count	1	8	31	40
		% within group	2.5%	20.0%	77.5%	100%
Total		Count	1	14	71	86
		% within group	1.2%	16.3%	82.6%	100%

Source: own elaboration.

When it comes to justifying why they considered that one text deserves a higher grade than another, students relied on aspects linked to the use of language, the better structure presented by the text and content depth found in them. Beginning with this last point, the students highlighted the ability to point out with greater wealth of information: "The second text [AI's text] because it has not only been based on basic data, but has inquired and has made a greater search for information, also has related more political and historical aspects of the two world wars" (S.3); greater precision when identifying the causes and consequences of the war: "The second text [AI's text] deserves better marks, since it shows the causes and consequences occurred, in a more descriptive way, touching all economic and political points, as the social". (S.6); and the use of keywords related to the subject treated: "The text that would deserve a higher mark for me would be text 2 [AI's text]. Since I think it explains the events better as well as its extension is longer, giving key words or important events and names" (S.51). In this sense, they emphasized how all this provides clarity in the presentation of content: "I think that the second text [AI's text] deserves a higher score because it is an elaborated commentary, with interesting and justified data in which events such as the Holocaust are mentioned. In addition, it relates the data to reality, thus showing its impact on the present day, for example by naming the people who lived through the event and are still living today" (S.23).

If we focus on the writing, they highlighted the greater cohesion of ideas and coherence in its organization: "The second text [AI's text] is better structured and argued, as it provides a series of consequences on the historical fact ordered chronologically and relates it to the impact on the current world without forgetting the past context". (S. 70); and a more concise expression and without repetition of ideas: "In my opinion the second text [AI's text] deserves a better grade because it seems better structured and written. In addition, the use of vocabulary is not so repetitive, there is vocabulary more in line with the topic and all the ideas are argued more. Along with this, the student demonstrates greater knowledge of the topic in relation to arguments and ideas used." (S.47). Finally, they discussed how these data were presented with the correct use of spelling rules: "In my opinion, the second text [AI's text] would deserve a better grade, it is clear that he knows what he is talking about and has been interested in the subject. It gives more elaborated data, the presentation is better and it does not have spelling mistakes like the first one [person's text]" (S.80).

3.2 Question 2 - Ability to Convey Emotions

As can be seen in Table 2, contrary to what happened in the previous case, there is a greater diversity of opinions. While in group EC1 the students considered that the text generated by the student was capable of transmitting more emotions, the participants of group EC3 considered in a higher percentage that it was the text generated by IA. However, the chi-square test does not show an interaction between this variable and group membership (X^2 (2) = 3.15; p = .207). On the other hand, although the percentage is not large, in both groups there are students who considered that both texts had a similar capacity to transmit those emotions or feelings.

Table 2. Crosstabs table of consideration of which text is able to convey more emotions depending on the group to which the students' response belongs

			Question 2			
			Both Equal	**Student text**	**AI text**	**Total**
Group	EC1	Count	2	26	18	46
		% within group	4.3%	56.5%	39.1%	100%
	EC3	Count	3	15	22	40
		% within group	7.5%	37.5%	55.0%	100%
Total		Count	5	41	40	86
		% within group	5.8%	47.7%	46.5%	100%

Source: own elaboration.

When justifying their choice, they focused on aspects linked to the critical reflection and personal opinion present in them: "The second text [AI's text], since it focuses more on ethical and critical thinking and makes more allusions to a personal opinion about the pain and emotions that it produced and can produce today."(S.9); and to the ability to empathize with the people who lived at that time: "The first one [person's text], since it is based on the causes of the war at a more social level, from the point of view of humanity, also empathizes from a point of view with the society of that time and with what happened to them in the war" (S.3), "The second commentary [AI's text] because it expresses his own point of vier and in an empathic way" (S.11). In fact, this ability to put oneself in the shoes of those people is one of the aspects most highlighted by the future teachers: "The first text [person's text] transmits more emotions because, in my opinion, you can see how you have described the situation with expressions that put you in the shoes of the people who lived through it and taking into account the personal problems, mental problems, etc., that this experience has caused them. That this experience may have caused them" (S.62). It enables a work in values through what is narrated in the text: "The second one [AI's text] because in the conclusion it shows the importance of emotions such as tolerance, respect and the feeling that the horror of this war caused" (S.14) or "The second text [AI's text] provides more data on emotions by naming human nature, talks about tolerance, dialogue and its importance to generate a fairer world" (S.20).

In addition, students related the evocation of all these emotions to greater subjectivity when writing the text: "I felt that the first text [person's text] was more subjective and sentimental than the second [AI's text] in the way it describes the events that took place" (S.49). In short, they intertwined it with the personal vision of the author of the text at the time of writing it: "The first one [person's text], since it speaks in a more informal way and without so much data. It is more personal and talks more about humans and their suffering" (S.38) or "In my opinion, the 1st text [person's text] is able to convey more emotions and/or feelings, because it expresses his opinion about the 2nd World War and also expresses his concern, especially about the deaths it entailed" (S.41).

3.3 Question 3 - Consider the AI Capable of Generating a Similar Text

Students in the first year, i.e., group EC1, were more reluctant to consider the AI capable of generating a text like the ones they had evaluated than students in group EC3. In fact, while there is a certain parity between the students who considered that the AI could do it and those who did not, in the case of the third-year students (EC3) two thirds considered that it is capable of doing it, while one third did not. In fact, it is in this question where the chi-square test reports the interaction closest to significance (X^2 (2) = 2.864; p = .091) (see Table 3).

Table 3. Crosstabs table of students' perception about the AI capacity to write a similar text according to the students' response in each group

			Question 3		Total
			Yes	**No**	
Group	EC1	Count	22	24	46
		% within group	47.8%	52.2%	100%
	EC3	Count	27	14	41
		% within group	65.9%	34.1%	100%
Total		Count	49	38	87
		% within group	56.3%	43.7%	100%

Source: own elaboration.

When arguing for this ability or inability of the AI to write a similar text, the students emphasized that it could write a text with a similar wealth of information: "Yes, because it will be based on the data you have on the computer or on the server with which you work, so even if it is an objective commentary, you will be able to contextualize" (S.2). However, they also highlighted its lack of personal assessment communication: "Yes, but focusing more on what happened in the war, leaving the personal part that both texts have when explaining the social causes, or to give a little opinion" (S.3) or "It could make a similar one but at the end it would be a more complete text, since in part the purpose of the AI is to obtain more elaborated documents. And besides, it would be a text in which only historical data would appear and the part of personal opinion would not be present" (S.26). In this sense, they alluded to the inability of empathy or to put themselves in the shoes of those people that they would value as something positive when choosing someone in question 2: "An AI system could give the same information about the event or even more, but it would lack the feelings that a person has when doing the job. In addition, perhaps, it doesn't take into account the emotions that this caused in people at the time" (S.44) or "From my point of view I think not, because it would tell the facts in an objective way, simply narrating them without expressing that sentimental or empathetic point, since they do not have feelings. It would be like reading it in a history textbook. Without emotions or feelings" (S.76).

However, it is necessary to highlight the better structuring and lexical richness that they attributed to the AI in comparison with texts such as those evaluated: "It could make an even better one (with more vocabulary and better structured) but it would not transmit the same thing that these two handmade ones transmit, nor in the same way" (S.46) or "No, I think it could make a great text, with an impeccable vocabulary, but it could not have the ethics and feeling, to capture the devastating consequences on people sentimentally, as any of some of these authors" (S.83). This shows the great potential that future teachers attribute to these generative AI tools due, among others, to technological progress: "I think so, since it has evolved over the years about technology, and today there is much more availability of the use of technologies than a few years ago and thanks to AI you can elaborate texts like these" (S.42). In fact, due to this their improvement over time, students consider that it will allow AI to even express an opinion. Solving this current deficiency highlighted by participants: "I think so, that with time this type of intelligence is capable of writing a text like this, because there is a lot of information on the Internet from where they can get it as well as opinions of people who lived through this situation" (S.77).

3.3 Question 4 - Identification of AI-Generated Text

Finally, through the last question, the students were asked to choose one of the two texts as one generated by AI. Although almost all students correctly identified the text generated by ChatGPT, there were also students who considered the text generated by the future teacher as the one made by a generative AI. As can be seen in Table 4, this situation was independent of the group to which the student belonged (X^2 (2) = 3.46; p = .177), although it is true that a higher percentage of the EC1 group erred in identifying the text than the EC3 group. Finally, it should be noted that, despite the fact that there were only three students from group EC1 and one from group EC3, even if the students were asked to choose one of the texts, these four students still considered it unfeasible that either of the two texts could have been generated by the AI.

Table 4. Crosstabs table of consideration of correctly identification or not of the AI-text by students of both groups

			Question 4			
			Neither	**Incorrect**	**Correct**	**Total**
Group	EC1	Count	3	9	34	46
		% within group	6.5%	19.6%	73.9%	100%
	EC3	Count	1	3	35	39
		% within group	2.6%	7.7%	89.7%	100%
Total		Count	4	12	69	85
		% within group	4.7%	14.1%	81.2%	100%

Source: own elaboration.

When it comes to justifying why they thought that text was the one written by AI, again, the participants argued their choice on aspects such as better structure: "I think the second one [AI's text] offers a greater variety of real data and a more structured organization of ideas than the first one [person's text]. Therefore, I think the second commentary would be the one made with AI" (S.24). Similarly, they highlighted the more formal language and the greater cohesion of the text: "The second one [AI's text], because it is the one that is written in a more formal and cohesive way" (S.18). In addition they pointed out wider vocabulary: "Yes, the second text [AI's text] because the vocabulary used is much more complex and it is noticeable that text 1 [person's text] has a simpler vocabulary with simpler knowledge of what a person is capable of learning" (S.72); and the use of technical terms that denote informative and objective content despite the emotional incapacity: "the second one [AI's text] because it uses more technical terms and is based on explaining the facts and consequences of the Second World War without speaking from the feeling as the first one [person's text] does" (S.75).

However, even some of the participants continued to deny that either of these texts had been generated by an AI: "I think not, that both have been created by people. The first one [person's text] clearly indicates subjectivity and has its emotional side, and the second one [AI's text] also expresses an opinion about the consequences of WWII and how it affects us today" (S.65), considering that they still need to be refined to reach this point.

4. DISCUSSION

Technology and, in particular, conversational AI tools have been progressing and improving over time. In recent years there has been a breakthrough with generative language models and the emergence of tools based on them such as ChatGPT created by the OpenAI, which are expected to play a relevant role in the near future (Pelletier et al., 2023; Pelletier et al., 2024). However, despite the fact that the potential of conversational AI in education is reflected in the existing literature (Hwang & Chang, 2021), it is considered that people are often not yet fully aware of how it works (Pelletier et al., 2024). Given the current scenario, this study sought to analyze the perceptions and knowledge of trainee teachers about this type of technology by making them compare an argumentative text composed by another pre-service teacher and one composed by ChatGPT-3.5.

Such is the progress reached in text generation through AI-based systems that studies like the one carried out by Herbold et al. (2023), specifically centered in comparing human and AI-written essays, suggest that tools like ChatGPT are able to produce texts of higher quality than a person. The results obtained in our investigation seem to align with this idea, since the vast majority of participants considered, without knowing the author of any of the essays evaluated, that the one generated by ChatGPT would outscore the one written by another pre-service teacher in an exam, regarding the first question posed to the students and the first objective of the investigation. Both the person and ChatGPT were asked to comment on a historical source from the Second World War, which means they were expected to analyze and reflect on to argue about it. Therefore, current AI seems to be capable of replicating a process that involves not only explaining data, but also reflecting on it better than a university student, as already seen in (Tirado-Olivares et al., 2023). When trainee teachers were asked to justify their decision, most of them identified better performance, on the one hand, with more precise information and content related to the topic of the source commented, and on the other hand, with better writing. This could serve as an exemplification of some of the characteristics that are shown in the literature about the recent advancements in AI field. One of the most prominent features of ChatGPT is the big amount of data that it is trained on, something common for LLMs (Chang et al., 2024; Janiesch et al., 2021), which might explain why students highlight the content of the text produced by this tool. At the same time, authors like Májovský et al. (2023) noted that AI can address writing aspects such as word usage and sentence structure with convincing results and Wilson and Czik (2016) acknowledged that AI could serve as a support for addressing especially superficial aspects of writing like spelling. While the pre-service teachers' opinion in our study supports these assumptions, it should be noted that their considerations about ChatGPT's ability write texts with better structure and more coherent suggest that current AI could also be capable of tackling more profound aspects of writing.

Focusing on the second question asked to the participants, in which they had to choose which text conveyed more emotions and justify their choice, the results reveal a clear division of opinions, as the number of pre-service teachers who selected each text is almost the same. Some studies show that conversational AI tools are less emotional than humans and see this as one of the main differences between AI and human beings (Guo et al., 2023; Mitrović et al., 2023). Nonetheless, our results seem to reflect that this distinction may not be completely clear when reading a text. In the qualitative data collected, those students who determined that the commentary created by ChatGPT conveyed more emotions tended to explain it by emphasizing, for example, the critical reflection and the personal opinion the text contained as well as empathy that the author —unknown for them— demonstrated with the people who faced the reality described in the historical source. Consequently, the constant progress in the generative AI field can be making this usual distinguishing criterion not so useful and reliable. Moreover, this could serve as evidence of how recent conversational AI systems seem to possess the ability to generate more human-like discourse (Dale, 2021; Pavilk, 2023), using the large amount of information they have, the patterns learnt and their ability to predict the next word (Chang et al., 2024; Janiesch et al., 2021). Therefore, there could be a need for better understanding of this and for the development of skills to critically analyze the use of this tools and their outcome and impact.

When the students were posed the third question, over half of them believed AI was capable of generating a text like the two they had read. However, as seen in the previous question, they claim that it would lack emotivity. Again, participants assumed that the AI could generate a text comparable to those in front of them in terms of the amount of information and data presented. However, they denied that it could achieve a similar level of empathy and offer a personal point of view. Furthermore, they were convinced an AI-based system would outperform both texts in terms of writing quality, exhibiting more and better vocabulary as well as a greater ability to structure its discourse. As can be seen, their reasoning is similar to that demonstrated in the prior questions and it is based on the common believe that AI can be better at a more objective level (data, good spelling, etc.) but it may not be capable of anything related to showing subjectivity (Guo et al., 2023; Mitrović et al., 2023). Besides, attention should be paid to the fact that more than a 43% of the pre-service teachers denied the possibility of an AI tool creating a similar text, something that reinforces the idea pointed out in the last *EDUCAUSE Horizon Report* (Pelletier et al., 2024), in which the lack of understanding and preparation to face the changes that generative AI could bring to the educational landscape is remarked.

In the last question, which raised the possibility that one of the two commentaries was written by ChatGPT instead of another pre-service teacher, the data show that the majority of the students correctly identified to whom belonged each one. Despite this, there were still few people who affirmed that is not possible for AI to execute an activity like this and some trainee teachers stated that the text that was actually written by ChatGPT was the one developed by the person. Once again, they associated greater subjectivity and emotivity with the human author, while data and better writing were linked to AI. The divergencies perceived in our results align with the idea that further investigation about chatbots and conversational AI introduction in education is still needed, as Labadze et al. (2023) suggest in their systematic review, especially since this technology is constantly evolving and teachers do not seem to be fully aware of current AI capabilities. Besides, the participants insistence on the relation between data and AI would also highlight this need, since it could mean they are not cognizant of the fact that AI-generated answers are not always be accurate (Labadze et al., 2023).

Finally, regarding the second objective of the study, no significant differences depending on the academic year of the participants were found, so there would be no differences in the need for more preparation to face the future implications of AI in education (Pelletier et al., 2024). The absence of considerable divergences between those students who have just started their degree and those who have already completed half of it could be linked to the shortage in preparation mentioned before. At the same time, it could be related to the deficiencies in teacher digital competence that many studies have acknowledged (e.g. Claro et al., 2024; Dias-Trindade et al., 2021; Gabarda Méndez et al., 2021; Gutiérrez & Cabero-Almenara, 2016). As they show, pre-service and in-service teachers usually have medium to low competence level when it comes to the integration of technology in their job. These means not only that they need more instruction on it in general but also that the training they receive during their university studies may not be enough to see a difference between undergraduate students in their first and third year of university.

5. CONCLUSION

The recent advancements in the field of artificial intelligence and the effects it can have on the way we interact with information in the post-truth era, where fake news and misinformation abound (Barzilai & Chinn, 2020; Chinn et al., 2021) make the development of critical thinking and reflection skills vital to navigate this world. At the same time, it makes necessary that teachers know the potential and limitations of this technology to help students acquire those skills. In line with this, the present study aimed to analyze if pre-service teachers could distinguish a text generated by ChatGPT from one written by a person, which features they attributed to the artificial intelligence's capability to compose a text and if there were any differences in their belief's according to their academic year. This intended to examine their perception and knowledge about current conversational artificial intelligence and to see if their progress in they degree they are studying could affect it.

Considering the results obtained, conversational systems like ChatGPT not only seem to be capable of producing an argumentative text based on analyzing and commenting on another source, but they also appear to be able to do it better than a university student. This would exemplify the progress that has made this kind of tools to generate and increasingly human-like discourse (Dale, 2021; Pavlik, 2023). Apart from this, while most of the participants could correctly tell the text created by ChatGPT from that written by a person, not all of them did. In addition, there were some divergences in relation to the criteria they used to decide it, which does not seem to be absolutely reliable. Thus, not all pre-service teachers are fully aware of the writing abilities that current artificial intelligence tools have, which adds to the shortage in teacher digital competence in general that educators usually have (Gabarda Méndez et al., 2021). It also demonstrates that more research about the application of conversational artificial intelligence in education needs to be done (Okwonko and Ade-Ibijola, 2021) and more preparation is required (Pelletier et al., 2024).

Educators should be provided with the training and knowledge to integrate technology in their labor. This includes the understanding of new tools that are expected to have a huge impact on education and society, such as generative artificial intelligence, and the fostering of associated skills such as critical thinking. Being cognizant of their potentialities and risks will make them able to critically evaluate its use and the information generated with it to avoid misinformation and biases. As a consequence, they will be able to promote the development of the skills needed in the 21st century world in their students.

5.1. Limitations and Future Research

The investigation carried out has raised new questions that should be addressed in the future. Firstly, since our study focuses on argumentative texts based on commenting on a source, including other types of text could be of interest, as it would allow for an exploration of ChatGPT's writing skills in different contexts. Another aspect worth considering is the comparison with other generative artificial intelligence systems, such as Microsoft Copilot or Gemini, or even with other large language models or versions of the one that has been used: GPT-3.5.

Lastly, it could also be valuable to increase the number and of participants and vary their conditions. This would allow for factors such as gender, age and teaching experience to be taken into account, and even for comparisons between teachers at different levels of education. Hence, a wider range of variables that could influence the results would be covered and a broader picture of the educational landscape would be provided.

ACKNOWLEDGEMENT

We would like to thank all the pre-service teachers that participated in this investigation for their collaboration.

FUNDINGS: This work was supported by MCIN/AEI/110.13039/501100011033 and by the European Union NextGenerationEU/PRTR under Project TED2021-131557B-I00; by the University of Castilla-La Mancha and by the European Regional Development Fund (ERDF) under Grant 2022-GRIN-34039; by the Ministry of Education, Culture and Sports of Spain under Grant FPU20/02375; the Regional Government of Castilla-La Mancha and the European Social Fund Plus (EFS+) under Grant number: 2023-PREJCCM-000054 and under the agreement with the Regional Government of Castilla-La Mancha (220520CONV).

REFERENCES

Ait Baha, T., El Hajji, M., Es-Saady, Y., & Fadili, H. (2024). The impact of educational chatbot on student learning experience. *Education and Information Technologies*, 29(8), 10153–10176. DOI: 10.1007/s10639-023-12166-w

Akgun, S., & Greenhow, C. (2022). Artificial intelligence in education: Addressing ethical challenges in K-12 settings. *AI and Ethics*, 2(3), 431–440. DOI: 10.1007/s43681-021-00096-7 PMID: 34790956

Barzilai, S., & Chinn, C. A. (2020). A review of educational responses to the "post-truth" condition: Four lenses on "post-truth" problems. *Educational Psychologist*, 55(3), 107–119. DOI: 10.1080/00461520.2020.1786388

Bishop, C. M. (2006). *Pattern recognition and machine learning (Information science and statistics)*. Springer-Verlag New York, Inc.

Buckingham, D. (2019). Teaching media in a 'post-truth' age: Fake news, media bias and the challenge for media/digital literacy education. *C&E, Cultura y Educación*, 31(2), 213–231. DOI: 10.1080/11356405.2019.1603814

Budzianowski, P., & Vulić, I. (2019). Hello, It's GPT-2-How Can I Help You? Towards the Use of Pretrained Language Models for Task-Oriented Dialogue Systems. In Birch, A., Finch, A., Hayashi, H., Konstas, I., Luong, T., Neubig, G., Oda, Y., & Sudoh, K. (Eds.). *Proceedings of the 3rd Workshop on Neural Generation and Translation* (pp. 15-22). Association for Computational Linguistics. DOI: 10.18653/v1/D19-5602

Carrión-Martínez, J. J., Luque-de la Rosa, A., Fernández-Cerero, J., & Montenegro-Rueda, M. (2020). Information and Communications Technologies (ICTs) in Education for Sustainable Development: A Bibliographic Review. *Sustainability (Basel)*, 12(8), 3288. DOI: 10.3390/su12083288

Chang, Y., Wang, X., Wang, J., Wu, Y., Yang, L., Zhu, K., Chen, H., Yi, X., Wang, C., Wang, Y., Ye, W., Zhang, Y., Chang, Y., Yu, P. S., Yang, Q., & Xie, X. (2024). A Survey on Evaluation of Large Language Models. *ACM Transactions on Intelligent Systems and Technology*, 15(3), 1–45. DOI: 10.1145/3641289

Chaudron, S., Di Gioia, R., & Gemo, M. (2018). Young children (0-8) and digital technology, a qualitative study across Europe. Retrieved January 15, 2021, from https://op.europa.eu/s/oNl9

Chinn, A., Barzilai, S., & Duncan, R. G. (2021). Education for a "Post-Truth" World: New Directions for Research and Practice. *Educational Researcher*, 50(1), 6–68. DOI: 10.3102/0013189X20940683

Claro, M., Castro-Grau, C., Ochoa, J. M., Hinostroza, J. E., & Cabello, P. (2024). Systematic review of quantitative research on digital competences of in-service school teachers. *Computers & Education*, 215, 105030. DOI: 10.1016/j.compedu.2024.105030

Cong-Lem, N., Soyoof, A., & Tsering, D. (2024). A Systematic Review of the Limitations and Associated Opportunities of ChatGPT. *International Journal of Human-Computer Interaction*, 1–16. DOI: 10.1080/10447318.2024.2344142

Dale, R. (2021). GPT-3: What's it good for? *Natural Language Engineering*, 27(1), 113–118. DOI: 10.1017/S1351324920000601

de Lázaro Torres, M. L. (2024). La utilidad de ChatGPT en la enseñanza universitaria a distancia de geoinformación para futuros docentes de enseñanza secundaria. *Interdisciplinary Journal of Didactics*, (1), 83–104. DOI: 10.14198/ijd.28204

Dias-Trindade, S., Moreira, J. A., & Gomes Ferreira, A. (2021). Evaluation of the teachers' digital competences in primary and secondary education in Portugal with DigCompEdu CheckIn in pandemic times. *Acta Scientiarum. Technology*, 43, e56383. DOI: 10.4025/actascitechnol.v43i1.56383

European Commission. (2020). *Digital education action plan 2021-2027: Resetting education and training for the digital age*. Education and Training.

European Commission, Directorate-General for Education, Youth, Sport and Culture. (2019). *Key competences for lifelong learning*, Publications Office. https://data.europa.eu/doi/10.2766/569540

Gabarda Méndez, V., García Tort, E., Ferrando Rodríguez, M. L., & Chiappe Laverde, A. (2021). Pre-school and primary school teachers: Technological training and digital competence. *Innoeduca.International Journal of Technology and Educational Innovation*, 7(2), 19–31. DOI: 10.24310/innoeduca.2021.v7i2.12261

Ghomi, M., & Redecker, C. (2019). Digital Competence of Educators (DigCompEdu): Development and Evaluation of a Self-assessment Instrument for Teachers' Digital Competence. In *Proceedings of the 11th International Conference on Computer Supported Education (CSEDU 2019)*, 541-549. https://doi.org/DOI: 10.5220/0007679005410548

Goodman-Deane, J., Mieczakowski, A., Johnson, D., Goldhaber, T., & Clarkson, P. J. (2016). The impact of communication technologies on life and relationship satisfaction. *Computers in Human Behavior*, 57, 219–229. DOI: 10.1016/j.chb.2015.11.053

Guo, B., Zhang, X., Wang, Z., Jiang, M., Nie, J., Ding, Y., Yue, J., & Wu, Y. (2023). How Close is ChatGPT to Human Experts? Comparison Corpus, Evaluation, and Detection. ArXiv:2301.07597. https://doi.org//arXiv.2301.07597DOI: 10.48550

Gutiérrez, J.J., & Cabero-Almenara, J. (2016). Estudio de caso sobre la autopercepción de la competencia digital del estudiante universitario de las titulaciones de grado de Educación Infantil y Primaria. *Profesorado. Revista de currículum y formación del profesorado, 20*(2), 180-199.

Herbold, S., Hautli-Janisz, A., Heuer, U., Kikteva, Z., & Trautsch, A. (2023). A large-scale comparison of human-written versus ChatGPT-generated essays. *Scientific Reports*, 13(1), 18617. DOI: 10.1038/s41598-023-45644-9 PMID: 37903836

Hirschberg, J., & Manning, C. D. (2015). Advances in natural language processing. *Science*, 349(6245), 261–266. DOI: 10.1126/science.aaa8685 PMID: 26185244

Hwang, G.-J., & Chang, C.-Y. J. I. L. E. (2021). A review of opportunities and challenges of chatbots in education. *Interactive Learning Environments, 31*(7) 4099–4112. DOI: 10.1080/10494820.2021.1952615

Janiesch, C., Zschech, P., & Heinrich, K. (2021). Machine learning and deep learning. *Electronic Markets*, 31(3), 685–695. DOI: 10.1007/s12525-021-00475-2

Kasneci, E., Sessler, K., Küchemann, S., Bannert, M., Dementieva, D., Fischer, F., Gasser, U., Groh, G., Günnemann, S., Hüllermeier, E., Krusche, S., Kutyniok, G., Michaeli, T., Nerdel, C., Pfeffer, J., Poquet, O., Sailer, M., Schmidt, A., Seidel, T., & Kasneci, G. (2023). ChatGPT for good? On opportunities and challenges of large language models for education. *Learning and Individual Differences*, 103, 102274. DOI: 10.1016/j.lindif.2023.102274

King, M. R. (2023). The future of AI in medicine: a perspective from a Chatbot. *Annals of Biomedical Engineering, 51*(2), 291-295. DOI: 10.1007/s10439-022-03121-w

Koehler, , M. JMishra, , PCain, , W. (2013). What is Technological Pedagogical Content Knowledge (TPACK)? *Journal of Education, 193*(3), 13-19. DOI: 10.1177/002205741319300303

Kozyreva, A., Wineburg, S., Lewandowsky, S., & Hertwig, R. (2023). Critical ignoring as a core competence for digital citizens. *Current Directions in Psychological Science*, 32(1), 81–88. DOI: 10.1177/09637214221121570 PMID: 37994317

Labadze, L., Grigolia, M., & Machaidze, L. (2023). Role of AI chatbots in education: Systematic literature review. *International Journal of Educational Technology in Higher Education*, 20(1), 56. Advance online publication. DOI: 10.1186/s41239-023-00426-1

Lazer, D. M. J., Baum, M. A., Benkler, Y., Berinsky, A. J., Greenhill, K. M., Menczer, F., Metzger, M. J., Nyhan, B., Pennycook, G., Rothschild, D., Schudson, M., Sloman, S. A., Sunstein, C. R., Thorson, E. A., Watts, D. J., & Zittrain, J. L. (2018). The science of fake news. *Science*, 359(6380), 1094–1096. DOI: 10.1126/science.aao2998 PMID: 29590025

Lee, Y. F., Hwang, G. J., & Chen, P. Y. (2022). Impacts of an AI-based *cha*bot on college students' after-class review, academic performance, self-efficacy, learning attitude, and motivation. *Educational Technology Research and Development*, 70(5), 1843–1865. DOI: 10.1007/s11423-022-10142-8

Lim, W. M., Gunasekara, A., Pallant, J. L., Pallant, J. I., & Pechenkina, E. (2023). Generative AI and the future of education: Ragnarök or reformation? A paradoxical perspective from management educators. *International Journal of Management Education*, 21(2), 100790. DOI: 10.1016/j.ijme.2023.100790

Lo, C. K. (2023). What Is the Impact of ChatGPT on Education? A Rapid Review of the Literature. *Education Sciences*, 13(4), 410. DOI: 10.3390/educsci13040410

Lund, B. D., & Wang, T. (2023). Chatting about ChatGPT: How may AI and GPT impact academia and libraries? *Library Hi Tech News*, 40(3), 26–29. DOI: 10.1108/LHTN-01-2023-0009

Májovský, M., Černý, M., Kasal, M., Komarc, M., & Netuka, D. (2023). Artificial Intelligence Can Generate Fraudulent but Authentic-Looking Scientific Medical Articles: Pandora's Box Has Been Opened. *Journal of Medical Internet Research*, 25, 46924. DOI: 10.2196/46924 PMID: 37256685

Martín-Párraga, L., Llorente-Cejudo, C., & Barroso-Osuna, J. (2023). Self-Perception of Digital Competence in University Lecturers: A Comparative Study between Universities in Spain and Peru According to the DigCompEdu Model. *Societies (Basel, Switzerland)*, 13(6), 142. DOI: 10.3390/soc13060142

McIntyre, L. (2018). *Post-Truth*. MIT Press. DOI: 10.7551/mitpress/11483.001.0001

Mishra, P., & Koehler, M. J. (2006). Technological Pedagogical Content Knowledge: A Framework for Teacher Knowledge. *Teachers College Record*, 108(6), 1017–1054. DOI: 10.1111/j.1467-9620.2006.00684.x

Mitrović, S., Andreoletti, D., & Ayoub, O. (2023) ChatGPT or Human? Detect and explain. Explaining Decisions of Machine Learning Model for Detecting Short ChatGPT-generated Text. *EdArXiv*. arXiv:2301.13852

Moreno-Vera, J. R. (2023). Epilogue: "We wanna learn like common people whatever common people did". In Gómez-Carrasco, C. J. (Ed.), *Re-Imagining the Teaching of European History. Promoting Civic Education and Historical Consciousness* (pp. 219–223). Routledge.

OECD. (2016). *Skills for a Digital World: 2016 Ministerial Meeting on the Digital Economy Background Report, OECD Digital Economy Papers, No. 250*. OECD Publishing., DOI: 10.1787/5jlwz83z3wnw-

OECD (2020), *TALIS 2018 Results (Volume II): Teachers and School Leaders as Valued Professionals*, TALIS, OECD Publishing. https://doi.org/DOI: 10.1787/19cf08df-en

Okonkwo, C. W., & Ade-Ibijola, A. (2021). Chatbots applications in education: A systematic review. *Computers and Education: Artificial Intelligence*, 2, 100033. DOI: 10.1016/j.caeai.2021.100033

Oxford University Press. (n.d.). Post-truth. *Oxford English Dictionary*. Retrieved October 4, 2024, from https://www.oed.com/search/dictionary/?scope=Entries&q=post-truth

Paul, R. (1992). Critical thinking: What, why, and how. *New Directions for Community Colleges*, (77), 324.

Pavlik, J. V. (2023). Collaborating with ChatGPT: Considering the implications of generative artificial intelligence for journalism and media education. *Journalism & Mass Communication Educator*, 78(1), 84–93. DOI: 10.1177/10776958221149577

Pelletier, K., McCormack, M., Muscanell, N., Reeves, J., Robert, J., & Arbino, N. (2024). *2024 EDUCAUSE Horizon Report, Teaching and Learning Edition*. Educause.

Pelletier, K., Robert, J., Arbino, N., Muscanell, N., McCormack, M., Reeves, J., McDonald, B., & Grajek, S. (2023). *2023 EDUCAUSE Horizon Report: Holistic Student Experience Edition*. EDUCAUSE.

Pelletier, K., Robert, J., Muscanell, N., McCormack, M., Reeves, J., Arbino, N., Grajek, S., Birdwell, T., Liu, D., Mandernach, J., Moore, A., Porcaro, A., Rutledge, R., & Zimmern, J. (2023). *2023 EDUCAUSE Horizon Report, Teaching and Learning Edition*. EDUCAUSE.

Radford, A., Narasimhan, K., Salimans, T., & Sutskever, I. (2018). *Improving Language Understanding by Generative Pre-Training*. Preprint.

Rapp, A., Curti, L., & Boldi, A. (2021). The human side of human-chatbot interaction: A systematic literature review of ten years of research on text-based chatbots. *International Journal of Human-Computer Studies*, 151, 102630. DOI: 10.1016/j.ijhcs.2021.102630

Redecker, C. (2017). European framework for the digital competence of educators: DigCompEdu. *Joint Research Centre (JRC). Social Policy Report*, 95. Advance online publication. DOI: 10.2760/159770

Rogers, E. M. (1986). *Communication technology*. Simon and Schuster.

Rouhiainen, L. (2018). *Inteligencia artificial. 101 cosas que debes saber hoy sobre nuestro futuro*. Alienta Editorial.

Russell, S. J., & Norvig, P. (2021). *Artificial intelligence: A modern approach* (4th ed.). Pearson.

Seguido, M. (2024). Competencia Digital Docente: Desafíos para la enseñanza de las Ciencias Sociales a partir de un proyecto de innovación educativa. *Interdisciplinary Journal of Didactics*, (1), 19–32. DOI: 10.14198/ijd.27976

Shawar, B. A., & Atwell, E. (2007). Chatbots: Are they Really Useful? *Journal for Language Technology and Computational Linguistics*, 22(1), 29–49. DOI: 10.21248/jlcl.22.2007.88

Tirado Olivares, S., & Cózar-Gutiérrez, R. (2024). Enseñar y evaluar ciencias sociales en la era digital: el uso del learning analytics para la evaluación formativa. Una revisión del estado de la cuestión. *Didáctica De Las Ciencias Experimentales Y Sociales*, (46), 21–36. https://doi.org/DOI: 10.7203/dces.46.28468

Tirado-Olivares, S., Navío-Inglés, M., O'Connor-Jiménez, P., & Cózar-Gutiérrez, R. (2023). From human to machine: Investigating the effectiveness of the conversational AI ChatGPT in historical thinking. *Education Sciences*, 13(8), 803. DOI: 10.3390/educsci13080803

Tzafilkou, K., Perifanou, M., & Economides, A. A. (2023). Assessing teachers' digital competence in primary and secondary education: Applying a new instrument to integrate pedagogical and professional elements for digital education. *Education and Information Technologies*, 28(12), 16017–16040. DOI: 10.1007/s10639-023-11848-9 PMID: 37361819

van Laar, E., van Deurse, A. J. A. M., van Dijk, A. A. G. M., & de Haan, J. (2017). The relation between 21st-century skills and digital skills: A systematic literature review. *Computers in Human Behavior*, 72, 577–588. DOI: 10.1016/j.chb.2017.03.010

Vuorikari, R., Punie, Y., Carretero Gómez, S., & Van Den Brande, G. (2016). *DigComp 2.0: The Digital Competence Framework for Citizens. Update Phase 1: The Conceptual Reference Model* (EUR 27948 EN). Publications Office of the European Union. DOI: 10.2791/11517

Watson, C. A. (2018). Information Literacy in a Fake/False News World: An Overview of the Characteristics of Fake News and its Historical Development. *International Journal of Legal Information*, 46(2), 93–96. DOI: 10.1017/jli.2018.25

Wilson, J., & Czik, A. (2016). Automated essay evaluation software in English Language Arts classrooms: Effects on teacher feedback, student motivation, and writing quality. *Computers & Education*, 100, 94–109. DOI: 10.1016/j.compedu.2016.05.004

Zhang, K., & Aslan, A. B. (2021). AI technologies for education: Recent research & future directions. *Computers and Education: Artificial Intelligence*, 2, 100025. DOI: 10.1016/j.caeai.2021.100025

Zhang, P., & Tur, G. (2024). A systematic review of ChatGPT use in K-12 education. *European Journal of Education*, 59(2), e12599. DOI: 10.1111/ejed.12599

Zheng, B., Warschauer, M., Lin, C.-H., & Chang, C. (2016). Learning in One-to-One Laptop Environments: A Meta-Analysis and Research Synthesis. *Review of Educational Research*, 86(4), 1052–1084. DOI: 10.3102/0034654316628645

KEY TERMS AND DEFINITIONS

Argumentative Text: A type of text whose main purpose is to persuade or convince the reader of a certain point of view by means of reasons, evidence and examples.

Commentary: Written analysis and critical interpretation of a text, event or theme.

Conception: The way in which a person understands, interprets or perceives an idea or phenomenon based on their own knowledge, experience and perspective.

Conversational Artificial Intelligence: Technology based on natural language processing techniques that allows machines to simulate an interaction with the user.

Critical Thinking: The ability to analyze and evaluate the veracity of information, identifying biases, in order to form a well-founded judgment in an autonomous manner.

Post-Truth Era: A context in which objective facts are less relevant than beliefs and emotions. It implies the distortion of reality to influence an opinion.

Teacher Digital Competence: Educators' capability to effectively integrate technology in their job. It implies a proficiency in the use of technologies for different tasks ranging from teaching and administrative tasks to assessment and promoting the development of students' digital competence.

Chapter 15
Impact of Oral Transmission and Geolocation on the Romance of El Señor Don Gato

Narciso José López García
https://orcid.org/0000-0002-5665-5263
University of Castilla-La Mancha, Spain

Raquel Bravo Marín
https://orcid.org/0000-0001-5809-7739
University of Castilla-La Mancha, Spain

ABSTRACT

The romance de El Señor Don Gato is a clear example of how oral transmission and geographical location can modify both the text and the melody of a song. The research presented below is an analysis of sound recordings and textual transcriptions from different parts of the world, especially Spain, which show how geographical location, cultural context and social interaction, among other aspects, have a significant impact on traditional music, the most tangible consequences of which are reflected in both literary and musical alterations which, nevertheless, have served to adapt and enrich this well-known piece of music, maintaining the essence and the main idea of its original message.

DOI: 10.4018/979-8-3693-7185-5.ch015

Copyright © 2025, IGI Global Scientific Publishing. Copying or distributing in print or electronic forms without written permission of IGI Global is prohibited.

1. INTRODUCTION

When discussing folk, popular, or traditional music, one of the primary concepts that arises is the way these musical styles have been handed down through the ages. A key feature of these genres, in all their forms, is their intrinsic link to oral transmission. This indicates that historically, music has been acquired, maintained, and shared not through written scores or formal education, but rather through verbal communication, personal engagement, and collective practice. Oral transmission serves not only as a method for preserving music; it is an essential element that defines the identity of folk and traditional music. It affects how musicians and communities perceive their music and cultural heritage, as it is deeply rooted in memory, improvisation, and individual interpretation. In contrast to written music, which seeks to maintain a composition in its original form, oral transmission fosters the emergence of variations and personal interpretations within a shared musical tradition. This adaptability and fluidity are vital to the ongoing vitality of these musical expressions. As noted by Keedan- Phipps (2013), the process of oral transmission encompasses more than merely conveying melodies or lyrics; it also involves the sharing of cultural values, historical narratives, and social significances that are interwoven with the music. Musicians and community members act as both preservers and innovators of their musical legacy, facilitating its continuous transformation. This renders oral transmission a dynamic and living practice, ensuring that folk and traditional music remains pertinent and impactful across generations, while also mirroring the shifts and consistencies of the societies that nurture it. Oral transmission transcends mere preservation; it is a crucial, creative process that influences the music itself. It enables folk and traditional music to stay fluid and responsive, while still deeply anchored in cultural history and communal identity.

This is a fact. Oral transmission is generally considered by most scholars (musicians, philologists, musicologists and ethnomusicologists, among others) to be a central characteristic of European traditional music, a characteristic that differentiates it from what is wrongly called 'classical' or 'cultured' music. Let us not forget that, in Europe, the first written musical documents related to 'cultured' music date from around the 9th century, while it was not until the 19th century, at the height of the Romantic spirit, that the first proverbs, romances, tales... related to folklore appeared in musical songbooks. Also, the late medieval songbooks collect popular compositions with variations to which they gloss. As Shanahan and Albrecht (2019) said the primary method through which songs are conveyed across generations is oral transmission.

The power of propagation of content offered by oral transmission is beyond doubt, although it is true that this phenomenon can and does directly influence the propagated work, to the extent that it has the capacity to modify it. One of the main effects of the oral transmission of songs, romances, proverbs, refrains... is that this material is very often modified by those who learn it and subsequently pass it on.

During this straightforward process of transmission - listening and then singing a song again - it is probable that the singer will incorporate some changes in the new rendition, whether by accident or intentionally. It would be simplistic to assume that oral transmission would merely add random noise to the sung production. However, it is thought that oral transmission molds musical systems in systematic ways that mirror the biases of human reproduction (Mehr *et al.* 2020; Savage et al., 2022).

This process is called *communal recreation*. This *communal recreation* has given rise to situations as curious as they are convoluted, such as finding the same song with its text lengthened or abbreviated depending on the geographical area where it was collected; the same happens with the melodies, similar or completely different; the rhythmic and metrical changes, etc., which is why it is common for *folk pieces* not to have established authentic forms.

The verbal exchange of music in early human societies likely contributed a significant role in developing common melodic features found in musical traditions worldwide (Savage et al., 2015). This method of sharing would have allowed communities to create unified musical patterns and repertoires, enhancing a sense of identity and unity both within and among groups. Through oral transmission, music functions not only as a source of entertainment but also as a vehicle for cultural preservation and social connection.

Numerous studies have addressed this modality. The research conducted by Anglada-Tort et al., (2022) revealed that oral transmission significantly impacts the development of melodies, transforming random sounds into more organized systems by utilizing fewer basic elements (such as intervals and contours) that are repeatedly combined. Notably, the structural characteristics that arose from our experiments closely resemble common melodic features found in various musical traditions worldwide (Savage et al., 2015; Mehr et al., 2019). Additionally, this study indicated that melodies often adhere to specific cultural norms of Western music (e.g., the use of harmonic intervals), reflecting the predominantly Western background of their participants.

It is common, therefore, to find many versions and variants of the same romance, of the same nursery rhyme, of the same proverb, story, couplet, joke... all of them of equal value and importance, simply by browsing through some of the hundreds of songbooks which, fortunately, are available today.

This is the case with the romance *La mala suegra* ('The Bad Mother-in-law'), the romance *La loba parda* ('The Brown She-wolf'), the romance of *Gerineldo*, the tunes *Paloma blanca* ('White Pigeon') and *Al salir de la enramada* ('Leaving the Arbor'), the song of *Las tres comadres* ('The Three Wives') and a long etcetera in which it is essential to highlight as one of the paradigms of this oral transmission, the romance of *El Señor Don Gato* ('There Was Mr. D. Cat'), capable of even crossing the borders of the Iberian Peninsula and travelling thousands of kilometers to settle in the Canary Islands, France, Morocco and Latin America, mainly.

But how can this fact be explained, what is the reason for this diversity of versions of the same romance with such a wide (extensive) radius of expansion? Here we come up against a major obstacle: which theory do we use to shed light on our doubt? The most frequent is the one based on the use of history. According to this theory, each of the different Peoples that have passed through the Peninsula have left important traces of their musical ideas and tastes which have become part of the musical traditions of the population already settled there (Damrosch, 2003). At the same time, the receiving population has impregnated these peoples with its musical culture, songs and dances, resulting in an interesting and enriching fusion. Added to this are the musical differences that have existed before and after this musical fusion, in the different geographical areas of our Peninsula and, in the case of the romance of *El Señor Don Gato*, the Arab and Jewish influences, their subsequent expulsions and, of course, the arrival of Europeans in the Americas, have been especially fundamental.

2. ORIGINS

The concept of 'Ethnomusicology' was introduced by Jaap Kunst in 1950, which gave birth to a scientific discipline focused on the analysis of music within its cultural context. This new branch of knowledge is characterised by the examination of music not only as an art form, but also as a cultural phenomenon. The work of Allan P. Merriam had a significant impact on this field, defining ethnomusicology as the study of music in its cultural dimension, emphasising the need to understand music in relation to the practices and beliefs of the communities that create it (Merriam, 1964). Similarly, John Blacking (1973) expanded this view by describing music as humanly organised sound, a definition that point out the social nature of music and its connection to cultural and social structures.

With the consolidation of ethnomusicology as a discipline deriving directly from comparative musicology, a concern emerged to address fundamental questions related to the 'who', 'when', 'how' and 'where' of music, particularly in studies involving oral traditions. This approach, while generating valuable knowledge on many occasions, has also posted significant methodological challenges. The need to answer these questions has led, in some cases, to fruitless efforts that have obscured the real keys to the phenomenon in question. According to Rice (2014), ethnomusicologists, in attempting to address these questions, have faced difficulties in confronting the complexity of musical cultures, which has sometimes resulted in interpretations that fail to fully capture the essence of the musical traditions studied. This problem is particularly evident in the analysis of oral cultures, where the unwritten nature of the tradition complicates thorough documentation and understanding (Nettl, 2005).

In this framework, it is essential that ethnomusicology progresses towards methodological approaches that are more flexible and that fit the particularities of the cultures being analysed. According to Seeger (2004), the challenge is to strike a balance between scientific precision and sensitivity to cultures, avoiding simplistic solutions to complex questions, and recognising the diversity and dynamism of musical traditions.

In the last two or three decades, ethnomusicology has undergone a remarkable transformation, moving away from the traditional approaches that previously dominated the discipline. Instead of focusing solely on the formal aspects of music, ethnomusicology has adopted a more holistic view that combines sound analysis with the cultural and social context in which it is expressed. This change of perspective has been crucial for the advancement of the discipline, which now finds itself at the confluence of musical and anthropological studies. This development has been facilitated by increasing collaboration with other areas of knowledge, such as linguistics, semiotics, sociology, history and cultural studies (Stobart, 2008; Feld, 2012).

This interdisciplinary approach has brought significant value to ethnomusicology, facilitating a more integrate understanding of musical phenomena. According to Feld (2012), current ethnomusicology highlights the relevance of considering cultural and social context to fully grasp the meaning and function of music in diverse communities. Furthermore, Stobart (2008) stresses that this disposition towards other disciplines has enabled ethnomusicologists to deal more effectively with the complexity of musical phenomena and their various dimensions.

It is from this holistic perspective that research into the possible origins of our romance must be approached. The aim is not to determine with certainty who the author of the romance is, when it was composed or who set it to music; such attempts are futile and unproductive. Instead, the focus should be on presenting the various existing hypotheses about the provenance of this romance, which will allow for an informed comparison with the different versions and variants, both textual and musical, that have emerged and been located throughout the research process (Suzel Ana, 2006; Rees, 2010). This method not only avoids unfounded speculation, but also provides a more solid basis for understanding the evolution and diversity of cultural manifestations related to romance.

In this context, it is pertinent to present and analyse the five main hypotheses that attempt to shed light on the origins of the romance of *El Señor Don Gato*:

1. Cuban Indigenous origin linked to Areíto of Anacaona. This hypothesis suggests that the romance could derive from the Areíto, a form of cultural expression of the Indigenous Cubans. According to the theory, experts such as Fernando Ortiz, Zoila Gómez García, Victoria Eli Rodríguez and Menéndez Pidal, who argue that the Areíto, which means 'rhyme' or 'romance', is simply one of the many melodies used over the centuries to interpret this romance (Ortiz, 1985; Gómez García, 1998).
2. Children's entertainment of French origin, like the 18th-century 'Mambrú'. According to this theory, the romance may have emerged as a children's song in France, contemporary to the popular 'Mambrú', and shares with it certain rhythmic and melodic patterns common to children's songs and rounds from the same period. Cuban writer and musician Alejo Carpentier (1983) supports this idea. Also, this popular song 'Mambrú' has roots that can be linked to Catalonia due to the influence of wars and the circulation of songs during the 18th century. The Catalan origin can be justified when considering the connection between war conflict (such as the War of the Spanish Succession) and popular narratives in the Iberian Peninsula (Balcells, 1991).
3. Parodic origin based on the Portuguese version of 'Doña Infanta' and the Spanish model of the 'Cid'. Francisco Adolfo Coelho, Portuguese philologist and folklorist, together with the Chilean researcher Julio Vicuña, defend the idea that the romance could have arisen as a parody of these two literary sources, adapting and transforming itself in the cultural context of the Iberian Peninsula (Coelho, 1893; Vicuña, 1911).
4. Presence in the Sephardic tradition after the expulsion of the Jews from Spain in the 15th century. This hypothesis holds that the Sephardic Jews, after their expulsion from Spain, already knew versions of this romance. Professor Susana Weich-Shahak collects and comments on two variants of the romance in her work 'Romancero Sefardí de Marruecos', which suggests a significant antiquity and diffusion within the Sephardic diaspora (Weich-Shahak, 1994).
5. Spanish song from the final period of Arab rule in Spain (14th century). This theory, supported by Rafael Salazar, a Venezuelan musicologist, and Ofelia Sangil, a Spanish researcher and professor of literature, postulates that the romance tells the story of a prince from Madrid who falls in love with a Moorish woman, thus confronting the religious prejudices of the time and the principles of blood purity defended by the Christian nobility. Salazar mentions that the Madrilenians are popularly known as 'cats', in reference to a legend that highlights the agility and bravery of the Madrilenian warriors who, like cats, stormed the city walls in a historic warlike action. According to this version, the prince died after falling from his horse, which prevented the marriage from

being consummated, to the satisfaction of the nobility (Sangil, 2006; Salazar, 2002).

The last two hypotheses, the fourth and the fifth, are the most widely accepted and discussed at present, although, like all the others, they must be considered with caution, due to the lack of conclusive documentation that would allow a definitive validation (Salazar, 2002; Weich-Shahak, 1994).

On the other hand, Tejero (2003) also supports hypotheses four and five when he refers to the number seven and points out that, throughout history, it has been venerated by Babylonian, Hebrew, Greco-Latin and Christian civilisations, among others. In this sense, he refers to different children's songs, one of them being the romance *El Señor Don Gato* and his seven lives.

On the subject at hand, both Jewish and Christian traditions have many encounters with and in the number seven.

1.1. Geolocations

One of the most striking aspects of the process of communal recreation of the romance of *El Señor Don Gato* has been its ability to cross borders and travel thousands of kilometres to settle in different geographical areas. The host population of these new locations has imbued this romance with its musical culture, resulting in an interesting and enriching fusion. It is worth noting that Arab and Jewish influences are clearly recognisable in the localised texts.

The following map (figure 1) shows the regions where the different geographical locations where variants and/or versions of the variations on the romance of *El Señor Don Gato* have been found.

Figure 1. Map of localised versions

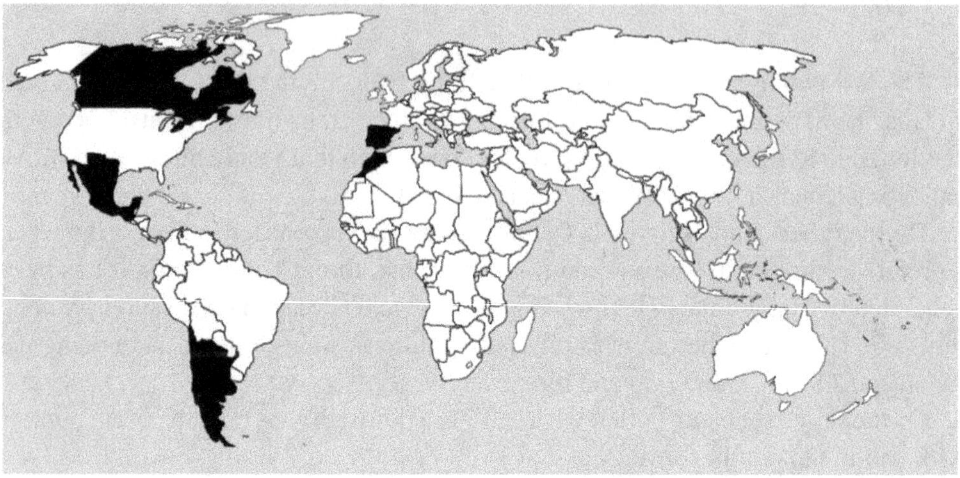

Source: Own elaboration

The map above (figure 1) includes how this Romance language has been able to travel from Spain to different geographical areas, especially the American continent.

Table 1 collects the number of documents located and the geographical area to which they belong.

Table 1. Localised versions

Source	Origin		Quantity	Text/ Lyric	Musical Transcription
Boiles (1973)	Spain	Extremadura	4	NO	4
		Castilla y León	4	NO	4
		Madrid	3	NO	3
Cabañas (2001)	Spain	Castilla-La Mancha	1	YES	1
Cárdenas & Cárdenas (1966)	Spain	-	2	YES	2
Córdova (1980)	Spain	Cantabria	2	YES	2
Díaz & Manzano (1989)	Spain	Castilla y León	1	YES	1
Echevarría (2005)	Spain	Castilla -La Mancha	2	YES	2
Fernández (1994)	Spain	Castilla -La Mancha	1	YES	1
Gil (1931)	Spain	Extremadura	1	YES	0
Gil (1987)	Spain	La Rioja	2	YES	2

continued on following page

Table 1. Continued

Source	Origin		Quantit y	Text/ Lyric	Musical Transcription
Katz & Tinnel (1999)	Spain		31	YES	31
	Jewish- Spanish		3	YES	3
	Portugal		2	YES	2
	Chile		2	YES	2
	Texas		1	YES	0
	New Mexico		1	YES	0
	Guatemala		1	YES	1
	Mexico		6	YES	6
https://www.kididdles.co m/	Canada		1	YES	1
Manzano (2001-2005)	Spain	Castilla y León	5	YES	5
Martí & Alonso (2006)	Spain	Comunidad de Madrid	1	YES	0
Mendoza (1990)	Spain	Castilla- La Mancha	2	YES	0
Montesdeoca (2005)	Spain	Islas Canarias	1	YES	0
Ruiz (2002)	Spain	Andalucía	2	YES	0
Seguí (1980)	Spain	Comunidad Valenciana	3	YES	3
Thomas (1966)	Mexico		4	YES	0
Weich-Shahak (1997)	Morocco	Sephardic culture	2	YES	2

Source: own elaboration.

3. METHODOLOGY

The working method used in this research was document analysis, understood as the systematic procedure of reviewing and evaluating information from a qualitative approach, with the aim of developing empirical knowledge (Bowen, 2009).

From this perspective, an exhaustive process of search and analysis of publications related to folklore, both national and international, has been carried out, with the aim of describing and interpreting the information provided by the documentation under study, in order to facilitate a cognitive approach to the original sources from a systematic and objective approach (Bardin, 2002; Peña and Pirela, 2007).

Documentary analysis is distinguished as a research tool that requires a systematic and rigorous review of the selected documents. This process involves detailed advance planning, which ensures that the selection of sources is appropriate and relevant to the specific objectives of the study. This approach allows researchers to gain a more accurate and comprehensive view of the issues being investigated, avoiding bias and ensuring that the selection of documents is as broad and representative as possible.

In contrast to other methods that focus on figures and statistical data, documentary analysis is based on the in-depth interpretation of textual content, emphasising the parameters of qualitative research. Researchers focus on understanding not only the explicit data presented in documents, but also the underlying meaning and context in which the information was produced.

Documentary analysis, on the other hand, is not limited to extracting information literally but, by interpreting the data within the social, cultural and historical context in which it was produced, attempts to go beyond this. This is crucial to capture how ideas, beliefs or narratives have evolved over time and how they have influenced the research topic.

From this point of view, contextual interpretation allows us to understand the value and relevance of documents in relation to the phenomenon under study, helping to identify the influences and factors that have shaped their development (Glenn, 2010).

Bowen (2009) establishes the following phases as milestones to consider in this type of process: the first stage of documentary analysis is the selection of documents. At this point, the inclusion criteria that will determine which documents will be studied are defined. These criteria may vary according to thematic relevance, historical period or type of source, and are essential to ensure that the documents chosen are aligned with the research objectives. The accurate and informed selection of sources is key to ensuring that the analysis is representative and based on an adequate sample of the phenomenon in question.

After the selection of the documents, the next step is the organisation of the information. The selected documents are processed in a way that makes the subsequent analysis more efficient. This process may include text coding, which consists of identifying relevant fragments and assigning them thematic codes, or the construction of categories of analysis that allow the information to be grouped in a coherent way. Proper organisation facilitates the interpretation and critical analysis of content, allowing researchers to visualise trends and patterns more clearly.

Once the information has been organised, the analysis and interpretation phase follow. Here the content of the documents is critically examined, identifying deeper meanings, recurring trends and relationships between the established categories. The analysis is carried out according to the theoretical framework of the research, which allows the findings to be interpreted in a broader academic context. Interpretation is not limited to mere description but seeks to understand how the documents reflect or influence the phenomenon under study.

Finally, the documentary analysis culminates in the synthesis and presentation of results. At this stage, the findings are condensed into a coherent and well-argued synthesis, which clearly reflects the conclusions drawn from the documentary analysis. This presentation should be based on the evidence gathered and provide a clear and comprehensible view of the results obtained. The final synthesis not only communicates the results, but also provides new perspectives or understandings of the research topic.

4. ANALYSIS OF TEXTUAL AND AUDIO DOCUMENTS

Comparing the localised texts, very different phraseological constructions and lexical uses can be observed. But, as López-García and De Moya (2019) point out, they all present the same argument and, therefore, a great similarity: the cat is in love and, therefore, a misfortune befalls it.

Additionally, in terms of musical transcriptions, López-García and De Moya (2019) present a classification model in which practically all the texts found fit, namely:

- Place where the cat sits before receiving news about the cat: on the roof; on a couch or chair made of gold, esparto grass or other materials.
- Description of the cat: white; black; brown; brown; blackberry.
- Provenance and/or parentage of the cat: daughter or niece of a brown cat; daughter or niece of a Roman cat; daughter of a wild cat; neighbour of a brown cat; neighbour of a Roman cat.
- Reason why the cat falls from the roof: when he is happy or joyful upon receiving the news, in a letter or from another cat, that he wants to marry the cat; when he receives a kiss from the cat or tries to give it to her; when he decides to go to see the cat.
- On the death of the cat: it does not die; it dies; it dies and, subsequently, is resurrected; it dies, is resurrected and, finally, dies again.

By way of example, two texts located in different geographical areas are transcribed below, highlighting some of the typological characteristics mentioned above:

Estando el señor don Gato ***sentadito en su tejado,***
cuando llegó la noticia *que si quería ser casado*
con una gatita parda ***sobrina de un gato pardo.***
Al oír esta noticia ***se ha caído del tejado,***
se ha roto siete costillas *y la puntita del rabo.*

Ya lo llevan a enterrar *por la calle del pescado:*
las gatitas visten luto, *los ratones de colorado.*
Al olor de las sardinas ***el gato ha resucitado:***
los ratones corren, corren, *las gatas se han desmayado.*
Por eso dice la gente: *"siete vidas tiene un gato".*
Estando el señor don Gato ***sentadito en silla de oro,***
estrenando medias blancas *y zapato acharolado.*
En eso pasó la gata *mostrándole indiferencia*
y el gato por darle un beso *se cayó desde el tejado.*
Llamaron al cirujano *y también al boticario,*
le recetaron un caldo *de longaniza y recado.*
Se murió el señor don Gato *a la mitad del puchero,*
la gata vistió de luto, *los gatos de pelo negro.*
Los ratones muy contentos *bailaban en el tejado,*
- se murió quien nos comía, *al caer descalabrado.*

Regarding sound documents, Manzano (1991) distinguishes between musical versions and musical variants, calling musical versions the different types or melodic inventions, unrelated in their features, with which the same text or its literary variants are sung. Regarding musical variants, he points out that they are those which present different physiognomy of the same melodic type, but which maintain the same defining features.

For this purpose, the text and melody proposed by Del Álamo and Gil (2000), a children's song with a greater number of coincidences with other texts and sound documents located, have been taken as standards:

EL SEÑOR DON GATO

Estando el señor Don Gato (a)
sentadito en su tejado, (b)
marramamiau, miau, miau, (i)
sentadito en su tejado. (b)
Ha recibido una carta (c)
que si quiere ser casado, (d)
marramamiau, miau, miau, (i)
que si quiere ser casado. (d)
Con una gata morena (e)
sobrina de un gato pardo, (f)
marramamiau, miau, miau, (i)
sobrina de un gato pardo. (f)
De contento que se ha puesto, (g)
se ha caído, se ha matado, (h)
marramamiau, miau, miau, (i)

se ha caído, se ha matado. (h)
Se ha roto siete costillas, (j)
el rabo y el espinazo, (k)
marramamiau, miau, miau, (i)
el rabo y el espinazo. (k)
Ya lo llevan a enterrar (l)
por la calle del Pescado, (m)
marramamiau, miau, miau, (i)
por la calle del Pescado. (m)
Al olor de las sardinas, (n)
el gato ha resucitado, (ñ)
marramamiau, miau, miau, (i)
el gato ha resucitado. (ñ)
Por eso dice la gente: (o)
"siete vidas tiene un gato", (p)
marramamiau, miau, miau, (i)
"siete vidas tiene un gato". (p)
Mr. Don Gato was sitting on his roof
Sitting on his roof,
marramameow, meow, meow, meow,
Sitting on his roof.
He has received a letter
That if he wants to be married,
marramameow, meow, meow, meow,
If he wants to be married.
With a brown cat
Niece of a brown cat,
marramameow, meow, meow, meow,
niece of a brown cat.
She's so happy she's been so happy,
she has fallen, she has killed herself,
marramameow, meow, meow, meow,
she's fallen, she's fallen, she's killed herself.
He has broken seven ribs,
the tail and the spine,
marramameow, meow, meow, meow,
his tail and his backbone.
They're already taking him to be buried
in the Fish Street,
marramameow, meow, meow, meow,

down Fish Street.
To the smell of sardines,
the cat has risen from the dead,
marramameow, meow, meow, meow,
the cat has risen.
That's why people say:
'a cat has seven lives',
marramameow, meow, meow, meow,
'seven lives a cat has'.

4.1. Structure and Form Analysis

The above text is characterised by its unique structure, where each of the 32 lines forms a quatrain stanza with an alternating rhyme. This arrangement, which combines an octosyllabic metre with a recurring onomatopoeic interjection, contributes significantly to the poem's musicality. The choice of octosyllabic verse, typical of popular poetry and traditional romances, facilitates its recitation and singing, making it an accessible and attractive text for the general public. The alternating assonant rhyme (abib, cdid, etc.) adds an additional layer of sonority which, together with the refrain, reinforces the rhythm and cyclical structure of the work.

The use of the refrain *marramamiau, miau, miau* at the end of each stanza plays a crucial role in the poem. This element not only adds a playful and humorous touch, but also acts as a rhythmic respite that separates and, at the same time, connects the different stanzas. The onomatopoeia, which imitates the sound of a cat's meow, underlines the relationship with the protagonist of the narrative and gives the poem an almost childlike tone, close to oral tradition and lullabies. The constant repetition of this refrain creates a hypnotic effect, capturing the listener´s attention and inviting him to follow the rhythm of the poem.

On the other hand, the structure of the poem, with its constant repetition of the refrain and its assonant rhyme, not only facilitates memorisation, but also invites collective participation. This type of composition was common in the folk tradition, where poetry was not only read, but also recited and sung in a group. The repetition of the refrain, easy to remember, allowed the audience to join in the recitation, making the poem a shared and dynamic experience.

Furthermore, *marramamiau, miau, miau, miau* can be interpreted as a stylistic device that plays with the idea of the everyday and the trivial. Although it apparently lacks deep meaning, its constant repetition turns it into a symbol which, as in popular sayings and proverbs, is integrated into the discourse to reinforce the sense of community and tradition. This resource gives the poem a playful and, at the same time, universal character, capable of transcending generations thanks to its simplicity and effectiveness.

The poem, with its octosyllabic form typical from the popular tradition, assonant rhyme, and the recurring onomatopoeic refrain, successfully embodies elements of popular tradition. Its melodic quality, combined with the straightforwardness of its structure and themes, renders it a piece that can be appreciated in both private reading and communal recitation, encapsulating the spirit of poetry as a vibrant, communal art form.

Below, as an example, table 2 shows other texts from different geographical locations in which the variations in content and style can be seen.

Table 2. Texts/ Lyrics from different countries

From Tetuan	Estaba 'l Señor don Gato sentadito `n su tejado Con las mano 'n la cintura y la otra 'n el costado. Por ahí pasó una gata con los ojo 'arrelumbrando, la gata, por darle un beso, se ha caído del tejado. Mandaron los ductores, ductores y cirujanos: uno le mira la pierna, otro le mira 'l costado. Señor Gato está muy malo, señor Gato ya se ha muerto, ya lo llevan a enterrar y a la plaza del pescado: al olor de las sardinas el gato ha resucitado! Hai, hai, hai!
From Guatemala	Estando el señor don Gato sentadito en silla de oro, estrenando medias blancas y zapato acharolado. En eso pasó la gata mostrándole indiferencia y el gato por darle un beso se cayó desde el tejado. Llamaron al cirujano y también al boticario, le recetaron un caldo de longaniza y recado. Se murió el señor don Gato a la mitad del puchero, la gata vistió de luto, los gatos de pelo negro. Los ratones muy contentos bailaban en el tejado. - se murió quien nos comía al caer descalabrado. Ya se murió el gato pinto, ya lo llevan a enterrar, entre cuatro gatos pardos que lo van a sepultar.
From Canada	Oh Señor Don Gato was a cat on a high red roof Don Gato sat he went there to read a letter, Meow, meow, meow where the reading light was better, meow, meow, meow `Twas a love note for Don Gato! 'I adore you', wrote the lady cat who was fluffy, white and nice and fat. There was not a sweeter kitty, meow, meow, meow inthe country or the city, meow, meow, meow and she said she´d wed Don Gato! Oh, Don Gato umped so happily! He fell off the roof and broke his knee. Broke his ribs and all his whiskers, meow, meow, meow and his little solar plexus, meow, meow, meow 'Ay Caramba!' cried Don Gato. Then the doctor all came on the run, just to see if something could be done. And they held a consultation, meow, meow, meow about how to save their patient, meow, meow, meow how to save Señor Don Gato. But despite everything they tried poor Señor Don Gato up and died. And it wasn´t very merry, meow, meow, meow for the ending of Don Gato. When the funeral passed the market square, such a smell of fish was in the air! Though his burial was slated, meow, meow, meow he became reanimated, meow, meow, meow he came back to life, Don Gato!

Source: Own elaboration (Wich-Shahak, 1997; Thomas, 1966; https://www.kididdles.com/lyrics/s038.html)

These three texts reflect the different models and classifications defined above, highlighting the specific features of communal recreation.

Figure 2 shows one of the more standardised scores of the romance *El Señor Don Gato.*

Figure 2. Score El Señor Don Gato

Source: (Del Álamo y Gil, 2000)

4.2. Brief Musical Analysis of the Score

The above score is in C major and consists of a sequence of 12 bars that are repeated on several occasions.

The score shows a simple and cheerful melody, with a 2/4 time signature, except for one bar that would change to a 3/4 ternary structure, which favours agile and continuous movement, a well-structured phrasing in two contrasting musical ideas and the use of figures that oscillate between quarter notes, eighth notes, sixteenth notes and their corresponding rests, providing dynamism and rhythmic variety. The presence of dotted notes and eighth-note rests, which contribute to a sense of light syncopation and rhythmic variety within the framework of 2/4 time signatures. The repetition of certain melodic motifs in each phrase also suggests the character of a popular round or theme, where repetition and melodic simplicity are fundamental.

The phrasing of the melody seems to be divided into two phrases of four bars each, each of them forming a complete musical idea. This can be seen in the structure of the bars, which culminate in the first phrase with a semi-cadence and in the second with a full cadence. This type of phrasing is typical in binary structures and provides a balance between question and answer within the piece.

In the following tables (figure 3, figure 4 and figure 5), different melodies are presented in which the differences between musical variant and musical version can be clearly observed.

Figure 3. Musical version found on the recording from Gerineldo (1994)

Transcription: Narciso José López García.

Figure 4. Musical version localised in León (1983)

Figure 5. Musical variant. Located on Voices of the Turtle (1990).

4.3. Identification of the Discovered Audio Documents

Table 3 manifests the sound documents that, from our point of view, have been most interesting for their ability to bring us closer to the characteristics, evolution and development of the oral transmission of the different groups and geographical areas to which these musical variants and versions belong.

All of them have been extracted after a search and analysis of sound recordings on different digital music streaming platforms, such as *Spotify* and *YouTube Music*.

Table 3. Sound documents located

Music album	Artist/Musical Group	Date of publication		Music
El gripau blau	Ara va de bo	1974	13	Version
La rueda 2. Canciones infantiles	Xesco Boix	1977	12	Version
Cantemos con el Pro Música de Rosario	Pro Música de Rosario	1981	15	Version
El Pro Música de Rosario le canta a los niños del mundo	Pro Música de Rosario	1982	4	Version
Con ton y son	Pro Música de Rosario	1986	17	Variant
Bridges of song: music of the Spanish Jews of Morocco	Voice of the turtle	1990	17	Variant
Music of New Mexico. Hispanic traditions	VVAA (Mercedes López)	1992	17	Version

continued on following page

Table 3. Continued

Music album	Artist/Musical Group	Date of publication		Music
Canciones para niños	Rosa León	1993	7	Version
En medio de aquel camino. Chansons judeo-espagnoles	Gerineldo	1994	5	Variant
Canciones de todo el mundo. La clave. música para niños	Resistencia	1996	6	Version
Con arte y con parte	Pro Música Niños Rosario	1998	8	Version
Canciones infantiles, vol. 3	VVAA	1999	4	Variant
Family Fiesta. Play songs from Northern Spain	Family Fiesta	2000	3	Variant
Con la música a todas partes	Pro Música Niños Rosario	2001	14	Version
Patitos feos	VVAA (Maita Vende Cá)	2002	7	Variant
Las canciones infantiles de siempre	Coro de las Escuelas Avemarianas	2005	4	Variant
100 temas infantiles, vol. 2	Joaquín Díaz	2006	2	Variant
Música para niños, vol. 2	Pro Música de Rosario	2007	7	Version
Canciones infantiles, vol. 5	Grupo Musical Ginesitos	2014	16	Variant
Trece romances y una cançó	Eliseo Parra	2018	9	Version

Source: Own elaboration

Eighteen themes listed in the previous table (Table 3) can be heard at the following link:

https://open.spotify.com/playlist/20ajBjsuZuvHOVJ11PmTgA?si=c078a9a06d8d463f

5. CONCLUSION

Oral transmission has been fundamental in the preservation and dissemination of folk, popular and traditional music through the generations. This aspect is considered a characteristic feature of European traditional music, in contrast to 'cultured' or 'classical' music, whose documentary development began much later. Oral transmission not only ensures the continuity of these songs, romances and sayings, but also plays an active part in their evolution, as performers, through their recreation in community, modify and adapt the content, generating different versions according to region and context. This phenomenon, far from being a simple process of deterioration, reflects human particularities in the reproduction and shaping of music in a systematic way, resulting in a rich variety of variants and non-authentic forms in the traditional musical repertoire.

This research has compiled a total of 91 variants and versions of the popular ballad *El Señor Don Gato* from 9 countries across three continents (Europe, Africa, and America). The versions and variants are presented in three languages (Spanish, English, and Portuguese), and some variations in the text's content have been observed throughout this study.

Collective recreation has led to a diversity of linguistic expressions and musical variations of the same work. A clear example of this phenomenon is the romance of *El Señor Don Gato*, a representative piece of Spanish folklore that evidences the effects of such recreation. Likewise, cultural interaction throughout the history of the Iberian Peninsula has favoured the appearance of these situations.

The documents examined reveal different versions and variants of romance, the result of the work of numerous folklorists and ethnomusicologists who have devoted themselves to the compilation and classification of musical elements of our tradition. This research work, as Ríos (1993) points out, has allowed the preservation of these elements and has reduced the risk of their being lost over time.

REFERENCES

Anglada-Tort, M., Harrison, P. M. C., & Jocoby, N. (2022). Studying the Effect of Oral Transmission on Melodic Structure using Online Iterated Singing Experiments. DOI: 10.1101/2022.05.10.491366

Balcells, A. (1991). *Catalonia and its History: A Survey of Modern History*. Palgrave Macmillan.

Bardin, L. (2002). *Análisis de contenido* (3rd ed.). Akal.

Blacking, J. (1973). *How musical is man?* University of Washington Press.

Bowen, G. A. (2009). Document analysis as a qualitative research method. *Qualitative Research Journal*, 9(2), 27–40. DOI: 10.3316/QRJ0902027

Carpentier, A. (1983). *La música en Cuba*. Fondo de Cultura Económica.

Coelho, F. A. (1893). *Cancioneiro popular português*. Typographia Castro Irmão.

Damrosch, D. (2003). *What is World Literature?* Princeton University Press. DOI: 10.1515/9780691188645

Del Álamo, L. y Gil, V. (2000). Taller de música 3. SM.

Feld, S. (2012). *Sound and Sentiment: Birds, Weeping, Poetics, and Song in Kaluli Expression*. Duke University Press.

Gerineldo (1994). El Señor don Gato. [canción] *En el medio de aquel camino: chansons judeo-espagnoles*. Colección de Oro Anahory-Librowicz.

Glenn, I. D. (2010). *Documentary research method. Palgrave Macmillan. Gómez García, Z. (1998). El areíto: Expresión de resistencia cultural*. Editorial de Ciencias Sociales.

Keegan-Phipps, S. (2013). An Aural Tradition with a Pause Button? The Role of Recording Technology in a North East English Folk Session. *Ethnomusicology*, 57(1), 34–56. DOI: 10.5406/ethnomusicology.57.1.0034

León, R. (1983). Estaba el Señor Don Gato [canción]. *Canciones para niños*. Fonomusic.

Mehr, S. A., Krasnow, M. M., Bryant, G. A., & Hagen, E. H. (2020). Origins of music in credible signaling. *Behavioral and Brain Sciences*, 44, e60. DOI: 10.1017/S0140525X20000345 PMID: 32843107

Mehr, S. A., Singh, M., Knox, D., Ketter, D. M., Pickens Jones, D., Atwood, S., Lucas, C., Jacoby, N., Egner, A. A., Hopkins, E. J., Howard, R. M., Hartshorne, J. K., Jennings, M. V., Simson, J., Bainbridge, C. M., Pinker, S., O'Donnell, T. J., Krasnow, M. M., & Glowacki, L. (2019). Universality and diversity in human song. *Science*, 366(6468), eaax0868. DOI: 10.1126/science.aax0868 PMID: 31753969

Merriam, A. P. (1964). *The Anthropology of Music*. Northwestern University Press.

Nettl, B. (2005). *The Study of Ethnomusicology: Thirty-One Issues and Concepts*. University of Illinois Press.

Peña, T., & Pirela, J. (2002). La complejidad del análisis documental. *Información, Cultura y Sociedad*, 6, 55–81. http://revistascientificas.filo.uba.ar/index.php/ICS/article/view/869

Rees, H. (2010). *Lives in Chinese Music*. University of Illinois Press.

Rice, T. (2014). *Ethnomusicology: A Very Short Introduction*. Oxford University Press.

Ríos, V. (1993). *Un muestreo en la poesía tradicional de la Mancha Baja*. Revista Zahora 33. Diputación Provincial de Albacete. Ortiz, F. (1985). *El contrapunteo cubano del tabaco y el azúcar.* Biblioteca Ayacucho.

Salazar, R. (2002). *Historia y leyenda de la música popular venezolana*. Monte Ávila Editores.

Sangil, O. (2006). *Literatura y leyendas de la España medieval*. Editorial Castalia.

Savage, P. E., Brown, S., Sakai, E., & Currie, T. E. (2015). Statistical universals reveal the structures and functions of human music. *Proceedings of the National Academy of Sciences of the United States of America*, 112(29), 8987–8992. DOI: 10.1073/pnas.1414495112 PMID: 26124105

Seeger, A. (2004). *Why Suyá Sing: A Musical Anthropology of an Amazonian People*. University of Illinois Press.

Shanahan, D., & Albrecht, J. (2019). Examining the effect of oral transmission on folksongs. *Music Perception*, 36(3), 273–288. DOI: 10.1525/mp.2019.36.3.273

Stobart, H. (2008). *The New (Ethno)musicologies*. The Scarecrow Press.

Suzel Ana, R. (2006). Intangible Cultural Heritage: Global Awareness and Local Interest. *In Yearbook for Traditional Music, 38,* 36-44.

Tejero, E. (2003). El siete, número cósmico y sagrado. Su simbología en la cultura y rendimiento en el Romancero. *Didáctica. Lengua y Literatura, 15,* 221-253. https://revistas.ucm.es/index.php/DIDA/article/view/DIDA0303110221A

Thomas, E. (1966). Three mexican indian carnaval songs. *Ethnomusicology*, 1(10), 58–69.

Vicuña, J. (1911). *El romancero de los sefardíes*. Imprenta Universitaria.

Voice of the Turtle (1990). Estaba Señor Don Gato [canción]. *Bridges of song: music of the Spanish Jews of Morocco*. Titanic Records.

Weich-Shahak, S. (1994). *Romancero Sefardí de Marruecos*. Instituto de Filología.

Weich-Shahak, S. (1997). *Romancero sefardí de Marruecos*. Alpuerto.

KEY TERMS AND DEFINITIONS

Community recreation: is a process closely related to the oral transmission of traditional music. In this context, individuals who learn a piece of music tend to modify it, either consciously or unconsciously, when performing it or teaching it to others. This transformation can manifest itself in different settings, such as popular festivities, rituals or in the everyday life of communities. Over time, as music is passed from one generation to the next, variations are introduced into the melody, lyrics, rhythm or structure of the original work. This may result in new versions of the same piece or even in the creation of completely different compositions. This phenomenon illustrates the dynamic and lively nature of folklore, where collective creativity plays a fundamental role in the evolution and preservation of musical traditions.

Cultural evolution: is the ongoing process through which a society's ideas, practices, technologies, and customs transform and advance over time. This process includes the transmission, modification, and adaptation of cultural elements across generations, often shaped by interactions and exchanges among various cultures. Innovation is a vital component of this evolution, as the emergence of new concepts and technologies can significantly impact societal norms and values. While cultural evolution has the potential to drive substantial social progress by promoting creativity, inclusivity, and adaptability to new conditions, it may also incite tensions and resistance, especially when such changes confront long-standing traditions or

disrupt established lifestyles. These tensions can surface in discussions regarding cultural identity: social cohesion, and the safeguarding of heritage, underscoring the intricate relationship between continuity and change within societies.

El Señor Don Gato: is a traditional children's tune of Spanish origin that narrates the tale of a cat who receives a letter informing him that his marriage proposal has been accepted by a female cat. However, upon reading the letter, he becomes so overwhelmed with joy that he falls from the roof and dies. This song, characterized by its humorous tone and catchy melody, has been sung by generations of children in Spain and Latin America. Despite the cat's demise, the conclusion of the song is cheerful, as the cat 'comes back to life' upon learning that his wedding is still on, transforming the narrative into a playful blend of tragedy and comedy, a hallmark of popular children's songs.

Ethnomusicology: is the examination of music within its cultural context. It can be described as a form of musical anthropology, as it seeks to understand the reasons and methods by which people create music across various regions of the world, and how this music mirrors their lives, traditions, beliefs, and environments. The study encompasses not only melodies and rhythms but also the entire framework surrounding musical creation and performance: who participates, when and where it occurs, the motivations behind it, and the significance it holds for those involved.

Oral transmission: refers to the process by which knowledge, customs, stories, songs, and traditions are conveyed from one generation to the next through spoken language, without the use of written forms. This method of transmission is essential in cultures where literacy is not widespread or utilized, relying on memorization and repetition. It enables the cultural heritage of a community to remain vibrant and evolve over time.

Popular songs: are compositions that are created and shared collectively within a community through oral tradition, reflecting the experiences, emotions, and values of the people. These texts are often anonymous and evolve over time as they are passed from person to person, adapting to cultural and social changes. Their content ranges from every day and romantic themes to epic tales and legends, serving as a vital means of preserving the identity and collective memory of a culture.

Variant vs musical version: the distinction between musical variant and musical version is based on the relationship that each establishes with the original melody. A musical version is defined as those melodic compositions which, although they can be used to accompany the same text or different literary adaptations of it, do not have common melodic characteristics. In this sense, a musical version represents a creation that separates itself from the original melody, offering a completely renewed perspective. In contrast, a musical variant refers to different interpretations of the same melody that preserve its distinctive elements, such as rhythm, structure or fundamental intervals. Although there may be variations in ornamentation or interpretative aspects, musical variants maintain an obvious connection with the original melody, which allows them to be identified as derivations of the same work. In conclusion, while musical versions are autonomous reinterpretations, musical variants are more subtle modifications within a shared melodic context.

Compilation of References

Abarca Salinas, M. E. (2017). El aprendizaje basado en proyectos, una estrategia didáctica desde historia y geografía para intencional la participación ciudadana en las comunidades educativas. *Revista Andamio, 4*(2), 53-71, https://revista-andamio.cl/index.php/revista/article/view/83

Abreu, C. (2014). Horacio Quiroga, a writer on the limits. *A contracorriente, 11* (2), 302-322.

Agencia Estatal Boletín Oficial del Estado del Gobierno de España. (2024). *Real Decreto 1630/2006, de 29 de diciembre, por el que se establecen las enseñanzas mínimas del segundo ciclo de Educación infantil.* https://www.boe.es/eli/es/rd/2006/12/29/1630

Agencia Estatal de Meterología (AEMET). (2021). *Proyecciones Climáticas para el siglo XXI en España.* https://www.aemet.es/es/serviciosclimaticos/cambio_climat

Aguilar, O., Rodríguez, E., Armiñana, R., & Padilla, A. (2023). Un itinerario didáctico, buenas prácticas para la formación de profesores de Geografía. *Revista Conrado, 19*(91), 320-331. 1990-8644-rc-19-91-320.pdf (sld.cu)

Aguilera, M. J., González, M. P., & Santos, J. M. (2009). *Geografía General I: Geografía Física.* UNED.

Aibar, E. A. (2023). *Programa educativo de conservación del medio ambiente para mejorar la formación de conciencia ambiental en los estudiantes de quinto y sexto grado de educación primaria de la institución educativa N°1152 - Dos de Mayo, Cercado de Lima, periodo 2019* [Tesis doctoral, Universidad Nacional Federico Villarreal]. Repositorio Institucional UNFV. https://hdl.handle.net/20.500.13084/7237

Aini, M., & Laily, P. (2010). Preparedness of Malaysian pre-school educators for environmental education. *Pertanika Journal of Social Science & Humanities*, 18(2), 271–283.

Ait Baha, T., El Hajji, M., Es-Saady, Y., & Fadili, H. (2024). The impact of educational chatbot on student learning experience. *Education and Information Technologies*, 29(8), 10153–10176. DOI: 10.1007/s10639-023-12166-w

Akçay, İ. (2006). Farklı ülkelerde okul öncesi öğrencilerine yönelik çevre eğitimi. [Environmental education for preschool students in different countries], (Master's Thesis), Uludağ University Institute of Social Sciences.

Akgun, S., & Greenhow, C. (2022). Artificial intelligence in education: Addressing ethical challenges in K-12 settings. *AI and Ethics*, 2(3), 431–440. DOI: 10.1007/s43681-021-00096-7 PMID: 34790956

Akınoğlu, O., & Sarı, A. (2009). İlköğretim programlarında çevre eğitimi. [Environmental Education in Prımary School Curriculum]. *Marmara University Atatürk Education Faculty Journal of Educational Sciences*, 30(30), 5–29.

Alacid de Pascual, S. (2021). La sostenibilidad dentro del aula de Educación Infantil. *Quest Journals. Journal of Research in Humanities and Social Science*, Volume 9-Issue 10, 74-83. https://www.questjournals.org/jrhss/papers/vol9-issue10/Ser-5/K09107483.pdf

Alacid, S. (2021). La sostenibilidad dentro del aula de Educación Infantil. *Journal of Research in Humanities and Social Science*, 9(10), 74–83. https://www.questjournals.org/jrhss/papers/vol9-issue10/Ser-5/K09107483.pdf

Alavés Ruíz, A. (2014). *Interculturalidad: conceptos, alcances y derecho*. Centro de Producción editorial del GCPRD. https://rm.coe.int/1680301bc3

Albaladejo, T. (1992). *Semántica de la narración: la ficción realista*. Taurus.

Alcántara, J., Rubio, S., & Mora, M. (2014). Google Earth como herramienta para formadores en la preparación de itinerarios didácticos. [Trabajo Académico, Universidad de Córdoba]. Universidad de Córdoba. https://helvia.uco.es/bitstream/handle/10396/15443/geografia6.pdf?sequence=1

Allen, D. E., & McAller, M. (2018). Fake news and indifference to scientific fact: President Trump's confused tweets on global warming, climate change and weather. *Scientometrics*, 117(1), 625–629. DOI: 10.1007/s11192-018-2847-y

Alomar-Garau, G., & Gómez-Zotano, J. (2022). Del paisaje en la Geografía: Perspectivas y actividades del grupo de paisaje de la Asociación Española de Geografía. *Cuadernos de Geografia*, 108-109(108-9), 857–874. DOI: 10.7203/CGUV.108-9.24067

Alonso Arenal, S., González Alonso, S., González Alonso, A. P., & González Alonso, M. (2010). *Didáctica de las Ciencias Sociales para la Educación Primaria*. Pirámide.

Alonso-Sainz, E. (2022). Las TIC en la etapa de educación infantil: Una mirada crítica de su uso y reflexiones para las buenas prácticas como alternativa educativa. *Vivat Academia.Revista de Comunicación*, 155, 241–263. DOI: 10.15178/va.2022.155.e1371

Alvarado-Herrera, A., Bigne, E., Aldas-Manzano, J., & Curras-Pérez, R. (2017). Una escala para medir las percepciones de los consumidores sobre la responsabilidad social corporativa siguiendo el paradigma del desarrollo sostenible. *Journal of Business Ethics*, 140, 243–262. DOI: 10.1007/s10551-015-2654-9

Alves, M. L., Brazão, M., & Martins, O. S. (2001). *Programa de Geografia A*. Ministério da Educação.

Amorós, A. (1994). Introducción. En Elena Quiroga, *La última corrida* (pp. 7-29). Editorial Castalia.

Amossy, R., & Herschberg Pierrot, A. (2010). *Estereotipos y clichés*. Eudeba.

Amsler, S., Kerr, J., & Andreotti, V. (2020). Interculturality in teacher education in times of unprecedented global challenges. *Education and Society (Melbourne)*, 38(1), 13–37. DOI: 10.7459/es/38.1.02

Anderson, B. (2023). Forms and scenes of attachment: A cultural geography of promises. *Dialogues in Human Geography*, 13(3), 392–409. DOI: 10.1177/20438206221129205

Anderson, M. P., Woessner, W. W., & Hunt, R. J. (2015). *Applied groundwater modeling: simulation of flow and advective transport*. Academic press.

Andreotti, V. (2006). Soft versus critical global citizenship *education. Policy & Practice: A Development.Education Review*, 3, 40–51.

Andújar Llosa, A., Sánchez Díaz, N., Pradillo Viñerta, Sh., & Sabín Galán, F. (2022). *Jóvenes y racismo. Estudio sobre las percepciones y actitudes racistas y xenófobas entre la población joven de España. Noviembre de 2022*. Centro Reina Sofía sobre Adolescencia y Juventud, Fundación Fad Juventud. DOI: 10.5281/zenodo.7268038

Anglada-Tort, M., Harrison, P. M. C., & Jocoby, N. (2022). Studying the Effect of Oral Transmission on Melodic Structure using Online Iterated Singing Experiments. DOI: 10.1101/2022.05.10.491366

Anon, S. (1992). *Çevre ve çocuk* [Environment and children]. Environment and Human Journal. Çevre Bakanlığı Yayınları.

Anyanwu, R., & Le Grange, L. (2017). The influence of teacher variables on climate change science literacy of Geography teachers in the Western Cape, South Africa. *International Research in Geographical and Environmental Education*, 26(3), 193–206. DOI: 10.1080/10382046.2017.1330039

Aral, N., Kandır, A. ve Can Yaşar, M. (2001). *Okul Öncesi Eğitim* [Preschool Education].Pre-School Education 1. Ya-Pa

Aranda Hernando, A. (2016). *Didáctica de las Ciencias Sociales en Educación Infantil*. Síntesis.

Aranda, A. M. (2016). *Didáctica de las ciencias sociales en Educación Infantil*. Síntesis.

Arango, M. A. (1982). Sobre dos cuentos de Horacio Quiroga: Correlación en el tema de la muerte, el ambiente y la estructura narrativa en "A la deriva" y "El hombre muerto". *Thesaurus*, XXXVII(1), 153–161.

Araya, F., & Cavalcanti, L. (2018). Desarrollo del pensamiento geográfico: Un desafío para la formación docente en Geografía. *Revista de Geografía Norte Grande*, 70(70), 51–69. DOI: 10.4067/S0718-34022018000200051

Ardoğan, R. (2012). Tüketim, Nüfus Ve Çevre Sorunları: Orantısız Denklem [Consumption, Population and Environment Problems: disproportionate equation]. *Tarih Kültür ve Sanat Arastirmalari Dergisi*, 1(4), 81–106.

Ardoin, N. M., & Bowers, A. W. (2020). Early childhood environmental education: A systematic review of the research literature. *Educational Research Review*, 31, 100353. DOI: 10.1016/j.edurev.2020.100353 PMID: 34173434

Arévalo Viveros, D. F. (2009). El cuento es la selva: lectura crítica-ecológica de *Los cuentos de la selva* de Horacio Quiroga. *452°F. Revista electrónica de teoría de la literatura y literatura comparada, 1*, 121-132

Arnell, N. W., Lowe, J. A., Challinor, A. J., & Osborn, T. J. (2019). Global and regional impacts of climate change at different levels of global temperature increase. *Climatic Change*, 155(3), 377–391. DOI: 10.1007/s10584-019-02464-z

Aslan Efe, H. (2015). The Effect of Anımatıon Supported Environmental Education on Academic Achievement, Retention and Attitude Towards the Environment. *Journal of Computer and Education Research*, 3(5), 130–143. DOI: 10.18009/jcer.90852

Associação de Professores de Geografia & Associação Portuguesa de Geógrafos. (2019). *Educação Geográfica e Ordenamento do Território*. https://www.aprofgeo.org/wp/?p=1361

Atasoy, E. (2005). *Çevre İçi Eğitim: İlköğretim Öğrencilerinin Çevresel Tutum ve Çevre Bilgisi Üzerine Bir Çalışma [Environmental Education: A Study on Environmental Attitude and Environmental Knowledge of Primary School Students]* (Doctoral dissertation, Bursa Uludag University Turkey).

Atasoy, E. (2005). *Çevre İçin Eğitim: İlköğretim Öğrencilerinin Çevresel Tutum ve Çevre Bilgisi Üzerine Bir Çalışma* [Education for the Envıronment:Primary Education Students' Environmental Attitudes And A Study On Environmental Knowledge]. (Doctoral dissertation) Uludag University.

Atasoy, E. (2010). Genel Coğrafya. [General Geography]. (2. Baskı). İstanbul: Ezgi Kitabevi Yayınları.

Ávila Dávalos, H. (2022). Multiculturalidad e interculturalidad: El papel de la educación superior para generación de competencias interculturales para el contexto organizacional. *Educación y Humanismo*, 24(43), 13–34. DOI: 10.17081/eduhum.24.43.4838

Aydın, M., Dündar, R. ve Korkut, Ş. (2016). Türkiye'de Ekolojik Okuryazarlık Eğitimine Yönelik Öğretmen Görüşleri. [Teacher Views Upon Ecological Literacy Education in Türkeye].*Abant İzzet Baysal University Journal of Faculty of Education,* Cilt 16, (USBES Özel Sayı II), 1160-1172.

Ayuso, S. (2006). Educación ambiental en espacios naturales protegidos: Aprendiendo en la naturaleza. *Medio Ambiente Castilla-La Mancha*, 13, 22–27.

Bailey, P. (1981). *Didáctica de la geografía*. Editorial Cincel.

Bailly, A., Beguin, H., & Scariati, R. (2016). *Introduction à la géographie humaine* (9th ed.). Armand Colin. DOI: 10.3917/arco.baill.2016.01

Balaguer Mora, P. A. (2018). Geografía crítica y pensamiento crítico. *Actualidades Pedagógicas*, (72), 73–95. DOI: 10.19052/ap.5232

Balcells, A. (1991). *Catalonia and its History: A Survey of Modern History*. Palgrave Macmillan.

Bale, J. (2012). *Didáctica de la Geografía en la Educación Primaria*. Morata.

Balkan Kıyıcı, F. (2009). Çevre eğitimi. [Environmental education].Sevinç, V. (Ed.), Eğitim fakülteleri için genel çevre bilimi içinde (s. 173- 183). Maya Akademi.

Balling, J. D., & Falk, J. H. (1982). Development of visual preference for natural environments. *Environment and Behavior*, 14(1), 5–28. DOI: 10.1177/0013916582141001

Baquedano, L. (2021). *Cinco panes de cebada*. SM.

Baquero Goyanes, M. (1951). *Problemas de la novela contemporánea*. Ateneo.

Baralo, M. (2003). Mestizaje e interculturalidad en la variación diatópica y su incidencia en español/ LE. En Pérez Gutiérrez, M. & Coloma Maestre, J. (eds.). *El español, lengua del mestizaje y la interculturalidad: actas del XIII Congreso Internacional de la Asociación para la Enseñanza del español como Lengua Extranjera (ASELE)*. ASELE, 152-164.

Bardin, L. (2002). *Análisis de contenido* (3rd ed.). Akal.

Barney, K. R. (1993). The Quiet Revolution of Elena Quiroga. *Revista Hispánica Moderna, 46*(1), 103–116. https://www.jstor.org/stable/30205865

Barnosky, A. D., & Hadly, E. A. (2016). *Tipping point for planet Earth: How close are we to the edge?* Editorial Macmillan.

Barrera, T. (2019). *Historia de la Literatura Hispanoamericana III: siglo XX*. Cátedra.

Barrett, P. M., Cooper, M., & Teoh, A. B. (2014). When time is of the essence: A rationale for earlier early intervention. *Journal of Psychological Abnormalities in Children*, 3(4), 1–9. DOI: 10.4172/2329-9525.1000133

Barzilai, S., & Chinn, C. A. (2020). A review of educational responses to the "post-truth" condition: Four lenses on "post-truth" problems. *Educational Psychologist*, 55(3), 107–119. DOI: 10.1080/00461520.2020.1786388

Basile, C. G. (2000). Environmental education as a catalyst for transfer of learning in young children. *The Journal of Environmental Education*, 32(1), 21–27. DOI: 10.1080/00958960009598668

Basile, C., & White, C. (2000). Respecting Living Thinks: Environmental Literacy for young Children. *Early Childhood Education Journal*, 28(1), 57–61. DOI: 10.1023/A:1009551705001

Baskarada, S. (2014). Qualitative case study guidelines. *The Qualitative Report*, 19(40), 1–25.

Bazolli, J. A. (2017). Nós Propomos! e a busca inovação no campo da extensão universitária. In J. A. Bazolli, S. Claudino, M. V. C. Silva, S. F. R. Viana, & W. C. Silva (Org.), *A Extensão Universitária como indutora à cidadania: a experiência do "Nós Propomos"* (pp. 13-27). EDUFT.

Beames, S., Higgins, P., Nicol, R., & Smith, H. (2023). *Outdoor learning across the curriculum: Theory and guidelines for practice*. Taylor & Francis. DOI: 10.4324/9781003010890

Bednarz, R., & Lee, J. (2019). What improves spatial thinking? Evidence from the Spatial Thinking Abilities Test. *International Research in Geographical and Environmental Education*, 28(4), 262–280. DOI: 10.1080/10382046.2019.1626124

Bellido, T. (1998). Lucía Baquedano, vida y obra. *Revista TK, 5*, 29-36. https://www.asnabi.com/revista/tk5/05bellido.pdf

Benavent, M. (2022). Una mirada constructiva a la realidad: los ODS en educación infantil. Una propuesta de promoción de la conciencia socio-ambiental. En D. García Monteagudo, N. Palacios, L.A. Rodríguez y C. Fuster (coords.), *La enseñanza de problemas socio-ambientales para una praxis escolar crítica* (pp. 37-52). Nau Llibres.

Benejam Arguimbau, P. (2002). La oportunidad de identificar conceptos clave que guíen la propuesta curricular de ciencias sociales, AAVV. *Las ciencias sociales: concepciones y procedimientos*, Graó, 11-19.

Benejam Arguimbau, P.; & Pagès Blanch, J. (coord.) (1997). *Enseñar y aprender Ciencias Sociales, Geografía e Historia en la Educación Secundaria.* ICE/HORSORI.

Benejam, P. (2003). Los objetivos de las salidas. *Didáctica de las Ciencias Sociales, Geografía e Historia* (7-12). Íber. Los objetivos de las salidas (1).pdf

Bennet, M. J. (1993). Toward ethnorelativism: A developmental model of intercultural sensitivity. In Paige, R. M. (Ed.), *Education for the intercultural experience* (pp. 21–71). Intercultural Press.

Bergua, S. B., Fields, D. L., & Gómez, N. B. (2024). *Geography and global change through interculturality and CLIL.* Universidad de Oviedo.

Berk, L. E. (2015). *Child Development* (9th ed.). Illinois State University: Pearson.

Bermúdez, M. (2010). Literatura infantil y juvenil hispanoamericana y educación intercultural: perspectivas de lectura. In Herrera, J., Abril, J. M., & Perdomo, C. (Eds.), *Estudios sobre didácticas de las lenguas y sus literaturas* (pp. 126–138). Universidad de la Laguna.

Besalú, X. (2002). *Diversidad cultural y educación.* Síntesis.

Binimelis, J., Usher, J., & Bagoly-Simó, P. (2023). How is Geography taught at Primary Education centers in Spain? Results of a state-wide study. *Boletín de la Asociación de Geógrafos Españoles*, 98.

Bishop, C. M. (2006). *Pattern recognition and machine learning (Information science and statistics).* Springer-Verlag New York, Inc.

Blacking, J. (1973). *How musical is man?* University of Washington Press.

Bladh, G. (2020). GeoCapabilities, didaktical analysis and curriculum thinking – Furthering the dialogue between Didaktik and curriculum. *International Research in Geographical and Environmental Education*, 29(3), 206–220. DOI: 10.1080/10382046.2020.1749766

Blanc, N. (13/08/2018). Cuentos de la selva: el centenario de un clásico salvaje. *La Nación*.

Blanco Lozano, P., De la Calle Carracedo, M., Fernández Rubio, C., Molero Otero, B., Ortega Sánchez, D., & Sánchez Agustí, M. (2008). La formación en didáctica de las Ciencias Sociales del maestro de Educación Infantil. Una propuesta adaptada al Espacio Europeo de Educación Superior. En R.M. Ávila Ruíz, M. Alcázar Cruz Rodriguez y M.C. Díez Bedmar (Coords.), *Didáctica de las Ciencias Sociales, currículo escolar y formación del profesorado. La didáctica de las Ciencias Sociales en los nuevos planes de estudio* (pp.113-134). Universidad de Jaén.

Bluestein, N., & Acredolo, L. (1979). Developmental Changes in Map-Reading Skills. *Child Development*, 50(3), 691–697. DOI: 10.2307/1128934

Boeve-de Pauw, J., Gericke, N., Olsson, D., & Berglund, T. (2015). The effectiveness of education for sustainable development. *Sustainability (Basel)*, 7(11), 15693–15717. DOI: 10.3390/su71115693

Bowen, G. A. (2009). Document analysis as a qualitative research method. *Qualitative Research Journal*, 9(2), 27–40. DOI: 10.3316/QRJ0902027

Boyd, K. J., Gold, A. U., & Littrell, M. K. (2022). Teaching practices around natural hazards and community resilience in Colorado. *Journal of Geoscience Education*, 70(4), 460–473. DOI: 10.1080/10899995.2021.1964319

Brent, A. (1959). The novels of Elena Quiroga. *Hispania*, 42(2), 210–213. DOI: 10.2307/335895

Briones Gómez, R. (2007). La integración del 'otro 'en la emergente España multicultural. *Gazeta de Antropología, 23*. https://docs.repositoriobiocultural.org/2021/08/03/la-integracion-del-otro-en-la-emergente-espana-multicultural/

Britland, J. (2024). El feminismo transtemporal y transnacional: mujeres que se unen y rebelan por la escritura y el *performance* en Elena Fortún y *Hermana (Placer)*, de María Folguera. En Á. Martín Pérez, S. Sevilla-Vallejo & J. Guzmán Mora (Eds.), *La dramaturgia femenina reclama su sitio* (pp. 345-362). Renacimiento.

Brooks, C. (2017). International differences in thinking geographically, and why "the local" matters. In Brooks, C., Butt, G., Fargher, M. (eds). *The Power of Geographical Thinking. International Perspectives on Geographical Education* (pp-169-179). Springer

Brugulat, M., Margalef, A., & Echeverria, A. (2021). Educación en riesgos naturales y autoprotección en Andorra. *Geotemas, 18* (ejemplar dedicado a: *X Congreso Geológico de España*), 631.

Bruner, J. S. (2018). *Desarrollo cognitivo y educación*. Ediciones Morata.

Buckingham, D. (2019). Teaching media in a 'post-truth' age: Fake news, media bias and the challenge for media/digital literacy education. *C&E, Cultura y Educación*, 31(2), 213–231. DOI: 10.1080/11356405.2019.1603814

Budzianowski, P., & Vulić, I. (2019). Hello, It's GPT-2-How Can I Help You? Towards the Use of Pretrained Language Models for Task-Oriented Dialogue Systems. In Birch, A., Finch, A., Hayashi, H., Konstas, I., Luong, T., Neubig, G., Oda, Y., & Sudoh, K. (Eds.). *Proceedings of the 3rd Workshop on Neural Generation and Translation* (pp. 15-22). Association for Computational Linguistics. DOI: 10.18653/v1/D19-5602

Butler, J. H., & Montzka, S. A. (2020). *The NOAA annual greenhouse gas index. Global Monitoring Laboratory*. https://www.esrl.noaa.gov/gmd/aggi/aggi.html

Cabello, A. (2023). Una habitación propia: El Premio Nadal y la escritora profesional en España. *Boletin de la Biblioteca de Menendez Pelayo*, 99(3), 21–47. DOI: 10.55422/bbmp.950

Cadová, R. (2007). La influencia de Edgard Allan Poe en Horacio Quiroga. *Études romanes de Brno, 37 (1)*, 149-157.

Çağlayan, Ç. (2015). Çocuk ve Çevre. [Child and Environment].Turk J Public Health is, 1(2), 32-38.

Calvo Buezas, T. (1989). *Los racistas son los otros: gitanos, minorías y derechos humanos en los textos escolares*. Editorial Popular.

Can, F. (2016). "Çevre Politikasının Ekonomik Araçları", [Economic Instruments of Envıronmental Policy]. Ömer Halisdemir *University* Faculty of Economics and Administrative Sciences Journal 9(3). 59-73.

Canal, M., Costa, D. & Santisteban, Al (2012). El alumnado ante problemas sociales relevantes: ¿Cómo los interpreta? ¿Cómo piensa la participación? En N. de Alba Fernández, F. F. García Pérez & A. Santisteban Fernández (Eds.), *Educar para la participación ciudadana en la enseñanza de las Ciencias Sociales,*.Vol. I (pp. 527-535). Díada Editora, S.L.

Canário, R. (2005). *O que é a Escola?* Porto Editora.

Canaza, F., Escobar-Mamani, F., & Huanca-Arohuanca, J. (2021). Reconocer a la bestia: Percepción de peligro climático en estudiantes de educación secundaria. *Revista de Ciencias Sociales*, XXVII(2), 417–434.

Cancer, L. A. (1999). *La degradación y la protección del paisaje.* Cátedra.

Canet, S., Morales, A. J., & García Monteagudo, D. (2019). Pensar geográficamente en la educación infantil: De la imaginación a la construcción social del espacio concebido. *Didáctica Geográfica*, 19(19), 23–46. DOI: 10.21138/DG.415

Canto, E. M. (2022). *El rol del maestro en la educación para la sostenibilidad ambiental* [Tesis de maestría, Universidad de Otavalo]. Repositorio Digital Universidad de Otavalo. https://repositorio.uotavalo.edu.ec/handle/52000/708

Capdevilla-Argüelles, N. (2005). Elena Fortún (1885-1952) y *Celia.* El *Bildungsroman* truncado de una escritora moderna. *Lectora: revista de dones i textualitat, 11*, 263-282. https://revistes.ub.edu/index.php/lectora/article/view/7127/9019

Capel, H.. (1983). *Ciencia para la burguesía: renovación pedagógica y enseñanza de la geografía en la revolución liberal española, 1814-1857.* Universitat de Barcelona, Edicions i Publicacions.

Cardona Andújar, J. (2016). *Saber docente desde la pedagogía poética.* Universidad Nacional de Educación a Distancia.

Caride, J. A., & Meira, P. Á. (2019). Educación, ética y cambio climático. *Innovación Educativa (México, D.F.)*, 29(29), 61–76. DOI: 10.15304/ie.29.6336

Carmona, R. (2012). The obligations of States Parties under the Convention on the rights of the child: a focus on rights in childhood policies in Spain. *Educatio Siglo XXI, 30*(2), 69-88. https://revistas.um.es/educatio/article/view/153691

Carpentier, A. (1983). *La música en Cuba.* Fondo de Cultura Económica.

Carrión, J. (2010). Las estructuras y el viaje. (Hacia un nuevo hispanismo). In Ortega, J. (Ed.), *Nuevos hispanismos interdisciplinarios y trasatlánticos* (pp. 239–251). Iberoamericana/Vervuert. DOI: 10.31819/9783954871537-014

Carrión-Martínez, J. J., Luque-de la Rosa, A., Fernández-Cerero, J., & Montenegro-Rueda, M. (2020). Information and Communications Technologies (ICTs) in Education for Sustainable Development: A Bibliographic Review. *Sustainability (Basel)*, 12(8), 3288. DOI: 10.3390/su12083288

Carter-Thuillier, B., López-Pastor, V., Gallardo-Fuentes, F., & Carter-Beltran, J. (2019). Inmigración y deporte escolar: Un caso de estudiantes latinoamericanos en España. *Journal of Sport and Health Research*, 11(2), 41–52.

Casanova Rodríguez, M. A., & Rodríguez, H. J. (2009). *La inclusión educativa, un horizonte de posibilidades*. Editorial La Muralla.

Case, R., & Wearing, J. (2016). *Teaching Geographical Thinking. Revised and expanded edition*. Royal Canadian Geographical Society.

Casper, A. M. A., Fernández-Giménez, M. E., & Balgopal, M. M. (2020). A tool for measuring ecological literacy: Coupled human-ecosystem interactions. *Journal of Agricultural Education and Extension*, 27(1), 21–34. DOI: 10.1080/1389224X.2020.1780139

Castellví, J., Diez-Bedmar, M.-C., & Santisteban, A. (2020). Pre-Service Teachers' Critical Digital Literacy Skills and Attitudes to Address Social Problems. *Social Sciences (Basel, Switzerland)*, 9(8), 134–145. DOI: 10.3390/socsci9080134

Castro, A., & Renés, P. (11 al 14 de noviembre de 2018). *La Educación Infantil y el desarrollo sostenible: pequeños pasos, grandes pisadas*. XXXVII Seminario Interuniversitario de Teoría de la Educación. Educación en la Sociedad de Conocimiento y el Desarrollo Sostenible. Universidad de la Laguna, San Cristóbal de La Laguna, España. https://riull.ull.es/xmlui/handle/915/11681

Castro, A., Patera, S. & Fernández, D. (2020). ¿Cómo aprenden las generaciones Z y Alpha desde la perspectiva docente? Implicaciones para desarrollar la competencia aprender a aprender. *Aula abierta, 49* (3), 279-285.

Catling: (2006). What Five-Year-Olds Know of the World? -Geographical Understanding and Play in Young Children's Early Learning.' Geography, C/S. 91 (1): 55-74.

Cavalcanti, L. S. (2019). *Pensar pela Geografia. Ensino e relevância social*. C&A Alfa Comunicação.

Centro para la Educación y Desarrollo de la Excelencia (CEDEC). (s. f.). *Banco de rúbricas y otros documentos*. Recuperado de Banco de rúbricas y otros documentos | Cedec (intef.es)

Cernev, T., & Fenner, R. (2020). The importance of achieving foundational Sustainable Development Goals in reducing global risk. *Futures*, 115, 102492. DOI: 10.1016/j.futures.2019.102492

Chams, N., & García-Blandón, J. (2019). On the importance of sustainable human resource management for the adoption of sustainable development goals. *Resources, Conservation and Recycling*, 141, 109–122. DOI: 10.1016/j.resconrec.2018.10.006

Chang, Y., Wang, X., Wang, J., Wu, Y., Yang, L., Zhu, K., Chen, H., Yi, X., Wang, C., Wang, Y., Ye, W., Zhang, Y., Chang, Y., Yu, P. S., Yang, Q., & Xie, X. (2024). A Survey on Evaluation of Large Language Models. *ACM Transactions on Intelligent Systems and Technology*, 15(3), 1–45. DOI: 10.1145/3641289

Chaudron, S., Di Gioia, R., & Gemo, M. (2018). Young children (0-8) and digital technology, a qualitative study across Europe. Retrieved January 15, 2021, from https://op.europa.eu/s/oNl9

Chawla, L. (2020). Childhood nature connection and constructive hope: A review of research on connecting with nature and coping with environmental loss. *People and Nature*, 2(3), 619–642. DOI: 10.1002/pan3.10128

Chevallard, Y. (2000). *La transposición didáctica: del saber sabio al saber enseñado*. Aique.

Chicano, C. (2015). Itinerario didáctico: la Catedral de Granada y su entorno. [Trabajo de Fin de Grado, Universidad de Granada]. Universidad de Granada. ITINERARIO DIDÁCTICO (ugr.es)

Chinn, A., Barzilai, S., & Duncan, R. G. (2021). Education for a "Post-Truth" World: New Directions for Research and Practice. *Educational Researcher*, 50(1), 6–68. DOI: 10.3102/0013189X20940683

Chitondo, L., & Chanda, C. T. (2023). Effects of ethnic diversity on the development of a country: A case of Zambia. *International Journal of Current Business and Social Sciences*, 9(5), 01-13.

Choulga, M., Janssens-Maenhout, G., Super, I., Solazzo, E., Agusti-Panareda, A., Balsamo, G., Bousserez, N., Crippa, M., Denier van der Gon, H., Engelen, R., Guizzardi, D., Kuenen, J., McNorton, J., Oreggioni, G., & Visschedijk, A. J. H. (2021). Global anthropogenic CO2 emissions and uncertainties as a prior for Earth system modelling and data assimilation. *Earth System Science Data*, 13(11), 5311–5335. DOI: 10.5194/essd-13-5311-2021

Cicin-Sain, B. (2015). Conserve and sustainably use the oceans, seas and marine resources for sustainable development. *UN Chronicle, 51*(4), 32-33. Decreto 81/2022, de 12 de julio, por el que se establece la ordenación y el currículo de Educación Primaria en la comunidad autónoma de Castilla-La Mancha. *Diario Oficial de Castilla-La Mancha, 134*, de 14 de julio de 2022.DOI: 10.18356/c5857454-fr

Cid, O. (1997). Educación ambiental en los espacios naturales protegidos. En R. Mendoza Castellón (Coord.), *Actas de la VIII aula de ecología en educación ambiental: Almería, 14 de febrero de 1994* (pp. 51-60). Instituto de Estudios Almerienses.

Ciplijauskaité, B. (1988). *La novela femenina contemporánea (1970-1985). Hacia una tipología de la narración en primera persona.* Anthropos.

Claro, M., Castro-Grau, C., Ochoa, J. M., Hinostroza, J. E., & Cabello, P. (2024). Systematic review of quantitative research on digital competences of in-service school teachers. *Computers & Education*, 215, 105030. DOI: 10.1016/j.compedu.2024.105030

Claudino Nunes, S., & Mendes, L. F. (2021). Project "We Propose!" Territorial citizenship and curricular innovation in portuguese geographical education. *Didáctica Geográfica*, 22(22), 47–71. DOI: 10.21138/DG.564

Claudino, S. (2019). Project We Propose! Building Territorial Citizenship from School. In J. A. Pineda-Alfonso, N. de Alba-Fernandez & E. Navarro-Medina, *Handbook of Research on Education for Participative Citizenship and Global Prosperity* (pp. 350-382). IGI Global. DOI: 10.4018/978-1-5225-7110-0

Claudino, S. (1990). Norman Graves: La Grande Patrie est la Communauté Européenne - un entretien avec Norman Graves. *The European Geographer*, 2, 51–54.

Claudino, S. (2015). A Educação Geográfica em Portugal e os Desafios Educativos. *Giramundo. Revista de Geografia do Colégio Pedro II*, 2(3), 7–19. DOI: 10.33025/grqcp2.v2i3.204

Claval, P. (2020). El papel del trabajo de campo en la geografía: de las epistemologías de la curiosidad a las del deseo. [Trabajo de campo, Universidad Nacional Autónoma de México]. Instituto de geografía. 2448-7279-igeo-103-e60283.pdf (scielo.org.mx)

Clements, R. (2004). An investigation of the state of outdoor play. *Contemporary Issues in Early Childhood*, 5(1), 68–80. DOI: 10.2304/ciec.2004.5.1.10

Coelho, F. A. (1893). *Cancioneiro popular português*. Typographia Castro Irmão.

Coménio, J. A. (1985). *Didáctica Magna - Tratado da Arte Universal de Ensinar Tudo a Todos* (3rd ed.). Fundação Calouste Gulbenkian.

Comes, P. (2002). La interpretación de las imágenes espaciales. En P. Martínez e I. González (coords.), *La geografía y la historia, elementos del medio* (pp. 209-244). Ministerio de Educación, Cultura y Deporte del Gobierno de España.

Comission on Geographical Education/International Geographical Education. (1992). *International Charter on Geographical Education*. Internacional Geographical Education. https://www.igu-cge.org/wp-content/uploads/2018/02/1.-English.pdf

Comission on Geographical Education/International Geographical Education. (2016). *International Charter on Geographical Education* (pp. 1–20). Internacional Geographical Education. https://www.igu-cge.org/wp-content/uploads/2019/03/IGU_2016_eng_ver25Feb2019.pdf

Cong-Lem, N., Soyoof, A., & Tsering, D. (2024). A Systematic Review of the Limitations and Associated Opportunities of ChatGPT. *International Journal of Human-Computer Interaction*, 1–16. DOI: 10.1080/10447318.2024.2344142

Consejería de Educación, Cultura y Deportes de la Junta de Comunidades de Castilla-La Mancha. (2022). Decreto 81/2022, de 12 de julio, por el que se establece la ordenación y el currículo de Educación Primaria en la comunidad autónoma de Castilla-La Mancha. *Diario Oficial de Castilla-La Mancha*, 135, 19860–19994.

Cordero, S. & Svarzman, J. (2007). *Hacer geografía en la escuela. Reflexiones y aportes para el trabajo en el aula.* Novedades educativas.

Corona, A. (2020). El sentido de pertenencia, una estrategia de mejora en el proceso formativo en las artes. Estudio de caso en Danza en una universidad mexicana. *Páginas de Educación*, 13(2), 59–79. DOI: 10.22235/pe.v13i2.2172

Corrales, M., & Garrido, J. (2024). Enseñanza de la geografía empleando como recursos los parques naturales de Extremadura: experiencia práctica con docentes en formación. En J. Mora, J. Garrido, & R. A. Castanho (Dirs.), *Estrategias públicas en la ordenación del espacio geográfico* (pp. 551–564). Aranzadi.

Cortizo Álvarez, T. (2020). La España vacía: De sintagma a geografísmo. *Polígonos. Revista de geografía*, 32, 159–178. https://revpubli.unileon.es/ojs/index.php/poligonos/article/view/6411/4933

Coscurão, R., & Claudino, S. (2022). O desenvolvimento de competências de pesquisa na disciplina de Geografia do ensino secundário: uma experiência didática. *Finisterra – Revista Portuguesa de Geografia*, 57(119), 193-211 DOI: 10.18055/Finis24884

Cox, C., Magendzo, A., Muñoz, G., & Redón, S. (2015). ¿Cómo debiera ser la formación ciudadana en la escuela? *Revista Docencia*, 58, 4–17.

Cramer, W., Guiot, J., Fader, M., Garrabou, J., Gattuso, J. P., Iglesias, A., Lange, M. A., Lionello, P., Llasat, M. C., Paz, S., Peñuelas, J., Snoussi, M., Toreti, A., Tsimplis, M. N., & Xoplaki, E. (2018). Climate change and interconnected risks to sustainable development in the Mediterranean. *Nature Climate Change*, 8(11), 972–980. DOI: 10.1038/s41558-018-0299-2

Crang, M. (2013). *Cultural geography*. Routledge. DOI: 10.4324/9780203714362

Crespo, J. M., Gómez, M. L., Cruz, L. A. (2018). Una aproximación a los Parques Nacionales y sus paisajes a través de itinerarios didácticos. *Revista Espacio, Tiempo y Forma, serie VI*(11), 121-140. Vista de Una aproximación a los Parques Nacionales y sus paisajes a través de itinerarios didácticos (uned.es)

Crespo, J. M. (2012). Un itinerario didáctico para la interpretación de los elementos físicos de los paisajes de la Sierra de Guadarrama. *Didáctica Geográfica*, 13, 15–34.

Crespo, J. M., Gómez, M. L., & Cruz, L. A. (2018). Una aproximación a los Parques Nacionales y sus paisajes a través de itinerarios didácticos. *Espacio, Tiempo y Forma. Serie VI, Geografia*, 11(11), 121–140. DOI: 10.5944/etfvi.11.2018.22359

Cristóvão, A., & Baptista, A. (2023). Desenvolvimento em contexto rural. In Monteiro, A. A., & Fragoso, A. (Eds.), *Desenvolvimento local em Portugal* (pp. 37–54). Edições Afrontamento.

Cruz, N., & Páramo, P. (2020). Educación para la mitigación y adaptación al cambio climático en América Latina. *Educación y Educadores*, 23(3), 469–489. DOI: 10.5294/edu.2020.23.3.6

Cuello, A., & García, F. F. (2019). ¿Ayudan los libros de texto a comprender la red fluvial de la ciudad? *Revista de Humanidades*, 37, 209–234.

Cullinan, B. E. (1992). *Read To Me: Raising Kids Who Love To Read*. Scholastic.

Da Rocha, Tibola, V., Brandli, L.L., Mazutti, J., Dal Moro, L., Gasperina, L. D. & Locatelli, R.M. (2020). Teacher's Approach on Climate Change Education a Case Study. *World Sustainability Series*, 617-642. DOI: 10.1007/978-3-030-30306-8_37

Dale, R. (2021). GPT-3: What's it good for? *Natural Language Engineering*, 27(1), 113–118. DOI: 10.1017/S1351324920000601

Damrosch, D. (2003). *What is World Literature?* Princeton University Press. DOI: 10.1515/9780691188645

Darby, W. J. (2020). *Landscape and identity: Geographies of nation and class in England*. Routledge. DOI: 10.4324/9781003135852

Davis, J. (1998). Young Children, Environmental Education and Future. *Early Childhood Education Journal*, 26(2), 117–123. DOI: 10.1023/A:1022911631454

De Brito Miranda, A. C., Jófili, Z., & Dos Anjos Carneiro-Leão, A. M. (2017). Ecological literacy – preparing children for the twenty-first century. *Early Child Development and Care*, 187(2), 192–205. DOI: 10.1080/03004430.2016.1226353

de Lázaro Torres, M. L. (2024). La utilidad de ChatGPT en la enseñanza universitaria a distancia de geoinformación para futuros docentes de enseñanza secundaria. *Interdisciplinary Journal of Didactics*, (1), 83–104. DOI: 10.14198/ijd.28204

De Luis, M., Brunetti, M., González-Hidalgo, J. C., Longares, L. A., & Martin-Vide, J. (2010). Changes in seasonal precipitation in the Iberian Peninsula during 1946–2005. *Global and Planetary Change*, 74(1), 27–33. DOI: 10.1016/j.gloplacha.2010.06.006

Decreto 81/2022, de 12 de julio, por el que se establece la ordenación y el currículo de Educación Primaria en la comunidad autónoma de Castilla-La Mancha.

Del Álamo, L. y Gil, V. (2000). Taller de música 3. SM.

del Molino, S. (2016). *La España vacía*. Turner.

Delors, J. (1996). de la publicación: La Educación Encierra un Tesoro. Informe a la UNESCO de la Comisión Internacional sobre la Educación para el Siglo XXI. *Laurus*, 14(26), 136–167.

Dewey, J. (1938). The philosophy of the arts. *John Dewey: The Later Works*, 13, 357–368.

Di Natale, A. F., Repetto, C., Riva, G., & Villani, D. (2020). Immersive virtual reality in K-12 and higher education: A 10-year systematic review of empirical research. *British Journal of Educational Technology*, 51(6), 2006–2033. DOI: 10.1111/bjet.13030

Dias-Trindade, S., Moreira, J. A., & Gomes Ferreira, A. (2021). Evaluation of the teachers' digital competences in primary and secondary education in Portugal with DigCompEdu CheckIn in pandemic times. *Acta Scientiarum. Technology*, 43, e56383. DOI: 10.4025/actascitechnol.v43i1.56383

Díaz Fernández, E., & Mérida Jiménez, R. (2021). María de la Encarnación Gertrudis Jacoba Aragoneses y de Urquijo. *Estudios Hispánicos*, 29, 55–63. DOI: 10.19195/2084-2546.29.6

Díaz Moreno, N., & Felices de la Fuente, M. M. (2017). Problemas sociales relevantes en el aula de primaria: la "cartografía de la controversia" como método, *REIDICS. Revista de Investigación en Didáctica de las Ciencias Sociales, nº 1*, 24-38, https://dehesa.unex.es/handle/10662/6610

Díaz Moreno, N., & Felices de la Fuente, M. D. M. (2017). Problemas sociales relevantes en el aula de primaria: La "cartografía de la controversia" como método. Reidics. *Revista de Investigación en Didáctica de las Ciencias Sociales.*, 1, 24–38.

Díaz-Barriga, A. (2014). Construcción de programas de estudio en la perspectiva del enfoque de desarrollo de competências. *Perfiles Educativos*, XXXVI(143), 142–162. DOI: 10.22201/iisue.24486167e.2014.143.44027

Díaz-Plaja Taboada, A. (2011). *Escrito y leído en femenino: novelas para niñas*. Ediciones de la Universidad de Castilla-La Mancha.

Díez Gutiérrez, E. J. (2020). Migraciones y refugiados: el papel de los centros educativos ante la situación actual. En M. Fernández-Hawrylaw, D. Heras Sevilla y J.A. Gómez Monedero. (Coords.), *Ampliando horizontes en Educación inclusiva* (pp. 81-88). Universidad de Burgos.

Dikmenli, Y. (2013). Geographic literacy perception scale (GLPS) validity and reliability study. [MIJE]. *Mevlana International Journal of Education*, 4(1), 1–15. DOI: 10.13054/mije.13.43.4.1

Dikmen, S. (1993). *İlköğretim Kurumlarında Çevre İçin Eğitim* [Education for Environment in Primary Schools]. Çevre Eğitimi, Türkiye Çevre Vakfı Yayını.

Direção-Geral de Educação. (2017). *Estratégia Nacional de Educação para a Cidadania*. República Portuguesa.

Doğan, D., & Seferoğlu, S. S. (2015). Mobile devices and digital transformation in education. In Akkoyunlu, B., İşman, A., & Odabaşı, H. F. (Eds.), *Educational Technologies Readings* (pp. 541–545). TOJET.

Domínguez Mujica, J., Díaz Hernández, R., & Parreño Castellano, J. M. (2002). Canarias, puerta de Europa. ¿Preferido destino de iberoamericano? *Coloquios de historia canario americana*, 57-77, https://accedacris.ulpgc.es/bitstream/10553/12033/4/Canarias_%20puerta_de_Europa.pdf

Domínguez Mujica, J. (2008). Un nuevo diálogo migratorio: Canarias, Latinoamérica y el Caribe en la Era de la Globalización. *Anuario de Estudios Atlanticos*, 1(54), 469–509. https://dialnet.unirioja.es/ejemplar/216156

Domínguez-Mujica, J., López de Lera, D., Ortega-Rivera, E., & Pérez-Caramés, A. (2020). El sistema migratorio de Latinoamérica-España: ¿Ha sido la crisis económica un paréntesis? *Cuadernos Geográficos*, 59(3), 37–57. DOI: 10.30827/cuadgeo.v59i3.9223

Downs, R. M., & Liben, L. S. (1991). The development of expertise in geography: A cognitive-developmental approach to geographic education. *Annals of the Association of American Geographers*, 81(2), 304–327. DOI: 10.1111/j.1467-8306.1991.tb01692.x

Dufresnoy, L. (1757). *Descripçao da Terra, ou Methodo Breve da Geographia, Dividido em Liçoes por perguntas e respostas*. Officina de Joseph da Costa Coimbra.

Eckardt, M. H. (2006). Fromm y su concepto de "biofilia". En J. Silva García (Coord.), *El humanismo de Erich Fromm: Actualidad del autor de "El arte de amar" y "El miedo a la libertad"* (pp. 75-84). Universidad Autónoma de México.

Eden, C. A., Chisom, O. N., & Adeniyi, I. S.Chima Abimbola EdenOnyebuchi Nneamaka ChisomIdowu Sulaimon Adeniyi. (2024). Cultural competence in education: Strategies for fostering inclusivity and diversity awareness. *International Journal of Applied Research in Social Sciences*, 6(3), 383–392. DOI: 10.51594/ijarss.v6i3.895

Egan, K. (1999). *Fantasía e imaginación, su poder en la enseñanza primaria: una alternativa a la enseñanza y el aprendizaje en la educación infantil y primaria.* Morata.

Eilam, E. (2022). Climate change education: The problem with walking away from disciplines. *Studies in Science Education*, 58(2), 231–264. DOI: 10.1080/03057267.2021.2011589

Elliott, S. (2003). Patches of Green EPA Social Research Series Early childhood environmental education in Australia: scope, status and direction Environment Protection Authority 59–61 Goulburn Street, http://www.penrithcity. nsw.gov.au/uploadedFiles/patchesofgreen.pdf

Equiza, P. (1976). Elena Quiroga y la novela española de posguerra. *Mundo Hispánico*, 336, 10–12.

Erkol, M. ve Erbasan, Ö. (2018). Öğretmenlerin çevre eğitimi öz-yeterliklerinin çeşitli değişkenler açısından incelenmesi. [Investigation of Teachers' Environmental Education Self-Efficacy in Terms of Different Variables]. *Erzincan University Journal of Education Faculty*, 20(3), 810-825. DOI: 10.17556/erziefd.446125

Erten, S. (2004) Çevre Eğitimi Ve Çevre Bilinci Nedir, Çevre Eğitimi Nasıl Olmalıdır? [What is Environmental Education and Environmental Awareness, How Should Environmental Education Be?]. Journal of Environment and Human, Publication of the Ministry of Environment and Forestry.. Sayı 65/66. 2006/25

Ertürk, R. (2017). İlkokul öğrencilerinin çevre sorunları ve çevre eğitimine yönelik algıları. [Environmental Problems Of The Primary School Students And Perceptions For Environmental Education]. *Inonu University Journal of the Faculty of Education*, 18(3), 12–24. DOI: 10.17679/inuefd.354142

Escoz, A., Gutiérrez-Pérez, J., & Meira, P. Á. (2020). Water and climate change, two key objectives in the agenda 2030: Assessment of climate literacy levels and social representations in academics from three climate contexts. *Water (Basel)*, 12(1), 1–33. DOI: 10.3390/w12010092

Escribano, C. & Tosello, J. (2020). ¿Qué sabe el profesorado español de educación infantil sobre la ciudadanía global? Una aproximación desde la teoría fundamentada.

Esho, E. O. (2024). The paradigm of peace enhancing peace culture. *Journal of Aggression, Conflict and Peace Research*, 16(1), 83–97. DOI: 10.1108/JACPR-12-2022-0765

Estrada, E. G., Mamani, H. J., & Huaypar, K. J. (2020). Eficacia del programa Cuidemos el ambiente en el desarrollo de la conciencia ambiental de estudiantes de educación primaria en Madre de Dios, Perú. *Ciencia Amazónica (Iquitos)*, 8(1), 85–98. DOI: 10.22386/ca.v8i1.282

European Comission. (2024). https://ec.europa.eu/info/index_es

European Commission, Directorate-General for Education, Youth, Sport and Culture. (2019). *Key competences for lifelong learning*, Publications Office. https://data.europa.eu/doi/10.2766/569540

European Commission. (2020). *Digital education action plan 2021-2027: Resetting education and training for the digital age*. Education and Training.

European Environment Agency (EEA). (2024). *Adaptación al cambio climático.* https://www.eea.europa.eu/es/themes/climate-change-adaptation

Evans, G. W., Otto, S., & Kaiser, F. G. (2018). Childhood origins of young adult environmental behavior. *Psychological Science*, 29(5), 679–687. DOI: 10.1177/0956797617741894 PMID: 29447064

Eyring, V., Gillett, N. P., Achuta Rao, K. M., Barimalala, R., Barreiro Parrillo, M., Bellouin, N., Cassou, C., Durack, P. J., Kosaka, Y., McGregor, S., Min, S., Morgenstern, O., & Sun, Y. (2021). Human influence on the climate system. In Masson-Delmotte, V., Zhai, P., Pirani, A., S.L. Connors, S.L., Péan, C., Berger, S., Caud, N., Chen, Y., Goldfarb, L., Gomis, M.I., Huang, M., Leitzell, K., Lonnoy, E., Matthews, J.B.R., Maycock, T.K., Waterfield, T., Yelekçi, O., Yu, R., & Zhou, B. (Eds.), *Climate change 2021: The physical science basis. Contribution of working group I to the sixth assessment report of the intergovernmental panel on climate change* (pp. 423–552). Cambridge University Press. DOI: 10.1017/9781009157896.005

Ezquerro, M. (1996). Los temas y la escritura quiroguianos. En Baccino Ponce de León, N. & Lafforgue, J. (coords.). *Horacio Quiroga. Todos los cuentos*. ALLCA XX, 1379-1414.

Farias, F. (2009). La epistemología de las Ciencias Sociales en la formación por competencias del pregrado. *Cinta de Moebio*, 34(34), 58–66. DOI: 10.4067/S0717-554X2009000100004

Feld, S. (2012). *Sound and Sentiment: Birds, Weeping, Poetics, and Song in Kaluli Expression*. Duke University Press.

Fernández Batanero, J. M. (2004). La presencia de alumnos inmigrantes en las aulas: un reto educativo. *Educación y educadores*, (7), 33-44, https://www.redalyc.org/pdf/834/83400705.pdf

Fernández Portela, J. (2016). El estudio del espacio geográfico cercano en educación infantil: itinerario didáctico para el conocimiento de la ciudad de Valladolid. In Alanís, L., Almuedo, J., de Oliveira, G., Iglesias, R., & Pedregal, B. (Eds.), *Nativos digitales y geografía en el siglo XXI: educación geográfica y sistemas de aprendizaje* (pp. 484–496). Grupo de Didáctica de la Geografía de la Asociación de Geógrafos Españoles y Universidad Pablo de Olavide.

Fernández, M., Valbuena, C., & Caro, R. (2017). *Evolución del racismo, la xenofobia y otras formas de intolerancia en España. Informe-encuesta 2017. Resumen*. Ministerio de Trabajo, Migraciones y Seguridad Social.

Ferrari, E., Ballegeer, A. M., Fuertes, M. Á., Herrero, P., Delgado, L., Corrochano, D., Andrés-Sánchez, S., Bisquert, K. M., García-Vinuesa, A., Meira, P. Á., Martínez, F., & Ruiz, C. (2019). Improvement on Social Representation of Climate Change through a Knowledge-Based MOOC in Spanish. *Sustainability (Basel)*, 11(22), 1–21. DOI: 10.3390/su11226317

Ferrer, M. J., & Riera, M. A. (2015). Relaciones y encuentros con las familias en Educación Infantil. *RELAdEI: revista latinoamericana de educación infantil, 4*(2), 27-42. https://revistas.usc.gal/index.php/reladei/article/view/4879

Ferrer, L. A., & Vivancos, A. E. (2022). *Didáctica de geografía e historia en educación primaria*. Editorial Síntesis.

Fleming, L. (1991). Introducción. En Quiroga, H. *Cuentos*. Cátedra.

Flores, C., De la Cruz, G., Domínguez, E., & Pérez, B. D. (2023). Revisión teórica acerca del conocimiento geográfico potente. Un nuevo enfoque conceptual en la educación geográfica. *Didáctica Geográfica*, 24(24), 17–38. DOI: 10.21138/DG.666

Foster, K. J. (2024). New classroom teachers' perceptions of multicultural teacher preparation. *Educational Leadership & Policy Studies Dissertations*, 4. https://mavmatrix.uta.edu/edleadershippolicy_dissertations/4

Fox, E., & Riconscente, M. (2008). Metacognition and self-regulation in James, Piaget, and Vygotsky. *Educational Psychology Review*, 20(4), 373–389. DOI: 10.1007/s10648-008-9079-2

Fraga Fernández-Cuevas, M.ª J. (2019). "Lo que Celia dice" en otros países: versiones de las novelas de Elena Fortún. *CLIJ: Cuadernos de literatura infantil y juvenil, 32*(290), 24-33.

Freire, P. (1970). Pedagogy of the oppressed. *Continuum : an Interdisciplinary Journal on Continuity of Care*.

Fuentes, A. F. (2015). Cuaderno de campo. En P. M. Jorgensen, A. F. Fuentes, T. Miranda, & L. Cayola (Eds.), *Manual de Trabajo, Proyecto Inventario Botánico de la Región del Madidi* (Cap. 5).

Funes, A. G., Cerdá, M. C., & Jara, M. A. (2019). Las finalidades de la enseñanza de las ciencias sociales y la historia para la educación ciudadana del siglo XXI. Experiencias en Argentina, Balbé, M., & González Monfort, N., Santisteban, A. (Eds). *Quin professorat, quina ciutadania, quin futur?* 227-231, https://ddd.uab.cat/record/220367

Gabarda Méndez, V., García Tort, E., Ferrando Rodríguez, M. L., & Chiappe Laverde, A. (2021). Pre-school and primary school teachers: Technological training and digital competence. *Innoeduca.International Journal of Technology and Educational Innovation*, 7(2), 19–31. DOI: 10.24310/innoeduca.2021.v7i2.12261

Gadamer, H. G. (2000). *La dialéctica de Hegel*. Ediciones Cátedra.

Galdona, R. I. (2018). Real Academia de la Historia. *Elena Quiroga de Abarca.* https://dbe.rah.es/biografias/10566/elena-quiroga-de-abarca

Gamalho, N. P. (2023). Juventudes e as periferias. In V. H. N. Oliveira (org.), Geografias das Juventudes (pp. 39-59). GEPJUVE.

García de la Vega, A. (2016). Análisis curricular, rigor científico en los textos literarios sobre las observaciones del tiempo y clima: Propuesta didáctica con fuentes primarias. In Sebastiá, R., & Tonda, E. M. (Eds.), *La investigación e innovación en la enseñanza de la Geografía* (pp. 211–236). Servicio de Publicaciones de la Universidad de Alicante. DOI: 10.14198/GeoAlicante2015.15

García Esteban, F.E.; & Murga Menoyo, M.A. (2015). El profesorado de Educación Infantil ante el Desarrollo Sostenible. Necesidades formativas, *Enseñanza &Teaching*, 33-1, 121-142. DOI: DOI: 10.14201/et2014331121142

García Monteagudo, D. (2022). De la representación del espacio a la construcción del pensamiento geográfico en Educación Infantil. En A. Paz, V. Figueora, E.J. Rodríguez y A.J. Montes (coords.), *Libro de Actas del 2º Congreso Caribeño de Investigación Educativa: Nuevos paradigmas y experiencias emergentes* (pp. 437-443). Instituto Superior de Formación Docente Salomé Ureña.

García Monteagudo, D. (2024). Formación inicial del profesorado de Educación Infantil en pensamiento geográfico: Paisaje e identidades territoriales. *ENSAYOS. Revista De La Facultad De Educación De Albacete*, 39(1), 57–73. DOI: 10.18239/ensayos.v39i1.3488

García Padrino, J. (2018). *Historia crítica de la Literatura Infantil y Juvenil en la España actual (1939-2015).* Marcial Pons.

García, A. (2004) El itinerario geográfico como recurso didáctico para la valoración del paisaje. *Didáctica Geográfica, 2ª época, 6*(1), 79-95. https://didacticageografica.age-geografia.es

García, A. L., & Martínez, J. M. (1988). Los itinerarios didácticos en la enseñanza de las Ciencias Sociales. *Revista del Departamento de Didáctica de las Ciencias Sociales de la Universidad de Granada, sep. 2º cuadernillo*, 107-113.

García, J. L., Serrano de la Cruz, M. A., & Jerez, Ó. (2022). *Guía del Parque Natural del Valle de Alcudia y Sierra Madrona (Ciudad Real).* 3d3 Editores.

García, A. (2004). El itinerario geográfico como recurso didáctico para la valoración del paisaje. *Didáctica Geográfica*, 6, 79–95.

García, A. (2012). El aprendizaje basado en problemas en el itinerario geográfico. *Educación y Futuro*, 27, 155–175.

García, A. (2023). Pensamento narrativo e construção de conceitos da paisagem. Oficina narrativa em Geografia. *Revista Signos Geográficos*, 5, 1–18. DOI: 10.5216/signos.v5.77428

García, A. L. (1997). El proceso de desarrollo de los Itinerarios Geográficos. *Didáctica Geográfica*, 2, 3–9.

García, C. (2017). *La educación intercultural: Claves para una enseñanza inclusiva*. Editorial Universitaria.

García, D. (2019). La construcción de pensamiento crítico en el aula de geografía escolar. *GeoGraphos: Revista Digital para Estudiantes de Geografía y Ciencias Sociales*, 10(118), 179–205.

García, J. L. (1995). *El medio natural en los Montes de Ciudad Real y el Campo de Calatrava*. Diputación Provincial.

García, M.. (2018). Se hace geografía al andar: La salida de campo itinerante y senderista. *Didáctica Geográfica*, 19, 103–125.

García-Monteagudo, D., Mendes, L. F. G., & Lastória, A. C. (2024). Percepciones docentes sobre ciencias sociales: El caso de Nosotros Proponemos. *Revista Colombiana de Educación*, 90(90), 9–34. DOI: 10.17227/rce.num90-14391

García-Vinuesa, A., Rui, S. A., Azeiterio, U. M., Meira, P. Á., & Pereira, M. (2020). Mozambican students' knowledge and perceptions about climate change: An exploratory study in Pemba City. *International Research in Geographical and Environmental Education*, 0(0), 1–17. DOI: 10.1080/10382046.2020.1863671

Garrido, A. (1992). *El texto narrativo*. Editorial Síntesis.

Gaudelli, W., & Heilman, E. (2009)... *Reconceptualizing Geography as Democratic Global Citizenship Education Teachers College Record*, 111(11), 2647–2677.

Gavari, E., & Pastrana, J. (2018). Evolución del caso japonés como referente internacional en la educación para la reducción del riesgo de desastres. *Revista Española de Educación Comparada*, 32(32), 52–67. DOI: 10.5944/reec.32.2018.22319

Gavidia, V., Solbes, J. A., & Souto, X. M. (2015). La supresión del área del conocimiento del medio en la educación primaria. *Aula*, 243-244, 31–36.

Gebhard, U. (2013). *Kind und Natur. Die Bedeutung der Natur für die psychische Entwicklung*. Westdeutscher Vlg. DOI: 10.1007/978-3-658-01805-4

Geray, C. (1995). Çevre koruma bilinci ve duyarlılığı için halkın eğitimi. [Public education for environmental protection awareness and sensitivity]. *Yeni Türkiye Özel Sayısı*, 1(5), 664–669.

Gerineldo (1994). El Señor don Gato. [canción] *En el medio de aquel camino: chansons judeo-espagnoles*. Colección de Oro Anahory-Librowicz.

Ghomi, M., & Redecker, C. (2019). Digital Competence of Educators (DigCompEdu): Development and Evaluation of a Self-assessment Instrument for Teachers' Digital Competence. In *Proceedings of the 11th International Conference on Computer Supported Education (CSEDU 2019)*, 541-549. https://doi.org/DOI: 10.5220/0007679005410548

Gil, C. G. (2018). Objetivos de Desarrollo Sostenible (ODS): una revisión crítica. *Papeles de relaciones ecosociales y cambio global, 140*, 107-118.

Glenn, I. D. (2010). *Documentary research method. Palgrave Macmillan. Gómez García, Z. (1998). El areíto: Expresión de resistencia cultural.* Editorial de Ciencias Sociales.

Glotfelty, C. (1996). Introduction: Literary Studies in an Age of Environmetal Crisis. In Glotfelty, C., & Fromm, H. (Eds.), *The Ecocriticism Reader: Landmarks in Literary Ecology* (pp. xv–xxxvii). University of Georgia Press.

Gobierno de Castilla-La Mancha. (2020). *Estrategia de Educación Ambiental de Castilla-La Mancha. Horizonte 2030 (2020-2025).* https://www.castillalamancha.es/sites/default/files/documentos/pdf/20200604/estrategia_ea-clm_h2030.pdf

Gökçe, N., Kaya, E., & Aktay, S. ve Özden, M. (2007). İlköğretim Öğrencilerinin Çevreye Yönelik Tutumları. [Elementary Students Attitutes Towards Environment]. *Elementary Education Online*, 6(3), 452–468.

Gomes, M. F. V. B. (2023). Contribuições do Projeto Nós Propomos! Guarapuava à formação inicial dos professores de Geografia. In M. F. V. B. Gomes, C. L. Silva, A. Roik & E. Y. Yamamoto (org.), *Formação de professores no Projeto Nós Propomos! Guarapuava. Contribuições pedagógicas para a Agenda da ONU 2030.* C&A Alfa Comunicação.

Gómez Gómez, M., & García Lázaro, D. (2023). Concienciación y conocimientos sobre los Objetivos de Desarrollo Sostenible en la formación del profesorado, *Profesorado. Revista de currículum y formación del profesorado, vol. 27, Nº 3*, 243-364. DOI: 10.30827/profesorado.v27i3.27948

Gómez Mendoza, J. (2013). Del patrimonio paisaje a los paisajes del patrimonio. *Documents d'Analisi Geografica*, 59(1), 5–20. DOI: 10.5565/rev/dag.48

Gómez- Saldivia. P. A. (2022). Formar al profesorado en educación para la ciudadanía a partir de problemas socialmente relevantes. En Bel-Martínez, J.C, Colomer-Rubio, J.C. y de Alba, N. *Repensar el currículum de Ciencias Sociales: prácticas educativas para una ciudadanía crítica* (pp. 323-331). Tirant Humanidades.

Gómez, A., Corrochano, D., Parra, G. (2017). Itinerarios didáctico-naturales en educación primaria: el noroeste de Zamora. *Didáctica Geográfica*, (18), 111-131. Vista de ITINERARIOS DIDÁCTICO-NATURALES EN EDUCACIÓN PRIMARIA: EL NOROESTE DE ZAMORA (age-geografia.es)

Gonçalves, C., & Claudino, S. (2021). Governance, Civic and Geographic Education and Local Development. In S. Saúde, M. A. Raposo, N. Pereira, A. I. Rodrigues, *Teaching and Learning Practices That Promote Sustainable Development and Active Citizenship* (pp. 236-253). IGI Global.

Gönen, M. ve Veziroğlu, M. (2013). Çocuk edebiyatının genel hedefleri. [General aims of children's literature].M. Gönen (Ed.), Çocuk Edebiyatı (ss. 1-11) içinde. Eğiten Kitap.

Gönen, M., Pektaş, D., Değirmenci, Y.G., Bülbün, B. & Ünal, Ş. (2016). Çevre konulu resimli çocuk kitaplarının resimleme ve içerik açısından incelenmesi. [Physical, Content and Content of Illustrated Children's Books Analyzing in Terms of Drawing Features].*Çocuk ve gençlik edebiyatında barış kültürü.*

González Iglesias, E. (2003). El trabajo con mapas en Educación Infantil. Un recurso para la educación en valores. En M.J. Marrón Gaite, C. Moraleda y H. Rodríguez (coords.), *La geografía ante las nuevas demandas sociales* (pp. 89-101). Asociación de Geógrafos Españoles.

González Monfort, N. (2011). Reflexions al voltant de les questions socialment vives i l'ensenyament de les ciències socials. Conclusiones de les VII Jornades Internacionals de Recerca en Didàctica de les Ciéncies Socials, Pagès Blanch, J., Santisteban, A. (Coords): *Les questions socialment vives i l'ensenyament de les ciències socials* (pp. 153-161). https://ddd.uab.cat/record/197111

González Picado, J. (1985). Teoría y práctica del cuento en Hispanoamérica: El caso de Horacio Quiroga. *Filología y Lingüística*, XI(1), 5–14.

González-Gaudiano, E., Meira, P. Á., & Gutiérrez-Pérez, J. (2020). ¿Cómo educar sobre la complejidad de la crisis climática? *Revista Mexicana de Investigación Educativa*, XXV(87), 843–872.

González, I. (2002). La formación inicial de los profesores de ciencias sociales de Educación Secundaria: necesidades y proyectos. In *VV. AA. Nuevos horizontes en la formación del profesorado de Ciencias Sociales* (pp. 429–451). Asociación Universitaria de Profesores de Didáctica de las Ciencias Sociales.

Goodman-Deane, J., Mieczakowski, A., Johnson, D., Goldhaber, T., & Clarkson, P. J. (2016). The impact of communication technologies on life and relationship satisfaction. *Computers in Human Behavior*, 57, 219–229. DOI: 10.1016/j.chb.2015.11.053

Graham, S., Kogachi, K., & Morales-Chicas, J. (2022). Do I fit in: Race/ethnicity and feelings of belonging in school. *Educational Psychology Review*, 34(4), 2015–2042. DOI: 10.1007/s10648-022-09709-x

Granados, J. (2019). Definición y justificación de un marco conceptual para la didáctica de la geografía. In Hortas, M. J., Dias, A., & De Alba, N. (Eds.), *Enseñar y aprender didáctica de las ciencias sociales: la formación del profesorado desde una perspectiva sociocrítica* (pp. 41–49). Asociación Universitaria de Profesorado de Didáctica de las Ciencias Sociales., https://dialnet.unirioja.es/servlet/articulo?codigo=7609709

Granados, J., & Medir, R. M. (2022). *Enseñar y aprender geografía para un mundo sostenible*. Octaedro.

Guardeño Juan, M., & Monsalve Lorente, L. (2024). Educación para el desarrollo sostenible en el currículum de España e Irlanda. *Universidad y Sociedad*, 16(1), 30–44. http://scielo.sld.cu/scielo.php?script=sci_abstract&pid=S221836202024000100030&lng=es&nrm=iso

Guerra, F. J. (2022). Medios para la interpretación del patrimonio. Planificación y gestión. *Servicios de Educación y Estudios Ambientales* (144-218). S.L. (SEEDA). capitulo_muestra_guia_digital_IP.pdf (juntadeandalucia.es)

Guijarro Arribas, D. (2021). De la circulation internationale à la globalisation des normes editoriales: Le cas des livres pour « adolescents » en France et en Espagne. *Reseaux (London)*, 226-227, 185–207. https://shs.cairn.info/revue-reseaux-2021-2-page-185?lang=fr&contenu=resume

Gülay, H. (2011). Ağaç yaş iken eğilir: Yaşamın ilk yıllarında çevre eğitiminin önemi. [You Cannot Teach an Old Dog New Tricks: The Importance of Environmental Education In The Early Years Of Life]. *TÜBAV Journal of Science*, 4(3), 240–245.

Güler, T. (2010). Çevre Eğitimi. [Environment Education].Berrin Akman, Gülden Uyanık Balat ve Güler, Tülin, (Ed.), *Okul Öncesi Dönemde Fen Eğitimi* (s.181-201). Pegem

Gülersoy, A. E., Dülger, İ., Dursun, E., & Ay, D. (2020). Nasıl bir çevre eğitimi? Çağdaş yaklaşımlar çerçevesinde bazı öneriler. [What Kind of Environmental Education? Some Suggestions in the Context of Contemporary Approaches]. *Turkish Studies (Elektronik)*, 15(5), 2357–2398. DOI: 10.7827/TurkishStudies.44074

Gül, F. (2013). İnsan-doğa ilişkisi bağlamında çevre sorunları ve felsefe. [Environmental Problems and Philosophy in The Context of Human-Nature Relationship]. *Pamukkale University Journal of Social Sciences Institute*, 14, 17–21. DOI: 10.5505/pausbed.2013.09609

Guo, B., Zhang, X., Wang, Z., Jiang, M., Nie, J., Ding, Y., Yue, J., & Wu, Y. (2023). How Close is ChatGPT to Human Experts? Comparison Corpus, Evaluation, and Detection. ArXiv:2301.07597. https://doi.org//arXiv.2301.07597DOI: 10.48550

Gürbüz, H., & Kışoğlu, M. (2011). Güncel çevre sorunları. [Current Environmental Problems]. H. Yazıcı & K. Arıbaş (Ed.), Günümüz dünya sorunları içinde (s. 153-190). Pegem Akademi.

Gutiérrez, J.J., & Cabero-Almenara, J. (2016). Estudio de caso sobre la autopercepción de la competencia digital del estudiante universitario de las titulaciones de grado de Educación Infantil y Primaria. *Profesorado. Revista de currículum y formación del profesorado, 20*(2), 180-199.

Gutierrez-Pérez, J., Meira, P. Á., & Gonzalez-Gaudiano, E. (2020). Presentación. *Revista Mexicana de Investigación Educativa*, 87, 820–842.

Güzelyurt, T., & Özkan, Ö. (2019). Okul Öncesi Dönemde Çevre Eğitimi: Çocuk Kitaplarına Yönelik Bir İnceleme1 [Environmental Education in Pre-School Period: An Examination of Children's Books]. *Elementary Education Online*, 18(1), 20–30.

Guzmán Mora, J., & Sevilla-Vallejo, S. (2019). Carmen Kurtz en el cine infantil de la década de 1970: intervenciones y adaptaciones. En M. Marcos Ramos (Coord.), *El cine como reflejo de la historia, de la literatura y del arte en la filmografía hispano-brasileña* (pp. 301-309). Universidad de Salamanca, Centro de Estudios Brasileños.

Hannoun, H. (1977). *El niño conquista el medio*. Kapelusz.

Haq, S. M. A., & Ahmed, K. J. (2020). Perceptions about climate change among university students in Bangladesh. *Natural Hazards*, 103(3), 3683–3713. Advance online publication. DOI: 10.1007/s11069-020-04151-0

Harvey, D. (2005). The sociological and geographical imaginations. *International Journal of Politics Culture and Society*, 18(3-4), 211–255. DOI: 10.1007/s10767-006-9009-6

Hedefalk, M., Almqvist, J., & Östman, L. (2015). Education for sustainable development in early childhood education: A review of the research literature. *Environmental Education Research*, 21(7), 975–990. DOI: 10.1080/13504622.2014.971716

Helldén, G., & Helldén, S. (2008). Students' early experiences of biodiversity and education for a sustainable future. *NorDiNa: Nordic Studies in Science Education*, 4(2), 123–131. DOI: 10.5617/nordina.286

Henares, I. (2015). La educación ambiental en los espacios naturales protegidos: retos. En *Blog Educación Ambiental 26 #EA26*. https://educacionambiental26.wordpress.com/2014/02/19/la-educacion-ambiental-en-los-espacios-naturales-protegidos-retos/

Herbold, S., Hautli-Janisz, A., Heuer, U., Kikteva, Z., & Trautsch, A. (2023). A large-scale comparison of human-written versus ChatGPT-generated essays. *Scientific Reports*, 13(1), 18617. DOI: 10.1038/s41598-023-45644-9 PMID: 37903836

Herbst, E., Cruz, T., Bower, C. A., Hirsh-Pasek, K., & Michnick, R. (2022). Playing for the Future: Spatial Thinking Belongs in Preschools and Home Environments. In Lynn, A., & Thai, K. P. (Eds.), *Handbook of Research on Innovative Approaches to Early Childhood Development and School Readiness* (pp. 416–451). IGI Global., DOI: 10.4018/978-1-7998-8649-5.ch017

Hernández Cervantes, L. (2021). La formación del profesorado como técnico y reproductor social. La práctica docente de Ana Lorenia Hernández Cervantes. En L.M. Aguayo Rendón, L.M. López Velarde, J.C. (Coords). *Ser profesor y doctorantes en tiempos de intertidumbre. Aportes al desarrollo profesional desde y sobre la pandemia* (pp. 175-202). Taberna Libaria editores.

Hernández Prados, M. A., Carbonell Bernal, N., & Polanco Mora, B. (2021). Diseño de una experiencia para trabajar la interculturalidad en Educación Infantil. *Revista Electrónica en Educación y Pedagogía*, 5(9), 165–178. DOI: 10.15658/rev.electron.educ.pedagog21.11050912

Hernández, A. M., Martínez, C. M., Torres, I. C., & Hernández, L. A. (2012). La excursión integradora en la enseñanza de aprendizaje de la carrera Biología-Geografía. [Redalyc.La excursión integradora en la enseñanza aprendizaje de la carrera Biología-Geografía]. *Revista Ciencias Holguín*, XVIII(2), 1–12.

Hernández, A., & García, A. (2022). Del análisis del paisaje a la emoción del paisaje. Aportaciones a su didáctica. *REIDICS.Revista de Investigación en Didáctica de las Ciencias Sociales*, 10, 6–23. DOI: 10.17398/2531-0968.10.6

Hernando, M. J., & Dikam, F. P. (2016). De los objetivos de desarrollo del milenio a los objetivos de desarrollo sostenible. *Libre pensamiento*, (88), 60-69.

Hirschberg, J., & Manning, C. D. (2015). Advances in natural language processing. *Science*, 349(6245), 261–266. DOI: 10.1126/science.aaa8685 PMID: 26185244

Hofferth, S. L., & Sandberg, J. F. (2001). How American children spend their time. *Journal of Marriage and Family*, 63(3), 295–308. DOI: 10.1111/j.1741-3737.2001.00295.x

Hopkins, C. (2012). Twenty Years of Education for Sustainable Development. *Journal of Education for Sustainable Development*, 6(1), 1–4. DOI: 10.1177/097340821100600101

Hueso, K. (2017). *Somos naturaleza. Un viaje a nuestra esencia*. Plataforma Editorial.

Hueso, K. (2021). *Educar en la naturaleza. Mejores personas para un planeta mejor*. Plataforma Editorial.

Hughes, J., Rogerson, M., Barton, J., & Bragg, R. (2019). Age and connection to nature: When is engagement critical? *Frontiers in Ecology and the Environment*, 17(5), 265–269. DOI: 10.1002/fee.2035

Hwang, G.-J., & Chang, C.-Y. J. I. L. E. (2021). A review of opportunities and challenges of chatbots in education. *Interactive Learning Environments*, *31*(7) 4099–4112. DOI: 10.1080/10494820.2021.1952615

IGOT-ULisboa. (2024). Regulamento do Projeto Nós Propomos! cidadania e Inovação na Educação Geográfica https://www.igot.ulisboa.pt/projeto-nos-propomos

Inarejos, J. A. (2018). *Guía docente para la enseñanza de la geografía y la historia en educación infantil.* Servicio de Publicaciones de la Universidad de Extremadura. https://dehesa.unex.es:8443/handle/10662/7023

Instituto Nacional de Estadística (INE). (2022). *Población extranjera por comunidades y provincias, nacionalidad y sexo*. Recuperado de https://bit.ly/3BtJiEK

Intergovernmental Panel on Climate Change. (2014). *Climate Change 2013 and Climate Change 2014* (3 vols.). https://www.ipcc.ch

Intergovernmental Panel on Climate Change. (2022). *Climate Change 2022: Impacts, Adaptation and Vulnerability*.https://www.ipcc.ch/report/ar6/wg2/

Izquierdo Escribano, A., & Martínez-Buján, R. (2014). From Favourites to 'With No Return': Permanence and Mobility of Latin American Immigration in Spain during the 21st Century. *Canadian Ethnic Studies*, 46(3), 103–130. DOI: 10.1353/ces.2014.0038

Jacobs, K., Holway, J., Hanak, E., Quay, R., Sternlieb, F., & Udall, B. (2022). Impacts and Opportunities at the Climate-Land Use-Energy-Water Interface: An Urgent Call for Dialogue. *Journal of the American Water Resources Association*, 58(3), 411–419. DOI: 10.1111/1752-1688.13014

Janiesch, C., Zschech, P., & Heinrich, K. (2021). Machine learning and deep learning. *Electronic Markets*, 31(3), 685–695. DOI: 10.1007/s12525-021-00475-2

Jaráiz, F. J., Hernández, A. M., & Hernández, A. (2016). La percepción del espacio: estudio de caso. En L. Alanís, J. Almuedo, G. de Oliveira, R. Iglesias & B. Pedregal (coords.), *Nativos digitales y geografía en el siglo XXI: educación geográfica y sistemas de aprendizaje* (pp. 601-616). Grupo de didáctica de la Geografía de la Asociación de Geógrafos Españoles, Universidad Pablo de Olavide y Universidad de Alicante. DOI: 10.14198/2016-nativos-digitales-y-geografia

Jardón, P. (2019). El sentido de la enseñanza de la Geografía y la Historia. 30 años del proyecto GEA-CLÍO. *Ar@cne: revista electrónica de recursos en internet sobre geografía y ciencias sociales*, 235, 1-16. https://raco.cat/index.php/Aracne/article/view/361146

JCCM. (1995). *Plan de Conservación del Medio Natural de Castilla-La Mancha.* Consejería de Agricultura y Medio Ambiente, Junta de Comunidades de Castilla-La Mancha.

JCCM. (2003). *Plan Regional de Educación Ambiental de Castilla-La Mancha.* Junta de Comunidades de Castilla-La Mancha.

Jeong, J. S., González-Gómez, D., Conde-Núñez, M. C., Sánchez-Cepeda, J. M., & Yllana-Prieto, F. (2021). Improving climate change awareness of preservice teachers (Psts) through a university science learning environment. *Education Sciences*, 11(2), 1–17. DOI: 10.3390/educsci11020078

Jerez, Ó. (2009a). *El Parque Nacional de las Tablas de Daimiel. Itinerario didáctico por un área protegida "en peligro de extinción"*. Materiales didácticos del Dpto. de Geografía en la E. U. de Magisterio de Ciudad Real. Carranchín.

Jerez, Ó. (2009b). *El Parque Nacional de Cabañeros. Cuaderno de campo etnobotánico.* Materiales didácticos del Dpto. de Geografía en la E. U. de Magisterio de Ciudad Real. Carranchín.

Jerez, Ó. (2019). La transferencia de conocimientos didáctico-geográficos en el medio rural. El senderismo didáctico como propuesta para un desarrollo rural sostenible. En X. C. Macía Arce, F. X. Armas Quintá, & F. Rodríguez Lestegás (Coords.), *La reconfiguración del medio rural en la sociedad de la información: nuevos desafíos en la educación geográfica* (pp. 1321–1333). Andavira Editora.

Jerez, Ó. (Dir.). (2014). *Itinerarios didácticos por La Mancha Húmeda. Libro digital bilingüe*. Óptima diseño e impresión.

Jerez, O., Serrano de la Cruz, M. A., & García, J. L. (2018). Propuesta metodológica para la planificación de un itinerario didáctico en un paisaje protegido. En A. García (Ed.), *Contribución didáctica al aprendizaje de la geografía* (717-733). https://www.researchgate.net/publication/340037981_Propuesta_metodologica_para_la_planificacion_de_un_itinerario_didactico_en_un_Paisaje_Protegio

Jerez, Ó., García, J. L., & Serrano de la Cruz, M. A. (2024). *Guía de excursiones por los espacios naturales de Castilla-La Mancha: Los Montes de Toledo*. Universidad de Castilla-La Mancha.

Jerez, Ó., Rodríguez, M. Á., & Escobar, E. (2014). El itinerario didáctico como recurso educativo para el desarrollo de valores ambientales. In Sáez-Martínez, F. J., Guadamillas, F., & Martín, R. (Eds.), *Experiencias docentes en Energía y Medio Ambiente: University Vocational Training Network* (pp. 60–73). Servicio de Publicaciones de la Universidad de Castilla-La Mancha.

Jerez, Ó., Serrano de la Cruz, M. A., & García, J. L. (2017). *Los espacios naturales protegidos en la provincia de Ciudad Real*. Diputación Provincial de Ciudad Real.

Jiménez, O., Villalón, G., & Évora, O. E. (2017). La educación para la percepción de riesgos de desastres como prioridad del trabajo educativo en la escuela cubana. *Educare (San José)*, 21(3), 1. Advance online publication. DOI: 10.15359/ree.21-3.20

Johnston, R. J., Gregory, D., & Smith, D. M. (2009). *The Dictionary of human geography* (3rd ed.). Blackwell Reference.

Juárez-Pulido, M., Rasskin-Gutman, I., & Mendo-Lázaro, S. (2019). El Aprendizaje Cooperativo, una metodología activa para la educación del siglo XXI: una revisión bibliográfica. *Revista Prisma Social*, (26), 200–210. https://revistaprismasocial.es/article/view/2693

Kandır, A., Can Yaşar, M., İnal, G., Yazıcı, E., Uyanık, Ö., & Yazıcı, Z. (2012). *Etkinliklerle Bilim Eğitimi* [Science Education with Activities]. Efil.

Karatekin, K., & Aksoy, B. (2012, Winter). Sosyal bilgiler öğretmen adaylarının çevre okuryazarlık düzeylerinin çeşitli değişkenler açısından incelenmesi, [Examination of Teacher Candidates Of Social Studies' Environmental Literacy Level in Terms of Various Variables]. Turkish Studies- International Periodical For The Languages. *Literature and History of Turkish or Turkic*, 7(1), 1423–1438.

Karsten, L. (2005). It all used to be better? Different generations on continuity and change in urban children's daily use of space. *Children's Geographies*, 3(3), 275–290. DOI: 10.1080/14733280500352912

Kasneci, E., Sessler, K., Küchemann, S., Bannert, M., Dementieva, D., Fischer, F., Gasser, U., Groh, G., Günnemann, S., Hüllermeier, E., Krusche, S., Kutyniok, G., Michaeli, T., Nerdel, C., Pfeffer, J., Poquet, O., Sailer, M., Schmidt, A., Seidel, T., & Kasneci, G. (2023). ChatGPT for good? On opportunities and challenges of large language models for education. *Learning and Individual Differences*, 103, 102274. DOI: 10.1016/j.lindif.2023.102274

Kayaer, M. (2012), Hukuksal Araçlarla Çevrenin Korunması [Protection of the Environment Through Legal Instruments]. (Doctoral dissertation) Sakarya University SBE.

Kaya, M. F. (2014). Sosyal bilgiler öğretmen adaylarının çevre sorunlarına ilişkin algıları: Metafor analizi örneği. [Social Studies Teachers' Perceptions Related to Environmental Problems: A Sample Analysis of Metaphors]. *Turkish Studies*, 9(2), 917–931. DOI: 10.7827/TurkishStudies.6308

Kaypak, Ş. (2013b). Çevre sorunlarının çözümünde küresel çevre politikalarının önemi. [The Global Environmental Politics for the Solution of Environmental Problems]. *Journal of the Institute of Social Sciences of Muğla University*, 31, 17–34.

Kazys, J. (2018). Climate change information on internet by different Baltic Sea Region languages: Risks of disinformation & misinterpretation. *Journal of Security and Sustainability Issues*, 7(4), 685–695. DOI: 10.9770/jssi.2018.7.4(6)

Keegan-Phipps, S. (2013). An Aural Tradition with a Pause Button? The Role of Recording Technology in a North East English Folk Session. *Ethnomusicology*, 57(1), 34–56. DOI: 10.5406/ethnomusicology.57.1.0034

Kefi, S., Çeliköz, N. ve Erişen, Y. (2013). Okulöncesi Eğitim Öğretmenlerinin Temel Bilimsel Süreç Becerilerini Kullanım Düzeyleri [Preschool Teachers' Levels of Using the Basic Science Process Skills].Journal of Research in Education and Teaching, C/S. 2 (2): 300-319.

Kellert, S. R. (2002). Experiencing nature: Affective, cognitive, and evaluative development in children. In Kahn, P. H.Jr, & Kellert, S. R. (Eds.), *Children and nature: Psychological, sociocultural, and evolutionary investigations* (pp. 117–151). MIT Press.

Kellert, S. R., & Wilson, E. O. (1993). *The biophilia hypothesis*. Island Press.

King, M. R. (2023). The future of AI in medicine: a perspective from a Chatbot. *Annals of Biomedical Engineering, 51*(2), 291-295. DOI: 10.1007/s10439-022-03121-w

Kırlıoğlu, H., & Zeytin, M. (2015). "Belediyelerde Çevre Muhasebesi ve Raporlama: Marmara Bölgesinde Bulunan Belediyeler Üzerine Bir Durum Analiz", [Environmental Accountıng and Reporting in Municipalities: A Case Analysis on the Municipalities Located in the Marmara Region].ASOS JOURNAL (The Journal of Academic Social Science), Yıl: 3. *Sayı*, 10, 753–777.

Kışoğlu, M., Yıldırım, T., Salman, M., & Sülün, A. (2016). İlkokul ve ortaokullarda çevre eğitimi verecek olan öğretmen adaylarında çevre sorunlarına yönelik davranışların araştırılması. [Investigation of Behaviors Towards Environmental Problems in Teacher Candidates Who Will Give Environmental Education at Primary and Secondary Schools].Erzincan University Journal of Education Faculty, 18 (1), 299-318. DOI: DOI: 10.17556/jef.93507

Koehler, , M. JMishra, , PCain, , W. (2013). What is Technological Pedagogical Content Knowledge (TPACK)? *Journal of Education, 193*(3), 13-19. DOI: 10.1177/002205741319300303

Kowalczyk, I. (2004). El Consejo de Europa: defensor de los derechos de las niñas y los niños. *Infancia en Europa: revista de una red de revistas europeas, 7*, 8-11.

Kozyreva, A., Wineburg, S., Lewandowsky, S., & Hertwig, R. (2023). Critical ignoring as a core competence for digital citizens. *Current Directions in Psychological Science*, 32(1), 81–88. DOI: 10.1177/09637214221121570 PMID: 37994317

Krauss, L. W. (2021). *El cambio climático. La ciencia ante el calentamiento global*. Ed. Pasado y Presente.

Kurup, P. M., Levinson, R., & Li, X. (2021). Informed-Decision Regarding Global Warming and Climate Change Among High School Students in the United Kingdom. *Canadian Journal of Science, Mathematics and Technology Education = Revue Canadienne de l'Enseignement des Sciences, des Mathématiques et de la Technologie*, 21(1), 166–185. DOI: 10.1007/s42330-020-00123-5

Kuzu, T. (2008). Aytül Akal'ın masallarıyla çocukta çevre bilinci geliştirme. [Environmental Awareness in Children With Aytül Akal's Fairy Tales Development]. *The Journal of Selcuk University Social Sciences Institute*, 19, 327–339.

Labadze, L., Grigolia, M., & Machaidze, L. (2023). Role of AI chatbots in education: Systematic literature review. *International Journal of Educational Technology in Higher Education*, 20(1), 56. Advance online publication. DOI: 10.1186/s41239-023-00426-1

Lacerda, D. S. (2023). Investigating the political economy of the territory: The contradictory responses of organisations to spatial inequality. *Organization*, 30(5), 1074–1093. DOI: 10.1177/13505084211061239

Lambert, D., & Morgan, J. (2010). *Teaching geography 11–18: A conceptual approach.* Open University Press.

Lane F. J. (2006), Environmental Education Implementation in Wisconsin: Conceptualizations and Practices, Wisconsiv-Madison Üniversitesi, yayımlanmamış doktora tezi.

Lazer, D. M. J., Baum, M. A., Benkler, Y., Berinsky, A. J., Greenhill, K. M., Menczer, F., Metzger, M. J., Nyhan, B., Pennycook, G., Rothschild, D., Schudson, M., Sloman, S. A., Sunstein, C. R., Thorson, E. A., Watts, D. J., & Zittrain, J. L. (2018). The science of fake news. *Science*, 359(6380), 1094–1096. DOI: 10.1126/science.aao2998 PMID: 29590025

Lee, Y. F., Hwang, G. J., & Chen, P. Y. (2022). Impacts of an AI-based *cha*bot on college students' after-class review, academic performance, self-efficacy, learning attitude, and motivation. *Educational Technology Research and Development*, 70(5), 1843–1865. DOI: 10.1007/s11423-022-10142-8

Lefebvre, H. (2000). La production de l'espace. *Anthropos: Studie z Oboru Anthropologie, Paleoethnologie, Paleontologie a Kvartérní Geologie.*

Legardez, A., & Simonneaux, L. (2006). *L'école à l'épreuve de l'actualité. Enseigner les questions vives.* ESF.

Legott, S. (2014). Negotiating censorship in the postwar spanish novel divorce and civil marriage in Elena Quiroga's "Algo pasa en la calle". *Revista de escritoras ibéricas, 2,* pp. 121-144. DOI: 10.5944/rei.vol.2.2014.12949

León, R. (1983). Estaba el Señor Don Gato [canción]. *Canciones para niños.* Fonomusic.

León, B., Boykoff, M. T., & Jordán, C. R. (2021). Climate change perception among Spanish undergraduates: A reception study on the combination of the local, global, gain and loss frames. *Communicatio Socialis*, 34(1), 57–75. DOI: 10.15581/003.34.1.57-75

Lewinsohn, T. M., Attayde, J. L., Fonseca, C. R., Ganade, G., Jorge, L. R., Kollmann, J., & da Rocha, P. L. (2015). Ecological literacy and beyond: Problem-based learning for future professionals. *Ambio*, 44(2), 154–162. DOI: 10.1007/s13280-014-0539-2 PMID: 24973054

LI, Y.Y. & Liu, S. C. (. (2021). Examining Taiwanese students' views on climate change and the teaching of climate change in the context of higher education. *Research in Science & Technological Education*, 1–14. DOI: 10.1080/02635143.2020.1830268

Liao, W., Yang, L., Zhong, S., Hess, J. J., Wang, Q., Bao, J., & Huang, C. (2019). Preparing the next generation of health professionals to tackle climate change: Are China's medical students ready? *Environmental Research*, 168, 270–277. DOI: 10.1016/j.envres.2018.10.006 PMID: 30342323

Li, C. J., Monroe, M. C., Oxarart, A., & Ritchie, T. (2021). Building teachers' self-efficacy in teaching about climate change through educative curriculum and professional development. *Applied Environmental Education and Communication*, 20(1), 34–48. DOI: 10.1080/1533015X.2019.1617806

Liceras, A. (2018) Los itinerarios didácticos en la enseñanza de la geografía. Reflexiones y propuestas acerca de su eficacia en educación. *Revista UNES. Universidad, Escuela y Sociedad*, (5). https://www.google.es/Los itinerarios didácticos en la enseñanza de la geografía. Reflexiones y propuestas acerca de su eficacia en educación

Liceras, A. (2022) Aportaciones de Giner de los Ríos y la ILE a la enseñanza del paisaje. Las excursiones. *Revista de Investigación en Didáctica de las Ciencias Sociales*, (10). Vista de Aportaciones de Giner de los Ríos y la ILE a la enseñanza del paisaje. Las excursiones (unex.es)

Li, J. Q., Xia, H. M., & Zhao, X. Y. (2022). Web GIS for sustainable education: Towards natural desastres education for high school students. *Sustainability (Basel)*, 14(5), 2694. Advance online publication. DOI: 10.3390/su14052694

Lim, W. M., Gunasekara, A., Pallant, J. L., Pallant, J. I., & Pechenkina, E. (2023). Generative AI and the future of education: Ragnarök or reformation? A paradoxical perspective from management educators. *International Journal of Management Education*, 21(2), 100790. DOI: 10.1016/j.ijme.2023.100790

Lo, C. K. (2023). What Is the Impact of ChatGPT on Education? A Rapid Review of the Literature. *Education Sciences*, 13(4), 410. DOI: 10.3390/educsci13040410

Loizaga, V. B. (2023). Ambitos sociolinguísticos en la ESO. El professorado ante el reto de la LOMLOE. *Íber.Didáctica de las Ciencias Sociales. Geografía e Historia*, 113, 29–33.

Lolo, E. (2006). *Cuentos de la selva* de Horacio Quiroga: La "otra" dicotomía civilizaciónVS barbarie. *Círculo: Revista de Cultura*, XXXV, 151–159.

LOMLOE. (2020). *Ley Orgánica 3/2020, de 29 de diciembre, por la que se modifica la Ley Orgánica 2/2006, de 3 de mayo, de Educación.*

Lomloe, U., & De La Democracia, L. E. (2020). Ley Orgánica 3/2020, de 29 de diciembre, por la que se modifica la Ley Orgánica 2/2006, de 3 de mayo, de Educación. *BOE*, 340, 122868–122953.

López Facal, R., & Santidrián Arias, V. M. (2011). Los "conflictos sociales candentes" en el aula. Iber:Didáctica de las Ciencias Sociales. *Geografía e Historia*, 69, 8–20.

López Gallego, M. (2013). Bildungsroman. Historias para crecer. *Tejuelo, 18*, 62-65. https://dehesa.unex.es/bitstream/10662/4521/1/1988-8430_18_62.pdf

López Martínez, M. J. (2017). Los problemas relevantes de la sociedad actual en el aula de 5 años: qué hacemos con las personas refugiadas sirias. *Investigación en la escuela nº 92*, 19-31, https://dialnet.unirioja.es/servlet/articulo?codigo=6275111

López Martínez, M. J. (2019). La enseñanza de las ciencias sociales a partir de problemas sociales relevantes en un aula de infantil. Estado de la cuestión y resultados de una investigación. *El futuro del Pasado: Revista electrónica de Historia*, nº 10, 57-59, https://dialnet.unirioja.es/servlet/articulo?codigo=7105038

López Torres, E., Carril Merino, M. T., & Alonso Neila, E. (2022). ApS con fuentes orales para hacer frente a la España vaciada en Educación Infantil. *Didáctica de las Ciencias Experimentales y Sociales*, (43), 53–70. DOI: 10.7203/dces.43.25270

López, C. (2021). María Folguera: "Venimos de una tradición en la que lo femenino se construía desde el sacrificio". *Eldiario.es*, 23 de mayo de 2021. https://www.eldiario.es/cultura/libros/maria-folguera-venimos-tradicion-femenino-construia-sacrificio_128_7950455.html

López, I. (2015). Sobre el desarrollo sostenible y la sostenibilidad: conceptualización y crítica. Barataria. *Revista Castellano-Manchega de Ciencias Sociales*, (20), 111-128. https://www.redalyc.org/articulo.oa?id=322142550007

López, J. A., Mora, M., Arrebola, J. C., Medina, S. (2017). Itinerarios didácticos interdisciplinares en el Grado de Educación Primaria: una propuesta en la ciudad de Córdoba. [Trabajo Académico, Universidad de Córdoba]. Didáctica de las Ciencias. 68_-_Itinerarios_didacticos_interdisciplinares_en_el_Grado_de_Educacion_Primaria.pdf (uab.cat)

López, M. J. (2017). Los problemas relevantes de la sociedad actual en un aula de 5 años: qué hacemos con las personas refugiadas sirias. *Investigación en la Escuela*, 92, 19-31. http://www.investigacionenlaescuela.es/articulos/R92/R92-2

Louv, R. (2005). *Last child in the woods: saving our children from nature-deficit disorder*. Algonquin Books of Chapel Hill.

Louv, R. (2018). *Los últimos niños en el bosque. Salvemos a nuestros hijos del trastorno por déficit de naturaleza*. Capitán Swing.

Luis Rico, M. I. (2020). La acogida del alumnado de familia inmigrada en la etapa de educación infantil. In Ampliando horizontes en educación inclusiva. [Recurso electrónico]: XVII Congreso Internacional y XXXVII Jornadas de Universidades y Educación Inclusiva. 21, 22 y 23 de septiembre de 2020 (aplazado en abril por Covid-19) (pp. 71-75). Servicio de Publicaciones e Imagen Institucional.

Lund, B. D., & Wang, T. (2023). Chatting about ChatGPT: How may AI and GPT impact academia and libraries? *Library Hi Tech News*, 40(3), 26–29. DOI: 10.1108/LHTN-01-2023-0009

Luzón-Trujillo, A., & Montes Moreno, S. (2018). Perspectiva histórica de la formación inicial del profesorado de Educación Infantil y Primaria en España. Una tarea inacabada. *Historia Caribe*, 13(33), 121–152. DOI: 10.15648/hc.33.2018.6

Mabratu, D. (1998). Sustainability and Sustainable Development: Historical and conceptual review. *Environmental Impact Assessment Review*, 18(6), 493–520. DOI: 10.1016/S0195-9255(98)00019-5

Maddux, W. W., Lu, J. G., Affinito, S. J., & Galinsky, A. D. (2021). Multicultural experiences: A systematic review and new theoretical framework. *The Academy of Management Annals*, 15(2), 345–376. DOI: 10.5465/annals.2019.0138

Maffi, L. (2005). Linguistic, cultural, and biological diversity. *Annual Review of Anthropology*, 34(1), 599–617. DOI: 10.1146/annurev.anthro.34.081804.120437

Májovský, M., Černý, M., Kasal, M., Komarc, M., & Netuka, D. (2023). Artificial Intelligence Can Generate Fraudulent but Authentic-Looking Scientific Medical Articles: Pandora's Box Has Been Opened. *Journal of Medical Internet Research*, 25, 46924. DOI: 10.2196/46924 PMID: 37256685

Malacarne, S. (2023). Mutual learning: A real way to cooperate? Evidence of more equal international development cooperation. *Italian Sociological Review*, 13(3), 347–369.

Mamedov, N. (1998). Environmental culture and education. Mamedov, N (Ed.). Environmental Education - Concept and Methodical in Approach. (pp.19-50). Izmir: Environmental Protection and Research Foundation Publications.

Mansilla Sepúlveda, J. M., Villalba, K., Mora García, J. P., & Turra Díaz, O. (2022). Tensión epistémica entre multiculturalismo anglosajón e interculturalidad latinoamericana: Sus proyecciones en educación. *Revista de Filosofía (Venezuela)*, 39(100), 132–143. DOI: 10.5281/zenodo.5979768

Marks, M. A. (1980). Elena Quiroga's Yo Voice and the Schism between Reality and Illusion. *Anales de La Narrativa Española Contemporánea*, 5, 39–55. https://www.jstor.org/stable/27741025

Marks, M. A. (1981). Time in the Novels of Elena Quiroga. *Hispania*, 64(3), 376–381. https://www.jstor.org/stable/342041. DOI: 10.2307/342041

Marques, E., Nossa, P., & de Guimarães, R. B. (2022). Saúde e território em contexto de pós-pandemia: Secção temática. *Finisterra, 57*(121) (AOP). DOI: 10.18055/Finis29127

Marrón Gaite, M. J. (1999). La geografía del comportamiento y de la percepción. Aportaciones a la investigación y a la enseñanza de la geografía. *Didáctica Geográfica*, 3, 85–108.

Marrón, M. J. (2013). Tratamiento interdisciplinar de la geografía, la historia y el arte con estudiantes del Grado de Magisterio. Una propuesta de enseñanza activa a través de un itinerario didáctico. [Trabajo Académico, Universidad Complutense de Madrid]. Innovación en la enseñanza de la geografía ante los desafíos sociales y territoriales. Tratamiento interdisciplinar de la Geografía, la Historia y el Arte con estudiantes del Grado de Magisterio. Una propuesta de enseñanza activa a través de un itinerario didáctico (dpz.es)

Marrón, M. J. (2000). *Enseñar geografía. De la teoría a la práctica.* Síntesis.

Marrón, M. J. (2011). Enseñar y aprender Geografía a partir de distintos lenguajes: Una experiencia de trabajo por proyectos para estudiar el clima de España a partir de los refranes. *Annales de Géographie*, 31(1), 107–123.

Martí, J., Amor, J. A., Giménez-García, R., Ruiz-Álvarez, V., & Biener, S. (2021). Episódios de chuvas torrenciais de 11 a 15 de setembro de 2019 no sudeste da Península Ibérica: Análise meteorológica e consequências das transformações nos usos do solo. *Finisterra*, 56(117), 151–174. DOI: 10.18055/Finis18993

Martín Rojo, L. (2004). Escuela y diversidad lingüística y cultural. In *Martín Rojo, L. (dir.). ¿Asimilar o integrar? Dilemas ante el multilingüismo en las aulas* (pp. 19–69). Ministerio de Educación y Ciencia.

Martín, C., & Navarro, J. I. (2015). *Psicología evolutiva en educación infantil y primaria*. Ediciones Pirámide.

Martínez Arnaldos, M. (1996). Deslinde teórico de la novela corta. *Monteagudo. Revista de Literatura Española, Hispanoamericana y Teoría de la Literatura, 1*, pp. 47–66. https://revistas.um.es/monteagudo/article/view/76881

Martínez Cachero, J. M. ª. (1979). *Historia de la novela española entre 1936 y 1975*. Castalia.

Martínez, J. (2000). Educación ambiental y conservación de paisajes frágiles. En *Libro de actas de la Reunión Internacional de Expertos en Educación Ambiental.*

Martínez, R., & Tonda, E. M. (2014). Nuevas perspectivas conceptuales y metodológicas para la educación geográfica. *Grupo de Didáctica de la Geografía de la Asociación de Geógrafos Españoles* pp. 227-229. INTERIOR VOLUMEN I.indd (age-geografia.es)

Martínez, C. (2020). *Infancia, pandemia y derechos. Treinta años de la Convención de los Derechos del Niño en España*. Thomson Reuters Aranzadi.

Martínez, J. M. (2012). De las escuelas al aire libre a las aulas de naturaleza. *Áreas. Revista Interna de Ciencias Sociales*, 20, 171–182.

Martínez-León, P., Ballester Roca, J., & Ibarra-Rius, N. (2017). Identidades fronterizas y pertenencias plurales: Diversidad e interculturalidad a través de la educación literaria. *Didáctica.Lengua y Literatura*, 29, 155–172. DOI: 10.5209/DIDA.57135

Martin, F., & Owens, P. (2005). Young children making sense of the world. In Scoffham, S. (Ed.), *Primary geography handbook* (pp. 63–73). Geographical Association.

Martín-Párraga, L., Llorente-Cejudo, C., & Barroso-Osuna, J. (2023). Self-Perception of Digital Competence in University Lecturers: A Comparative Study between Universities in Spain and Peru According to the DigCompEdu Model. *Societies (Basel, Switzerland)*, 13(6), 142. DOI: 10.3390/soc13060142

Massey, D. (2005). *For space*. Sage Publications.

Mastrángelo, M. E., Pérez-Harguindeguy, N., Enrico, L., Bennett, E., Lavorel, S., Cumming, G. S., Abeygunawardane, D., Amarilla, L. D., Burkhard, B., Egoh, B. N., Frishkoff, L., Galetto, L., Huber, S., Karp, D. S., Ke, A., Kowaljow, E., Kronenburg-García, Á., Locatelli, B., Martín-López, B., & Zoeller, K. (2019). Key knowledge gaps to achieve global sustainability goals. *Nature Sustainability*, 2(12), 1115–1121. DOI: 10.1038/s41893-019-0412-1

Maters, M. (2020). *123 curiosidades que todo el mundo debería conocer sobre el clima*. Geoplaneta.

Maude, A., & Caldis, S. (2019). Teaching Higher-Order Thinking and Powerful Geographical Knowledge Through the Stage 5 Biomes and Food Security Unit. *Geographical Education, 32*, 30-39. https://eric.ed.gov/?id=EJ1238503

Mbaye, L. M. (2014). "Barcelona or die": Understanding illegal migration from Senegal. *IZA Journal of Migration*, 3(1), 1–19. DOI: 10.1186/s40176-014-0021-8

McCrindle, M. (2014). *The ABC of XYZ: understanding the global generations*. UNSW Press.

McIntyre, L. (2018). *Post-Truth*. MIT Press. DOI: 10.7551/mitpress/11483.001.0001

McWhirther, N., & Shealy, T. (2018). Case-based flipped classroom approach to teach sustainable infrastructure and decision-making. *International Journal of Construction Education and Research*, 1–21. DOI: 10.1080/15578771.2018.1487892

Medina-Nicolalde, M. A., & Tapia-Calvopina, M. P. (2017). El aprendizaje basado en proyectos una oportunidad para trabajar interdisciplinariamente. OLIMPIA. Revista de la Facultad de Cultura Física de la Universidad de Granma, 14(46), 236-246. https://dialnet.unirioja.es/servlet/articulo?codigo=6220162

Mehr, S. A., Krasnow, M. M., Bryant, G. A., & Hagen, E. H. (2020). Origins of music in credible signaling. *Behavioral and Brain Sciences*, 44, e60. DOI: 10.1017/S0140525X20000345 PMID: 32843107

Mehr, S. A., Singh, M., Knox, D., Ketter, D. M., Pickens Jones, D., Atwood, S., Lucas, C., Jacoby, N., Egner, A. A., Hopkins, E. J., Howard, R. M., Hartshorne, J. K., Jennings, M. V., Simson, J., Bainbridge, C. M., Pinker, S., O'Donnell, T. J., Krasnow, M. M., & Glowacki, L. (2019). Universality and diversity in human song. *Science*, 366(6468), eaax0868. DOI: 10.1126/science.aax0868 PMID: 31753969

Memişoğlu, H. (2017). Opinions of Teachers and Preservice Teachers of Social Studies on Geo-Literacy. *Educational Research Review*, 12(19), 967–979. DOI: 10.5897/ERR2017.3315

Mendes, L. (2024). Pensamento espacial crítica, activismo e cidadania territorial: introdução a uma trialéctica. In M. J. M. Gaité (Ed.) *Geografía, Educación y Innovacionnes y Didácticas* (pp. 146-153). Asociación Española de Geografía (AGE). Grupo de Didáctica.

Méndez, M., Fernández, A., Cruz, A., & Bueno, P. (2020). De la representación social del cambio climático a la acción: El caso de estudiantes universitarios. *Revista Mexicana de Investigación Educativa*, 87, 1043–1068.

Meredith, J., Cantrell, D., Conner, M., Evener, B., Hunn, D., & Spector, P. (2000). Best Practices for Environmental Education: Guidelines for Success.

Merino, C., Pérez, M., & Calaghan, P. (1998). Espacios naturales protegidos y educación ambiental. En P. Aznar (Coord.), *La educación ambiental en la sociedad global: actas* (pp. 292-298). Universitat de València, Servei de Publicacions.

Merriam, A. P. (1964). *The Anthropology of Music*. Northwestern University Press.

Mínguez, Mª del C. (2010). El paisaje como objeto de estudio de la Geografía. Un Itinerario Didáctico en el Marco de la Semana de la Ciencia de la Comunidad de Madrid. *Revista Didáctica Geográfica* (11), 37-62. Vista de EL PAISAJE COMO OBJETO DE ESTUDIO DE LA GEOGRAFÍA. UN ITINERARIO DIDÁCTICO EN EL MARCO DE LA SEMANA DE LA CIENCIA DE LA COMUNIDAD DE MADRID. (age-geografia.es)

Miró-Pérez, J., & Estrela, M. (2016). Reconstrucción de la señal térmica local en la Comunidad Valenciana entre 1948 y 2011 a partir de un downscaling estadístico mediante una red neuronal artificial: Detección de patrones locales de cambio. *Boletín de la Asociación de Geógrafos Españoles*, 70(70), 113–147. DOI: 10.21138/bage.2165

Mishra, P., & Koehler, M. J. (2006). Technological Pedagogical Content Knowledge: A Framework for Teacher Knowledge. *Teachers College Record*, 108(6), 1017–1054. DOI: 10.1111/j.1467-9620.2006.00684.x

Mitrović, S., Andreoletti, D., & Ayoub, O. (2023) ChatGPT or Human? Detect and explain. Explaining Decisions of Machine Learning Model for Detecting Short ChatGPT-generated Text. *EdArXiv*. arXiv:2301.13852

Molina, M. P. (2018). *La Educación Patrimonial en la Didáctica de las Ciencias Sociales en Primaria*. [Trabajo Académico, Universidad de Córdoba]. Contextos Educativos. bibybape,+12_MolinaTorres.pdf

Mon, P. R. (2021). An introduction to the concept of landscape in geography. *The Asian Review of Civil Engineering*, 10(1), 20–25. DOI: 10.51983/tarce-2021.10.1.2941

Montenegro, P. G. (2006). Símbolos inadvertidos en *El almohadón de plumas* de Horacio Quiroga. *El cuento en red: estudios sobre la ficción breve, 14* (2).

Montero Caro, M. D. (2021). Educación, gobierno abierto y progreso: Los Objetivos de Desarrollo Sostenible (ODS) en el ámbito educativo. Una visión crítica de la LOMLOE. *Revista de Educación y Derecho*, (23), 1–26. DOI: 10.1344/REYD2021.23.34443

Montes Pedrajo, C. (2023). *Actitudes lingüísticas de estudiantes cántabros de bachillerato hacia las variedades americanas del español: diagnóstico e intervención* [TFG inédito]. Universidad de Cantabria.

Montoya, G., Valencia, L., Vargas, L., García, J., Franco, J., & Calderón, H. (2022). Ruralidad, educación rural e identidad profesional de maestras y maestros rurales. *Praxis & Saber*, 13(34), e13323. DOI: 10.19053/22160159.v13.n34.2022.13323

Monzón Perdomo, M. E. (2019). «Mujeres solas» luces y sombras de la emigración canaria a América (siglos XVIII-XIX). *Anuario de Estudios Atlánticos*, (65), 1-24. Cabildo Insular de Gran Canaria.

Morales Hernández, A. J., & López Martínez, M. J. (2022). Didáctica de las Ciencias Sociales en Educación Infantil: Expectativas curriculares y formación docente. *Revista Educatio Siglo XXI*, 40(3), 61–82. DOI: 10.6018/educatio.487691

Mora, V. L. (2018). Líneas de fuga neorrurales de la literatura española contemporánea. *Tropelías.Revista de Teoría de la Literatura y Literatura Comparada*, 4(4), 198–221. DOI: 10.26754/ojs_tropelias/tropelias.201843071

Moreno Fernández, O. (2013). *Educación ambiental y educación para la ciudadanía desde una perspectiva planetaria. Estudio de experiencias educativas en Andalucía* [Environmental education and education for citizenship from a planetary perspective. Study of educational experiences in Andalucía]. Universidad Pablo de Olavide.

Moreno, L. (2024). Recorriendo Nuestras Calles. [Storyjumper]. "Recorriendo nuestras calles" - Read stories online. Create books for kids | StoryJumper

Moreno, M. C., & Gallego, C. I. (2013). "Egipto en mi aula": didáctica de la geografía para educación infantil. En R. de Miguel, M.L. de Lázaro y M.J. Marrón (coords.), *Innovación en la enseñanza de la geografía ante los desafíos sociales y territoriales* (pp. 353-366). Diputación provincial de Zaragoza.

Moreno-Vera, J. R. (2023). Epilogue: "We wanna learn like common people whatever common people did". In Gómez-Carrasco, C. J. (Ed.), *Re-Imagining the Teaching of European History. Promoting Civic Education and Historical Consciousness* (pp. 219–223). Routledge.

Morón, H. (2016). *El itinerario didáctico como herramienta para la activación patrimonial: una experiencia docente desde el parque María Luisa*. [Trabajo académico, Universidad de Sevilla]. EL ITINERARIO DIDÁCTICO COMO HERRAMIENTA PARA LA ACTIVACIÓN PATRIMONIAL: UNA EXPERIENCIA DOCENTE DESDE EL PARQUE MARÍA LUISA (us.es)

Morote, Á. F. (2021). Investigación y enseñanza en Didáctica de la Climatología: Estado de la cuestión (1980-2020) (España). *Espacio, Tiempo y Forma. Serie VI, Geografía*, 14(14), 89–120. DOI: 10.5944/etfvi.14.2021.31332

Morote, Á. F., Campo, B., & Colomer, J. C. (2021). Percepción del cambio climático en alumnado de 4º del grado en Educación Primaria (Universidad de Valencia, España) a partir de la información de los medios de comunicación. *Revista Electrónica Interuniversitaria de Formación del Profesorado*, 24(1), 131–144. DOI: 10.6018/reifop.393631

Morote, Á. F., & Gómez-Trigueros, I. M. (2023). La brecha digital de género y enseñanza de los riesgos naturales en la formación del profesorado de Ciencias Sociales. *Research in Education and Learning Innovation Archives*, 30(30), 60–75. DOI: 10.7203/realia.30.24712

Morote, Á. F., & Hernández, M. (2020). Social Representations of Flooding of Future Teachers of Primary Education (Social Sciences): A Geographical Approach in the Spanish Mediterranean Region. *Sustainability (Basel)*, 12(15), 1–14. DOI: 10.3390/su12156065

Morote, Á. F., & Hernández, M. (2024). Knowledge and perception of Spanish school children of climate change. *Children's Geographies*, 22(3), 465–479. DOI: 10.1080/14733285.2024.2303581

Morote, Á. F., & Olcina, J. (2021). La importancia de la enseñanza del cambio climático: Propuestas didácticas para la geografía escolar. *Estudios Geográficos*, 82(291), e078. DOI: 10.3989/estgeogr.202189.089

Morote, Á. F., & Olcina, J. (2023a). Cambio climático y educación. Una revisión de la documentación oficial. *Documents d'Analisi Geografica*, 69(1), 107–134. DOI: 10.5565/rev/dag.749

Morote, A. F., & Olcina, J. (2023b). Educando en el riesgo. Formación y propuestas didácticas para incrementar la resiliencia socio-territorial de la población escolar. *Anales de Geografía de la Universidad Complutense*, 43(2), 413–434. DOI: 10.5209/aguc.90582

Morote, Á. F., Olcina, J., & Hernández, M. (2022). How is flood risk explained in the subject of Geography in Spanish schools? An approach based on Social Science textbooks (Primary Education). *International Research in Geographical and Environmental Education*, 32(2), 124–139. DOI: 10.1080/10382046.2022.2133955

Morote, Á. F., & Souto, X. M. (2020). Educar para convivir con el riesgo de inundación. *Estudios Geográficos*, 81(288), 1–14. DOI: 10.3989/estgeogr.202051.031

Moseley, C. (2000). Teaching for environmental literacy. *The Clearing House: A Journal of Educational Strategies, Issues and Ideas*, 74(1), 23–24. DOI: 10.1080/00098655.2000.11478633

Muñoz Jumilla, A. R. (2002). Efectos de la globalización en las migraciones internacionales. *Papeles de Población*, 8(33), 9–45. https://www.scielo.org.mx/pdf/pp/v8n33/v8n33a2.pdf

Muñoz, E. M., & García Monteagudo, D. (2021). El agua como propuesta didáctica piloto en el grado de maestro/a en Educación Infantil. En J. Granados y R.M. Medir (coords.), *Enseñar y aprender geografía para un mundo sostenible* (pp. 361-377). Octaedro.

Muñoz, E., & Jerez, Ó. (2018). El trabajo de campo como herramienta didáctica y su relación con el medio en el proyecto ¡Nosotros proponemos! En M. A. Rodríguez & S. Claudino (Coords.), *¡Nosotros proponemos! Ciudadanía, sostenibilidad e innovación geográfica ante los desafíos educativos de la sociedad* (pp. 41-65). Graó.

Muñoz, C., Schultz, D., & Vaughan, G. (2020). *A Midlatitude Climatology and Interannual Variability of 200- and 500-hPa Cut-Off Lows*. American Meteorological Society.

Muñoz, J. A., & Otaño, Y. (2014). Software educativo: "a prepararnos para enfrentar los desastres naturales". *Ejemplar dedicado a:Memorias del XXVII Fórum Científico Nacional de Estudiantes Universitarios de Ciencias Pedagógicas, 14*, (39), 1-13.

Murga Menoyo, M.A. (2020). El camino hacia los ODS: Conformar una ciudadanía planetaria mediante la educación. *Comillas Jorunal of International Relations*, núm 19: Los desafíos de la Agenda 2030. Educación y ciudadanía global, 01-11, DOI: 10.14422/cir.i19.y2020.001

Murga-Menoyo, M. Á., & Novo, M. (2017). Sostenibilidad, desarrollo glocal y ciudadanía planetaria: referentes de una Pedagogía para el desarrollo sostenible. *Teoría de la Educación: Revista Interuniversitaria, 29*(1), 55-78. https://digital.casalini.it/10.14201/teoredu20172915578

Musgrave, R. A. (2020). Los niños no pueden explorar el mundo, pero sí cartografiarlo. *National Geographic*. https://www.nationalgeographic.es/family/2020/06/los-ninos-no-pueden-explorar-el-mundo-pero-si-cartografiarlo

Negrín, M. Á., & Marrero, J. J. (2021). La nueva Ley de Educación (LOMLOE) ante los Objetivos de Desarrollo Sostenible de la Agenda 2030 y el reto de la COVID-19. *Avances En Supervisión Educativa*, (35). Advance online publication. DOI: 10.23824/ase.v0i35.709

Negrín, O. (2005). *Teorías e instituciones contemporáneas de educación*. Editorial Universitaria Ramón Areces.

Nelles, D., & Serrer, C. (2020). *El pequeño manual del cambio climático*. Grijalbo.

Nelson, T. A., Goodchild, M. F., & Wright, D. J. (2022). Accelerating ethics, empathy, and equity in geographic information science. *Proceedings of the National Academy of Sciences of the United States of America*, 119(19), e2119967119. DOI: 10.1073/pnas.2119967119 PMID: 35507875

Neto, M. S., Silva, F. D. P. R., Alencar, J. J., & Araújo, R. L. (2024). Transformações espaciais e o Projeto Nós Propomos! em Campo Maior/Piauí/Brasil. In M. A. Rodríguez-Domenech (Ed.), *El pensamiento geográfico y su enseñanza en el siglo XXI: Tendencias y perspectivas a través del proyecto de participación ciudadana ¡Nosotros Proponemos!* (pp. 49-59). Dykinson Ebook.

Nettl, B. (2005). *The Study of Ethnomusicology: Thirty-One Issues and Concepts*. University of Illinois Press.

Nguyen, C. V., & Pham, N. M. (2018). The impact of natural disasters on children's education: Comparative evidence from Ethiopia, India, Peru, and Vietnam. *Review of Development Economics*, 22(4), 1561–1589. DOI: 10.1111/rode.12406

Normatova, A. (2024). Five ways to build intercultural competence and confidence in your classroom. *Educational Research in Universal Sciences*, 3(4 SPECIAL), 512-514.

Normatova, A. (2024). Five ways to build intercultural competence and confidence in your classroom. *Educational Research in Universal Sciences*, 3(4), 512–514.

Núñez Romero, V. (2017) Las salidas escolares como recurso motivador para los niños [Trabajo Fin de Grado, Universidad de Valladolid]. Universidad de Valladolid. Las salidas escolares como recurso motivador para los niños

Núñez, M. ª P. (2005). Literatura juvenil y educación intercultural: ¿estética vs. axiología? En Herrera, F. et alii (eds.): *Inmigración, interculturalidad y convivencia, IV*. Instituto de Estudios Ceutíes / Ciudad Autónoma de Ceuta, 609-618.

Nussbaum, M. C. (2019). *The monarchy of fear: A philosopher looks at our political crisis*. Simon & Schuster.

O'Flaherty, J., & Liddy, M. (2018). The impact of development education and education for sustainable development interventions: A synthesis of the research. *Environmental Education Research*, 24(7), 1031–1049. DOI: 10.1080/13504622.2017.1392484

Observatorio Español del Racismo y la Xenofobia (OBERAXE). (2022). *Informe sobre la integración de los estudiantes extranjeros en el sistema educativo español. Resumen*. Ministerio de Inclusión, Seguridad Social y Migraciones.

Ocampo Ospina, L. F., & Valencia Carvajal, S. (2019). Los problemas sociales relevantes: enfoque interdisciplinar para la enseñanza integrada de las ciencias sociales, *REIDICS. Revista de Investigación en Didáctica de las Ciencias Sociales* n º 4, 60-75, https://dehesa.unex.es/bitstream/10662/9328/1/2531-0968_2019_4_60.pdf

OECD (2020), *TALIS 2018 Results (Volume II): Teachers and School Leaders as Valued Professionals*, TALIS, OECD Publishing. https://doi.org/DOI: 10.1787/19cf08df-en

OECD. (2016). *Skills for a Digital World: 2016 Ministerial Meeting on the Digital Economy Background Report, OECD Digital Economy Papers, No. 250*. OECD Publishing., DOI: 10.1787/5jlwz83z3wnw-

Okonkwo, C. W., & Ade-Ibijola, A. (2021). Chatbots applications in education: A systematic review. *Computers and Education: Artificial Intelligence*, 2, 100033. DOI: 10.1016/j.caeai.2021.100033

Okyay, Ö., Demir, Z. G., Sayın, A., & Özdemir, K. (2021). Ekolojik okuryazarlık eğitiminin okul öncesi öğretmenlerinin ekolojik farkındalığı ve çevreye yönelik motivasyonlarına etkisi. [The effect of ecological literacy training on preschool teachers' ecological awareness and motivation towards the environment]. *Başkent University Journal of Education*, 8(1), 129–146.

Okyay, Ö., Sayın, A., Demir, Z. G., & Özdemir, K. (2022). Ekolojik Okuryazarlık Eğitiminin Okul Öncesi Çocukların Çevreye Yönelik Farkındalık ve Tutumlarına Etkisi. [The Impact of Ecological Literacy Training on Environmental Awareness and Attitudes of Preschool Childrens]. *Milli Eğitim Dergisi*, 51(236), 2851–2870. DOI: 10.37669/milliegitim.944271

Olave, D. (2005). El itinerario didáctico: Una propuesta metodológica para el análisis geo-histórico local. [Redalyc.El itinerario didáctico: una propuesta metodológica para el análisis geo-histórico local]. *Revista Geoenseñanza*, 10(2), 197–208.

Olawumi, T. O., & Chan, D. W. M. (2018). A scientometric review of global research on sustainability and sustainable development. *Journal of Cleaner Production*, 183, 231–250. DOI: 10.1016/j.jclepro.2018.02.162

Olazagasti, E. (1988). "Plácida, la joven", síntesis de la narrativa quiroguiana. *Anales de La Literatura Española Contemporánea*, 13(1/2), 135–146. https://www.jstor.org/stable/27741844

Olcina, J. (2017). Incremento de episodios de inundación por lluvias de intensidad horaria en el sector central del litoral mediterráneo español: Análisis de tendencias en Alicante. *Rev. Semata*, 29, 143–163.

Olcina, J. (2020). Clima, cambio climático y riesgos climáticos en el litoral mediterráneo español: Oportunidades para la geografía. *Documents d'Analisi Geografica*, 66(1), 159–182. DOI: 10.5565/rev/dag.629

Olcina, J., & Biener, S. (2019). Está cambiando el clima valenciano? Realidades e incertidumbres. [Is the Valencian climate changing? Realities and uncertainties] In Olcina, J., & Moltó, E. A. (Eds.), *Climas y Tiempos en el País Valencià* (pp. 162–170). Publicaciones de la Universidad de Alicante.

Olcina, J., & Moltó, E. A. (Eds.). (2019). *Climas y Tiempos del País Valenciano*. Publicaciones de la Universidad de Alicante.

Olcina, J., & Morote, A. F. (2023). Cambio climático y Objetivos de Desarrollo Sostenible (ODS). Adaptación y enseñanza como retos principales en el ámbito mediterráneo. *Finisterra. Revista Portuguesa de Geografía, 58* (122), 39-61. DOI: 10.18055/Finis28898

Olcina, J., Morote, Á. F., & Hernández, M. (2022). Teaching Floods in the Context of Climate Change with the Use of Official Cartographic Viewers (Spain). *Water (Basel)*, 14(21), 1–20. DOI: 10.3390/w14213376

Oller, M. (1999). Trabajar problemas sociales en el aula, una alternativa a la transversalidad. En García Santa María, T. (Coord.). *Un currículum de ciencias sociales para el siglo XXI: qué contenidos y para qué* (pp. 123-131). Díada Editora, S.L.

Oller, M. (1999). Trabajar problemas sociales en el aula, una alternativa a la transversalidad. En T, García Santa María. (Coord.), *Un currículum de ciencias sociales para el siglo XXI: qué contenidos y para qué*, (pp. 123-131)., Díada Editora S.L.

Oltman, M. (2002). Natural Wonders: A Guide to Early Childhood for Environmental Educators.

Önder, A., Özkan, B. (2013). *Sürdürülebilir Çocuk Gelişimi.* [Sustainable Child Development]. *Environmental Education with Activities in Preschool.* Anı Yayıncılık

Onur, B. (2016). *Child environment nature: Environment and citizenship education.* İmge Bookstore.

Opuni-Frimpong, N. Y., Essel, H. B., Opuni-Frimpong, E., & Obeng, E. A. (2022). Sustainable Development Goal for Education: Teachers' Perspectives on Climate Change Education in Senior High Schools (SHS). *Sustainability (Basel)*, 14(13), 8086. DOI: 10.3390/su14138086

Ordóñez Díaz, L. (2021). *Ríos que cantan, árboles que lloran: Imágenes de la selva en la narrativa hispanoamericana.* Universidad del Rosario. Universidad de los. *Andes (Salta).*

Organización de las Naciones Unidas [ONU] (2015). *Objetivo 14: Conservar y utilizar sosteniblemente los océanos, los mares y los recursos marinos.*https://www.un.org/sustainabledevelopment/es/oceans/

Organización de las Naciones Unidas [ONU] (2023). *Acción por el clima.*https://www.un.org/es/climatechange

Organización de las Naciones Unidas [ONU] (2024). *Objetivos y metas de desarrollo sostenible.*https://www.un.org/sustainabledevelopment/es/sustainable-development-goals/

Örnek, G. (1994). *Çevre eğitimi ve lise eğitim programlarındaki yeri.* Yayınlanmamış Yüksek Lisans Tezi, Hacettepe Üniversitesi Fen Bilimleri Enstitüsü.

Orr, D. W. (1992). *Ecological Literacy: Education and ahe Transition to a Postmodern World.* State University of New York Press.

Ortega Rocha, E. (2020). Concepciones del profesorado chileno sobre la conciencia geográfica. *REIDICS.Revista De Investigación En Didáctica De Las Ciencias Sociales*, 6, 126–143. DOI: 10.17398/2531-0968.06.126

Ortega Sánchez, D., & Olmos Vila, R. (2018). Los problemas sociales relevantes o las cuestiones socialmente vivas en la enseñanza de las ciencias sociales. En M.A. Jara y A. Santisteban. (Coords). *Contribuciones de Joan Pagés al desarrollo de las ciencias sociales, la historia y la geografía en Iberoamérica* (pp. 203-214). Universidad Nacional del Comahue. https://ddd.uab.cat/pub/llibres/2018/196869/jara-santisteban-2018-bariloche.pdf

Ortiz, A., Prats, M., & Baylina, M. (2012). Métodos visuales y geografías de la infancia: dibujando el entorno cotidiano. *Scripta Nova, Revista Electrónica de Geografía y Ciencias Sociales*, XVI (400). https://www.ub.edu/geocrit/sn/sn-400.htm

Osborne, M., Kearns, P., & Yang, J. (2013). Learning cities: Developing inclusive, prosperous and sustainable urban communities. *International Review of Education*, 59(4), 409–423. DOI: 10.1007/s11159-013-9384-y

Oswin, N. (2020). An other geography. *Dialogues in Human Geography*, 10(1), 9–18. DOI: 10.1177/2043820619890433

Oxford University Press. (n.d.). Post-truth. *Oxford English Dictionary.* Retrieved October 4, 2024, from https://www.oed.com/search/dictionary/?scope=Entries&q=post-truth

Özdemir, O. (2007). A new environmental education perspective: "Education for sustainable development". *Education in Science*, 32(145), 23–38.

Özdoğan, B. (2001). Altı-on iki yaşlarındaki çocukların eğitimi ve okul başarıları. [Education and School Success of Children Aged Between Six and Twelve]. *Education in Science*, 26(120), 3–7.

Özgen, N., ve Kahyaoğlu, M. (2011). Farklı Fonksiyonel Özeliğe Sahip Yerleşim Ünitelerinde İkamet Eden İlköğretim Öğrencilerinin Çevre Sorunlarını Algılama ve Çözüm Önerileri: Fenomenografik Bir Araştırma [Perception of Primary School Students Residing in the Settlement Units With Diverse Functional Features on Environmental Problems and Solution Recommendations: A Phenomenographic Research]. *Electronic Journal of Social Sciences*, 10(38), 136–157.

Özkan-Kılıç, Ö. (2012). Okul Öncesi Dönem Resimli Öykü Kitaplarının Coğrafi Kavramları İçermesi Yönünden İncelenmesi, [Analyze whether picture story books published for preschoolers include geographic concepts]. (Master's Thesis), Çanakkale Onsekiz Mart University.

Pagès, J. (1998). La formación del pensamiento social. En P. Benejam y J. Pagès, J. (Coords.). *Enseñar y aprender Ciencias Sociales, Geografía e Historia en la Educación Secundaria* (pp.152-164). Horsori.

Pagès, J. (2009). Enseñar y aprender ciencias sociales en el siglo XXI: reflexiones casi al final de una década. In Investigación en Educación, Pedagogía y Formación Docente, II Congreso Internacional. Libro (Vol. 2).

Pagès, J. (2019). La ciudadanía global y la enseñanza de las Ciencias Sociales: retos y posibilidades para el futuro. *Reidics:Revista de Investigación en didáctica de las ciencias sociales*. 5, 5-22.

Pagès, J., & Santisteban, A. (2011). *Les qüestions socialment vives i l'ensenyament de les ciències socials*. Universitat Autónoma Barcelona.

Palacios, D. (1988). La aplicación del método indagatorio a la enseñanza de la Geografía a través del trabajo de campo. En *Actas de las Primeras Jornadas de Didáctica de la Geografía* (pp. 124-130). A.G.E-Grupo de Didáctica.

Palacios, D. (1990). *Propuesta de un método para la didáctica de la geografía física a través del trabajo de campo. La indagación* (Tesis doctoral). Universidad de La Laguna.

Pallares Jiménez, M. A., & Lleida Añón, A. (2023). Diseño de actividades en el rincón de ODS en el aula de Educación Infantil. En M.E. Cambil Hernández, A.R. Fernández Paradas, N. De Alba Fernández. (Coords). *La didáctica de las Ciencias Sociales ante el reto de los ODS* (pp. 675-682). Narcea S.A. Ediciones.

Palma, A. (2023). Didáctica de las Ciencias Sociales y Educación para la Paz. Valoraciones desde la práctica docente. *Revista UNES. Universidad. Escuela y Sociedad*, 14(14), 5–25. DOI: 10.30827/unes.i14.27121

Palmer, J., & Birch, J. (2004). *Geography in the early years*. Routledge.

Pan, Z. (2021). *Cuentos de la selva* of Horacio Quiroga: A regionalism adapted for children readers. *International Journal of Humanities and Social Science*, 11(7), 96–104.

Parkinson, A. (2020). *Why study geography?* London Publishing Partnership.

Parra, M., Domínguez, G., & Caballero, P. J. (2008). El cuaderno de campo: Un recurso para dinamizar senderos desde la educación en valores. *Ágora para la Educación Física y el Deporte*, (7-8), 145–158.

Parreño-Castellano, J. M., & Rodríguez Rodríguez, M. A. (2021). *Aprendizaje basado en proyectos colaborativos. Una experiencia con estudiantes del Grado en Geografía y Ordenación del Territorio* (No. COMPON-2021-CINAIC-0060). Universidad de Zaragoza, Servicio de Publicaciones. DOI: 10.26754/CINAIC.2021.0060

Pascual, J., & Navío, A. (2018). Concepciones sobre innovación educativa. ¿Qué significa para los docentes en Chile? Profesorado. *Revista de Currículum y Formación del Profesorado*, 22(4), 71–90. DOI: 10.30827/profesorado.v22i4.8395

Paul, R. (1992). Critical thinking: What, why, and how. *New Directions for Community Colleges*, (77), 324.

Pausas, J. G., & Millán, M. (2019). Greening and browning in a climate change hotspot: The Mediterranean Basin. *Bioscience*, 67(2), 143–151. DOI: 10.1093/biosci/biy157

Pavlik, J. V. (2023). Collaborating with ChatGPT: Considering the implications of generative artificial intelligence for journalism and media education. *Journalism & Mass Communication Educator*, 78(1), 84–93. DOI: 10.1177/10776958221149577

Pedraza Jiménez, F. B. (2002). *Manual de historia de la literatura hispanoamericana*. Cénlit.

Pedretti, E., & Nazir, J. (2014). Tensions and opportunities: A baseline study of teachers' views of environmental education. *International Journal of Environmental and Science Education*, 9(3), 265–283.

Pelletier, K., McCormack, M., Muscanell, N., Reeves, J., Robert, J., & Arbino, N. (2024). *2024 EDUCAUSE Horizon Report, Teaching and Learning Edition*. Educause.

Pelletier, K., Robert, J., Arbino, N., Muscanell, N., McCormack, M., Reeves, J., McDonald, B., & Grajek, S. (2023). *2023 EDUCAUSE Horizon Report: Holistic Student Experience Edition*. EDUCAUSE.

Pelletier, K., Robert, J., Muscanell, N., McCormack, M., Reeves, J., Arbino, N., Grajek, S., Birdwell, T., Liu, D., Mandernach, J., Moore, A., Porcaro, A., Rutledge, R., & Zimmern, J. (2023). *2023 EDUCAUSE Horizon Report, Teaching and Learning Edition*. EDUCAUSE.

Pena, M., & Caamaño, M. J. (2019). El libro informativo y juvenil: un recurso para la enseñanza y el aprendizaje de la geografía. En X.C. Macía, F.X. Armas y F.R. Lestegás (coords.), *La reconfiguración del medio rural en la sociedad de la información: nuevos desafíos en la educación geográfica* (pp. 581-590). Andavira.

Peña, T., & Pirela, J. (2002). La complejidad del análisis documental. *Información, Cultura y Sociedad*, 6, 55–81. http://revistascientificas.filo.uba.ar/index.php/ICS/article/view/869

Pérez Alonso, J. E. (2015). Las que cuentan. Narradoras en la literatura infantil y juvenil española. *Espéculo.Revista de Estudios Literarios*, 55, 11–28.

Pérez, A. I., Soto, E., & Serván, M. J. (2010). Participatory action research and reconstruction of teachers' practical thinking: Lesson studies and central reflection. An experience in Spain. *Educational Action Research*, 18(1), 73–87. DOI: 10.1080/09650790903484533

Pérez-delHoyo, R., Mora, H., Martí-Ciriquián, P., Pertegal-Felices, M. L., & Mollá-Sirvent, R. (2020). Introducing innovative technologies in higher education: An experience in using geographic information systems for the teaching-learning process. *Computer Applications in Engineering Education*, 28(5), 1110–1127. DOI: 10.1002/cae.22287

Pérez-Morales, A., Gil-Guirado, S., & Olcina, J. (2022). La geografía de los riesgos en España (1992-2022). Cambios y oportunidades en una temática de trabajo consolidada y en alza. In *La Geografía española actual. Estado de la cuestión* (pp. 183–202). Comité Español de la UGI.

Pfautsch, S., & Gray, T. (2017). Low factual understanding and high anxiety about climate warming impedes university students to become sustainability stewards: An Australian case study. *International Journal of Sustainability in Higher Education*, 18(7), 1157–1175. DOI: 10.1108/IJSHE-09-2016-0179

Piaget, J., & Inhelder, B. (1947). *La représentation de l'espace chez l'enfant*. Presses Universitaires de France.

Pickles, L. (2012). Geography and intercultural understanding. *Interaction*, 40(1), 9–11. https://search.informit.org/doi/10.3316/ielapa.038913886192132

Piñeiro, M. R. (1987). La lectura del mapa en el niño de preescolar. *Revista interuniversitaria de formación del profesorado, 1*, 66-76. https://digitum.um.es/digitum/handle/10201/124224

Piñeiro, M.R. (1996). Representación del concepto de pueblo y ciudad en niños asturianos de 4 a 7 años. *Iber: didáctica de las ciencias sociales, geografía e historia, 9*, 27-36.

Pinillas, S., & Torralba-Burrial, A. (2021). El cuaderno de campo como eje del aprendizaje de naturaleza cercana en Educación Infantil. *Revista Eureka sobre Enseñanza y Divulgación de las Ciencias, 18*(3), 3202.

Piulats, O. (2017). *La ciencia espiritual de la naturaleza: Seis meditaciones sobre ecología*. Ediciones Carena.

Portela, M. O. B. (2024). A cidade e o urbano no ensino de Geografia: possibilidades temáticas para o trabalho em sala de aula. In G. A. Teles, J. F. Sobrinho & S. Claudino, *Educação goegráfica e formação de professores no Brasil e Portugal: trajetórias e atualidades* (pp. 71-88). Sertão Cult.

Portera, A. (2023). Global versus intercultural citizenship education. *Prospects*, 53(3), 233–248. DOI: 10.1007/s11125-021-09577-3

Priest, S., & Gass, A. M. (1997). *Effective leadership in adventure programming*. Human Kinetics.

Prieto Rosas, V., & López-Gay, A. (2015). Push and Pull Factors of Latin American Migration. In Domingo, A., Sabater, A., & Verdugo, A. (Eds.), *Demographic Analysis of Latin American Immigrants in Spain: From Boom to Boost* (pp. 1–28). Springer., DOI: 10.1007/978-3-319-12361-5_1

Pruzzo, V. (2014). La enseñanza entre la praxis y la poiesis. Aportes a la didáctica general. *Revista Pilquen. Sección Psicopedagogía, 11*(1), 1–13. https://revele.uncoma.edu.ar/index.php/psico/article/view/2103

Puig, J. M., & Martín, X. (2015). Para un currículum de educación en valores. *Folios*, 41(41), 7–22. DOI: 10.17227/01234870.41folios7.22

Pujante, C. (2010). Notas para una recreación de la teoría sobre la novela corta española del siglo XX: A propósito de las escritoras en La Novela del Sábado (1953-1955). *Castilla: Estudios de Literatura, 1,* pp. 38-59. https://uvadoc.uva.es/handle/10324/12182

Pujolas, P. (2004). *Aprender juntos, alumnos diferentes. Los equipos de aprendizaje cooperativo en el aula*. Editorial Octaedro.

Pulgarín Silva, R., Ramírez Acosta, J. I., & Vélez Venegas, C. (2019). La espacialidad y los derechos básicos de aprendizage de Ciencias Sociales en Colombia. In L. A. Rodríguez Pizzinato, N. Palacios Mena & X. M. Souto González, *La construcción global de una enseñanza de los problemas sociales desde el Geoforo Iberoamericano* (pp. 21-34). Nau Llibres.

Pulgarín, R. (1998). La excursión escolar como estrategia didáctica en la enseñanza de la geografía. [Trabajo académico, Universidad de Antioquia]. La Gaceta Didáctica nº 2. La_salida_de_campo_como_estrategia_didactica-libre.pdf (d1wqtxts1xzle7.cloudfront.net)

Pulido-Capurro, V., & Olivera-Carhuaz, E. (2018). Aportes pedagógicos a la educación ambiental: una perspectiva teórica. *Revista de investigaciones Altoandinas, 20*(3), 333-346. DOI: 10.18271/ria.2018.397

Puttick, S., Chandrachud, P., Chopra, R., Robson, J., Singh, S., & Talks, I. (2022). Climate change education: Following the information. *Children. Education and Geography: Rethinking Intersections*, 30, 168–181. DOI: 10.4324/9781003248538-15

Quiles Cabrera, M.ª del, & Sánchez García, M.ª R. (2017). Educación medioambiental a través de la LIJ: Claves para la formación del profesorado en los grados de magisterio. *RIFOP: Revista interuniversitaria de formación del profesorado: continuación de la antigua Revista de Escuelas Normales, 31*(90), 65-77.

Quiroga, H. (1917). *Cuentos de amor, de locura y de muerte*. [(2019). Alma].

Quiroga, H. (1918). *Cuentos de la selva*. [(2011). Anaya].

Quiroga, E. (1985). *Plácida, la joven y otras narraciones*. Plaza y Janés.

Quiroga, E. (2013). *La enferma*. Cátedra.

Quiroga, E. (2022). *Tristura*. Bamba.

Quiroga, H. (2005). *Cuentos* (de Leonor, F., Ed.). Cátedra.

Quiroga, H. (2008). *Anaconda y otros cuentos* (de Consuelo, L., & Bueno, J. A., Eds.). Edaf.

Ra, S., Jagannathan, S., & Maclean, R. R. (Eds.) (2021) - Powering a Learning Society During an Age of Disruption. Springer. DOI: 10.1007/978-981-16-0983-1

Radford, A., Narasimhan, K., Salimans, T., & Sutskever, I. (2018). *Improving Language Understanding by Generative Pre-Training*. Preprint.

Ramiro, E. (2005). Ideas para preparar itinerarios de una forma amena y con buen humor. *Didáctica Geográfica*, 7, 463–472.

Ramos, A., & Torralba, A. (2020). Uso y potencial del Programa LIFE para la Educación Ambiental en educación formal, no-formal e informal, y especialmente en Educación Primaria. *Revista Eureka sobre Enseñanza y Divulgación de las Ciencias, 17*(3), 3501. https://digibuo.uniovi.es/dspace/handle/10651/54889

Rapp, A., Curti, L., & Boldi, A. (2021). The human side of human-chatbot interaction: A systematic literature review of ten years of research on text-based chatbots. *International Journal of Human-Computer Studies*, 151, 102630. DOI: 10.1016/j.ijhcs.2021.102630

Redecker, C. (2017). European framework for the digital competence of educators: DigCompEdu. *Joint Research Centre (JRC). Social Policy Report*, 95. Advance online publication. DOI: 10.2760/159770

Rees, H. (2010). *Lives in Chinese Music*. University of Illinois Press.

Requena, M. (2000). La protección jurídica del menor en el ámbito del consejo de Europa: el Convenio Europeo sobre el ejercicio de los derechos del niño. En M. Calvo y N. Fernández (coords.), *Los derechos de la infancia y la adolescencia* (pp. 187-207). Editores Mira.

Resa Ocio, A., Lucas Palacios, L. & Diez Ros, R (2023). Hacia unas competencias coeducativas en Educación Infantil para la formación docente desde la Didáctica de las Ciencias Sociales. *Clío. History and History teaching 49*, 209-233, DOI: 10.26754/ojs_clio/clio.2023499540

Ricardo, C., Peñuela Epalza, M., Lombana Coy, J., Guevara Novoa, C. C., Ramírez, M., & Cano, J. (2024). *Transformaciones de la competencia intercultural. Prácticas de una comunidad de aprendizaje docente (CAD).* Editorial Uninorte.

Rice, T. (2014). *Ethnomusicology: A Very Short Introduction.* Oxford University Press.

Richmond, K. (2004). *Las mujeres en el fascismo español: la Sección Femenina de la Falange, 1934-1959.* Alianza.

Rieckmann, M. (2018). Learning to transform the world: Key competencies in Education for Sustainable Development. In Leicht, A., Heiss, J., & Byun, W. J. (Eds.), *Issues and trends in education for sustainable development* (pp. 39–59). UNESCO publishing.

Rigal, R. (2006). *Educación motriz y educación psicomotriz en preescolar y primaria.* Inde Publicaciones.

Ríos, V. (1993). *Un muestreo en la poesía tradicional de la Mancha Baja.* Revista Zahora 33. Diputación Provincial de Albacete. Ortiz, F. (1985). *El contrapunteo cubano del tabaco y el azúcar.* Biblioteca Ayacucho.

Rivera, J. A. S. (2016). La acción didáctica de las Ciencias Sociales y el desarrollo del pensamiento crítico. *Educación y Humanismo*, 18(31), 241–256. DOI: 10.17081/eduhum.18.31.1377

Rivero, M. ª.P., y Gil. A. (2011). Pensar y expresar el espacio en el aula de Infantil. En Mª P. Rivero (Coord.), *Didáctica de las Ciencias Sociales para Educación Infantil* (pp. 31-47). Editores Mira.

Roa, P. A., & Vargas, C. (2009). El cuaderno de campo como estrategia de enseñanza en el Departamento de Biología de la Universidad Pedagógica Nacional. *Bio-grafía: Escritos sobre la Biología y su Enseñanza, 2*(3), 1-41.

Rodríguez García, A. M., Hinojo-Lucena, F. J., & Ágreda-Montoro, M. (2019). Diseño e implementación de una experiencia para trabajar la interculturalidad en Educación Infantil a través de realidad aumentada y códigos QR. *Educar, 55*(1), 59-77, https://raco.cat/index.php/Educar/article/view/348881

Rodríguez Moreno, E. A., & Souto González, X. M. (2019). Síntesis del Geoforo en los ãnos 2008 a 2018. In L. A. Rodríguez Pizzinato, N. Palacios Mena & X. M. Souto González, *La construcción global de una enseñanza de los problemas sociales desde el Geoforo Iberoamericano* (pp. 9-20). Nau Llibres.

Rodríguez Rodríguez, M. Á., & Parreño Castellano, J. M. (2023). Aprendizaje activo en el aula universitaria actual: una experiencia de aprender haciendo. *Didáctica geográfica.* nº 24-2023, 39-61. 39. DOI: 10.21138/DG.663

Rodríguez, F. (2000). *La actividad humana y el espacio geográfico*. Síntesis.

Rogers, E. M. (1986). *Communication technology*. Simon and Schuster.

Romero Claudio, C., Heredia Ponce, H., & Romero Oliva, M. F. (2023). Formación inicial de las docentes y los docentes y biografías de mujeres en libros ilustrados de no ficción, *Cadernos CEDES*, Vol. 44. *Cadernos CEDES*, 122(122), 106–120. DOI: 10.1590/cc271187

Romero Morante, J. (2021). Apuntes para una sociogénesis del currículum centrado en problemas sociales. *REIDICS.Revista de Investigación en Didáctica de las Ciencias Sociales*, 9, 14–40. DOI: 10.17398/2531-0968.09.14

Romero, J., & Luis, A. (2008). El conocimiento socio-geográfico en la escuela: Las tensiones inherentes a la transmisión institucionalizada de cultura y los dilemas de la educación para la democracia en este mundo globalizado. *Scripta Nova, Revista Electrónica de Geografía y Ciencias Sociales, 12*(270). https://www.ub.edu/geocrit/sn/sn-270/sn-270-123.htm

Romero, J., & Olcina, J. (Eds.). (2021). Cambio climático en el Mediterráneo. Procesos, riesgos y políticas. Ed. Tirant lo Blanc.

Roth, C. E. (1992). *Environmental literacy: Its roots, evolution and directions in the 1990s.* ERIC Clearinghouse for Science, Mathematics, and Environmental Education. https://files.eric.ed.gov/fulltext/ED348235.pdf

Rouhiainen, L. (2018). *Inteligencia artificial. 101 cosas que debes saber hoy sobre nuestro futuro*. Alienta Editorial.

Rubio, M. (2023). Los itinerarios didácticos virtuales como herramienta pedagógica complementaria a los itinerarios didácticos reales en geografía e historia en 1° de ESO. [Trabajo de Fin de Máster, Universidad Católica de Murcia]. UCAM. https:// *Rubio_García_Marina.pdf (ucam.edu)

Rudd, J. A. (2021). From Climate Change Ignorant to Climate Change Educator. *Chemistry (Weinheim an der Bergstrasse, Germany)*, 27(20), 6107–6111. DOI: 10.1002/chem.202004414 PMID: 33704844

Ruggerio, C. A. (2021). Sustainability and sustainable development: A review of principles and definitions. *The Science of the Total Environment*, 786, 147481. DOI: 10.1016/j.scitotenv.2021.147481 PMID: 33965820

Ruiz de Velasco, A., & Abad, J. (2011). *El juego simbólico*. Graó.

Russell, S. J., & Norvig, P. (2021). *Artificial intelligence: A modern approach* (4th ed.). Pearson.

Sachs, J. D., Kroll, C., Lafortune, G., Fuller, G., & Woelm, F. (2022). *Sustainable development report 2022*. Cambridge University Press., DOI: 10.1017/9781009210058

Sadık, F., Çakan, H., & Artut, K. (2011). Çocuk resimlerine yansıyan çevre sorunlarının sosyo-ekonomik farklılıklara göre analizi [Analysis of the Environmental Problems Pictures of Children from Different Socio-economical Level]. *Elementary Education Online*, 10(3), 1066–1080.

Sağsöz, G. (2018). Resimli çocuk kitaplarının barış eğitimine ilişkin bileşenler açısından incelenmesi. [Investigating illustrated children's books from the point of the components related to peace education]. (Doctoral dissertation), Ankara *University* Eğitim Bilimleri Enstitüsü

Sáiz-Linares, A., & Ceballos-López, N. (2022). Una mirada a las geografías de la infancia desde el análisis de los patios escolares. *Educación Básica, Cultura, Currículo, 52*, e09886, 1-20. DOI: 10.1590/198053149886

Sala, M., & Batalla, R. J. (1996). *Teoría y Métodos en Geografía Física*. Síntesis.

Salazar, R. (2002). *Historia y leyenda de la música popular venezolana*. Monte Ávila Editores.

Salgueiro, T. B. (2023). Pensar o território através do território. In T. B. Salgueiro, *Temas e reflexões. Breve Antologia* (pp. 57-67). Associação Portuguesa de Geógrafos.

Sammut, G., Andreouli, E., Gaskell, G., & Valsiner, J. (2015). Social representations: a revolutionary paradigm? In Sammut, G., Andreouli, E., Gaskell, G., & Valsiner, J. (Eds.), *Cambridge Handbook of Social Representations* (pp. 3–11). Cambridge University Press. DOI: 10.1017/CBO9781107323650.003

San Román Gago, S. (2015). Madres sociales de la aldea: La maestra española durante el nacional-catolicismo. *Educação e Pesquisa*, 41(2), 359–373. DOI: 10.1590/s1517-97022015041806

Sanahuja, J. A., & Tezanos, S. (2017). Del milenio a la sostenibilidad: Retos y perspectivas de la Agenda 2030 para el desarrollo sostenible. *Política y Sociedad (Madrid)*, 54(2), 521–543. DOI: 10.5209/POSO.51926

Sánchez Fernández, M. P. (1982). La educación preescolar en España en el momento actual (I). *Vida escolar, 220-221*, 78-89. http://hdl.handle.net/11162/79094

Sánchez Ogallar, A. (2010). El trabajo de campo y las excursiones. *Parte II: Estrategias y Recursos Didácticos*, 159-184. A S Ogallar Itinerarios.pdf

Sánchez, E. (2010) Innovación e interdisciplinariedad en los itinerarios didácticos de ciencias sociales. *Revista Electrónica de Didácticas Específicas,* (3). https://www.google.es/Innovación e interdisciplinariedad en los itinerarios didácticos de ciencias sociales

Sánchez, L. (2005). Enseñar Geografía: De la significatividad psicológica local a la significatividad lógica global. En E. Sande Lemos (Coord.), *Ensinar Geografia na sociedade do conhecimento* (pp. 429-451). APG-AGE.

Sánchez-Almodóvar, E., Gómez-Trigueros, I. M., & Olcina, J. (2023). Climate Change and Extreme Weather Events in the Education of the Citizens of the Twenty-First Century: The Perception of Secondary Education Students. *Social Sciences (Basel, Switzerland)*, 12(1), 27. DOI: 10.3390/socsci12010027

Sánchez, L., & Jerez, Ó. (2004). El lugar en las ciencias sociales. In *Métodos para enseñar ciencias sociales: interacción, cooperación y participación* (pp. 69–80). Íber, Didáctica de las Ciencias Sociales, Geografía e Historia, Graó.

Sánchez, M. C., Campillo, J. M., & Vivas, V. (2021). *La formación del profesorado en didáctica de las ciencias sociales en el ámbito Iberoamericano*. EDITUM- Ediciones de la Universidad de Murcia., DOI: 10.6018/editum.2919

Sandahl, J. (2023). Engaging with scale and place: Geographical thinking on migration in middle-school. [ZGD]. *Zeitschrift für Geographiedidaktik*, 51(2), 55–69.

Sangil, O. (2006). *Literatura y leyendas de la España medieval*. Editorial Castalia.

Sanmartí, N. (2018). *La finalidad principal de la evaluación es la regulación tanto de la enseñanza como del aprendizaje. 10 ideas clave*. Evaluar para Aprender. Graó.

Santana, D., Morales, A. J., Colomer, J. C., Campo Pais, B., & Caurín Alonso, C. (2016). Parques naturales: La necesaria conceptualización transformadora en la educación primaria y secundaria. *Didáctica Geográfica, 16*, 73–94. https://didacticageografica.age-geografia.es/index.php/didacticageografica/article/view/296

Santos, M. (1995). *De la totalidad al lugar*. Editorial Oikos-Tau.

Santos, M. (2000). La naturaleza del espacio. *Ariel*.

Santos, M. (2000). Por uma outra globalização: Do pensamento único à consciência universal. *Record*.

Sanz Simón, C., & Sonlleva Velasco, M. (2020). La maestra franquista en el imaginario del alumnado femenino. Un estudio a través de testimonios orales. *Revista Diversidade e Educaçao, 8*, 213-237. https://repositorio.furg.br/handle/123456789/10820

Sanz, C. (2002). El paisaje como recurso. In *Estudios sobre el paisaje* (pp. 75–84). Fundación Duques de Soria. UAM.

Saregar, A., Sunyono, H. E. Y., Hasan, H., Putra, F. G., & Diani, R. (2022). Natural disaster Education in School: A bibliometric analysis with a detailed future insight overview. *International Journal of Educational Methodology*, 8(4), 743–757. DOI: 10.12973/ijem.8.4.743

Sauer, C. O. (1974). The fourth dimension of geography. *Annals of the Association of American Geographers*, 64(2), 189–192. DOI: 10.1111/j.1467-8306.1974.tb00969.x

Savage, P. E., Brown, S., Sakai, E., & Currie, T. E. (2015). Statistical universals reveal the structures and functions of human music. *Proceedings of the National Academy of Sciences of the United States of America*, 112(29), 8987–8992. DOI: 10.1073/pnas.1414495112 PMID: 26124105

Sbardelotto, V. S., & Francischett, M. N. (2011). "Nós Propomos!" O conhecimento geográfico e a formação pelo ensino. In M. N. Francischett, A. C. Biz & G. Toffolo (orgs), "Nós Propomos!" Ressignficar o ensino na educação geográfica (pp. 23-37). Edunioeste.

Schultz, P. W. (2002). In Dietz, T., & Stern, P. C. (Eds.), *Knowledge, information and household recycling: Examining the knowledge-deficit model of behavior change. New Tools for Environmental Protection- Education, Information, and Voluntary Measures* (pp. 67–82). National Academy Press.

Sebastià, R., & Tonda, E. M. (2015). El concepto y representación del espacio geográfico en la enseñanza de la Geografía en los niveles educativos no universitarios. Análisis bibliométrico. In de la Riva, J., Ibarra, P., Montorio, R., & Rodrigues, M. (Eds.), *Análisis espacial y representación geográfica: innovación y aplicación* (pp. 1505–1514). Universidad de Zaragoza-Asociación Española de Geografía.

Sebba, R. (1991). The landscapes of childhood: The reflection of childhood surroundings in adult memory and children's attitudes. *Environment and Behavior*, 23(4), 395–422. DOI: 10.1177/0013916591234001

Seeger, A. (2004). *Why Suyá Sing: A Musical Anthropology of an Amazonian People*. University of Illinois Press.

Seguido, M. (2024). Competencia Digital Docente: Desafíos para la enseñanza de las Ciencias Sociales a partir de un proyecto de innovación educativa. *Interdisciplinary Journal of Didactics*, (1), 19–32. DOI: 10.14198/ijd.27976

Serrano de la Cruz, M. A., & Sánchez, L. (Coords.). (2007). *Educación geográfica a través de los paisajes de la provincia de Ciudad Real*. Ediciones de la Universidad de Castilla-La Mancha y Asociación de Geógrafos Españoles.

Serrano de la Cruz, M. A., García, J. L., & Jerez, Ó. (2016). Propuesta preliminar para la identificación de lugares de interés didáctico en espacios naturales protegidos. *Didáctica Geográfica*, 17, 159–176.

Serrano Pastor, F. J., Martínez España, A., & Montesinos Navarro, A. M. (2015). Construcción de la identidad docente de los futuros maestros de Educación Infantil y Educación Primaria: entre la vocación y la profesionalización. In Mirete Ruiz, A. B., & Nortes Martínez-Artero, R. (Eds.), *Investigación e innovación: una constante necesaria en la formación del profesorado* (pp. 125–136). Servicio de Publicaciones de la Universidad de Murcia.

Serrano, R. (2017). *Reconstrucción climática instrumental de la precipitación diaria en España: ensayo metodológico y aplicaciones*. Tesis doctoral. Universidad de Zaragoza. http://geografia.unizar.es/noticias/tesisrobertoserrano

Seyman, N., & Gülaçtı, İ. E. (2022). Mobile Application Designed to Raise Environmental Awareness in Children and the Results of the Application. [SSSJournal]. *Social Sciences Studies Journal*, 8(93), 32–40. DOI: 10.26449/sssj.3762

Sezen-Barrie, A., & Marbach-Ad, G. (2021). Cultural-Historical Analysis of Feedback from Experts to Novice Science Teachers on Climate Change Lessons. *International Journal of Science Education*, 43(4), 497–528. DOI: 10.1080/09500693.2020.1870759

Shanahan, D., & Albrecht, J. (2019). Examining the effect of oral transmission on folksongs. *Music Perception*, 36(3), 273–288. DOI: 10.1525/mp.2019.36.3.273

Shawar, B. A., & Atwell, E. (2007). Chatbots: Are they Really Useful? *Journal for Language Technology and Computational Linguistics*, 22(1), 29–49. DOI: 10.21248/jlcl.22.2007.88

Sheehy, B., & Farneti, F. (2021). Corporate social responsibility, sustainability, sustainable development and corporate sustainability: What is the difference, and does it matter? *Sustainability (Basel)*, 13(11), 5965. DOI: 10.3390/su13115965

Shin, K. H. (2008). Development of environmental in the Korean kindergarten context. Unpublished Doctoral Thesis, University of Victoria, Canada.

Sikorska, L. (2024). Different meanings of "diversity". *Linguistic, Literary, and Cultural Diversity in a Global Perspective*, 21, 8–26. DOI: 10.1075/fillm.21.01sik

Şimşekli, Y. (2004). Çevre bilincinin geliştirilmesine yönelik çevre eğitimi etkinliklerine ilköğretim okullarının duyarlılığı [Sensitivity of Elemantary Schools to the Environmental Education Activities For Increasing Environmental Knowledge]. *Journal of Faculty of Education Uludag University*, 17(1), 83–92.

Smith, A. (2001). Early childhood a wonderful time for science learning. *Australian Primary and Junior Science Journal*, 17(2), 18–20.

Sobejano, G. (1975). *Novela española de nuestro tiempo*. Prensa Española.

Sobejano, G. (2006). *Novela española de nuestro tiempo. (En busca del tiempo perdido)*. Marenostrum Comunicación.

Sobel, D. (2004). Place-based education: Connecting classroom and community. Nature and listening, 4(1), 1-7.

Soldevila, I. (1980). *La novela desde 1936*. Alhambra.

Soloman, J. M. (2023). Embracing Diversity to Increase Belongingness and Foster a Culture of Success. *The Organizational Improvement Plan at Western University*, 321. https://ir.lib.uwo.ca/oip/321

Sonlleva Velasco, M., Martínez Scott, S., & Barba Martín, R. (2019). Las marcas del género en las trayectorias biográficas de las maestras de educación infantil. Analizando nuestra historia para desmontar el patriarcado. *Cabás, 22*, 47-60. https://doi.org/DOI: 10.35072/CABAS.2019.17.63.006

Sotelo, M., & Ripoll, B. (2023). El Premio Nadal y las narradoras de posguerra. *Boletin de la Biblioteca de Menendez Pelayo*, 99(3). Advance online publication. DOI: 10.55422/bbmp.958

Sousa, S. A., García Monteagudo, D., & Souto, X. M. (2016). Educación geográfica y las salidas de campo como estrategia didáctica: un estudio comparativo desde el Geoforo Iberoamericano. *Biblio3W. Revista Bibliográfica de Geografía y Ciencias Sociales, 21*(1155). https://www.ub.edu/geocrit/b3w-1155.pdf

Sousa, S. A., García, D., & Souto, X. M. (2016). Educación geográfica y las salidas de campo como estrategia didáctica: un estudio comparativo desde el Geoforo Iberoamericano. *Revista Bibliográfica de Geografía y Ciencias Sociales, vol. XXI*(1.155), 1-19. Educación Geográfica y las salidas de campo como estrategia didáctica (uv.es)

Souto Alabarce, A. (2003). Introducción. In Quiroga, H. (Ed.), *Más cuentos, Cuentos de la selva*. Porrúa. VII-XV.

Souto, X. M. (2017). Propuestas para innovar e investigar sobre Conocimiento del medio en Educación Infantil y Primaria. *Biblio 3w: revista bibliográfica de geografía y ciencias sociales*, 22. https://raco.cat/index.php/Biblio3w/article/view/329017

Souto, X. M., & Claudino, S. (2019). Construímos uma Educação Geográfica para a Cidadania Participativa. O caso do Projeto Nós Propomos! *Revista Signos Geográficos – Boletim NEPEG de Ensino de Geografia*, 1, 16 p.

Souto, X. M., & García-Monteagudo, D. (2023). Formación docente y cultura profesional. In I. Bellatti, C. Fuentes Moreno, P. Miralles Martínez & L. Sánchez (coord.), *Las ciencias sociales y su didáctica pensamiento histórico y educación democrática: Homenaje a Joaquín Prats Cuevas*. Horizontes Universidad, Octaedro.

Souto, X. M., García-Monteagudo, D., & Campo País, B. (2024). Formando ciudadanos críticos en el Geoforo Iberoamericano. *Geo UERJ*, 46, 26 p. DOI: 10.12957/geouerj.2024.86466

Souto, X.M. (2011). Identidades espaciales y territorios políticos. *Iber, didáctica de las ciencias sociales, geografía e historia,* 69, 55-63.

Souto, X. M. (1999). *Didáctica de la geografía. Problemas sociales y conocimiento del medio*. Serbal.

Souto, X. M. (2013). Investigación e innovación educativa: El caso de la Geografía escolar. *Scripta Nova. Revista Electrónica de Geografía y Ciencias Sociales*, 17, 1–36.

Souto, X. M. (2018). La geografía escolar: Deseos institucionales y vivencias de aula. *Boletín de la Asociación de Geógrafos.*, 2757(79), 1–31. DOI: 10.21138/bage.2757

Stenhouse, L. (1987). *Investigación y desarrollo del currículum*. Morata.

Stobart, H. (2008). *The New (Ethno)musicologies*. The Scarecrow Press.

Suárez Hernán, C. & Bermúdez Martínez, M. (2024). La presencia de la literatura hispanoamericana en Educación Secundaria Obligatoria y Bachillerato en España. Concepciones del profesorado. *América sin Nombre, 31,* 147-164. DOI: 10.14198/AMESN.24540

Suárez-Eiroa, B., Fernández, E., Méndez-Martínez, G., & Soto-Oñate, D. (2019). Operational principles of circular economy for sustainable development: Linking theory and practice. *Journal of Cleaner Production*, 214, 952–961. DOI: 10.1016/j.jclepro.2018.12.271

Suazo, L. E., & Torres, A. (2022). Perceptions, Knowledge, and Teaching of Climate Change and Natural Disaster Risk at Honduran Universities. *Journal of Higher Education Theory and Practice*, 22(3), 118–130. DOI: 10.33423/jhetp.v22i3.5090

Suzel Ana, R. (2006). Intangible Cultural Heritage: Global Awareness and Local Interest. *In Yearbook for Traditional Music, 38,* 36-44.

Tadaki, M., Salmond, J., Le Heron, R., & Brierley, G. (2012). Nature, culture, and the work of physical geography. *Transactions of the Institute of British Geographers*, 37(4), 547–562. DOI: 10.1111/j.1475-5661.2011.00495.x

Talay, İ., Aslan, F., & Belkayalı, N. (2010). Okul Öncesi Eğitim Kurumlarında Doğa Dostu ve Çocuk Katılımı Temelli Dış Mekan Tasarım Yaklaşımları Bir Proje Önerisi. [Nature Friendly And Child Participation Based Outdoor Design Approaches in Preschool Institutions: A Project Proposal]. *Kastamonu Education Journal*, 18(1), 317–322.

Taşkın, Ö. ve Şahin B. (2008). Çevre kavramı ve altı yaş okul öncesi çocuklar. *Pamukkale University Journal of Education*, 1(23), 1–14.

Tebár Cuesta, F. (2018). Autonomía de los centros educativos (intervención de la Inspección de Educación). *Avances en supervisión educativa*, (29). DOI: 10.23824/ase.v0i29.617

Tedesco, J. C. (2008). *O novo pacto educativo. Educação, competitividade e cidadania na sociedade moderna* (3rd ed.). Fundação Manuel Leão.

Tejero, E. (2003). El siete, número cósmico y sagrado. Su simbología en la cultura y rendimiento en el Romancero. *Didáctica. Lengua y Literatura, 15*, 221-253. https://revistas.ucm.es/index.php/DIDA/article/view/DIDA0303110221A

TEMA (2011-2015) Türkiye Erozyonla Mücadele Vakfı Genel Müdürlüğü Arşivi,

Thomas, E. (1966). Three mexican indian carnaval songs. *Ethnomusicology*, 1(10), 58–69.

Thomas, J. W. (2000). *A review of research on project-based learning*. The Autodesk Foundation.

Tirado Olivares, S., & Cózar-Gutiérrez, R. (2024). Enseñar y evaluar ciencias sociales en la era digital: el uso del learning analytics para la evaluación formativa. Una revisión del estado de la cuestión. *Didáctica De Las Ciencias Experimentales Y Sociales*, (46), 21–36. https://doi.org/DOI: 10.7203/dces.46.28468

Tirado-Olivares, S., Navío-Inglés, M., O'Connor-Jiménez, P., & Cózar-Gutiérrez, R. (2023). From human to machine: Investigating the effectiveness of the conversational AI ChatGPT in historical thinking. *Education Sciences*, 13(8), 803. DOI: 10.3390/educsci13080803

Tomar, A. (2009). *Toprak ve su kirliliği ve su havzalarının korunması* [Soil and water pollution and protection of watersheds.]. TMMOB İzmir Kent Sempozyumu.

Tonda, E. M. (2001). *La didáctica de las ciencias sociales en la formación del profesorado de Educación Infantil*. Servicio de Publicaciones de la Universidad de Alicante.

Tonda, E. M., & Sebastiá, R. (2003). Las dificultades en el aprendizaje de los conceptos de tiempo atmosférico y clima: La elaboración e interpretación de climogramas. *Revista de Educación de la Universidad de Granada*, 16, 47–69.

Torres Nebrera, G. (2013). Introducción. En Elena Quiroga, *La enferma* (p. 9-91). Cátedra.

Torres, B. (2001). *Estudio de los modos narrativos en el discurso autobiográfico de* Tristura, *de Elena Quiroga*. Estudios y ensayos 50. Universidad de Málaga.

Torres, M. (1999). Mi hada madrina. *Cuadernos de estudio y cultura, 11*, 21-24. https://acec-web.org/PDF/CUAD/11.pdf

Trepat, C., & Comes, P. (2000). *El tiempo y el espacio en la didáctica de las ciencias sociales*. Graó.

Tsalis, T. A., Malamateniou, K. E., Koulouriotis, D., & Nikolaou, I. E. (2020). New challenges for corporate sustainability reporting: United Nations' 2030 Agenda for sustainable development and the sustainable development goals. *Corporate Social Responsibility and Environmental Management*, 27(4), 1617–1629. DOI: 10.1002/csr.1910

Tuan, Y. F. (2007). *Topofilia*. Editorial Melusina.

Tuğrul, B. (2007). Etkili okul öncesi eğitim programlarının özellikleri [Elektronik versiyon] [Characteristics of effective preschool education programs]. Symposium on preschool education (s.50-93). Antalya.

Türkoğlu, B. (2019). Opinions of preschool teachers and pre-service teachers on environmental education and environmental awareness for sustainable development in the preschool period. *Sustainability (Basel)*, 11(18), 4925. Advance online publication. DOI: 10.3390/su11184925

Türksoy, Ö. (1991). Çevre duyarlılığı eğitiminde bilişsel yaklaşım: Çocuk ve temel çevre özellikleri. [Cognitive approach in environmental sensitivity education: The child and basic environmental characteristics.]. *Education in Science*, 15(80).

Türküm, A. S. (1998). *Çağdaş toplumda çevre sorunları ve çevre bilinci* [Environmental problems and environmental awareness in contemporary society]. Çağdaş Yaşam Çağdaş İnsan. Anadolu Üniversitesi Açık Öğretim Fakültesi İlköğretim Öğretmenliği Lisans Tamamlama Programı.

Turner, S., & Leydon, J. (2012). Improving Geographic Literacy among First-Year Undergraduate Students: Testing the Effectiveness of Online Quizzes. *The Journal of Geography*, 111(2), 54–66. DOI: 10.1080/00221341.2011.583263

Tzafilkou, K., Perifanou, M., & Economides, A. A. (2023). Assessing teachers' digital competence in primary and secondary education: Applying a new instrument to integrate pedagogical and professional elements for digital education. *Education and Information Technologies*, 28(12), 16017–16040. DOI: 10.1007/s10639-023-11848-9 PMID: 37361819

Ünal, S., Mançuhan, E. ve Sayar, A. (2001). Çevre bilinci, bilgisi ve eğitimi. [Environmental awareness, knowledge and education]. Marmara Üniversitesi Teknik Eğitim Fakültesi Matbaası

Unattributed (1992). Entrevista a Lucía Baquedano. *Peonza, 22*, 23-27. https://www.cervantesvirtual.com/obra-visor/peonza-revista-de-literatura-infantil-y-juvenil--60/html/0272fb60-82b2-11df-acc7-002185ce6064_21.htm

Unattributed (2010). Lucía Baquedano: "Después de los padres, es el maestro quien debe encauzar bien a un niño". *Nuestro Tiempo, 660.* https://nuestrotiempo.unav.edu/es/grandes-temas/despiece/lucia-baquedano

UNEP. (1972). United Nations Conference on the Human Environment, United Nations: https://www.unep.org/Documents.Multilingual/Default.asp?DocumentID=97

UNESCO & UNFCCC. (2016). *Action for CLIMATE EMPOWERMENT. Guidelines for accelerating solutions through education, training and public awareness* (Issue October). UNESCO. https://unfccc.int/files/cooperation_and_support/education_and_outreach/application/pdf/action_for_climate_empowerment_guidelines.pdf

UNESCO. (1977). Intergovernmental Conference on Environmental Education organized by Unesco in co-operation with UNEP, Final Report, Tbilisi (USSR).

UNICEF. (1989). *Convención sobre los Derechos del Niño.* https://www.unicef.org/

United Nations (UN). (2015a). *Objetivos de Desarrollo Sostenible.* https://www.un.org/sustainabledevelopment/es/objetivos-de-desarrollo-sostenible/

United Nations (UN). (2015b). *Oficina de Naciones Unidas para la Reducción del Riesgo de Desastres. 2015. Marco de Sendai (2015-2030).*https://www.undrr.org/es/implementando-el-marco-de-sendai/que-es-el-marco-de-sendai-para-la-reduccion-del-riesgo-de

United Nations Educational, Scientific and Cultural Organization [UNESCO] (2012). ESD—*Building a better, fairer world for the 21st century.* https://unesdoc.unesco.org/ark:/48223/pf0000216673

United Nations Educational, Scientific and Cultural Organization [UNESCO] (2020). *Education for sustainable development. A roadmap.* UNESCO.

United Nations General Assembly. (2015). *Transforming Our World: the 2030 Agenda for Sustainable Development.* https://sdgs.un.org/2030agenda

United Nations. (2024). *Programa de trabajo de Doha (2013-2020).*https://www.bing.com/search?q=organización+de+las+naciones+unidas&form=ANNTH1&refig=4e33c3212b664ce6ad8bfec203d421b4&sp=1&qs=AS&pq=organización+de+las+&sk=PRES1&sc=8-20&cvid=4e33c3212b664ce6ad8bfec203d421b4

Ünlü, M., Üçışık, S. ve Özey, R. (2002) . Coğrafya eğitim ve öğretiminde haritaların yeri ve önemi. [The Importance of othe Maps in Geography Education and Instruction].lnternational Journal of Geography and Geography Education, 5, 9-25.

Unwin, T. (1995). *El lugar de la Geografía.* Cátedra.

Uribarri, R. (1969). *Elena Quiroga habla para el*. Diario de Navarra. Diario de Navarra.

Uz, C., Anteneh, M. R., & Thompon, M. (2021). What's So Special About Spatial?: A Review Study Joining Virtual Reality and Spatial Ability. In Russell, D. (Ed.), *Implementing Augmented Reality Into Immersive Virtual Learning Environments* (pp. 56–73). IGI GLOBAL., DOI: 10.4018/978-1-7998-4222-4.ch003

Vakfı, T. E. M. A. (2016), https://www.tema.org.tr/Sayfalar/Calismalarimiz/ com

Valbuena, M., & Valverde, J. Á. (2006). La climatología local: Procedimiento para su enseñanza y aprendizaje. *Didáctica Geográfica*, 8, 93–108.

Van Dijk, T. (2010). Racismo y la Prensa: Análisis Crítico del Discurso. Ponencia en el Auditórium del Edificio Centro de Innovación Docente de la Facultad de Filosofía y Humanidades de la Universidad Austral de Chile el 3 de enero de 2010.

van Laar, E., van Deurse, A. J. A. M., van Dijk, A. A. G. M., & de Haan, J. (2017). The relation between 21st-century skills and digital skills: A systematic literature review. *Computers in Human Behavior*, 72, 577–588. DOI: 10.1016/j.chb.2017.03.010

Vañó, C. (2021). *Las emociones del espacio vivido. La identidad desde el ámbito local*. Trabajo Final de Grado inédito. Grado en maestro/a en educación infantil. Universitat de València.

Vara Muñoz, J. L. (2008). Cinco décadas de Geografía de la percepción. *Eria*, 77, 371–384. DOI: 10.17811/er.0.2008.371-384

Vargas Silva, F. A. (2019). *Tendencias, corrientes y modelos pedagógicos: relaciones, semejanzas y diferencias (una mirada desde occidente)*. Sello Editorial Universidad del Tolima.

Vázquez Fernández, M.ª I. (2007). *Miguel Delibes, el camino de sus héroes*. Pliegos.

Veiga, F. H., García, O. F., & García, F. (2019). Envolvimento dos alunos na escola: conceptualização, avaliação e promoção. In *F. H. Veiga, Psicologia da Educação. Temas de aprofundamento científico para a Educação XXI* (pp. 37–58). Climepsi Editores.

Velásquez, N. (2017). *Efecto del programa "Cuidando el medio ambiente desde mi aula", para mejorar la responsabilidad ambiental en los estudiantes del primer grado de educación primaria de la IE N° 88320, San Bartolo, Santa, Áncash, Perú, 2014*. [Tesis doctoral, Universidad Nacional del Santa Chimbote - Perú]. Repositorio Institucional Digital UNS. https://repositorio.uns.edu.pe/handle/20.500.14278/3096

Vergara Ramírez, J.J. (2016). Aprendo porque quiero. El Aprendizaje Basado en Proyectos (ABP), paso a paso. SM.

Vicuña, J. (1911). *El romancero de los sefardíes*. Imprenta Universitaria.

Vidal, J.M. (2022). *Plácida, la joven*, de Elena Quiroga. Entre la denuncia y la autoficción. *Hécula: revista de la Fundación Castillo-Puche, 9*, pp. 95-106.

Vidal, J. M. (2006). *Trayecto uno,* novela corta de Elena Quiroga. *Montearabí*, 36, 63–76.

Vilarrasa, A. (2003). Salir del aula. Reapropiarse del contexto. *Íber*, (36), 13-25.

Vilches, A., Gil, D., & Cañal, P. (2010). Educación para la sostenibilidad y educación ambiental. *Investigación en la Escuela*, (71), 5–15. DOI: 10.12795/IE.2010.i71.01

Villanueva, D. (2019). *La trayectoria novelística de Elena Quiroga*. Volvoreta. Revista de Literatura.

Villena-Taranilla, R., Tirado-Olivares, S., Cózar-Gutiérrez, R., & González-Calero, J. A. (2022). Effects of virtual reality on learning outcomes in K-6 education: A meta-analysis. *Educational Research Review*, 35, 100434. DOI: 10.1016/j.edurev.2022.100434

Voice of the Turtle (1990). Estaba Señor Don Gato [canción]. *Bridges of song: music of the Spanish Jews of Morocco.* Titanic Records.

Vuorikari, R., Punie, Y., Carretero Gómez, S., & Van Den Brande, G. (2016). *DigComp 2.0: The Digital Competence Framework for Citizens. Update Phase 1: The Conceptual Reference Model* (EUR 27948 EN). Publications Office of the European Union. DOI: 10.2791/11517

VV. AA. (1998). La educación ambiental en los espacios naturales protegidos. *Aula Verde: Revista de Educación Ambiental*, 16, 3–12.

Vygotsky, L. S. (1978). *Mind in society: The development of higher psychological processes*. Harvard University Press.

Wass, S. (1992). *Salidas escolares y trabajo de campo en la educación primaria*. Ediciones Morata.

Watson, C. A. (2018). Information Literacy in a Fake/False News World: An Overview of the Characteristics of Fake News and its Historical Development. *International Journal of Legal Information*, 46(2), 93–96. DOI: 10.1017/jli.2018.25

Weich-Shahak, S. (1994). *Romancero Sefardí de Marruecos*. Instituto de Filología.

Weich-Shahak, S. (1997). *Romancero sefardí de Marruecos*. Alpuerto.

Wells, N. M., & Lekies, K. S. (2006). Nature and the life course: Pathways from childhood nature experiences to adult environmentalism. *Children, Youth and Environments*, 16(1), 1–24.

Wells, N., & Evans, G. (2003). Nearby nature. A buffer of life stress among rural children. *Environment and Behavior*, 35(3), 311–330. DOI: 10.1177/0013916503035003001

Wilde, G. (2007). De la depredación a la conservación: Génesis y evolución del discurso hegemónico sobre la selva misionera y sus habitantes. *Ambiente & Sociedade*, 10(1), 87–106. DOI: 10.1590/S1414-753X2007000100006

Wilson, E. O. (2021). *Biofilia: El amor a la naturaleza o aquello que nos hace humanos*. Errata Naturae Editores.

Wilson, J., & Czik, A. (2016). Automated essay evaluation software in English Language Arts classrooms: Effects on teacher feedback, student motivation, and writing quality. *Computers & Education*, 100, 94–109. DOI: 10.1016/j.compedu.2016.05.004

Wilson, R. A. (1996). *Starting Early: Environmental Education During The Early Childhood Years*. ERIC Clearinghouse for Science Mathematics and Environmental Education.

Winther, A. A., Volk, T. L., & Shrock, S. A. (2002). Teacher decision making in the 1st year of implementing an issues-based environmental education program: A qualitative study. *The Journal of Environmental Education*, 33(3), 27–33. DOI: 10.1080/00958960209600812

World Commission on Environment and Development [WCED] (1987). *Our common future. The Brundtland Repor*t. Oxford University Press.

Yalçınkaya, E. (2013). İlköğretim 8. Sınıf öğrencilerine göre çevre sorunları: Nitel bir çalışma. [Environmental Issues to 8th Grades Elementary Education Students: A Qualitative Study]. *lnternational Journal of Geography and Geography Education*, 27, 416-439

Yaman, K., & Gül, M. (2018). Kuruluşundan Günümüze Avrupa Birliği'nin Çevre Politikası [Environmental Policy of the European Union since its foundation]. *Journal of Economy Business and Management*, 2(2), 198–217.

Yayla Ceylan, Ş. ve Ülker, P. (2014). Okul öncesi dönem çocuklarında çevre eğitimi neden önemlidir? [Why is environmental education important for preschool children?]. (Ed.: H. Gülay Ogelman). *Çocuk ve çevre. Küçük Çocuklar ve Çevre Eğitimi El kitabı*. Eğiten Kitap, s. 37-58. http//tez2.yok.gov.tr

Yoleri, S. (2012). Çocuk ve Çevre: Okul Öncesi Çocuklar Arasında Çevre Bilinci Oluşturma. [Children and the Environment: Creating Environmental Awareness Among Preschool Children]. *Buca Faculty of Education Journal*, 34, 100–111.

Yraola, A. (2014). Actitudes lingüísticas en España. In Chiquito, A. B., & Quesada Pacheco, M. Á. (Eds.), *Actitudes lingüísticas de los hispanohablantes hacia el idioma español y sus variantes. Bergen Language and Linguistic Studies (BeLLS), 5, 551-636*. Disponible., DOI: 10.15845/bells.v5i0.685

Zalli, E. (2024). Globalization and education: Exploring the exchange of ideas, values, and traditions in promoting cultural understanding and global citizenship. *Interdisciplinary Journal of Research and Development*, 11(1 S1), 55–55. DOI: 10.56345/ijrdv11n1s109

Zaragoza, A. (2022). *La importancia del factor educación como medida preventiva ante el riesgo de inundación. Análisis y propuestas para la Educación Secundaria en Orihuela*. Trabajo de Fin de Máster. Máster en Planificación de Riesgos Naturales, Universidad de Alicante.

Zatlin, P. (1992). Introducción. En Elena Quiroga, *Escribo tu nombre*, (pp. 9-21). Austral.

Zatlin, P. (1977). *Elena Quiroga*. Twayne Publishers.

Zayimoğlu Öztürk, F., & Öztürk, T. (2015). "Öğretmen Adaylarının Çevre ve Çevre Eğitimi ile İlgili Görüşleri (Ordu Üniversitesi Örneği) " [Opinions of Pre-Service Teachers About Environment and Environmental Education Ordu University Sample]. Balıkesir University Journal of Social Sciences Institute, 18(33). 115-132

Zeichner, K. (2010). Nuevas epistemologías en formación del profesorado, Repensando las conexiones entre las asignaturas del campus y las experiencias de prácticas en la formación del profesorado en la universidad, *Revista Interuniversitaria de Formación del Profesorado*, Vol. 24, núm. 2, 123-149, https://www.redalyc.org/articulo.oa?id=27419198007

Zhang, K., & Aslan, A. B. (2021). AI technologies for education: Recent research & future directions. *Computers and Education: Artificial Intelligence*, 2, 100025. DOI: 10.1016/j.caeai.2021.100025

Zhang, P., & Tur, G. (2024). A systematic review of ChatGPT use in K-12 education. *European Journal of Education*, 59(2), e12599. DOI: 10.1111/ejed.12599

Zheng, B., Warschauer, M., Lin, C.-H., & Chang, C. (2016). Learning in One-to-One Laptop Environments: A Meta-Analysis and Research Synthesis. *Review of Educational Research*, 86(4), 1052–1084. DOI: 10.3102/0034654316628645

About the Contributors

Cristina Honrubia-Montesinos holds a Ph.D. with International Mention in Education from the University of Castilla La Mancha. She is an accredited contracted doctor. Additionally, she has various scientific publications in high-impact journals, the latest of which was published in 2023 in the journal Early Education and Development. Furthermore, she has participated in national and international conferences. She has undertaken stays at the University of Edinburgh and the University of Minho, as well as two Erasmus+ teaching assignments at the Middle East Technical University (Turkey). In terms of professional experience, she has worked as a predoctoral researcher at the University of Castilla La Mancha and has taught at the International University of La Rioja and the Isabel I University. Her contact email is cristina.honrubia@uclm.es.

Antonio Morcillo-Martínez holds a Graduate's Degree in Primary Education and a Master's Degree in Educational Research and Innovation (UCLM). He is currently pursuing a PhD in Research in Humanities, Arts and Education with a thesis attached to the departments of Pedagogy and Didactics of Physical Education, Art and Music at the Faculty of Education of Albacete (University of Castilla-La Mancha), being the beneficiary of a predoctoral contract for university teacher training (reference: FPU21/06754) funded by the Ministry of Universities of Spain. He has various scientific publications in high-impact journals and has also participated in national and international conferences.

***

Lorena Alleres Otero is a Graduate in Early Childhood Education, holding a Master's degree in Attention to Diversity, and currently pursuing a Ph.D. Her expertise focuses on inclusive education and fostering diversity in early learning environments

Sérgio Claudino is a Researcher of the Center for Geographical Studies and Professor of the Institute of Geography and Spatial Planning of the University of Lisbon, whose Center for Teacher Training is coordinator. He also coordinates the Master's Degree in Geography Teaching at the University of Lisbon. Member of the Board of Directors of GEOFORO Iberoamerican Forum of Education, Geography and Society. Founder and coordinator of the international Project We Proposed! Citizenship and Innovation in Geographic Education.

Nuria Cuartero Ruiz is a Graduate in Primary Education Teacher.

Carmen Escribano-Muñoz is a Professor of Social Sciences Teaching. Her lines of research focus on the teaching of Social Sciences and the area of professional practices. Academic coordinator of the Degrees of Teacher in Early Childhood Education and Primary Education at the International University of La Rioja. Author of different university teaching materials, she has developed professional application projects for the Pompeu Fabra University in the field of Training in University Teaching and has collaborated with continuing education projects for the Center for Learning Innovation and Knowledge (CLIK) of this university. She has also conducted online training courses in General Didactics for teachers at the National University of Villarrica in Paraguay and in Social Sciences for rural teachers in El Salvador. She has also conducted training and lectures on ICT and its educational application at the Villanueva University Center attached to the Complutense University of Madrid and has conducted training in globalizing methodologies at educational centers in Catalonia. She is currently part of the Internationalization Project of the History Didactics Chair in collaboration with the National University of La Pampa (Argentina), the University of Tolima (Colombia) and the Academic University of Christian Humanism (Chile).

Diego García Monteagudo is Permanent Labor Professor. Graduate in Geography. Doctor in Specific Didactics (specialty in Social Science didactics).

María Isabel Gil-García is a Graduate in Spanish Language and Literature. She is currently studying the Master's Degree in Teacher of Compulsory Secondary Education and Baccalaureate, Vocational Training and Language Teaching.

Isabel María Gómez Barreto has a PhD of Education. UNED, Madrid. Spain. Program: Teacher Training in the Curriculum Areas and Evaluation of Educational Institutions. Master in Management in Education. Degree in Psychopedagogy. Specialist teacher in Special Education. Research professor in the department of Pedagogy Knowledge area: Teaching and School Organization. Education Faculty. University Castilla La Manca. Coordinator of the degree of the Teacher Degree in

Early Childhood Education and Coordinator of the Teacher Training Master on the Albacete campus of the University of Castilla La Mancha. Spain. Research lines: Initial training and professional development of teachers - Thought and learning - Intercultural education. Publications in national and international indexed scientific journals, participation and direction in national and foreign educational innovation and research projects. Research stays at Florida International University and Harvard University. USA. Development of teacher training activities in Spanish universities, in North America and South America.

Manuel Ramón González Herrera has 40 years of academic experience in university teaching and research in subjects such as tourism geography, landscapes, environment and sustainability, tourism planning and management, tourism destinations, tourism research and scientific publication, in countries such as Spain, Cuba, Peru, Mexico, Chile, Brazil, Nicaragua, Canada and the United States. Participation in 45 research projects, and more than 20 consultancies related to the research line Sustainable Tourism Management, as well as 22 academic and research stays. Director of 2 post-doctoral stays, 8 doctoral theses and 19 master's theses. Participation in 106 international scientific events and 55 national events as speaker. Author of 17 books, 30 book chapters, 68 articles in scientific journals, and 46 articles in memoirs of events, he has also collaborated in different editorial committees of scientific journals and events.

Jesús Guzmán Mora works as Interim Assistant Professor in the Department of Hispanic and Classical Philology (Area of Language and Literature Didactics) at the Faculty of Education in Albacete, University of Castilla-La Mancha. He holds a PhD in Spanish Literature from the University of Salamanca and a Bachelor's Degree in Humanities from the University of Castilla-La Mancha. He has been a postdoctoral fellow of the Deutscher Akademischer Austauschdienst at the Universität Rostock (Germany) and has taught at the International University of La Rioja and the University of Valladolid.

Laura Hernández González holds a degree in Hispanic Philology (First National Prize for Final Year) and a PhD in Spanish Literature from the University of Valladolid, with a thesis on Golden Age Theatre. A former Secondary School teacher on leave of absence, she currently works as a professor in the Department of Hispanic and Classical Philology at the University of Castilla-La Mancha. Her main research fields are Golden Age Theatre and Literature Didactics, areas in which she has published two books and numerous chapters and articles in specialized journals.

Óscar Jerez García is an Associate Professor in the Department of Geography and Land Planning.

Narciso Jose Lopez Garcia has a Bachelor´s Degree in Primary Education: Human Sciences (University of Castilla-La Mancha), Bachelor of History and Music Sciences (University of La Rioja), Doctor in Specific Didactics-Music Education (University of Málaga), Staff Officer of the Body of Teachers-Specialty of Musical Education and member of the Research Group LABINTIC, at the Faculty of Education of Albacete (University of Castilla - La Mancha). Master´s Degree of Music of the Public Centers of Pre Primary Education and Primary Cervantes School (Albacete - Spain) and San Isidro Labrador (Aguas Nuevas-Albacete-Spain). Associate Lecture of the Department of Didactics of Musical Expression at the Faculty of Education in Albacete (UCLM)

María Yazmina Lozano Mas has a PhD in Geography and History from the University of Las Palmas de Gran Canaria. Lecturer in the area of Social Sciences Didactics at the University of Las Palmas de Gran Canaria (Canary Islands, Spain). She has worked on issues related to gender equality in the framework of social sciences, citizen participation in the framework of non-formal education, and education for sustainable development.

Raquel Bravo Marín is teacher of Music Education and Early Childhood Education. Graduate in Humanities, Graduate in History and Science of Music and PhD in Humanities, Arts and Education. Professional Degree in the speciality of clarinet. She is an Assistant Lecturer in the Department of Didactics of Physical, Artistic and Music Education of the UCLM in the Faculty of Education of Albacete. She is also a lecturer in the Master's Degree in Teacher Training at the Alfonso X El Sabio University and in the Master's Degree in Music Pedagogy at the UCLM. Master's Degree in Music Pedagogy at the UNIR. Author of several research articles, speaker and presenter at numerous national and international congresses related to music and music education. and international congresses related to Music and Education. She has taught at the Universities of Bologna, Lisbon, Ankara, Daugavpils and the Colegio Español Federico García Lorca in Paris. Federico García Lorca School in Paris as part of the Erasmus+ Teacher Mobility programmes.

Laura Moreno Macías is a Graduate in Primary Education Teacher.

Alvaro-Francisco Morote is a lecturer at the Department of Experimental and Social Sciences Education (Faculty of Teaching Training of the University of Valencia, Spain). His research work, in relation to the Didactics of Geography and Social Sciences, has to do with topics on the teaching-learning of Climatology, climate change, natural risks (floods, droughts, etc.), natural resources (water), landscape and field trips. This line of work is carried out in the school stages (Primary, Secondary and Baccalaureate Education) as well as in university stages (undergraduate and postgraduate) and from 3 points of analysis: 1) social representations of school students, teachers in training and in service; 2) textbook analysis; and 3) didactic proposals.

Mercedes de los Ángeles Rodríguez Rodríguez is an Associate Professor at the ULPGC. PhD in Teacher Training since 2016. She has directed and collaborated in several projects of Educational Innovation and has obtained the 2024 award for Educational Innovation from the Association of Geographers of Spain, she has published in several journals related to the didactics of geography.

Maria Navío-Inglés is a PhD student at the University of Castilla-La Mancha and a member of the GITCE research group. She holds a degree in Hispanic Philology and a master's degree in Secondary Education Teaching, as well as a master's degree in educational research and innovation. Her research interests include language teaching and learning, artificial intelligence, and educational technology.

Paula O'Connor-Jiménez is a student in the Primary Education Teaching degree (Bilingual programme, English) at the University of Castilla-La Mancha. She has a contract as an investigator's assistant in the LabinTic research group. Her reseach interests include computational thinking, STEM disciplines' motivation, personalized learning and educational technology.

Manuel Antonio Serrano de la Cruz Santos-Olmo is an Associate Professor in the Department of Geography and Land Planning at the University of Castilla La Mancha.

Paula Tarancón Sánchez is a Student of double degree in Primary Education Teacher and Early Childhood Education Teacher.

Sergio Tirado Olivares is R&D hired staff in the consolidated research group LabinTic (Laboratory for the Integration of Technologies in the Classroom). His lines of research focus on the application of playful-active learning methodologies with Primary Education students and pre-service teachers in STEAM disciplines, the development of active methodologies in the school setting and the development of new formative assessment approaches.

Clara Vañó Albert is a Graduate in Early Childhood Education infantil. Universitat de València.

Bedia Yıldırım is part of the Teaching and Research Staff at the University of Aksaray (Turkey), assigned to the Child Care and Youth Services Department

Index

I

L

M

N

O

P

R

S

T

W

www.ingramcontent.com/pod-product-compliance
Lightning Source LLC
LaVergne TN
LVHW080352011026
849161LV00009B/2278